RECENT TRENDS IN OPTIMIZATION THEORY AND APPLICATIONS

World Scientific Series in Applicable Analysis
Volume 5

Editor

R. P. Agarwal

Department of Mathematics
National University of Singapore

RECENT TRENDS IN OPTIMIZATION THEORY
— AND —
APPLICATIONS

Published by

World Scientific Publishing Co. Pte. Ltd.

P O Box 128, Farrer Road, Singapore 9128

USA office: Suite 1B, 1060 Main Street, River Edge, NJ 07661

UK office: 57 Shelton Street, Covent Garden, London WC2H 9HE

British Library Cataloguing-in-Publication Data
A catalogue record for this book is available from the British Library.

ISBN 981-02-2382-X

This book is printed on acid-free paper.

Printed in Singapore by Uto-Print

Contributors

B.D.O.Anderson (Australia)

M.Bertaja (Italy)

O.J.Boxma (Holland)

O.Burdakov (France)

A.Cantoni (Australia)

D.J.Clements (Australia)

B.D.Craven (Australia)

J.B.Cruz, Jr. (U.S.A.)

P.Diamond (Australia)

S.V.Drakunov (U.S.A.)

Y.G.Evtushenko (Russia)

N.M.Filatov (Germany)

I.Galligani (Italy)

J.C.Geromel (Brazil)

F.Giannessi (Italy)

M.J.Grimble (U.K.)

G.O.Guardabassi (Italy)

D-W Gu (U.K.)

C.H.Houpis (U.S.A.)

D.G.Hull (U.S.A.)

C.Itiki (U.S.A.)

X.Jian (P.R.China)

M.A.Johnson (U.K.)

R.E.Kalaba (U.S.A.)

J.C.Kalkkuhl (U.K.)

M.R.Katebi (U.K.)

T.J.Kim (U.S.A.)

P.Kloeden (Australia)

T.Kobylarz (U.S.A.)

A.J.Laub (U.S.A.)

C.S.Lee (Singapore)

G.Leitmann (U.S.A.)

B-G Liu (P.R.China)

J.Liu (U.S.A.)

Z-Q Luo (Canada)

K.A.Lurie (U.S.A.)

P.Maponi (Italy)

J.B.Matson (Australia)

A.Mees (Australia)

G.Pacelli (Italy)

M.Pachter (U.S.A.)

I.Postlethwaite (U.K.)

T.Rapcsak (Hungary)

M.C.Recchioni (Italy)

Y.Sakawa (Japan)

S.V.Savastyuk (U.S.A.)

K.Schittkowski (Germany)

Y.Shi (U.S.A.)

M.A.Sikora (U.S.A.)

D.D.Siljak (U.S.A.)

K.L.Teo (Australia)

C.Tovey (Australia)

P.Tseng (U.S.A.)

F.E.Udwadia (U.S.A.)

H.Unbehauen (Germany)

A.Vladimirov (Australia)

B.Vo (Australia)

J.F.Whidborne (U.K.)

R.Xu (U.S.A.)

P.L.Yu (U.S.A.)

V.G. Zhadan (Russia)

F.Zirilli (Italy)

Preface

World Scientific Series in Applicable Analysis (WSSIAA) aims at reporting new developments of high mathematical standard and current interest. Each volume in the series shall be devoted to the mathematical analysis that has been applied or is potentially applicable to the solutions of scientific, engineering, and social problems. This volume contains 30 research articles on the theory of optimization and its applications by the leading scientists in the field. It is hoped that the material in the present volume will open new vistas in research. The editor is grateful to his colleague Sheng Qin for his help in the preparation of this volume.

R.P.Agarwal

CONTENTS

WSSIAA 5 (1995) pp. 1–16

Static Optimization of Queueing Systems

Onno J. Boxma

CWI

P.O. Box 94079, 1090 GB Amsterdam, The Netherlands;

Tilburg University, Faculty of Economics

P.O. Box 90153, 5000 LE Tilburg, The Netherlands

Abstract

This paper discusses some recent developments in the static optimization of queueing systems. Special attention is given to three problem classes: (i) the optimal allocation of servers, or service capacity, to queues in a network; (ii) the optimal allocation of the visits of a single server to several queues (a polling system); (iii) the optimal allocation of a single arrival stream to several single server queues.

1. INTRODUCTION

When several users compete for the use of a common resource, the limited capacity of the resource can give rise to congestion. This situation occurs in a plethora of everyday situations: people queue at a counter in a bank or supermarket, congestion occurs in road traffic, products encounter delays at machines during their production process, messages wait for access to a common transmission channel and computer jobs for the use of a set of processors.

Queueing occurs even when the service capacity of the resource strongly exceeds the demand. This is due to the fact that the interarrival times of the users, and their required service times, are generally not fixed. A mathematical model of congestion phenomena therefore usually represents interarrival and service times of users by random variables. The resources are called service facilities, with a single server or multiple servers, and the users are called customers. Customers often visit a number of service facilities, encountering several queues during their stay in the system.

Queueing theory is devoted to the description, analysis and optimization of such *queueing systems*. It concentrates on a few key *performance measures*, like queue lengths and waiting times. Due to the stochastic nature of the arrival and service processes, and of the routing process of customers through a network of queues, the main performance measures are also random variables (or moments thereof). Generally, costs are in a natural way associated with these performance measures. The ultimate goal of performance analysis is optimization - and that is the subject of this paper.

While an enormous amount of literature has been devoted to the probabilistic analysis of queueing systems, their optimization is somewhat lagging behind. This is partly due to the mathematical complexity of queueing systems: only rarely does one find nice structural properties or simple explicit expressions that allow straightforward optimization. Still, a sizable literature discusses the optimization and control of queues. One possible classification of this literature is according to the aspect of the queueing system to which it refers: (i) facility lay-out, (ii) admission control, (iii) customer routing, (iv) processing capacity ((re-)allocation of numbers of servers, and also control of the speed of servers), (v) service order, and (vi) buffer allocation. Some of these problems (facility lay-out, buffer allocation) typically occur in the design and fine-tuning phases of a service facility, whereas other problems mainly arise in its daily operation.

This brings us to another classification: *static* versus *dynamic* queueing system optimization. Consider for example a stream of jobs that have access to N parallel processors, possibly with different processing speeds. Suppose that the jobs must be allocated to the processors in a way that minimizes the mean waiting cost, different costs being assigned to one unit of delay at each of the processors. An important element in this customer assignment problem is the available information; this can range from 'complete observation', i.e., total knowledge about the system at any point in time (including exact queue lengths and service times), to only information about some basic characteristics like arrival rates or mean service times. In general, the term *dynamic* is used for policies which operate under time-dependent information, whereas policies operating under time-independent characteristics are called *static*.

Clearly, the more information is available for making decisions, the better the allocation can be. Dynamic policies in general perform better than static policies. However, static allocation policies are also of considerable interest. First of all, the situation of total knowledge at all times is unrealistic. From a viewpoint of costs, overhead grows as the amount of information to be exchanged, stored and processed increases. Furthermore, dynamic policies are not always that effective: there will always be some delay between updates of the system's current state, and this may have a considerable effect upon the quality of the policy. Moreover, it may be extremely difficult or time-consuming to solve a control problem under time-dependent information, while only rarely the structure of the optimal policy can be fully determined. Static policies, which often lend themselves more easily to performance analysis, can then be employed to provide performance indications (e.g., bounds) for dynamically controlled systems.

In this paper we restrict ourself to (a selection of) *static* queueing optimization problems. We refer to Stidham and Weber [29] for a survey of dynamic control problems in queueing networks, with an emphasis on models based on Markov decision theory; Chapter 8 of Walrand [32] is also highly recommended, for the structural insight it provides in the dynamic control of queues.

In Section 2 we discuss the optimal (re-)allocation of servers to queues in a network; we also pay attention to the assignment of service capacity, in the form of service speeds, to the single server queues of a network. Section 3 is devoted to the optimization of a *polling* system, i.e., a multiqueue system with only one server who moves from queue to queue. The service disciplines of the server at the various queues are studied, as well as the optimal route of the server along the queues. Section 4 considers a problem that is in some sense dual to the latter problem: the earlier mentioned optimal allocation of customers to several queues in parallel. We always assume that buffer capacity is unlimited; for buffer allocation problems we refer to the survey [34].

Remark 1.1
Most of the optimization problems that are discussed in this paper have the following property: the objective function to be minimized by choosing a set of parameters $v_1, \ldots, v_N$ can be separated into N terms, the ith being a function of v_i only, that is convex in v_i, $i = 1, \ldots, N$. This is characteristic of a class of resource allocation problems discussed in the book of Ibaraki and Katoh [18]. They present several algorithms for such problems. The required convexity/concavity properties of the performance measures of queueing systems have only recently been studied systematically; see Liyanage and Shanthikumar [25] and Buzacott and Shanthikumar [11] and references therein. Important references are in particular a series of papers by Shaked, Shanthikumar and Yao that develop a sample-path based approach to obtain structural properties of queueing systems; see e.g. Shaked and Shanthikumar [28].

2. STATIC SERVICE CAPACITY ALLOCATION

Consider an open Jackson network of M/M/. queues $Q_1, Q_2, \ldots, Q_N$. We discuss the following problems. (i) The server reallocation problem: how should a pool of M servers be distributed over the queues such as to minimize a weighted sum of the mean numbers of customers? (ii) The server allocation problem: how many servers should be allocated to each station, such that a weighted sum of the mean numbers of customers is below a certain level while minimizing costs? Or, dually: how many servers should be allocated to each station, such that server investment costs are kept below a certain level while minimizing a weighted sum of the mean numbers of customers? (iii) Kleinrock's capacity assignment problem: allocate server speeds to N single-server stations such that investment costs are kept below a certain level while minimizing the mean sojourn time of a customer in the network. The dual problem is also considered.

(i) *The server reallocation problem*
Let λ_i be the total arrival rate (external plus internal, the latter being determined by the Markovian routing matrix) at Q_i of the Jackson network, and let μ_i be the service rate of each server at Q_i. Hence $m_i^L := \lfloor \lambda_i/\mu_i \rfloor + 1$, with $\lfloor \ \rfloor$ the integer rounddown operation, denotes the minimal number of servers at Q_i such that the traffic intensity at Q_i is less than one. The server reallocation problem (**SR**) for the open Jackson network is formulated as follows:

SR

$$\text{Min}_{m_1,\ldots,m_N} \sum_{i=1}^{N} c_i \text{EL}_i(m_i) \tag{2.1}$$

$$s.t. \sum_{i=1}^{N} m_i = M, \quad m_i \geq m_i^L, \quad i = 1, \ldots, N.$$

Here $\mathrm{EL}_i(m_i)$ denotes the mean number of customers at Q_i when this queue has m_i servers, and c_i is a cost factor. It is proven in [9] that the following greedy, or marginal allocation, algorithm is optimal for the **SR** problem. Start by allocating m_i^L servers to Q_i, $i = 1, \ldots, N$. At each iteration step add one server to that queue where the greatest decrement in the objective function is achieved. Repeat this procedure until all M servers have been allocated. The optimality of this marginal allocation algorithm is typical for resource allocation problems in which the objective function is separable into convex terms while the constraints are *linear*.

(ii) *The server allocation problem*
Again consider the open Jackson network. The server allocation problem (**SA**) is formulated as follows:

SA

$$\mathrm{Min}_{m_1,\ldots,m_N} \sum_{i=1}^{N} F_i(m_i), \tag{2.2}$$

$$s.t. \sum_{i=1}^{N} c_i \mathrm{EL}_i(m_i) \leq W, \qquad m_i \geq m_i^L, \quad i = 1, \ldots, N.$$

Here $F_i(m_i)$ is a convex and decreasing function of m_i, that denotes the investment costs involved in allocating m_i servers to Q_i; W is a given number. If, e.g., $c_i \equiv 1/\mu_i$ then W indicates an upper bound on the mean total workload in the system.
Problem **SA** can be regarded as a generalization of the knapsack problem. Hence it is NP-complete. In [9] a simple greedy heuristic is proposed that represents a useful approach to the solution of problem **SA**: Start by allocating m_i^L servers to Q_i. At each iteration step, allocate one server to the queue for which the ratio of the increment of the objective function and the decrement of the weighted sum of mean queue lengths is the smallest. Stop as soon as adding a server makes the allocation feasible.
In [14] two algorithms are proposed that build upon this algorithm. They lead to substantially better results, at the expense of the complexity increasing by a factor N respectively N^2. For those two algorithms also worst-case performance ratios of 2 respectively 3/2 are proven in [14], whereas for the greedy heuristic it has only been proven that the minimal value of the objective function lies in between the values of the one-but-last (infeasible) and last allocations. Van Vliet and Rinnooy Kan [31] extend the greedy heuristic of [9] in another way. They allow general external interarrival time distributions and general service time distributions at the queues, and they use a two-moment parametric decomposition approach to estimate the first two moments of all interarrival times; subsequently they approximate the mean numbers of customers

in the queues, based on the first two moments of interarrival and service times at each queue in isolation. For the latter approximation they use a known approximation formula that is convex in the number of servers.

Aarts [1] considers the dual server allocation problem (**DSA**):

DSA

$$\text{Min}_{m_1,\ldots,m_N} \sum_{i=1}^{N} c_i \text{EL}_i(m_i), \tag{2.3}$$

$$s.t. \ \sum_{i=1}^{N} F_i(m_i) \leq F, \qquad m_i \geq m_i^L, \quad i = 1,\ldots,N,$$

with F some constant. He discusses the greedy heuristic and two refinements, similar to the three algorithms mentioned for (2.2). He observes that if $F_i(m_i) = dm_i$, $i = 1,\ldots,N$, then problem (2.3) amounts to the server reallocation problem (2.1), which is solved exactly by the greedy heuristic.

(iii) *The capacity assignment problem*

Consider an open Jackson network of M/M/1 queues, with service capacity μ_i at queue Q_i. Let λ_i denote the total (external plus internal) arrival rate at Q_i. The mean total sojourn time ET of an arbitrary customer in the network is given by

$$\text{ET} = \frac{1}{\gamma} \sum_{i=1}^{N} \frac{\mu_i}{\lambda_i - \mu_i}.$$

Here γ denotes the total external arrival rate in the network. Kleinrock [20] has posed and solved the following capacity assignment problem (**CA**); he has formulated it in the framework of assigning channel capacities in a communication network.

CA

$$\text{Min}_{\mu_1,\ldots,\mu_N} \text{ET}, \tag{2.4}$$

$$s.t. \ \sum_{i=1}^{N} d_i \mu_i \leq D, \quad \mu_i > \lambda_i, \quad i = 1,\ldots,N.$$

Again the separability/convexity structure appears. This time the problem even allows an explicit solution, that is easily obtained using the Lagrange multiplier technique. The optimal service rates turn out to be [20]:

$$\mu_i^* = \lambda_i + \frac{D - \sum_{j=1}^{N} \lambda_j d_j}{d_i} \frac{\sqrt{\lambda_i d_i}}{\sum_{j=1}^{N} \sqrt{\lambda_j d_j}}, \quad i = 1,\ldots,N. \tag{2.5}$$

After having allocated λ_i service capacity to Q_i, the remaining funds are invested proportionally to the square root of λ_i and d_i.

Remark 2.1
The dual capacity assignment problem (**DCA**) reads:
DCA

$$\text{Min}_{\mu_1,\dots,\mu_N} \sum_{i=1}^{N} d_i \mu_i, \tag{2.6}$$

$$s.t.\ \text{ET} \leq T, \quad \mu_i > \lambda_i, \quad i = 1,\dots,N,$$

with T some positive constant. It is solved in exactly the same way, yielding a very similarly structured solution. One easily verifies [1] that, if T equals the optimal value in problem **CA**, then the optimal value of $\sum_{i=1}^{N} d_i \mu_i$ in the dual problem equals D.

Remark 2.2
Problems **CA** and **DCA** have been extended in several directions. We refer to Bitran and Tirupati [2] for a heuristic approach (related to those for **SA**) to **DCA** in the case of general service time distributions.

3. STATIC POLLING OPTIMIZATION

The standard polling system is a queueing system in which a single server, S, visits N queues $Q_1,\dots,Q_N$ in some order. Polling systems presently receive much attention, partly because of their ability to model many resource allocation phenomena in computer-communications. After a brief model description, we discuss: (i) optimal server routing and (ii) optimal service behaviour at the queues.

Model description
S serves N infinite-capacity queues $Q_1,\dots,Q_N$. Customers arrive at all queues according to independent Poisson processes. The arrival intensity at Q_i is λ_i, $i = 1,\dots,N$. Customers arriving at Q_i are called class-i customers; their service times are independent random variables $\mathbf{B}_i$ with mean β_i and second moment $\beta_i^{(2)}$, $i = 1,\dots,N$. After their service at Q_i they leave the system. The offered traffic load, ρ_i, at Q_i is defined as $\rho_i := \lambda_i \beta_i$, $i = 1,\dots,N$, and the total offered load is called ρ. When swapping out of Q_i, the server incurs a switchover period of type i; the switchover durations of type i are independent random variables $\mathbf{S}_i$ with mean s_i and second moment $s_i^{(2)}$, $i = 1,\dots,N$. All interarrival, service and switchover processes are independent stochastic processes. The *service discipline* at Q_i determines how many customers are served when S visits Q_i. Important disciplines are:

- Exhaustive (E): S serves Q_i until it has become empty.

- Gated (G): S serves exactly the customers that were present at the beginning of the visit.

- k-limited ($k - L$): S serves k customers or empties Q_i, whichever happens first.

Furthermore, for some applications (traffic lights, timed-token protocol in a local area network), the visit period is restricted by a fixed time limit. In the application of a signalized traffic intersection, it may be natural to leave the server (green light) at a queue even if it is empty; but in the now following discussion of polling optimization we shall restrict ourself to the case in which S never resides at an empty queue. In the sequel we assume the traffic parameters to be such that the polling system is ergodic. A necessary condition for this to hold is $\rho < 1$; it is known to be also sufficient when the service discipline at all queues is E or G, but not when, e.g., a queue is $k - L$.

(i) *The optimal server routing problem*
We start with a simple result for a *probabilistic polling model*, viz., a polling model in which S visits the queues according to a probabilistic routing mechanism: with probability p_i it chooses Q_i for its next visit, $i = 1, \ldots, N$. Suppose that all switchover times between the various queues have mean σ and second moment $\sigma^{(2)}$, and that service at a set of queues e is exhaustive while it is gated at the remaining set, g.
Let $\mathbf{EW}_i$ denote the mean waiting time at Q_i, $i = 1, \ldots, N$. Consider the following optimal server routing problem (**OSR**):
OSR

$$\mathrm{Min}_{\mathrm{P1},\ldots,\mathrm{PN}} \sum_{i=1}^{N} \rho_i \mathbf{EW}_i \tag{3.1}$$

$$s.t. \quad \sum_{i=1}^{N} p_i = 1, \quad p_1 \geq 0, \ldots, p_N \geq 0.$$

It is easily seen that minimizing this objective function is equivalent to minimizing the mean *workload*. In this model

$$\sum_{i=1}^{N} \rho_i \mathbf{EW}_i = \rho \frac{\sum_{i=1}^{N} \lambda_i \beta_i^{(2)}}{2(1-\rho)} - \frac{\sigma}{1-\rho} \sum_{k \in e} \frac{\rho_k^2}{p_k} + \frac{\sigma}{1-\rho} \sum_{k=1}^{N} \frac{\rho_k}{p_k} - \rho\sigma + \rho\frac{\sigma^{(2)}}{\sigma}. \tag{3.2}$$

We now have a classical non-linear optimization problem with linear constraints. Using standard Lagrange multiplier techniques we find the following solution (cf. [7]):

$$k \in e: \quad p_k = \frac{\sqrt{\rho_k(1-\rho_k)}}{\sum_{j \in e} \sqrt{\rho_j(1-\rho_j)} + \sum_{j \in g} \sqrt{\rho_j}}; \tag{3.3}$$

$$k \in g: \quad p_k = \frac{\sqrt{\rho_k}}{\sum_{j \in e} \sqrt{\rho_j(1-\rho_j)} + \sum_{j \in g} \sqrt{\rho_j}}. \tag{3.4}$$

We refer to [27] for an extensive study of the optimization of *Markovian* server routing (i.e., server routing probability p_{ij} from Q_i to Q_j). Below we turn to the following related problem, again for a mixture of E and G queues and i.i.d. switchover times. S is allowed to visit the queues according to a fixed pattern, a *polling table* (like $Q_1Q_2Q_3\ldots Q_N$, which is called cyclic polling, or $Q_1Q_2Q_1Q_3Q_1\ldots Q_1Q_N$, which is called star polling). We seek the polling table that minimizes the mean workload, or equivalently, $\sum_{i=1}^{N}\rho_i\mathbf{EW}_i$. An implicit expression for this sum is known, but it does not allow a straightforward optimization. If an upper bound on the table size is given, then an integer programming problem results; but without such a restriction it is not clear a priori whether a given 'good' table cannot be improved by taking a much larger table with a very similar structure. In [7] an approximate approach is proposed to the problem of choosing an optimal polling table, and shown to perform well. It consists of three steps.

Step 1. Choose the occurrence frequencies of the queues in the table by taking the optimal frequencies p_k obtained in (3.3) and (3.4) for the probabilistic polling model with the same traffic parameters.

Step 2. Based on those occurrence frequencies, determine a 'good' table size M (take the smallest possible M such that for all k, Mp_k is within a predetermined small distance from a positive integer).

Step 3. Given the round-off occurrence numbers, say n_i, find a table such that Q_i occurs n_i times with the visits to each particular queue as evenly spaced as possible. This spacing problem is handled using the so-called *Golden Ratio* policy, cf. Itai and Rosberg [19].

Remark 3.1

Comparison of the optimal mean workloads in the cases of probabilistic polling and polling tables reveals that the latter minimum is considerably lower. The explanation is that the variance of the time between successive visits to a particular queue is much smaller in a polling table; this more regular behaviour leads to a smaller mean workload. Still, the occurrence frequencies of the queues in the optimal polling table are well predicted by the occurrence frequencies in the probabilistic polling model.

Remark 3.2

The Golden Ratio policy has recently been applied to several 'even spacing' problems. In [5], in the context of optimizing the *time limits* in a polling model, an algorithm is proposed which seems to outperform Golden Ratio; it is based on a neat result of Hajek [16] concerning the optimal splitting of point processes.

Let us now consider the somewhat more general polling table optimization problem with objective function:

$$\text{Min} \sum_{i=1}^{N} c_i \lambda_i \mathbf{EW}_i. \tag{3.5}$$

Here c_i are nonnegative constants that reflect the cost of waiting one unit of time at Q_i. The choice $c_i \equiv \beta_i$ again yields the objective function of (3.1), while $c_i \equiv 1$ leads to the minimization of the mean total number of waiting customers in the system (note that Little's formula implies that $\lambda_i \mathbf{EW}_i$ equals the mean number of waiting customers at Q_i). Let us also allow $1 - L$ service at one or more of the queues, and non-identically distributed switchover times. Even probabilistic polling no longer yields an explicit analytic solution when one of these changes is introduced. For the corresponding (and practically more important) polling table minimization problem, step 1 above now has to be adapted after which steps 2 and 3 can again be used. In [8] some approaches are developed for this adaptation. One of them is based on the following approximations for $\mathbf{EW}_i$, for a polling table with n_i evenly spaced occurrences of Q_i:

$$i \in e : \quad \mathbf{EW}_i \approx A(1 - \rho_i)\frac{\sum_{j=1}^{N} n_j s_j}{n_i}; \tag{3.6}$$

$$i \in g : \quad \mathbf{EW}_i \approx A(1 + \rho_i)\frac{\sum_{j=1}^{N} n_j s_j}{n_i}; \tag{3.7}$$

$$i \in 1 - L : \quad \mathbf{EW}_i \approx A\frac{1 - \rho + \rho_i}{1 - \rho - \lambda_i \sum_{j=1}^{N} n_j s_j/n_i}\frac{\sum_{j=1}^{N} n_j s_j}{n_i}. \tag{3.8}$$

A denotes some positive constant which we don't need to specify, as only the ratio of the various mean waiting times matters for the optimization. Minimization of $\sum_{i=1}^{N} c_i \lambda_i \mathbf{EW}_i$ easily yields the optimal n_i values, up to a multiplicative constant:

$$i \in e : \quad n_i \sim \sqrt{c_i \lambda_i (1 - \rho_i)/s_i}; \tag{3.9}$$

$$i \in g : \quad n_i \sim \sqrt{c_i \lambda_i (1 + \rho_i)/s_i}; \tag{3.10}$$

$$i \in 1 - L : \quad n_i \sim \lambda_i + (1 - \rho - \sum_{k=1}^{N} \lambda_k s_k)\frac{\sqrt{c_i \lambda_i (1 - \rho + \rho_i)/s_i}}{\sum_{j=1}^{N} s_j \sqrt{c_j \lambda_j (1 - \rho + \rho_j)/s_j}}. \tag{3.11}$$

If all queues are $1 - L$ and we add the constraint $\sum_{j=1}^{N} n_j s_j/(1 - \rho) = C^*$, which amounts to prescribing that the mean cycle time of the polling table equals C^*, then the resulting 1-limited polling table problem (**LPT**) is:

LPT

$$\mathrm{Min}_{n_1,\dots,n_N} \sum_{i=1}^{N} \frac{c_i \lambda_i (1 - \rho + \rho_i)}{n_i - \lambda_i C^*} \tag{3.12}$$

$$s.t. \sum_{j=1}^{N} n_j s_j / (1 - \rho) = C^*.$$

It has a similar form as problem **CA** of Section 2. The solution is given by (3.11) when its righthand side is multiplied by C^*. Its interpretation is that Q_i should be visited at least $\lambda_i C^*$ times per cycle, as this is the mean number of arrivals at Q_i during a cycle of mean C^*; the remaining 'capacity' is allocated according to a square root rule. In [8] the resulting rule is shown to perform quite satisfactorily.

Remark 3.3
An interesting application of polling table optimization occurs in stochastic economic lot scheduling, where several items need to be produced in a common facility. A. Federgruen (personal communication) studies the problem of determining an optimal production sequence (with a somewhat more involved objective function), using a three-step procedure as outlined above.

Remark 3.4
We refer to Yechiali [35] for an interesting overview of semi-dynamic control of the server route in a polling system. Yechiali (see also Browne and Yechiali [10]) considers one-cycle look-ahead policies. For various service disciplines, he determines the visit order of the queues in the next cycle - in which all queues are visited exactly once - that minimizes the expected duration of that cycle.

(ii) *Optimal service behaviour at the queues*
Given a route of S along the queues, the question arises which service discipline should be used at each queue. Borst et al. [6] consider a cyclic polling model, viz., a polling model in which S visits the queues in cyclic order. The service discipline at Q_i is k_i-limited, $i = 1, \ldots, N$. They consider the following unconstrained k-limited problem (**UkL**) for this cyclic polling model:
UkL

$$\text{Min}_{k_1,\ldots,k_N} \sum_{i=1}^{N} c_i \lambda_i \mathbf{EW}_i. \tag{3.13}$$

Next to this *unconstrained* optimization problem, they also consider the *constrained* k-limited problem (**CkL**) where the objective function in (3.13) must be minimized under the additional condition that $\sum_{i=1}^{N} \gamma_i k_i \leq K$ for some nonnegative parameters γ_i and constant K. Such a condition could reflect a limit on the (expected) cycle time of the server, which is relevant in some access protocols in local area networks.
Problem **CkL** is tackled in two different ways (both necessarily approximative: an exact analysis of polling models with k-limited service seems to be prohibitive in all but a few exceptional cases). One approach is to take such k_i values that the ratios $k_i : k_j$ agree with the ratios of the visit frequencies for a corresponding polling table with 1-limited service at all queues (see (3.11)). A minor adaptation of (3.8) has to be

made to incorporate the fact that switchover times out of Q_i occur once instead of k_i times. That results in the following mean waiting time approximation, with D_1 some constant and $s := \sum_{i=1}^{N} s_i$:

$$\mathbf{EW}_i \simeq D_1 \frac{1 - \rho + \rho_i}{k_i(1 - \rho) - \lambda_i s}. \tag{3.14}$$

A second approach is to use the Fuhrmann-Wang approximation [15] for $\mathbf{EW}_i$ in the cyclic polling model with k-limited service:

$$\mathbf{EW}_i \simeq D_2 \frac{(1 - \rho_i)(1 - \rho) + (2 - \rho)\rho_i/k_i}{1 - \rho - \lambda_i s/k_i}; \tag{3.15}$$

here D_2 is some constant. There is no need to specify the constants D_1 and D_2, as only the ratio of the mean waiting times plays a role in the optimization. Taking into account the constraint $\sum_{i=1}^{N} \gamma_i k_i \leq K$, the optimal solution k_i^*, $i = 1, \ldots, N$ of (3.13) using either (3.14) or (3.15) is found in a straightforward way. We mention only the one based on (3.15), as it is slightly better:

$$k_i^* = \frac{\lambda_i s}{1 - \rho} + (K - \sum_{j=1}^{N} \gamma_j \frac{\lambda_j s}{1 - \rho}) \frac{\sqrt{c_i \lambda_i [\rho_i(2 - \rho) + \lambda_i s(1 - \rho_i)]/\gamma_i}}{\sum_{j=1}^{N} \gamma_j \sqrt{c_j \lambda_j [\rho_j(2 - \rho) + \lambda_j s(1 - \rho_j)]/\gamma_j}}. \tag{3.16}$$

Again the resemblance with problem **CA** of Section 2 should be noted.
In [6] slightly better results for **CkL** are obtained by sharpening (3.15) somewhat, at the expense of no longer obtaining explicit expressions for the k_i; that improved approximation also yields good results for **UkL**.

Remark 3.5
An unproven conjecture in [6], that is used in the optimization procedure, is that $\mathbf{EW}_i$ is decreasing in k_i and nondecreasing in k_j for all $j \neq i$. Using a sample-path argument, Levy et al. [22] show that the service discipline that minimizes the mean workload, or equivalently $\sum_{i=1}^{N} \rho_i \mathbf{EW}_i$, in the unconstrained case is to serve all queues exhaustively (i.e., $k_i \equiv \infty$). Liu et al. [23], also allowing idling policies, in more generality discuss the server behaviour that stochastically minimizes the unfinished work and the number of customers in the system.

Remark 3.6
In the so-called Bernoulli service discipline, S serves after each service completion yet another customer with probability q_i and leaves with probability $1 - q_i$. This discipline can be viewed as the stochastic counterpart of the k-limited discipline. Analogous to **UkL**, Blanc and Van der Mei [3] try to find those q_i that minimize the objective function (3.5). Their main approach is a numerical one, based on the use of the so-called power series algorithm.

12

Remark 3.7
A pioneering polling optimization study is due to Klimov [21]. It combines elements of both problem types studied in this section. Klimov studies a polling model without switchover times, but with probabilistic customer routing along the queues. After each service completion a decision is made as to which queue S should visit next. Klimov shows that the optimal server routing policy has an index structure: the optimal policy corresponds to an ordering of the queues such that S always chooses to serve a customer from the first nonempty queue in that order.

4. STATIC TRAFFIC ALLOCATION

Consider the following situation. Customers arrive according to a Poisson process with rate Λ. At the instance of arrival, a customer has to be assigned to one of N parallel single servers $Q_1, \ldots, Q_N$. The service time of a customer that is assigned to Q_i has distribution $B_i(\cdot)$ with mean β_i and second moment $\beta_i^{(2)}$, $i = 1, \ldots, N$. All service times are independent. Let P denote an allocation policy. Our aim is to minimize

$$\text{Min}_P \ \sum_{i=1}^{N} c_i f_i(P) \mathbf{EW}_i(P). \tag{4.1}$$

The notation is as before, with now (P) indicating a dependency on the allocation policy P; the factors $f_i(P)$ are load-dependent weight factors. For example, if under P a fraction p_i of the customers is assigned to Q_i, then $f_i(P) = \Lambda p_i$ yields the objective function in (3.5).

We refer to Wang and Morris [33] for a survey on load balancing, including numerical comparisons of various dynamic and static allocation policies. We restrict ourself again to static policies. Buzen and Chen [12] have studied the *probabilistic* allocation, in which a customer is sent to Q_i with probability p_i, $i = 1, \ldots, N$. The arrival process at each queue now is a Poisson process, and hence each queue is an M/G/1 queue. They have used mathematical programming techniques to minimize the mean *sojourn* time of a customer. Let us also consider the class of probabilistic allocation policies, but with objective function (4.1). First take $f_i \equiv p_i$, and put $\lambda_i := \Lambda p_i$. The probabilistic allocation problem (**PA**) now reduces to:

PA

$$\text{Min}_{\lambda_1, \ldots, \lambda_N} \ \sum_{i=1}^{N} c_i \lambda_i \frac{\lambda_i \beta_i^{(2)}}{2(1 - \lambda_i \beta_i)} \tag{4.2}$$

$$s.t. \ \sum_{i=1}^{N} \lambda_i = \Lambda, \quad 0 \leq \lambda_i \leq \frac{1}{\beta_i}, \quad i = 1, \ldots, N.$$

It can also be verified that this optimization problem has a feasible solution provided that $\sum_{i=1}^{N} 1/\beta_i > \Lambda$, i.e., the arrival rate does not exceed the total service capacity.

Here and in the remainder of the section that is assumed to be the case.

Note that the objective function in (4.2) is separable into N terms, the ith term being strictly convex in λ_i; cf. Remark 1.1, where the reader is referred to the book of Ibaraki and Katoh [18]. Just like **CA**, this case is so simple that it can be solved explicitly, using Lagrange multiplier techniques; we find that the unique optimal rates λ_i^* are (cf. [13]):

$$\lambda_i^* = \frac{1}{\beta_i} - \frac{1}{\beta_i}[\sqrt{1 + \frac{2\beta_i\delta}{c_i\beta_i^{(2)}}}]^{-1}, \quad i = 1, \ldots, N, \tag{4.3}$$

in which the Lagrange multiplier δ is determined by the constraint $\sum_{i=1}^N \lambda_i = \Lambda$. If in (4.1) $f_i \equiv 1$ instead of $f_i \equiv p_i$, then problem **PA** has the same separable structure, with the control variables only interacting through the linear restriction $\sum_{i=1}^N \lambda_i = \Lambda$. In this case an explicit analytic solution is not obtained (in fact some of the λ_i^* now are equal to zero), but one can easily solve the problem numerically, using for example the algorithm RANK in [18], p. 19.

Algorithm RANK strongly depends on the strict convexity of the N terms. The convexity property implies that there is only one local minimum, which consequently has to be the optimal solution for the allocation problem. One of the cases in which the property of strict convexity may not hold is the probabilistic traffic allocation problem with a *general* arrival process, as studied in Tang and Van Vliet [30]. Their method involves the Frank-Wolfe algorithm, which was originally developed for quadratic programming and which provides a local minimum; they claim that it should be close to the global minimum.

Just like in polling optimization, in customer allocation one may expect to improve considerably upon a probabilistic allocation by allocating according to a fixed pattern. This expectation is based on the reduced variability of the arrival processes at the queues. In the sequel we assume that allocation is being done according to a fixed pattern, and we consider the problem of finding a 'good' pattern (w.r.t. the objective function (4.1)). Only in exceptional cases optimality of a pattern can be proven. We refer to Liu and Towsley [24] for such a result and further references. Liu and Towsley [24] consider the case of identical service time distributions at all queues, with an increasing failure rate. They show that the round-robin policy (send customers consecutively to $Q_1, Q_2, \ldots, Q_N, Q_1, \ldots$) minimizes, in the sense of a separable increasing convex ordering, the customer sojourn times and the numbers of customers in the queues.

Let $(a_1, a_2, \ldots, a_M)$ denote an allocation table, a_i indicating the number of the queue to which every $(i + kM)$-th customer is being sent, $k = 1, 2, \ldots; i = 1, \ldots, M$. Combé and Boxma [13] observe that the resulting arrival processes at the queues fall into the class of Markovian Arrival Processes (MAP), cf. [26]. Subsequently they present a three-step algorithm, similar to the three-step polling optimization algorithm of Section 3, to construct a 'good' pattern. The occurrence frequencies of the queues in the allocation table are estimated in step 1, after which a suitable table size M and an even spacing are determined. It turns out that the optimal probabilistic allocation again gives a reasonable indication for the occurrence frequencies. However, the optimal probabilistic allocation seems to underestimate the traffic assignment to the queues with relatively high mean service time. The explanation is that the effect of 'regularizing' the arrival stream in pattern allocation is strongest for the queues with relatively large service times, so relatively small assignment probabilities. E.g., if the optimal probabilistic allocation fractions in a two-queue case are 8/9 and 1/9, then Q_2 would receive an Erlang-9 arrival process under pattern allocation, and Q_1 something close to Poisson (far less regular than Erlang-9). In step 1, instead of using the optimal probabilistic allocation, it is better to approximate the interarrival time distributions to the queues by Gamma distributions, subsequently approximate the mean waiting time in a Gamma/G/1 queue in a suitable way (cf. [13]) and solve the resulting non-linear optimization problem. That problem again has the separability/convexity structure as found in (4.2), and can be easily solved numerically (cf. [13]).

Remark 4.1
The approach using MAP and Gamma approximations seems applicable to several generalizations of the problem described above, like: the original arrival process is not a Poisson process; some of the queues also receive a 'dedicated' arrival stream; the arrival process must be allocated to *multiserver* queues in parallel.

Remark 4.2
Hordijk, Koole and Loeve [17] study the class of pattern allocation policies, for the special case of exponentially distributed service times, by using Markov Decision theory. Their algorithm leads to results of similar quality as those of [13].

Remark 4.3
Borst [4] considers the probabilistic allocation of not one but *several heterogeneous customer classes* to a set of parallel servers with different speeds. Each customer class has a Poisson arrival process and generally distributed service requirements. He studies the minimization of a weighted sum of the mean waiting times, exposing the structure of the optimal allocation. [11] considers a version of this problem in which all servers have the same speed, and in which the overall mean waiting time is the objective function. It would be interesting to study allocation patterns for this type of model.

Acknowledgement
The author's research has been supported by the European Grant BRA-QMIPS of CEC DG XIII.

REFERENCES
1. M.A.J. Aarts (1992). Production logistics - A queueing theoretic approach, *M.Sc. Thesis, Tilburg University*, in Dutch.

2. G.R. Bitran and D. Tirupati (1989). Trade-off curves, targeting, and balancing in

manufacturing networks, *Operations Research* **37**, 547-564.

3. J.P.C. Blanc and R.D. van der Mei (1992). Optimization of polling systems with Bernoulli schedules, *Report FEW 563, Tilburg University*, to appear in Performance Evaluation (1995).

4. S.C. Borst (1994). Optimal probabilistic allocation of customer types to servers, *CWI Report BS-R9415, Amsterdam*; to appear in the Proceedings of ACM Sigmetrics/Performance '95.

5. S.C. Borst, O.J. Boxma, J.H.A. Harink and G.B. Huitema (1994). Optimization of fixed time polling schemes, *Telecommunication Systems* **3**, 31-59.

6. S.C. Borst, O.J. Boxma and H. Levy (1993). The use of service limits for efficient operation of multi-station single-medium communication systems, *CWI Report BS-R9312, Amsterdam*.

7. O.J. Boxma, H. Levy and J.A. Weststrate (1990). Optimization of polling systems, In: *Performance '90*, eds. P.J.B. King, I. Mitrani and R.J. Pooley (North-Holland, Amsterdam) pp. 349-361.

8. O.J. Boxma, H. Levy and J.A. Weststrate (1991). Efficient visit frequencies for polling tables: minimization of waiting cost, *Queueing Systems* **9**, 133-162.

9. O.J. Boxma, A.H.G. Rinnooy Kan and M. van Vliet (1990). Machine allocation problems in manufacturing networks, *Eur. J. Oper. Res.* **45**, 47-54.

10. S. Browne and U. Yechiali (1989). Dynamic priority rules for cyclic-type queues, *Adv. Appl. Probab.* **21**, 432-450.

11. J.A. Buzacott and J.G. Shanthikumar (1992). Design of manufacturing systems using queueing models, *Queueing Systems* **12**, 135-213.

12. J.P. Buzen and P.P.-S. Chen (1974). Optimal load balancing in memory hierarchies, in: *Proc. IFIP 1974*, ed. J. Rosenfeld (North-Holland, Amsterdam) pp. 271-275.

13. M.B. Combé and O.J. Boxma (1994). Optimization of static traffic allocation policies, *Theoretical Computer Science* **125**, 17-43.

14. H. Frenk, M. Labbé, M. van Vliet and S. Zhang (1994). Improved algorithms for machine allocation in manufacturing systems, *Operations Research* **42**, 523-530.

15. S.W. Fuhrmann and Y.T. Wang (1988). Analysis of cyclic service systems with limited service: bounds and approximations, *Performance Evaluation* **9**, 35-54.

16. B. Hajek (1985). Extremal splittings of point processes, *Math. Oper. Res.* **10**, 543-556.

17. A. Hordijk, G.M. Koole, J.A. Loeve (1994). Analysis of a customer assignment model with no state information, *Prob. in the Eng. and Inform. Sciences* **8**, 419-429.

18. T.I. Ibaraki and N. Katoh (1988). *Resource Allocation Problems.* MIT Press, Cam-

bridge.

19. A. Itai and Z. Rosberg (1984). A Golden Ratio control policy for a multiple-access channel, *IEEE Trans. Autom. Control* **29**, 712-718.

20. L. Kleinrock (1964). *Communication Nets - Stochastic Message Flow and Delay.* McGraw-Hill, New York, 1964 (republished by Dover, 1972).

21. G.P. Klimov (1974). Time-sharing service systems I, *Theory of Prob. and its Appl.* **19**, 532-551.

22. H. Levy, M. Sidi and O.J. Boxma (1990). Dominance relations in polling systems, *Queueing Systems* **6**, 155-171.

23. Z. Liu, P. Nain and D. Towsley (1992). On optimal polling policies, *Queueing Systems* **11**, 59-83.

24. Z. Liu and D. Towsley (1994). Optimality of the round-robin routing policy, *J. Appl. Probab.* **31**, 466-475.

25. L. Liyanage and J.G. Shanthikumar (1992). Second-order properties of single-stage queueing systems, in: *Queueing and Related Models*, eds. U.N. Bhat and I.V. Basawa (Oxford Univ. Press, Oxford), pp. 129-160.

26. D.M. Lucantoni (1991). New results on the single server queue with a batch Markovian arrival process, *Stochastic Models* **7**, 1-46.

27. R.D. van der Mei (1995). Polling systems with Markovian server routing, *Report CentER, Tilburg.*

28. M. Shaked and J.G. Shanthikumar (1988). Stochastic convexity and its applications, *Adv. Appl. Probab.* **20**, 427-446.

29. S. Stidham, Jr. and R. Weber (1993). A survey of Markov decision models for control of networks of queues, *Queueing Systems* **13**, 291-314.

30. C.S. Tang and M. van Vliet (1994). Traffic allocation for manufacturing systems, *Eur. J. Oper. Res.* **75**, 171-185.

31. M. van Vliet and A.H.G. Rinnooy Kan (1991). Machine allocation algorithms for job shop manufacturing, *J. Intell. Manufacturing* **2**, 83-94.

32. J. Walrand (1988). *An Introduction to Queueing Networks.* Prentice-Hall, Englewood Cliffs (NJ).

33. Y.-T. Wang and R.J.T. Morris (1985). Load sharing in distributed systems, *IEEE Trans. Comp.* **C-34**, 204-217.

34. H. Yamashita and R.O. Onvural (1994). Allocation of buffer capacities in queueing networks with arbitrary topologies, *Annals of Operations Research* **48**, 313-332.

35. U. Yechiali (1991). Optimal dynamic control of queueing systems, in: *Queueing, Performance and Control in ATM*, eds. J.W. Cohen and C.D. Pack (North-Holland, Amsterdam) pp. 205-217.

WSSIAA 5 (1995) pp. 17–24

ON PROPERTIES OF NEWTON'S METHOD
FOR SMOOTH AND NONSMOOTH EQUATIONS

OLEG BURDAKOV

Parallel Algorithms Team, CERFACS

42, Avenue Gustave Coriolis, 31057 Toulouse Cedex, France

Abstract

Variational inequalities, nonlinear programming, complementarity problems and other problems can be reduced to nonsmooth equations, for which some generalizations of Newton's method are known. The Newton path, as a natural generalization of the Newton direction, was suggested by D. Ralph for enlarging the convergence region (globalization) of Newton-Robinson's method in the nonsmooth case. We investigate some properties of both the Newton direction and the Newton path, which seem to be basic for various globalization strategies. In particular, a simple formula for the derivative of an arbitrary norm of residuals along the Newton direction, derived earlier by the author for the smooth equations, is generalized here for the derivative along the Newton path.

1. Introduction

Consider the problem of solving a nonlinear equation

$$f(x) = 0, \tag{1}$$

where the mapping $f : D \subset X \to Y$, and X and Y are Banach spaces. For brevity we will use the same notation $\| \cdot \|$ both for $\| \cdot \|_X$ and $\| \cdot \|_Y$.

Nonlinear programming problems, variational inequalities, complementarity problems and other problems can be reduced to a smooth or nonsmooth equation Eq. 1. Some generalizations of the Newton method were suggested[1-7] for solving such equations.

The classic Newton iteration[8,9] is based upon a linear approximation

$$A_c(x) = f_c + f'_c(x - x_c) \tag{2}$$

of the smooth mapping f constructed for the current iterate x_c (here and henceforth we use the notation $f_c = f(x_c)$ and $f'_c = f'(x_c)$). The next iterate results from solving the equation

$$A_c(x) = 0. \tag{3}$$

In the nonsmooth case, the generalized Newton iteration is based upon a nonlinear, nonsmooth, approximation A_c of f. The main practical requirements for this

18

approximation are that, first, it must take into account the nature of nonsmoothness of f as much as possible; second, Eq. 3 must be easily solved.

It is known[8,9] that, although it has a high rate of convergence, the classical Newton's method is guaranteed to converge only in the local vicinity of the solution. One of the possibilities to enlarge the region of its convergence is to apply a line search[9] along the vector connecting x_c and the solution to Eq. 3, called the Newton direction. In this line search, a norm of residuals $\| f(x) \|$ (or some other merit function[10]) is reduced. In the local vicinity of the solution, these modifications coincide automatically with the classical Newton's method. Such globalization strategies apply the Armijo-Danilin-Goldstein step-size rules[11–13], well known in the unconstrained minimization. The basic property of the Newton direction enabling this application is that the function $\| f(x) \|$ is differentiable along the Newton path, and that the formula for this derivative[14,15] is very simple.

For globalization in the nonsmooth case, Ralph[7] suggested a generalization of the Newton direction, called the Newton path, and a search along this nonsmooth path, which can be also regarded as a generalization of the line search.

The purpose of this work is to present new properties of the Newton search direction for the smooth Eq. 1 (Section 2) and of the Newton path for the nonsmooth case (Section 3). These properties are expected to be useful for globalization. We investigate them at an arbitrary point $x_c \in \text{int}(D)$ without any assumption on the existence of a solution to Eq. 1.

2. Smooth Equations

Consider the smooth case and suppose that the following assumption holds.

Assumption 1. *The mapping* $f : D \subset X \to Y$ *is Fréchet-differentiable at* $x_c \in \text{int}(D)$, *and the first derivative* f_c' *is invertible.*

Let p^N denote the Newton search direction $-(f_c')^{-1}f_c$ and let the scalars $\alpha_- < 0$ and $\alpha_+ > 0$ be such that $x_c + \alpha p^N \in D$ for all $\alpha \in (\alpha_-, \alpha_+)$.

Suppose that $f_c \neq 0$ and consider the behavior of the curve $f(x_c + \alpha p^N)$ in the space Y for $\alpha \in (\alpha_-, \alpha_+)$. The tangent to this curve at the point f_c is given by the derivative

$$(f(x_c + \alpha p^N))'_{\alpha=0} = f_c' p^N = -f_c. \tag{4}$$

Thus, the tangent vector connects the point f_c with the origin $0 \in Y$. Since the level sets of any norm are convex and concentric at 0, this tangent vector is a descent direction for $\| f \|$ at f_c. It implies for the curve that

$$\| f(x_c + \alpha p^N) \| < \| f_c \| \tag{5}$$

for all sufficiently small $\alpha > 0$. Hence the Newton direction is a descent direction of $\| f(x) \|$ for any norm. This fact is well known[8], but a stronger statement involving the derivative of $\| f(x) \|$ along p^N will be further discussed.

The reasoning above enables us to formulate the following property of the Newton direction, that can be used as a basis for various generalizations of the Newton's method for various problems.

Proposition 1. *Suppose that Assumption 1 holds and $f_c \neq 0$. Then the Newton direction is the unique direction which produces descent in $\| f(x) \|$ for all norms $\| \cdot \|$ in Y.*

Proof. Suppose, to the contrary, that there exists a vector $d \in X$, which is not collinear to p^N, but it is such that all norms of $f(x)$ decrease along this direction. Consider the curve $f(x_c + \alpha d)$ in the space Y. The tangent to this curve at the point f_c is a vector

$$\tau = (f(x_c + \alpha d))'_{\alpha=0} = f'_c d.$$

By assumption, it is not collinear to the tangent $-f_c$ to the curve generated by p^N. Hence the line $L = \{f_c + t\tau : t \in R\}$ does not contain 0. Denote for some norm in Y the closed ball of radius ϵ centered at 0 by ϵB. Then there exists a sufficiently small scalar $\epsilon > 0$, such that ϵB has an empty intersection with L. Denote the convex closed hull of f, $-f$ and ϵB by B'. Consider the Minkowski function[16]

$$\mu(y) = \min\{t \geq 0 : y \in tB'\}. \tag{6}$$

Since, by construction, B' is a restricted, closed, convex and equilibrated ($y \in B'$ implies $-y \in B'$) set containing 0 as an interior point, the functional $\mu(y)$ is a norm in Y. Note that the point f_c is the only intersection of L with B', and

$$(\mu(f_c + t\tau))'_{t=+0} > 0.$$

This implies that $\mu(f(x_c + \alpha d)) > \mu(f_c)$ for all sufficiently small $\alpha > 0$, because L is a tangent line to $f(x_c + \alpha d)$. This inequality contradicts the assumption that d is a descent direction of $\| f(x) \|$ for any norm, including the one given by Eq. 6. $\quad\square$

Though the norm $\| f(x) \|$ is, generally speaking, not differentiable with respect to x, it turns out that the function $\| f(x_c + \alpha p^N) \|$ of the argument α is differentiable at $\alpha = 0$. Moreover, in the case of arbitrary norm the formula

$$\| f(x_c + \alpha p^N) \|'_{\alpha=0} = - \| f_c \| \tag{7}$$

holds independent of the distance of x_c from the solution of Eq. 1, indeed, independent of the existence of such a solution. This formula was published[14] earlier and its complete proof was given[15] for the finite dimensional case. By Proposition 3 of the next section, it is valid also for the infinite dimensional case.

The Armijo-Danilin-Goldstein line search algorithms are used widely for globalization of unconstrained minimization methods[8,9]. They use the derivative of the smooth minimized function along the search direction. Formula Eq. 7 provides a basis for generalizing them for the case of nonlinear equations, if one regards $\| f(x) \|$ as a function (nondifferentiable, in general) to be decreased along the Newton direction or its approximation. The formula was applied in the case of smooth Eq. 1 for constructing various globalization strategies for the Newton's[14,15] and quasi-Newton's[17] methods. Moreover, this property of the Newton direction seems to be fundamental for other globalization strategies (see[9,10] for the references). Formula Eq. 7 and Proposition 4 of Section 3 (appeared earlier[18]) were also used[19] in developing of a global convergence theory for arbitrary norm trust-region methods for smooth equations.

3. Nonsmooth Equations

Supposing here that the mapping f is nonsmooth, we regard the smooth equations as a special case. Consider the following generalization of Newton's method suggested by Robinson[4]. Assume that at the current iteration there are x_c and an approximation $A_c : D_c \subset X \to Y$ $(D_c \subset D)$ of the mapping f. The solution of Eq. 3 gives a new approximation to the solution of Eq. 1. Though the mapping A_c is in general nonsmooth, Eq. 3 is supposed to be much easier to solve than Eq. 1. Under some reasonable assumptions on the mappings f and A_c, the iterations based on this idea have a local superlinear (or quadratic) rate of convergence to the solution of Eq. 1.

In order to enlarge the convergence region of the Newton-Robinson method, Ralph[7] introduced a so-called Newton path

$$p(\alpha) = A_c^{-1}((1 - \alpha)f_c)$$

and suggested a search along the path, which is generally nonsmooth because of the nonsmoothness of f. The iterations based on this approach coincide automatically with the Newton-Robinson iterations in the local vicinity of the solution of Eq. 1. The characteristic feature of the Newton path is that the value of A_c goes from f_c to zero "rapidly" on the path $p(\alpha)$. Note, that in the smooth case, where the Newton iterations are based on the linear mapping Eq. 2, the Newton path is given by the formula

$$p(\alpha) = x_c + \alpha p^N.$$

In our analysis of the Newton path we will use the following

Assumption 2. *There exists a mapping $A_c : D_c \subset X \to Y$ $(D_c \subset D)$ that approximates $f : D \subset X \to Y$ for $x_c \in \mathrm{int}(D_c)$, such that*

$$\| f(x) - A_c(x) \| = o(\| x - x_c \|). \tag{8}$$

For some scalars $\alpha_- < 0 < \alpha_+ \leq 1$ there exists a mapping $p : (\alpha_-, \alpha_+) \to X$, such that $p(\alpha) \in D_c$ and

$$A_c(p(\alpha)) = (1 - \alpha)f_c \tag{9}$$

for all $\alpha \in (\alpha_-, \alpha_+)$, and also such that

$$\| p(\alpha) - x_c \| = O(\alpha). \tag{10}$$

Note that this assumption is less restrictive than those used by Ralph[7]. In particular, it requires neither the invertibility nor the continuity of the mappings f and A_c. Moreover, the continuity of the mapping p is assumed only at $\alpha = 0$.

Consider properties of the Newton path $p(\alpha)$. In the following proposition we show that in spite of the possible nondifferentiability of $f(x)$ at x_c and $p(\alpha)$ at $\alpha = 0$, the curve $f(p(\alpha))$ is smooth at $\alpha = 0$, and give a formula for the tangent line similar to Eq. 4.

Proposition 2. *Suppose Assumption 2 holds. Then the mapping $\alpha \mapsto f(p(\alpha))$ is differentiable at $\alpha = 0$, and the first derivative is given by the formula*

$$(f(p(\alpha)))'_{\alpha=0} = -f_c. \tag{11}$$

Proof. It is necessary to show that

$$\| f(p(\alpha)) - f_c + \alpha f_c \| = o(\alpha). \tag{12}$$

Eqs. 8 and 10 give

$$\| f(p(\alpha)) - A_c(p(\alpha)) \| = o(\| p(\alpha) - x_c \|) = o(\alpha). \tag{13}$$

Substituting here Eq. 9 we finally prove Eq. 12. $\square$

Now we can derive a formula similar to Eq. 7. We will show that the function $\alpha \mapsto \| f(p(\alpha)) \|$ is differentiable at $\alpha = 0$, in spite of the possible nondifferentiability of both $\| \cdot \|$ and the other mappings that compose the function.

Proposition 3. *Suppose Assumption 2 holds. Then the function $\alpha \mapsto \| f(p(\alpha)) \|$ is differentiable at $\alpha = 0$ and the first derivative is given by the formula*

$$\| f(p(\alpha)) \|'_{\alpha=0} = - \| f_c \| . \tag{14}$$

Proof. Denote $\phi(\alpha) = \| f(p(\alpha)) \|$. It is necessary to show that

$$\phi(\alpha) - \phi(0) - \alpha\phi'(0) = o(\alpha), \tag{15}$$

where $\phi'(0) = -\phi(0)$. Indeed, from Eqs. 8 and 10 we have

$$| \| f(p(\alpha)) \| - \| f(p(0)) \| - \alpha (- \| f_c \|) | = | \| f(p(\alpha)) \| - \| (1 - \alpha) f_c \| |$$

$$= | \| f(p(\alpha)) \| - \| A_c(p(\alpha)) \| | \leq \| f(p(\alpha)) - A_c(p(\alpha)) \| .$$

Applying here Eq. 13, we prove Eq. 15. $\square$

It is possible to show by simply modifying the proofs of Propositions 2 and 3, that Eqs. 11 and 14 remain true, if one substitutes Eqs. 8 and 10 by

$$\| f(x) - A_c(x) \| = o(\| f_c - A_c(x) \|).$$

In this case the discontinuity of p at $\alpha = 0$ becomes admissible as well.

Note that if in Assumption 2 $p(\alpha)$ is defined for $\alpha \geq 0$ only, then $\alpha = 0$ should be substituted in Eqs. 11 and 14 by $\alpha = +0$.

Definition. *We call $x_c \in \mathrm{int}(D)$ a degenerate point of the mapping $f : D \subset X \to Y$, if Assumption 2 does not hold for it.*

By this definition, a point $x_c \in \mathrm{int}(D)$ with $f_c = 0$ is never degenerate, because in this case Assumption 2 holds obviously for $A_c \equiv f$ and $p(\alpha) \equiv x_c$.

Proposition 4. *Let $f : D \subset X \to Y$. Suppose $x_c \in \mathrm{int}(D)$ is a local minimizer for the function $\| f(x) \|$ in some norm. Then either $f_c = 0$ or x_c is a degenerate point.*

Proof. Suppose, to the contrary, that $f_c \neq 0$ and x_c is not a degenerate point. Then by Assumption 2 and Proposition 3, there exists a Newton path $p(\alpha)$, such that

$$\| f(p(\alpha)) \|'_{\alpha=0} < 0. \tag{16}$$

According to Eqs. 9 and 10, $p(0) = x_c$, $p(\alpha) \neq x_c$ for all $\alpha \neq 0$, and the mapping $p(\alpha)$ is continuous at $\alpha = 0$. Hence, Eq. 16 means that in any neighborhood of x_c there is a point x such that $\| f(x) \| < \| f_c \|$. This contradicts the assumption that x_c is a local minimizer for $\| f(x) \|$. $\qquad\square$

Note, that if $x_c \in \mathrm{int}(D)$ is a degenerate point of a mapping $f : D \subset X \to Y$, which is Fréchet-differentiable at x_c, then f'_c is not invertible. For this smooth case, Proposition 4 implies the following

Corollary. *Let the mapping $f : D \subset X \to Y$ be Fréchet-differentiable at $x_c \in int(D)$. Suppose x_c is a local minimizer for the function $\| f(x) \|$ in some norm. Then $f_c = 0$ or f'_c is not invertible.*

One can see that the properties of the Newton path presented in this section are very similar to those of the Newton direction (compare, e.g., Eq. 4 and Eq. 11, and also Eq. 7 and Eq. 14). It confirms that the Newton path is a natural generalization of the Newton direction.

In conclusion, we note that it is possible to use formula Eq. 14 for constructing various globalization strategies for solving nonsmooth equations based on the Armijo-Danilin-Goldstein step-size rules. All the strategies suggested by the author[14,15] for smooth equations are directly applicable to the nonsmooth case with the use of the approach developed by Ralph[7]. Though the latter paper[7] focuses on finite dimensional

Euclidean spaces X and Y, the main results of that work remain valid also when X and Y are Banach spaces with arbitrary norm.

Acknowledgements

The author would like to thank Natalia Alexandrov, Yuri Ledyaev and Danny Ralph for their helpful suggestions, Anatoli Belyankov for suggesting Proposition 1 and asking if it is true, and also Georgi Smirnov, who simplified the initial proof of this proposition.

References

[1] O.L. Mangasarian, Equivalence of the complementarity problem to a system of nonlinear equations, *SIAM J. Appl. Math.* **31** (1976), 89–91.

[2] N.H. Josephy, Newton's method for generalized equations, Technical Summary Report #1965, Mathematics Research Center, University of Wisconsin-Madison (1979).

[3] S.M. Robinson, Generalized equations, in *Mathematical Programming: The State of the Art*, ed. by A. Bachem *et al.*, Springer-Verlag, Berlin (1983), 346–367.

[4] S.M. Robinson, Newton's method for a class of nonsmooth functions, manuscript (Department of Industrial Engineering, University of Wisconsin-Madison, Madison, WI 53706, August 1988).

[5] B. Kummer, Newton's method for non-differentiable functions, in *Mathematical Research. Advances in Mathematical Optimization*, ed. by J. Guddat *et al.*, Akademie Verlag, Berlin (1988), 114–125.

[6] P.T. Harker and J.-S. Pang, Finite-dimensional variational inequality and nonlinear complementarity problems: A survey of theory, algorithms and applications, *Math. Prog.* **48** (1990), 161-220.

[7] D. Ralph, Global convergence of damped Newton's method for nonsmooth equations, via the path search, *Math. of Oper. Res.* **19** (1994), 352–389.

[8] J.M. Ortega and W. Rheinboldt, *Iterative Solution of Nonlinear Equations in Several Variables*, Academic Press, New York, 1970.

[9] J.E. Dennis and R.B. Schnabel, *Numerical Methods for Nonlinear Equations and Unconstrained Optimization*, Prentice Hall, Englewood Cliffs, New Jersey, 1983.

[10] M.C. Ferris and S. Lucidi, Nonmonotone stabilization methods for nonlinear equations, *J. of Optim. Theory and Applics* **81** (1994), 53–71.

[11] L. Armijo, Minimization of functions having continuous partial derivatives, *Pacific J. Math.* **16** (1966), 1–3.

[12] Ju. M. Danilin, On an approach to minimization problems, *Soviet Math. Dokl.* **10** (1969), 1274-1275.

[13] A. Goldstein, On steepest descent, *SIAM J. Control* **3** (1965), 147–151.

[14] O.P. Burdakov, Some globally convergent modifications of Newton's method for solving systems of nonlinear equations, *Soviet Math. Dokl.* **22** (1980), 376–378.

[15] O.P. Burdakov, *Numerical Methods of Saddle Point Search, Dissertation Thesis*, Moscow Inst. of Engineering Physics (MPhTI), Dolgoprudnyj (1980), Chap. 2, 57–69. (in Russian)

[16] A. N. Kolmogorov and S. V. Fomin, *Elements of the Theory of Functions and Functional Analysis*, Graylock Press, Rochester, N.Y., 1961.

[17] O.P. Burdakov, A curvilinear search for globalization of quasi-Newton methods, Preprint No. 5/91, Technische Hochschule Köthen, Inst. für Mathematik, Köthen (1991).

[18] O.P. Burdakov, On some properties of Newton's method for solving smooth and nonsmooth equations, Preprint No. 07–13–91, Technische Univ. Dresden, Inst. für Numerische Mathematik, Dresden (1991).

[19] M. El Hallabi and R.A. Tapia, A global convergence theory for arbitrary norm trust-region methods for nonlinear equations, Tech. Rep. TR93-41, Dept of Computational and Applied Mathematics, Rice Univ., Houston (1993).

WSSIAA 5 (1995) pp. 25–34
© World Scientific Publishing Company

RELATIONS BETWEEN INVEX PROPERTIES

B. D. Craven, University of Melbourne.
*Mathematice Dept., University of Melbourne,
Parkville, Victoria 3052, Australia*

ABSTRACT

Optimization problems under invex hypotheses preserve many properties of convex problems. However a vector function may be invex at a point without being convexifiable by a transformation. It is shown that a generalization of Weir and Mond's preinvex property, not requiring derivatives, characterizes invex at a point, and a further generalization characterizes the convexifiable property. The results are applied to Lagrangian duality.

0. Introduction

It is well known that many properties of convex optimization problems are preserved when the vector of functions involved has a weaker *invex* property. However, the properties of invex at one point, invex at each point in a region (essentially the original definition), or transformable to a convex function (the property that gave rise to the term invex) must be distinguished, since they are not equivalent. Moreover, while *convex* can be defined without reference to any derivative, *invex* appears to need some sort of derivative to describe it. Weir and Mond [6] defined *pre-invex* without derivatives, ande *pre-invex* implies *invex*, though not conversely.

It is now shown that a generalization of *pre-invex* characterizes *invex* at a point, and a further generalization characterizes the transformable property in a region. These necessary and sufficient properties do not use any derivatives. Other conditions are found, under which invex at each point implies the transformable property.

1. Invex and convexifiable

Let $X = R^n$, $X_0 \subset X$ a convex open domain, $x_0 \in X_0$, $Y = R^m$, $K \subset Y$ a closed convex order cone, with int $K \neq \varnothing$, and $\geq_K$ the ordering defined by K. The notation f' is used for Fréchet derivative of a function f. Let $E \subset X_0$. A differentiable function $\varphi : X_0 \to Y$ is *convex on* E with respect to K, if

$$(\forall x \in X_0, \forall u \in E) \qquad \varphi(x) - \varphi(u) \geq_K \varphi'(u)(x - u) . \qquad (1)$$

Note that E may be chosen as a singleton $\{x_0\}$, or some neighbourhood of $x_0 \in X_0$, or as the whole domain. The analogous *invex* properties are the following.

Definition 1 A differentiable function $f : X_0 \to Y$ is *invex on* E with respect to the order cone K if a continuous map $\omega : X_0 \times X_0 \to X_0$ exists such that

$$(\forall x \in X_0)(\forall u \in E) \; f(x) - f(u) \geq_K f'(u)\omega(x-u,u) . \qquad (2)$$

Hanson [3] gave thus definition for real functions f, writing $\eta(x,u)$ for the present $\omega(x-u,u)$; his assumption that the same $\eta(.,.)$ applied to each component of f is more conveniently expressed as (2). The function $\omega(.,.)$ is called the *scale function* for f. Note that *invex on* a singleton $\{x_0\}$ does not imply invex on $\{u\}$ when $u \neq x_0$. It was shown in [2] that if $g : X_0 \to Y$ is *convex on* E with respect to K, where E is $\{x_0\}$ or a sufficiently small neighbourhood of x_0, and if $\psi : X_0 \to X_0$ is differentiable, with $\psi(x_0) = x_0$ and the gradient $\psi'(x_0)$ invertible, then the composition $g \circ \psi$ is invex on E.

Definition 2 A differentiable function $f : X_0 \to Y$ is *convexifiable on* E with respect to the cone K if there exists a differentiable map $\varphi : X_0 \to X_0$ with $(\forall u \in E) \varphi'(u); X \to X$ invertible, such that the composition $f \circ \varphi$ is *convex on* E with respect to K.

If $h = f \circ \varphi$ and $\psi = \varphi^{-1}$, then $f = h \circ \psi$. If f is also directionally differentiable at x_0, then

$$\lambda^{-1}[f(x_0 + \lambda(x-x_0)) - f(x_0)] \leq f(x) - f(x_0) ,$$

which implies, as $\lambda\downarrow 0$, that the directional derivative
$$f'(x_0 ; x-x_0) \leq_K f(x) - f(x_0) . \tag{3}$$
Thus f is *invex on* $\{x_0\}$ with $(\forall z)\ \omega(z;x_0) = z$. Conversely, if f is invex on $\{x_0\}$ with $\omega(0)=0$ and $\omega'(0) = 1$ (identity map), let $q := f\circ\omega^{-1}$; then
$$q(z) - q(0) = f(\omega^{-1}(z)) - 0 \geq_K f'(0)\omega(\omega^{-1}(z)) = q'(0)z .$$
This q is *convex at* 0, but not generally convex at other points.

The term *invex* arose (Craven [2]) from the following construction. Let $h : R^n \to R^r$ and $\varphi : R^n \to R^n$ be continuously differentiable; let h be K-convex; assume that $(\forall u \in E)\ \varphi'(u); X \to X$ is invertible. Then $f := h\circ\varphi$ satisfies
$$(\forall u, u+z \in E)\ f(u+z)-f(u) \geq_K h'(\varphi(u))\ [\varphi(u+z)-\varphi(u)]$$
$$= f'(u)\varphi'(u)^{-1}[\varphi(u+z)-\varphi(u)]$$
$$\equiv f'(u)\omega(z;u) \tag{4}$$
where
$$\omega(z;u) := \varphi'(u)^{-1}[\varphi(u+z)-\varphi(u)] = z + o(\|z\|). \tag{5}$$
Thus invexity on E follows, but not conversely.

2. Extending *pre-invex*

Weir and Mond [7] defined $f : X \to R$ to be *pre-invex* at $x_0 \in X$ if
$$(\forall\lambda\in(0,\delta))\ (\forall x)\ f(x_0 + \lambda\omega(x-x_0)) \leq (1-\lambda)f(x_0) + \lambda f(x) ,$$
for some $\delta > 0$. This implies, similarly to (3), that f is *invex on* $\{x_0\}$ with *scale function* $\omega(.) \equiv \omega(.;x_0)$, namely
$$(\forall x)\qquad f'(x_0;\omega(x-x_0)) \leq f(x) - f(x_0). \tag{6}$$
Since this deduction depends on letting $\lambda\downarrow 0$, (6) still follows if terms with higher powers of λ are added to $\lambda\omega(x-x_0)$.

Definition 3 The function $f : X_0 \to Y$ is *generalized pre-invex on* $E\subset X_0$ with respect to the order cone K if
$$(\forall u \in E, \forall x \in X_0, \forall\lambda\in(0,1))$$
$$f(u + \Delta(\lambda,x-u,u)) \leq_K (1-\lambda)f(u) + \lambda f(x), \tag{7}$$
where $\varphi : X_0 \to X_0$ is an invertible function, and
$$\Delta(\lambda,x-u,u) := -u + \varphi((1-\lambda)\varphi^{-1}(u) + \lambda\varphi^{-1}(x))) . \tag{8}$$

28

Remarks If, at $x_0 \in E$, there hold $\varphi^{-1}(x_0) = 0$ and the gradient $\varphi'(0) = 1$ (the identity map), and if f is differentiable, then (7) gives

$$f(x) - f(x_0) \geq_K f'(u).(\partial/\partial\lambda)\Delta(\lambda,x-u,u)|_{\lambda=0} = \varphi^{-1}(x), \tag{9}$$

thus f is invex on $\{x_0\}$ with scale function $\omega(.,x_0) = \varphi^{-1}(x)$. At another point $u \in E$, $u \neq x_0$, supposing φ differentiable and φ invertible, and denoting $\theta := \varphi^{-1}$,

$$(\forall x \in X) \; f(x) - f(u) \geq_K f'(u)\varphi'(\theta(u))[\theta(x) - \theta(u)]$$
$$= f'(u) [(x-u) + o(\|x-u\|)] ; \tag{10}$$

thus f is invex on $\{u\}$.

Theorem 1 The function $f : X_0 \to Y$ is *generalized pre-invex on* $E \subset X_0$ if and only if $f \circ \varphi$ is convex on E, thus if f is *convexifiable on* E.

Proof Let f be convexifiable on E; let $F := f \circ \varphi$, with φ from Definition 2. Let $u \in E$, $x \in X_0$, $0 < \lambda < 1$, $u = \varphi(v)$ and $x = \varphi(y)$. Since F is convex,

$$f(\varphi((1-\lambda)\varphi^{-1}(u) + \lambda\varphi^{-1}(x))) = F((1-\lambda)v + \lambda y)$$
$$\leq_K (1-\lambda)F(v) + \lambda F(y)$$
$$= (1-\lambda)f(u) + \lambda f(x), \tag{11}$$

so that f is generalized pre-invex on E.

Conversely, let f be generalized pre-invex on E, with φ from Definition 3; let $F := f \circ \varphi$. Then

$$F((1-\lambda)v + \lambda y) = f(\varphi((1-\lambda)\varphi^{-1}(u) + \lambda\varphi^{-1}(x)))$$
$$\leq_K (1-\lambda)f(u) + \lambda f(x)$$
$$= (1-\lambda)F(v) + \lambda F(y),$$

so that F is convex.

$$\square$$

Remark The scale function at x_0, $\omega(.;x_0)$, determines the mapping φ, which in turn determines the function Δ. If $\omega(x-u;u) = M(u)(x-u) + o(\|x-u\|)$, where $M(u)$ is an invertible mapping, and if φ is a C^2-function, then, considering only terms up to second order,

$$\varphi^{-1}(z) = z + \tfrac{1}{2}x \backslash z^T M z + \ldots , \tag{12}$$

then $\varphi(w) = w - \tfrac{1}{2}w^T M_{\cdot} w$. Substituting into $\Delta(.)$ and simplifying gives

$$\Delta(\lambda,x-u,u) = \lambda(x-u) + \tfrac{1}{2}L(1-\lambda)(x-u)^T M_{\cdot}(x-u) + \dots . \tag{13}$$

Assume now f is locally Lipschitz. Define

$$\rho(w,\lambda) := \lambda^{-1}\Delta(\lambda,w,u).$$

Then

$$f(x) - f(u) \geq \lambda^{-1}[f(u+\Delta(\lambda,x-u,u) - f(u)]$$
$$= \lambda^{-1}[f(u+\lambda\rho(x-x_0,0)) - f(u)]$$
$$+ \lambda^{-1}[f(u+\lambda\rho(x-x_0,\lambda)-f(u+\lambda\rho(x-x_0,0)] . \tag{14}$$

If f is real-valued, the first term tends to $f^{\circ}(x_0;\rho(x-x_0,0)$ for suitable sequences $\{u_k\}\to x_0$ and $\{\lambda_k\}\to 0$. If f is vector-valued, the first term tends to an element of the Clarke generalized Jacobian $\partial f(x_0)$ of f at x_0, for suitable sequences $\{u_k\}\to x_0$ and $\{\lambda_k\}\to 0$. Given f locally Lipschitz, the second term tends to 0, if also $\rho(.,.)$ is Lipschitz. These results do not follow from (7), since the Clarke derivative and Jacobian require neighbouring points to x_0. This has proved:

Theorem 2   Assume that f is locally lipschitz, and *generalized pre-invex on* a neighbourhood E of x_0, with $\lambda^{-1}\Delta(.,.,.)$ a Lipschitz. function. Then

(f real-valued): $\qquad (\forall x)\ f(x) - f(x_0) \geq f^{\circ}(x_0;\omega(x-x_0)\ ;$ $\qquad$ (15)

(f vector-valued): $\quad (\forall x)\ (\forall \zeta \in \partial f(x_0))\ f(x) - f(x_0) \geq\ <\zeta,\ \omega(x-x_0);$ $\quad$ (16)

where $\omega(.) := \varphi(.)^{-1}$, with $\varphi(.)$ as in Theorem 1.

Remark The version of generalized invex proved here was introduced in [1], and was used there for sufficient conditions for an optimum.

Remark It is well known that a local minimum of a real-valued invex function is necessarily a global minimum. The domain can be the whole space, or a convex subset. There is a similar result for *generalized pre-invex*, which does not require any kind of derivative for the function.

Theorem 3 Let f be a real-valued function, which is *generalized pre-invex on* $\{x_0\}$. If x_0 is a local minimum of f, then x_0 is a global minimum of f.

Proof Suppose x_0 is not a global minimum; then $c := f(x_0) - f(p) > 0$ for some point p. From *generalized pre-invex on* $\{x_0\}$,

$$(\forall \lambda \in (0,1)) \quad f(x_0 + \Omega(\Lambda, p - x_0)) - f(x_0) \leq -\lambda c, \tag{17}$$

contradicting the local minimum, as $\lambda \downarrow 0$. $\square$

3. Lagrangian duality with invex

Consider the optimization problem:

$$\text{MIN } f(x) \text{ subject to } -g(x) \in S, x \in \Gamma , \tag{18}$$

where Γ is a convex set, and S is a convex cone. Assume that a minimum is reached at $x = x_0$. If f is convex, g is S-convex, and a Slater constraint qualification is assumed, then a well known duality result holds :

$$\text{MAX}_{v \in S*} \text{ MIN}_{x \in \Gamma} \; f(x) + vg(x) = f(x_0) , \tag{19}$$

where $S*$ denotes the dual cone of S.

Suppose now instead that Γ is the whole space, and that the vector function consisting of f(.) and g(.) is *generalized pre-invex on* a neighbourhood N of x_0, with respect to the cone $R_+ \times S$. Note that no differentiability has been assumed. By Theorem 1, this vector function is *convexifiable on* N. Hence, for some invertible mapping φ, with $\varphi(x_0) = x_0$ and $\varphi'(x_0) = 1$, the vector function consisting of $f \circ \varphi$ and $g \circ \varphi$ is convex with respect to $(R_+ \times S)$ on N, and the transformed problem (with LMIN denoting *local minimum*)

$$\text{LMIN } f \circ \varphi(z) \text{ subject to } -g \circ \varphi(z) \in S \tag{20}$$

reaches a local minimum at x_0. Aseume now a Slater constraint qualification, that $-g(q) \in \text{int } S$ for some q in a sufficiently small neighbourhood of x_0, thus with $q \in \varphi^{-1}(N)$. Then a Slater qualification holds for the transformed problem. Hence

$$\text{MAX}_{v \in S*} \text{ LMIN}_z \; f \circ \varphi(z) + vg \circ \varphi(z) = f(x_0) , \tag{21}$$

Hence the following duality relation holds for the original problem:

$$\text{MAX}_{v \in S*} \text{ LMIN}_x \; f(x) + vg(x) = f(x_0) . \tag{22}$$

4. When does *invex on E* imply *convexifiable on* E ?

If f is *convexifiable on* a neighbourhood of 0 then, from (5),
$$\omega(z;x) := \varphi'(x)^{-1}[\varphi(x+z)-\varphi(x)] = z + o(\|z\|). \tag{23}$$
Note that, if $\varphi(x_0)=x_0$ and $\varphi'(x_0)=1$ (the identity map), then φ is determined by ω, from $\omega(.;x_0) = \varphi(.)$. In order to deduce *convexifiable* from *invex on* E, some hypothesis is needed, about the dependence of $\omega(.;.)$ on its second argument. Note that *invex on* E does not specify $\omega(.;.)$ uniquely; what is required is that some scale function ω exists, with a suitable property. Denote by ω_1 the partial derivative of ω with respect to its first argument.

Theorem 4 Let f be differentiable and invex on a neighbourhood N of $x_0 \in X_0$, with scale function $\omega(.;.)$ such that $\omega(.,x_0)$ is bijective on N of x_0, $\omega(z,x)=z+o(\|z\|)$ for each $x \in X_0$, and
$$\omega_1(x,x_0)\omega(z,x) = \omega(x+z,x_0) - \omega(x,x_0) \tag{24}$$
Then $f = \theta \circ \varphi$, where θ is differentiable and convex on N with respect to K, and φ is differentiable and bijective, with $\varphi'(a)=1$.

Remark If $\Phi(x_0)=x_0$ and $\Phi'(x_0)=1$ are assumed, then (5) gives $\varphi(.)=\omega(.;x_0)$. Substituting this expression for $\varphi(.)$ into (5) gives (24). Thus the condition (264 is necessary as well as sufficient.

Proof Define $\varphi(.) := \omega(.,x_0)$. Then $\varphi'(x_0)=1$, and $\varphi(.)$ is bijective on a neighbourhood of x_0. Let $\psi:=\varphi^{-1}$, and let $\theta:=f \circ \psi$. Let $\xi:=\varphi(x)$ and $\sigma:=\varphi(x+z)$. From hypothesis (24),
$$\varphi'(x)\omega(z,x) = \varphi(x+z) - \varphi(x). \tag{25}$$
Then
$$\theta(\sigma) - \theta(\xi) = f(\psi(\sigma)) - f(\psi(\xi))$$
$$\geq_K f'(\psi(\xi))[\omega(\psi(\sigma) - \psi(\xi)), \psi(\xi)]$$
$$= \theta'(x) \circ \psi'(\xi)^{-1}\omega(z,x)$$
$$= \theta'(x)[\varphi(x+z) - \varphi(x)] . \tag{26}$$
Thus θ is convex at each point $x \in N$, and $f = \theta \circ \varphi$. $\qquad\square$

Remark If $x=0$ and $\varphi(0)=0$, then
$$\theta(\sigma) - \theta(0) = f(\psi(\sigma)) - f(\psi(0)) . \tag{27}$$

Substitute $\sigma := \beta(s)$, where $\beta(s) := sv$, $0 \leq s \leq t$, with v a fixed vector. Now $f \circ \psi$ is convex if f is r-invex, and the other hypotheses of Theorem 5 apply, and then

$$\theta(\beta(s)) - \theta(\beta(0)) = (f \circ \psi)(\beta(s))) - (f \circ \psi)(\beta(0)). \qquad (28)$$

Remark   This may be compared with Pini's [5] definition of an *integrable* function $\eta(.,.)$ on a manifold, in the special case when the manifold is R^n. This property requires the existence of a function $\alpha : [0,1] \to R^n$ for which (in present notation) $\alpha(0) = \xi = 0$, $\alpha(1) = \sigma$, and

$$(\forall t, u \in [0,1]) \qquad \alpha'(u)(t-u) = \eta(\alpha(t), \alpha(u)). \qquad (29)$$

Under this hypothesis, applied to a real-valued r-invex function f, Pini obtains $f \circ \alpha(.)$ to be convex.   Thus Pini's *invex* is effectively *invex on* a neighbourhood, with the integrability condition (29) added. Note that substitution of $\rho := \varphi^{-1}$, $t := \varphi(x+z)$ and $u := \varphi(u)$ into (8) gives

$$\omega(\rho(t)-\rho(u)), \rho(u)) = \rho'(u)(t-u) \ , \qquad (30)$$

which closely resembles (29), except that in (30) t and u are vectors.

Remark   Expanding (24) in powers of z, assuming now that $\omega(.;.)$ is a C^2 function, and equating terms of first order in z, then terms of second order in z, shows (setting $x_0 = 0$ for simplicity) that

$$\text{(linear terms):} \qquad \omega_1(y;0)\omega_1(0;y) = \omega_1(y;0) \qquad (31)$$
$$\text{(This is fulfilled since } \omega_1(0;y) = 1)$$
$$\text{(quadratic terms):} \qquad \omega_1(y;0)\omega_{11}(0;y) = \omega_{11}(y;0) \qquad (32)$$

Here ω_{11} denotes the second derivative of ω with respect to its first argument; and composition by matrix multiplication is assumed.

Since (32) is a differential equation involving the gradient of a function on R^n, some compatibility conditions (analogs of $\nabla h = q \Rightarrow$ curl $q = 0$ in R^3) must be satisfied. So (32) cannot hold trivially, in general. It remains to find whether (32) is sufficient, as well as necessary.

5. Relation to other generalizations

In [6] and [1], several different kinds of generalized convex were defined, which relate to those introduced here. In the present notation, a function f is

B-vex at x_0 if
$$(\forall x)\,(\forall \lambda \in (0,1))\quad f(x_0 + \lambda(x-x_0)) \le (1-b)f(x_0) + bf(x) \qquad (33)$$
in which $v \equiv b(x_0,x,\lambda) \in (0,1)$.

This definition allows extension in the manner of Definition 3. Define f to be

generalized B-vex at x_0 if
$$(\forall x)\,(\forall \lambda \in (0,1))\quad f(x_0 + \Omega(\lambda,x-x_0)) \le (1-b)f(x_0) + bf(x) \qquad (34)$$
in which $v \equiv b(x_0,x,\lambda) \in (0,1)$, and $\Omega(.,.)$ satisfies (8), with $\omega(.,x_0)$ invertible. If $\Omega(.,.)$ and $f(.)$ are differentiable, then (34) implies that
$$(\forall x)\quad f'(x_0)\omega(x-x_0) \le v(x,x_0)[f(x) - f(x_0)] , \qquad (35)$$
where $\omega(x-x_0) = \Omega_\lambda(0,x_0)$ and $v(x,x_0) = b_\lambda(x_0,x,0)$. This may be compared with the V-invex property of Jeyakumar and Mond [5], namely
$$(\forall x)\quad f_i(x) - f_i(x_0) \ge \alpha_i(x,x_0)f_i{}'(x_0)\eta(x,x_0) , \qquad (36)$$
with positive $\alpha_i(x,x_0)$. Clearly, (35) and (36) agree if $v(x,x_0) > 0$, the order cone is an orthant, and $\eta(x,x_0) = \omega(x-x_0)$, $\alpha_i(x,x_0) = 1/v_i(x,x_0)$.

If (34) is assumed, and the origins are shifted so that $x_0=0$, $f(x_0)=0$, then a similar proof to that of Theorem 1 shows that there is an invertible function $\varphi(.) = \omega(.)^{-1}$ such that $\varphi(x_0)=x_0$ and
$$(\forall y)\quad f\circ\varphi(\lambda y) \le b(\varphi(x_0),\varphi(y),\lambda)f\circ\varphi(y) . \qquad (37)$$
Thus generalized B-invex at x_0 is tramsformed to B-vex at x_0.

Analogous definitions could be given, with at x_0 replaced by on E.

6. References

1. B.D. Craven, Nondifferentiable optimization by smooth approximations, Optimization <u>17</u> (1986), 3-17.
2. B. D. Craven, Duality for generalized convex fractional programs,

pages 473-489 of *Generalized Concavity in Optimization and Economics*, S. Schaible and W.T. Ziemba (eds.), Academic Press, New York (1981).

3. M.A. Hanson, On sufficiency of the Kuhn-Tucker conditions, J. Math. Anal. Appl. 80 (1981), 545-550

4. V. Jeyakumar and B. Mond, Om generalized convex mathematical programming, J. Austral. Math. Soc., Series B, **34** 91992), 43-53.

5. R. Pini, Convexity along curves and invexity, Optimization, to appear.

6. T. Weir and B. Mond, Pre-invex functions in multiple objective optimization, J. Math. Anal. Appl. 126 (1988), 29-38.

WSSIAA 5 (1995) pp. 35–49

DUAL CONES IN SPACES OF CONVEX SETS AND FUZZY SETS

PHIL DIAMOND, A. VLADIMIROV

Mathematics Department, University of Queensland, Australia 4072

and

P. KLOEDEN

School of Computing and Mathematics, Deakin University,
Geelong, Australia 3217.

ABSTRACT

Kuhn–Tucker conditions for mathematical programming problems with fuzzy objective functions or fuzzy constraints require the characterization of dual cones of spaces of convex homogeneous functions. This is trivial when the fuzzy sets are all defined on the real line. Here the characterization is established for fuzzy sets defined on a two–dimensional base space.

1. Introduction

Increasing interest in fuzzy control has motivated, in recent years, many publications concerned with fuzzy optimization problems. See, as a representative example, [6, 9, 10, 13, 18, 19, 20, 22, 24, 25, 26].

The fuzziness may be interpreted as a lack of precise knowledge about the actual behavior of the system under consideration. In these settings, it is natural to consider a problem of constrained optimization:

$$\text{maximize } F(x) \quad \text{subject to} \quad G(x) \le 0, \tag{1}$$

with fuzzy or crisp objective function $F(x)$ and fuzzy constraints $G(x)$. This actually was done both for the cases of linear [6, 13, 9, 18, 20] and nonlinear [4, 10, 19, 22, 25, 26] programming problems. A convenient technique for the problems of type Eq. 1 is the method of Lagrange multipliers and Kuhn–Tucker theorems [1, 2, 7, 12, 21]. These are based on the representation of dual cones in various spaces.

Here we address only the problem of characterization of certain dual cones in the space of convex homogeneous functions. It is related to the problem Eq. 1 for the class of convex fuzzy functions; this characterization will be used elsewhere to derive, in certain cases, explicit necessary and sufficient conditions of optimality in fuzzy programming problems.

2. Statement of the problem

Definition 1 *Let Q be a positive real cone. The dual positive cone $Q^{\oplus}$ is defined to be*

$$Q^{\oplus} = \{\mu \in Q^* : \langle q, \mu \rangle \geq 0 \text{ for all } q \in Q\},$$

where Q^* is the dual of Q.

The measure derivative Df of a locally integrable function f is defined in the distributional sense. Characterization of the positive dual cones will involve solutions of measure differential equations. Further details on these may be found in [14, 16].

Definition 2 *Let μ be a discrete measure on an interval $I \subseteq \mathfrak{R}$. The expression*

$$DY = F(Y) + \mu \tag{2}$$

denotes a measure differential equation. A function $Y(t)$ is said to be a solution of Eq. 2 if it is a right continuous function of bounded variation on I and the measure derivative of Y satisfies Eq. 2.

Consider the space $\mathcal{K}_C$ of nonempty compact convex subsets of $\mathfrak{R}^n$ with the Hausdorff metric D_H. Let $s_A(x)$ be the support function of a compact set $A \subset \mathfrak{R}^n$,

$$s_A(x) = \max_{a \in A} \langle a, x \rangle,$$

where $\langle a, x \rangle$ is the inner product in $\mathfrak{R}^n$. Denote the set of all the convex homogeneous functions defined on all $\mathfrak{R}^n$ by CH. A function $s : \mathfrak{R}^n \to \mathfrak{R}$ is a support function of some compact convex set in $\mathfrak{R}^n$ iff it belongs to CH [1]. Clearly, CH is a convex cone.

Denote by S^{n-1} the unit sphere in $\mathfrak{R}^n$:

$$S^{n-1} = \{x \in \mathfrak{R}^n : \|x\| = 1\}.$$

The space of continuous functions on the unit sphere will be denoted $C(S^{n-1})$. Being homogeneous, any support function s_A, $A \in \mathcal{K}_C$ (and hence A itself) is well defined by its restriction to S^{n-1}. The restriction of any support function $s_A(x)$ to S^{n-1}, clearly, belongs to $C(S^{n-1})$. Moreover, for any $A, B \in \mathcal{K}_C$,

$$D_H(A, B) = \max_{x \in S^{n-1}} |s_A(x) - s_B(x)| = \|s_A - s_B\|_{C(S^{n-1})}.$$

Consequently, we see that the space $\mathcal{K}_C$ is metrically isomorphic to the subspace of $C(S^{n-1})$ consisting of the restrictions of support functions from CH

to S^{n-1}. Where no confusion can occur, this subspace is also denoted by CH. This isomorphism also preserves positive convex combinations:

$$s_{\alpha A + \beta B}(x) = \alpha s_A(x) + \beta s_B(x),$$

where the set operations are Minkowski addition and scalar multiplication by non-negative numbers.

Let P be the positive cone of nonempty compact convex sets in the positive octant $\mathfrak{R}^n_+$ of $\mathfrak{R}^n$, $P = \{A \in \mathcal{K}_C : A \subset \mathfrak{R}^n_+\}$. Let SP be the cone induced by P in the space of continuous functions $C(S^{n-1})$. That is, SP is the set of all support functions $s_A(\cdot)$, $A \in P$.

The following problem arises naturally in the optimization theory for convex fuzzy valued functions (see [3],[4]):

Problem. What is the dual positive cone $SP^\oplus$ of SP?

By the Riesz representation theorem [23], $C^*(S^{n-1})$ can be interpreted as the set of all signed Borel measures μ on S^{n-1}. We shall use the notation

$$\int_{S^{n-1}} f(s)d\mu(s) = \langle f, \mu \rangle = \langle \mu, f \rangle.$$

It is important to characterize $SP^\oplus$ by explicit representations of $\mu \in SP^\oplus$, in order to write down optimality criteria for fuzzy programming problems.

As is shown in [3], the case $n = 1$ is trivial. Here we derive the dual cone representation for $n = 2$. The general case will be treated elsewhere.

3. Decomposition of dual cones

Lemma 1 *A support function $s(x)$ belongs to SP iff it is negative on the negative octant,*

$$s(x) \leq 0 \quad \text{whenever} \quad x \in -\mathfrak{R}^n_+.$$

Proof. First, let $A \subset \mathfrak{R}^n_+$ and $p \in -\mathfrak{R}^n_+$. Then $\langle p, x \rangle \leq 0$ for any $x \in A$; hence $s_A(p) = \max_{x \in A}\langle p, x \rangle \leq 0$. Conversely, let $x \in A$ and $x \notin \mathfrak{R}^n_+$. Suppose, for definitiveness, that $x_1 < 0$. Then, for $p = (-1, 0, \ldots, 0) \in -\mathfrak{R}^n_+$, we have

$$s_A(p) = \max_{a \in A}\langle p, a \rangle \geq \langle p, x \rangle = -x_1 > 0. \qquad \square$$

Therefore, $SP = H \cap N$, where the set H is the restriction to S^{n-1} of the set of all continuous convex homogeneous functions on $\mathfrak{R}^n$ and N is the set of all continuous functions on S^{n-1} that are negative on the negative octant.

It is not immediately apparent whether $SP^{\oplus} = H^{\oplus} + N^{\oplus}$, so we shall look for another representation.

Denote by L the set of all $f \in C(S^{n-1})$ such that $f(-e_i) \leq 0$, $i = 1, \ldots, n$ (here $e_i = (0, \ldots, 0, 1, 0, \ldots, 0)$).

Lemma 1 implies the following

Corollary 1 *For H and L as above,*

$$-L^{\oplus} \bigcap H^{\oplus} = \{0\}.$$

Obviously,

$$L^{\oplus} = \left\{ -\sum_{i=1}^{n} \alpha_i \mu(-e_i) : \alpha_i \leq 0, \ i = 1, \ldots, n \right\}.$$

The cone $SP^{\oplus}$ is a cone in the space of signed Borel measures on S^{n-1}.

The following two known results will be needed.

Lemma 2 ([7]) *For any pair A, B of closed convex cones in $C(S^{n-1})$, the cone $(A \bigcap B)^{\oplus}$ is the weak-$*$ closure of the cone $A^{\oplus} + B^{\oplus}$ and will be denoted by $\overline{A^{\oplus} + B^{\oplus}}$.*

We shall prove below that, in fact, $(H \bigcap L)^{\oplus} = H^{\oplus} + L^{\oplus}$.

Lemma 3 ([23]) *Let a space Z be dense in C and let $\{\mu_i\}$ be a given sequence of bounded measures on C. If for all $z \in Z$*

$$\lim_{i \to \infty} \langle \mu_i, z \rangle = 0,$$

then $\{\mu_i\}$ is weakly-$*$ convergent to zero as $i \to \infty$.

The following result reduces the original problem to that of explicit representation of the cone $H^{\oplus}$.

Lemma 4

$$SP^{\oplus} = H^{\oplus} + L^{\oplus}. \tag{3}$$

Proof. First, we show that $SP = H \cap L$. Indeed, a homogeneous convex function is non-positive on the whole negative octant iff it is nonpositive at the extreme rays of this octant which are exactly the rays $-\alpha e_i$, $\alpha \geq 0$. By virtue of Lemma 2, it thus suffices to show that the cone $H^{\oplus} + L^{\oplus}$ is weak-$*$ closed. Suppose it is not. Then there exists $\nu \notin H^{\oplus} + L^{\oplus}$ such that each of its weak-$*$ neighborhoods contains points of $H^\oplus + L^\oplus$. The space $C(S^{n-1})$ is separable. Choose a dense countable subset $\{x_i\}$ in $C(S^{n-1})$ and consider the sequence of weak-$*$ neighborhoods of ν:

$$V_m = \left\{ \mu : |\langle \nu - \mu, x_i \rangle| \le \frac{1}{m}, \quad 1, \ldots, m \right\}.$$

Now take a sequence of $\mu_m \in V_m$, $m = 1, \ldots$. From Lemma 3, the sequence $\{\mu_m\}$ is weak-$*$ convergent to ν. Now μ_i can be represented as $\mu_i = \sigma_i + \lambda_i$ where $\sigma_i \in H^\oplus$, $\lambda_i = -\sum_{j=1}^n \alpha_i^j \mu_j \in L^\oplus$, and

$$\langle \nu, f \rangle = \lim_{i \to \infty} \langle \sigma_i + \lambda_i, f \rangle$$

for any continuous f on S^{n-1}. Consider the alternatives:

1) Each sequence $\{\alpha_i^j : i = 1, 2, \ldots\}$, $j = 1, \ldots, n$, is bounded. Extracting, if necessary, converging subsequences, we can assume $\alpha_i^j \to \alpha^j \le 0$ as $i \to \infty$, $j = 1, \ldots, n$. Thus, the sequence $\{\lambda_i\}$ is weak-$*$ convergent to some $\lambda = -\sum_{j=1}^n \alpha^j \mu_j$ which implies weak-$*$ convergence of $\sigma_i \in H^\oplus$ to $\nu - \lambda \notin H^\oplus$. This is impossible because of the weak-$*$ closedness of $H^\oplus$.

2) There is an unbounded subsequence $\{\alpha_i^j : i = 1, 2, \ldots\}$, i.e., $\|\lambda_i\| \to \infty$ as $i \to \infty$. Since $\sigma_i + \lambda_i \to \nu$ $*$-weakly as $i \to \infty$, we have

$$\frac{\sigma_i}{\|\lambda_i\|} + \frac{\lambda_i}{\|\lambda_i\|} \to 0$$

in the norm of $C^*(S^{n-1})$ as $i \to \infty$. Now, extracting a subsequence, if necessary, we have

$$\frac{\alpha_i^j}{\|\lambda_i\|} \to \alpha^j \le 0 \quad \text{as } i \to \infty,$$

and hence

$$\frac{\lambda_i}{\|\lambda_i\|} \to \gamma \in L^\oplus,$$

which implies

$$\frac{\sigma_i}{\|\lambda_i\|} \to -\gamma \in -L^\oplus.$$

Now, we have $-L^\oplus \bigcap H^\oplus \ne \{0\}$, but this is impossible due to Corollary 1.
$\square$

4. Triangular measures

Now that the problem is reduced to that of describing the dual cone $H^\oplus$, let us turn to the case $n = 2$. The general case $n > 2$ will be studied elsewhere. Here we shall exploit the Lebesgue-Stieltjes representation of signed measures on S^1.

The circle S^1 will be identified with the interval $[0, 2\pi)$ and functions $f \in C(S^1)$ are then 2π-periodic continuous functions on $\mathfrak{R}$. As is well-known, each measure in $C^*(S^1)$ has a distribution function $F(t) = F_\mu(t)$, that is a right-continuous function of locally bounded variation on $\mathfrak{R}$ such that $\mu(t) = dF(t)$ [5]. The function F is unique up to an additive constant. We shall always assume that $F_\mu(t) = \mu((0, t]), \ 0 < t \le 2\pi$. Then, because of the right-continuity, $F(0) = 0$.

Lemma 5 *A sequence of measures $\{\mu_i\}$ is weak-$*$ convergent to μ if and only if the corresponding distribution functions $F_i = F_{\mu_i}$ converge to F at each point of continuity of F.*

Proof. See Exercise 4.37 in [8]. $\square$

Next, we describe a special subclass of discrete measures in $C^*(S^1)$. Denote by δ_p the unit measure concentrated at $p \in S^1$. The discrete measures are of the form

$$\delta = \sum_{i=1}^{m} \alpha_i \delta_{p_i}, \quad p_i \in S^1, \quad \alpha_i \in \mathfrak{R}.$$

We shall also use the notation $\delta(t)$ for $t \in \mathfrak{R}$ as

$$\delta(t) = \delta_p, \qquad p = (\sin t, \cos t) \in S^1.$$

Definition 3 *A discrete measure μ is called triangular if*

$$\mu = \delta(t_1) + \delta(t_2) - 2\cos\left(\frac{t_2 - t_1}{2}\right) \delta\left(\frac{t_2 + t_1}{2}\right),$$

where $0 < t_2 - t_1 \le \pi$. Denote this by $\mu(t_1, t_2)$.

The set of triangular measures will be denoted by T. Note that any values of t_1, t_2 are allowed, and not only those in $[0, 2\pi)$.

Special attention is accorded the triangular measures because in a sense they are the simplest measures in the cone $H^\oplus$. Note that $\langle \mu, f \rangle \ge 0$ for any $f \in CH$ and thus μ is indeed in $H^\oplus$.

Lemma 6 *For any $f \in C(S^1)$ and any triangular measure $\mu = \mu(t_1, t_2)$,*

$$\langle \mu, f \rangle = f(x_1) + f(x_2) - \|x_1 + x_2\| f\left(\frac{x_1 + x_2}{\|x_1 + x_2\|}\right),$$

where

$$x_i = (\sin t_i, \cos t_i).$$

Proof. This follows directly from

$$\|x_1 + x_2\| = 2\cos\left(\frac{t_2 - t_1}{2}\right). \qquad \square$$

Thus, for any homogeneous convex f and any triangular measure μ, we have

$$\langle \mu, f \rangle = f(x_1) + f(x_2) - 2f\left(\frac{x_1 + x_2}{2}\right) \geq 0.$$

Therefore,

Lemma 7

$$T \subseteq H^\oplus.$$

Next, we introduce the trigonometric representation of triangular measures. This representation will be later generalized to the explicit description of the cone $H^\oplus$.

Lemma 8 *For any triangular measure $\mu(t_1, t_2)$ there exists a continuous non-negative piecewise-smooth 2π-periodic function $X(t)$ on $\mathfrak{R}$ such that*

$$\ddot{X}(t) + X(t) = 0$$

for any $t \neq t_1 \pm 2\pi k, t_2 \pm 2\pi k, (t_1 + t_2)/2 \pm 2\pi k, \ k = 0, 1, 2, \ldots$, and the jump of $\dot{X}(t)$ at each point t of concentration of measure μ equals the value of μ at this point. This will be written symbolically as a measure differential equation

$$\ddot{X} + X = \mu. \tag{4}$$

Proof. Construct X as follows: $X(t) \equiv 0$ outside the $\bigcup_{k=-\infty}^{\infty}[t_1 + 2k\pi, t_2 + 2k\pi]$. In the segment $[t_1, (t_1 + t_2)/2]$ take $X(t) = \mu(t_1)\sin(t - t_1)$, in $[(t_1 + t_2)/2, t_2]$ take $X(t) = \mu(t_2)\sin(t_2 - t)$, and put $X(t + 2k\pi) = X(t), \ k = 0, \pm 1, \pm 2, \ldots$. $\quad \square$

It will be shown in Theorem below that $H^\oplus$ consists exactly of all the measures μ satisfying Eq. 4.

Now, the triangular measures are weak-$*$ dense in the cone $H^\oplus$.

Theorem 1 *The weak-* closure $\overline{T}$ of the convex hull of the set of triangular measures T coincides with $H^{\oplus}$.*

Proof. From Lemma 7, $T \subseteq H^{\oplus}$. Suppose that $\nu \in C(S^{n-1})^*$ and $\nu \notin \overline{T}$. Then there exists a continuous function f separating ν and $\overline{T}$, that is,

$$\langle \nu, f \rangle < 0 \quad \text{and} \quad \langle \mu, f \rangle \geq 0 \quad \text{for any } \mu \in T.$$

If f is not homogeneous convex, then there are points t_1 and t_2 in $\mathfrak{R}$ such that $0 < t_2 - t_1 < \pi$ and

$$f(x_1) + f(x_2) - 2f\left(\frac{x_1 + x_2}{2}\right) < 0$$

where

$$x_i = (\sin t_i, \cos t_i).$$

Thus, $\langle \mu(t_1, t_2), f \rangle < 0$, which is impossible. So, f is a homogeneous convex function, and hence $\nu \notin H^{\oplus}$. $\square$

5.   Convex positive measures

The derivative $\dot{X}(t)$ of an absolutely continuous function $X(t)$ is said to be of bounded variation and right-continuous if there exists a a function $h(t)$ with these properties and $\dot{X}(t) = h(t)$ almost everywhere.

Definition 4 *A measure $\mu \in C^*(S^1)$ is said to be convex positive if there exists an absolutely continuous non-negative 2π-periodic function $X(t)$ on $\mathfrak{R}$ such that its derivative $\dot{X}(t)$ is of locally bounded variation on $\mathfrak{R}$, right-continuous, and X satisfies the measure differential equation*

$$D\left[\dot{X}(t) + \int_0^t X(\tau)d\tau\right] = \mu(t), \quad 0 \leq t < 2\pi. \tag{5}$$

Denote the set of all convex positive measures on S^1 by CP. By Lemma 8, $T \subseteq CP$. Since CP is a convex cone, from Theorem 1, the following holds:

Lemma 9 *In terms of weak-* closure, $\overline{CP} \supseteq \overline{T} = S^{\oplus}$.*

It will be shown below that, in fact, $CP = \overline{CP} = S^{\oplus}$.

Denote by $\mathcal{D}$ the set of all discrete convex positive measures.

Lemma 10 *For any $\mu \in \mathcal{D}$, there is a positive 2π-periodic, continuous function $X \in C(S^1)$ composed of pieces of $\alpha_i \sin(t_i + t)$ such that the jump of its derivative at each point t_i equals the value $\mu(t_i)$ of μ at this point.*

Proof. Indeed, if the measure

$$D\left[\dot{X}(t) + \int_0^t X(\tau)d\tau\right] \equiv 0 \quad \text{on } [s_1, s_2],$$

then there exist $\alpha, t^* \in \mathfrak{R}$ such that $X(t) = \alpha\sin(t^* + t)$ on $[s_1, s_2]$. The rest of the proof follows in much the same way.

An upper bound for absolute values of X and $\dot{X}$ in Eq. 5 can be found in terms of the norm of μ.

Lemma 11 *Suppose $\mu \in CP$ is represented as in Eq. 5. Then*

$$|\dot{X}(t)| \le \|\mu\| \tag{6}$$

and

$$|X(t)| \le \left(\pi + \frac{1}{2\pi}\right)\|\mu\| \tag{7}$$

for any $t \in \mathfrak{R}$.

Proof. First, by definition, $\|\mu\|$ is the total variation of the function $\dot{X}(t) + \int_0^t X(s)ds$ on $[0, 2\pi)$; hence

$$\|\mu\| = V_- + V_+ \ge V_+,$$

where V_- and V_+ are the negative and positive variation of this function on the same interval. Note that

$$Y(t) = \int_0^t X(s)ds$$

is a non-decreasing function, thus $V_+ \ge V_+(\dot{X})$.

Being a derivative of 2π-periodic function, $\dot{X}(t)$ satisfies the following:

$$V_+(\dot{X}) \ge \max_{t\in[0,2\pi)} \dot{X}(t) - \min_{t\in[0,2\pi)} \dot{X}(t) \ge \max_{t\in[0,2\pi)} |\dot{X}(t)|.$$

Taking into consideration $\int_0^{2\pi} \dot{X}(t)dt = 0$, we get Eq. 6. This implies

$$|X(t_2) - X(t_1)| \le \|\mu\| \quad t_1, t_2 \in \mathfrak{R}. \tag{8}$$

Since $\int_0^{2\pi} X(s)ds \le \|\mu\|$, we have

$$\min_{0\le t<2\pi} X(t) \le \frac{\|\mu\|}{2\pi},$$

and hence, from Eq. 8,

$$\max_{0 \le t < 2\pi} X(t) \le \|\mu\| \left(\pi + \frac{1}{2\pi} \right). \qquad \square$$

6. Smooth measures

A function $Z(t)$ on the open interval $I \subseteq \Re$ is said to be *smooth* if it has a continuous second derivative on I. A measure μ is said to be a *smooth measure* if, for any $G \in C(S^1)$,

$$\langle \mu, G \rangle = \int_0^{2\pi} Z(t)G(t)dt,$$

where $Z(t)$ is a smooth 2π-periodic function on $\Re$.

Smooth functions on S^1 are dense in the cone H.

Lemma 12 *The cone $H \subset C(S^1)$ is the closure of the set of all smooth functions in H.*

Proof. Take an arbitrary function $f \in H$ and consider it as a 2π-periodic continuous function on $\Re$. Next, choose a smooth positive "cap" $g(t)$ over the origin such that $\int_{-\infty}^{+\infty} g(t)dt = 1$ and the support of g is bounded. Define $g_i(t) = ig(it)$, $i = 1, 2, \dots$. The convolution

$$(f * g_i)(t) = \int_{-\infty}^{+\infty} f(t - s)g_i(s)ds$$

is a smooth 2π-periodic function. Clearly, that it is still convex homogeneous if considered on S^1 since it is an integral of convex functions with respect to a positive measure g_i. Finally, the sequence $f * g_i$ converges to f with respect to the norm of $C(S^1)$ and the lemma is proved. $\square$

Furthermore, the smooth measures are dense in CP. Let $\mathfrak{X}$ be the totality of smooth measures which are representable as

$$\mu = \ddot{X} + X,$$

where X, $\ddot{X}$ are both smooth functions.

Lemma 13 *$\mathfrak{X}$ weak-$*$ dense in CP.*

Proof. In much the same way as in the previous lemma, any measure $\mu \in C^*(S^1)$ can be approximated by smooth measures $\mu_i = \ddot{X}_i + X_i$, where $X_i = X * g_i$, that is, μ_i weak-$*$ converges to μ as $i \to \infty$. Such a smoothification preserves the convex positivity of the measure on the cone H, because each relevant property of the corresponding function $X(t)$ is invariant under the convolution with a positive smooth function g_i. $\quad\square$

Several further statements are required.

Lemma 14 *A smooth 2π-periodic function $X(t)$ belongs to H iff $\ddot{X}(t) + X(t) \geq 0$ for any $t \in \mathfrak{R}$.*

Proof. It is sufficient to show that the homogeneous function $z(p)$ on $\mathfrak{R}^2$

$$z(p) = \|p\| X(\mathrm{Arg}(p))$$

is convex. We only need to prove that its restriction to any straight line is convex. From the homogeneity of $z(p)$, it is sufficient to consider only the lines tangent to the unit circle. Because of the symmetry, it is sufficient to prove it only for the line $\{(x, y) : x = 1\}$. On this line,

$$z(1, y) = z(y) = \sqrt{1 + y^2} X(\arctan(y)).$$

For notational clarity, the argument $\arctan(y)$ of X, $\dot{X}$ and $\ddot{X}$ is omitted. Then

$$z'(y) = \frac{y}{\sqrt{1 + y^2}} X + \frac{\sqrt{1 + y^2}}{1 + y^2} \dot{X} = \frac{yX + \dot{X}}{\sqrt{1 + y^2}}$$

and

$$z''(y) = \frac{(X + y\dot{X}/(1 + y^2) + \ddot{X}/(1 + y^2))\sqrt{1 + y^2} - y(yX + \dot{X})/\sqrt{1 + y^2}}{1 + y^2}.$$
$$(9)$$

Multiplying Eq. 9 by the positive number $\sqrt{1 + y^2}(1 + y^2)$, obtain

$$X(1 + y^2) + y\dot{X} + \ddot{X} - y^2 X - y\dot{X} = X + \ddot{X} \geq 0.$$

This implies the convexity of $z(y)$. $\quad\square$

Lemma 15 *A smooth measure $\mu = \ddot{Y} + Y$ belongs to $H^\oplus$ whenever $Y(t) \geq 0$, $t \in \mathfrak{R}$.*

Proof. Indeed, integrating by parts, we get

$$\langle \mu, X \rangle = \int_0^{2\pi} (Y(t)X(t) - \dot{Y}(t)\dot{X}(t))dt = \int_0^{2\pi} Y(t)(\ddot{X}(t) + X(t))dt \quad (10)$$

for any smooth $X(t) \in S$. By Lemma 14, the integral in Eq. 10 is non-negative for any X iff $Y(t)$ is non-negative for any t. By Lemma 12, $\langle \mu, X \rangle \geq 0$. $\square$

We have proved that any smooth measure in CP belongs to $H^{\oplus}$ and that smooth measures are weak-$*$ dense in CP (Lemma 13). Thus, $\overline{CP} \subseteq H^{\oplus}$.

Theorem 2

$$H^{\oplus} = \overline{CP}.$$

Proof. This follows from the above remark, together with Lemma 9. $\square$

7. The main theorem

We shall need the two following lemmas.

Lemma 16 *Let ν be a Borel measure on $[0, a] \subset \mathfrak{R}$. Then*

$$P(t) = \int_0^t \nu(s)ds, \qquad 0 \leq t \leq a,$$

is a right-continuous function of locally bounded variation and

$$\int_0^p \left(\int_0^t \nu(s)ds \right) dt = \int_0^p \nu(t)(p - t)dt \qquad (11)$$

for any $p \in [0, a]$.

Proof. This immediately follows from the Fubini theorem. $\square$

Lemma 17 *The cone CP is weak-$*$ closed.*

Proof. Since $C(S^1)$ is separable, it suffices to prove that the limit of any weak-$*$ converging sequence of elements of CP belongs to CP. Let $\mu_i \to \mu$ weakly-$*$ as $i \to \infty$

and for each i let $\ddot{X}_i + X_i = \mu_i$. The sequence $\|\mu_i\|$ is clearly bounded. Hence, by Lemma 11, the sequences $\{\dot{X}_i(t)\}$ and $\{X_i(t)\}$ are uniformly bounded. Moreover, the sequence $X_i(t)$ is equicontinuous and hence relatively compact in $C(S^1)$. Extracting, if needed, a subsequence, we can consider a corresponding sequence $X_i(t)$ uniformly converging to some continuous 2π-periodic non-negative function $X(t)$. It remains to show that $\mu = \ddot{X} + X$. In terms of measure derivatives, $\mu_i(t) = D\dot{X}_i(t) + X_i(t)$. Write $\nu_i = D\dot{X}_i$. Since $\|X_i - X\|_C \to 0$, we have

$$\nu_i \to \nu = \mu - X(t) \quad \text{weakly-$*$ as } i \to \infty.$$

Now put

$$\dot{X}_i(t) = \int_0^t \nu_i(s)ds; \qquad X_i(p) = X_i(0) + \int_0^p \int_0^t \nu_i(s)dsdt \qquad (12)$$

for any $0 \le t \le p < 2\pi$. We can suppose that $X_i(0) \to \tilde{X}_0$ as $i \to \infty$. From Eq. 11 and the weak-$*$ convergence of ν_i,

$$\lim_{i \to \infty} |X_i(p) - \tilde{X}(p)| = 0$$

for any $p \in [0, 2\pi)$, where

$$\tilde{X}(p) = \tilde{X}_0 + \int_0^p \int_0^t \nu(s)dsdt.$$

Thus, $\dot{X}(t) = \int_0^t \nu(s)ds$, $0 \le t < 2\pi$ and so

$$\mu = D\left[\dot{X} + \int_0^t X(s)ds\right]. \quad \square$$

Theorem 2 implies

Corollary 2

$$H^\oplus = CP.$$

Now, our main result follows immediately:

Theorem 3 *A measure μ on S^1 belongs to $SP^\oplus$ if and only if it can be represented as*

$$\begin{aligned} \mu(t) &= \nu - a_1\mu(e_1) - \alpha_2\mu(e_2), \quad \alpha_1, \alpha_2 \ge 0, \\ \nu &= \ddot{X}(t) + X(t), \end{aligned}$$

where $X(t)$ is a non-negative 2π-periodic function on $\mathfrak{R}$ with a right-continuous derivative $\dot{X}$ of locally bounded variation.

8.   Acknowledgement

This research has been supported by the Australian Research Council Grant A 49330974.

References

[1] J.-P. Aubin and H. Frankowska, *Set–Valued Analysis*, Birkhäuser, Basel, 1990.

[2] F.H. Clarke, *Optimization and Nonsmooth Analysis*, SIAM, Philadelphia, 1990.

[3] P. Diamond and P. Kloeden, *Metric Spaces of Fuzzy Sets: Theory and Applications*, World Scientific Publishers, Singapore, 1994.

[4] P. Diamond and P. Kloeden, Robust Kuhn–Tucker conditions and optimization under imprecision, in *Fuzzy Optimization. Recent Advances*, Editors: M. Delgado *et al*, Physica–Verlag, Heidelberg, 1994, pages 61–66.

[5] J.L. Doob, *Measure Theory*, Springer-Verlag, New York, 1993.

[6] D. Dubois, Linear programming with fuzzy data, in *Analysis of fuzzy information*, CRC, Boca Raton, Fla., 1987, Volume **3**, 241–263.

[7] I.V. Girsanov, *Lectures on Mathematical Theory of Extremum Problems*, Springer Lecture Notes in Mathematics Volume **67**, Springer-Verlag, Berlin, 1972.

[8] R.B. Holmes, *Geometric Functional Analysis and its Applications*, Springer-Verlag, New York, 1975.

[9] Margit Kovacs, Linear programming with centered fuzzy numbers, *Ann. Univ. Sci. Budapest. Sect. Comput.*, **12** (1991), 159–165.

[10] Y. Jou Lai and C. Lai Hwang, *Fuzzy mathematical programming*, Springer Lecture Notes in Economics and Mathematical Systems Volume **394**, Springer-Verlag, Berlin, 1992.

[11] S.R. Lay, *Convex Sets and Their Applications*, John Wiley and Sons, New York, 1982.

[12] D.G. Luenberger, *Linear and Nonlinear Programming*, 2nd ed., Addison–Wesley, Reading, Mass., 1984.

[13] Marian Matloka, Fuzzy parameters in linear programming, *J. Fuzzy Math.*, **1** (1993), 509–515.

[14] J.J. Moreau, P.D. Panagiotopolous and G. Strang, *Topics in Nonsmooth Mechanics*, Birkhuser Verlag, Basel, 1988.

[15] C.V. Negoita and D.A. Ralescu, *Applications of Fuzzy Sets to Systems Analysis*, John Wiley And Sons, New York, 1975.

[16] S.G. Pandit and S.G. Deo, *Differential Systems Involving Impulses*, Springer Lecture Notes in Mathematics Volume **954**, Springer–Verlag, Berlin, 1982.

[17] R.T. Rockafellar, *Convex Analysis*, Princeton University Press, Princeton, N.J., 1970.

[18] H. Rommelfanger, R. Hanuscheck, and J. Wolf, Linear programming with fuzzy objectives, *Fuzzy Sets and Systems*, **29** (1989), 31–48.

[19] H. Tanaka, T. Okuda, and K. Asai, On fuzzy–mathematical programming, *J. Cybernet.*, **3**(1973), 37–46.

[20] H. Tanaka and K. Asai, Fuzzy linear programming based on fuzzy functions, *Bull. Univ. Osaka Prefect. Ser. A*, **29** (1980), 113–125.

[21] J. Werner, *Optimization Theory and Applications*, Friedr. Vieweg, Braunschweig, Wiesbaden, 1984.

[22] R. R. Yager, Mathematical programming with fuzzy constraints and a preference on the objective, *Kybernetes*, **8**(1979), 285–291.

[23] K. Yoshida, *Functional Analysis*, Springer-Verlag, Berlin, 1965.

[24] H. J. Zimmermann, Fuzzy programming and linear programming with several objective functions, *Fuzzy Sets and Systems*, **1** (1978), 45–55.

[25] H. J. Zimmermann, Fuzzy mathematical programming, *Comput. Oper. Res.*, **10** (1983), 291–298.

[26] H. J. Zimmermann, Applications of fuzzy set theory to mathematical programming, *Inform. Sci.*, **36** (1985), 29–58.

WSSIAA 5 (1995) pp. 51–66

51

DUAL BARRIER-PROJECTION METHODS IN LINEAR PROGRAMMING*

Yuri G. Evtushenko and Vitali G. Zhadan

March 15, 1995

*Computing Centre, Russian Academy of Sciences, 40 Vavilov Str.,
117967 Moscow GSP-1, Russia*

Abstract

A surjective space transformation technique is used to convert an original dual
linear programming problem with equality and inequality constraints into a problem
involving only equality constraints. Continuous and discrete versions of the stable
gradient projection method are applied to the reduced problem. The numerical methods involve performing inverse transformations. The convergence rate analysis for
dual linear programming methods is presented. By choosing a particular exponential
space-transformation function we obtain the dual affine scaling algorithm. Variants of
methods which have linear local convergence are given.

1 INTRODUCTION

Since 1973, we have developed a family of numerical methods based on space transformation techniques. Using a space transformation, we convert the original problem with
equality and inequality constraints to a problem with equality constraints only. This is
an old notion commonly used in the optimization literature. Numerous variants of this
basic idea exist. In [4]-[12] we used a surjective space transformation and then applied the
gradient projection method and Newton's method to solve the reduced nonlinear programming problem. After an inverse transformation to the original space a family of numerical
methods for solving optimization problems with equality and inequality constraints was
obtained. The proposed algorithms are based on the numerical integration of systems of
ordinary differential equations. As a result of a space transformation the vector fields of
differential equations are changed and additional terms are introduced which serve as a
barrier preventing the trajectories from leaving the feasible set. In our algorithms the
barrier functions are continuous and equal zero on a boundary. The space transformations
are carried out without using conventional barrier or penalty functions and this feature
provides a high rate of convergence.

Different numerical methods are obtained by different choices of the space transformations. For example, if we choose an exponential space transformation in the linear

---

*Research supported by the grant N 94-01-01379 from Russian Scientific fund

programming case, we obtain the Dikin algorithm [3] from the family of primal barrier-projection methods. This algorithm, however, does not posses local convergence properties and it converges only for starting points inside the feasible set. Furthermore, the discrete version has a less than linear rate of convergence. In [8]-[10] it was shown that if we apply stable versions of the gradient projection algorithm and use the quadratic space transformation, then we obtain local linear convergence. A survey of our results in this field is given in [11] and applications to linear programming are presented in [12].

The content of this paper is similar to that of [12], but in contrast to [12], we focus our attention on the solution of the dual linear programming problem, though the methods which we propose permit us to obtain the solution of the primal problem simultaneously with the solution of the dual problem.

For the local convergence analysis we use the Lyapunov linearization principle of determining the stability from the equatio

n of the first approximation about an equilibrium state [2]. Nonlocal convergence is investigated by using the second (direct) method of Lyapunov.

In Section 2, we describe a family of dual barrier-projection methods. These methods are described by systems of ordinary differential equations. Numerical algorithms are obtained as discretizations of dynamical systems. Sufficient conditions for local convergence of continuous and discrete versions of numerical methods are given. We show that if the quadratic space transformation is used then we obtain an exponential rate of convergence for continuous methods and a linear rate of convergence for discrete versions.

In Section 3, we use a nonconventional representation of the dual linear programming problem and we propose a different set of algorithms. After some simplification and after choosing a particular exponential space transformation function we obtain the dual affine scaling method proposed by I.Adler, N.Karmarkar, M.Resende and G.Veiga [1]. In Section 4, we investigate nonlocal convergence properties.

2 BASIC APPROACH AND OUTLINE OF THE METHODS

Consider a linear programming problem in standard form

$$\text{minimize } c^T x \text{ subject to } x \in X = \{x \in R^n : b - Ax = 0_m, \ x \geq 0_n\} \qquad (2.1)$$

and its dual problem

$$\text{maximize } b^T u \text{ subject to } u \in U = \{u \in R^m : v = c - A^T u \geq 0_n\}, \qquad (2.2)$$

where $A \in R^{mn}(m < n); c, x, v \in R^n; b, u \in R^m$ and $\text{rank}(A) = m; 0_s$ is the $s-$dimensional null vector.

We define the interior set of U as:

$$U_0 = \{u \in R^m : v = c - A^T u > 0_n\},$$

and assume that this set is nonempty.

Throughout the paper we assume that the problems have nonempty solution sets X_* and U_*, respectively. We also introduce the following sets:

$$V = \{v \in R^n : \text{there exists } u \in R^m \text{ such that } v = c - A^T u\},$$

$$V_U = \{v \in R^n : \text{there exists } u \in U \text{ such that } v = c - A^T u\}.$$

Here v is the $n-$vector of slack variables. The set V_U is the image of U under the mapping $v(u) = c - A^T u$. Therefore $V_U = V \cap R^n_+$, where R^n_+ is the nonnegative orthant of R^n.

We denote the components of a vector by using superscripts and the iterate numbers by using subscripts; $D(z)$ denotes the diagonal matrix whose entries are the components of z. The dimensionality of this matrix is determined by the dimensionality of z.

We now introduce a new n-dimensional space with the coordinates $[w^1, \ldots, w^n]$ and define a differentiable transformation from this space to the original one: $v = \varphi(w)$. This surjective transformation maps R^n onto R^n_+ or $\text{int} R^n_+$, i.e. $R^n_+ = \overline{\varphi(R^n)}$, where $\bar{B}$ denotes the closure of B. In this case every vector in R^n_+ is the image of at least one vector in R^n or it is the image of a limit point of a sequence in R^n. In other words for each $v \in R^n_+$ there exists a $w \in R^n$ such that $v = \varphi(w)$ or $v = \lim_{i \to \infty} \varphi(w_i)$, where $w_i \in R^n$.

For the sake of simplicity we use a componentwise space transformation where

$$\varphi(w) = [\varphi^1(w^1), \varphi^2(w^2), \ldots, \varphi^n(w^n)].$$

Let $w^i = \psi^i(v^i)$ denote the inverse transformation of $\varphi^i(w^i)$. This transformation exists at least in a neighborhood of a point $v_0^i = \varphi^i(w_0^i)$, as long as $\dot{\varphi}^i(w_0^i) \neq 0$.

We introduce a n-vector $\theta(v)$ and a $n \times n$ matrix $G(v)$:

$$\theta(v) = [\theta^1(v^1), \theta^2(v^2), \ldots, \theta^n(v^n)], \qquad G(v) = D(\theta(v)),$$

where $\theta^i(v^i) = (\gamma^i(v^i))^2, \gamma^i(v^i) = \dot{\varphi}^i(\psi^i(v^i)), 1 \leq i \leq n$.

We impose the following conditions on the space transformation $\varphi(v)$:

C_1. The functions $\theta^i(v^i)$ are defined and continuous in some neighborhood of R^1_+ and $\theta^i(v^i) = 0$ if and only if $v^i = 0$, where $1 \leq i \leq n$.

C_2. The functions $\theta^i(v^i)$ are continuously differentiable in some neighborhood of R^1_+ and $\dot{\theta}^i(0) > 0, 1 \leq i \leq n$.

Different numerical methods with various convergence properties can be obtained from different choices of the space transformation functions. Here we consider only two simple surjective transformations

$$v = \frac{1}{4}D(w)(w), \quad v = e^{-w}, \tag{2.3}$$

where the $i-$th component of the $n-$vector e^{-w} is e^{-w^i}. We shall refer to these transformations as to quadratic and exponential space transformations, respectively.

For the transformations (2.3) we obtain, respectively

$$\theta(v) = v, \quad G(v) = D(v); \qquad \theta(v) = D(v)v, \quad G(v) = D^2(v).$$

In both cases the Jacobian matrix is singular on the boundary of the set R^n_+. These transformations satisfy C_1. Condition C_2 holds only for the first quadratic transformation (2.3).

By extension of the space and by converting the inequality constraints to equalities, we transform the original dual problem (2.2) into the following equivalent problem

$$\text{maximize } b^T u \text{ with respect to } u \text{ and } w \text{ subject to } \varphi(w) + A^T u - c = 0_n. \qquad (2.4)$$

The Lagrangian associated with this problem is defined by

$$\tilde{L}(u, w, x) = b^T u + x^T [\varphi(w) + A^T u - c].$$

For solving problem (2.4) we use the stable version of the gradient projection method which is described in [16]. The method is stated as an initial-value problem involving the following system of ordinary differential equations

$$\frac{du}{dt} = \tilde{L}_u(u, w, x(u, w,)), \qquad \frac{dw}{dt} = \tilde{L}_w(u, w, x(u, w,)). \qquad (2.5)$$

The function $x(u, w)$ is chosen to satisfy the following condition

$$\tilde{L}_{xu}(u, w, x)\dot{u} + \tilde{L}_{xw}(u, w, x)\dot{w} = -\tau \tilde{L}_x(u, w, x). \qquad (2.6)$$

Since $\dot{v} = \varphi_w \dot{w}$, (2.5) can be rewritten in terms of u and v as follows

$$\frac{du}{dt} = b - Ax(u, v), \qquad \frac{dv}{dt} = -G(v)x(u, v), \qquad (2.7)$$

where $\Phi(v)x(u, v) = A^T b + \tau(v + A^T u - c)$ and $\Phi(v) = G(v) + A^T A$.

We say that an extreme point u of the feasible set U is nondegenerate if the vector $v(u)$ has only m zero components.

Lemma 1 *Let the space transformation $\varphi(w)$ satisfy C_1. Then the matrix $\Phi(v(u))$ is positive definite for all $u \in U_0$.*

Lemma 2 *Let the space transformation $\varphi(w)$ satisfy C_1. Assume that a point $u \in U$ can be represented as*

$$u = \sum_{j=1}^{s} \alpha^j u_j, \quad \alpha^j > 0, \quad \sum_{j=1}^{s} \alpha^j = 1,$$

where $u_j, 1 \le j \le s$ are extreme points of U, and at least one point u_j is nondegenerate. Then the matrix $\Phi(v(u))$ is positive definite.

Proof. To show the nonsingularity of $\Phi(v)$, where $v = v(u)$ it suffices to show that the nullspace of $\Phi(v)$ contains only the point 0_n. Consider the linear system of equations

$$\Phi(v)\bar{x} = G(v)\bar{x} + A^T A\bar{x} = 0_n. \qquad (2.8)$$

By multiplying (2.8) on the left by $\bar{x}^T$, we obtain

$$\bar{x}^T G(v)\bar{x} + \bar{x}^T A^T A\bar{x} = 0.$$

Both expressions are nonnegative and therefore

$$\bar{x}^T G(v)\bar{x} = 0, \quad \bar{x}^T A^T A\bar{x} = 0. \tag{2.9}$$

Let $S_j = \{1 \le i \le n : a_i^T u_j = c^i\}, S = \bigcap_{j=1}^{s} S_j$, where a_i is the i-th column of the matrix A. Without loss of generality we assume that $S = \{1, 2, \ldots, k\}$. We select the first k columns and denote them by B. We partition $A, \bar{x}$ and v as

$$A = [B \mid N], \quad \bar{x} = \begin{bmatrix} \bar{x}^B \\ \bar{x}^N \end{bmatrix}, \quad v = \begin{bmatrix} v^B \\ v^N \end{bmatrix}.$$

Since at least one extreme point is nondegenerate, it follows that $k \le m$ and B has full rank. Since $v^B = 0_k$ and $v^N > 0_{n-k}$, we obtain from (2.9) that $\bar{x}^N = 0_{n-k}$. Hence $B\bar{x}^B = 0_m$ and we conclude that $\bar{x} = 0_n$. $\square$

Corollary 1. If an extreme point u of a polytope U is nondegenerate, then $\Phi(v(u))$ is positive definite.

Corollary 2. If all extreme points of the bounded set U are nondegenerate, then $\Phi(v(u))$ is positive definite for all $u \in U$.

According to Corollary 2, if $v \in V_U$, then the matrix $\Phi(v)$ is invertible. Because of the continuity it is also invertible in some neighborhood of V_U. For all points from this set we get

$$x(u, v) = \left(G(v) + A^T A\right)^{-1} \left(A^T b + \tau(v + A^T u - c)\right).$$

Substituting this formula into the right-hand side of (2.7) we find that the system takes the explicit form

$$\begin{aligned}
du/dt &= b - A\left(G(v) + A^T A\right)^{-1} \left(A^T b + \tau(v + A^T u - c)\right), \\
dv/dt &= -G(v)\left(G(v) + A^T A\right)^{-1} \left(A^T b + \tau(v + A^T u - c)\right).
\end{aligned} \tag{2.10}$$

Let $[u(t, z_0), v(t, z_0)]$ denote the solution of the Cauchy problem (2.10) with initial conditions $u(0, z_0) = u_0, v(0, z_0) = v_0, z_0^T = [u_0^T, v_0^T]$. Let $y(u, v) = c - A^T u - v$. The condition (2.6) can be written as

$$\frac{dy(u, v)}{dt} = y_u^T(u, v)\dot{u} + y_v^T(u, v)\dot{v} = -\tau y.$$

Hence, the system of ordinary differential equations (2.10) has the first integral

$$c - A^T u(t, z_0) - v(t, z_0) = \left(c - A^T u_0 - v_0\right) e^{-\tau t}.$$

This implies that $c - A^T u(t, z_0) - v(t, z_0) \to 0_n$ as $t \to +\infty$. Moreover, along the trajectories of the system (2.10) we get

$$b^T \frac{du}{dt} = b^T\left(b - Ax(u, v)\right) = \|b - Ax(u, v)\|^2 + x^T(u, v)A^T\left(b - Ax(u, v)\right) =$$

$$= \|b - Ax(u, v)\|^2 + x^T(u, v)G(v)x(u, v) + \tau x^T(u, v)\left(c - A^T u - v\right).$$

From the second equation of (2.10) it follows that if the transformation $\varphi(v)$ satisfies condition C_1, then each component of the vector $v(t, z_0)$ does not change its sign. Hence, if $v_0 \geq 0$, then $v(t, z_0) \geq 0$ on the entire trajectory. We obtain this important property because of the matrix $G(v)$ in the right-hand side of (2.10), which plays the role of a "barrier", preventing the trajectory $v(t, z_0)$ from passing through the boundary of R_+^n. Hence, we call (2.10) a "dual barrier-projection method".

Note that $y(u(t, z_0), v(t, z_0)) \equiv 0_n$ if $y(u_0, v_0) = 0_n$. We conclude that if $u_0 \in U$, then we can get rid of the equation for v and this way simplify systems (2.7) and (2.10). In this case, (2.7) can be expressed as

$$\frac{du}{dt} = b - Ax(u), \quad \left(G(v(u)) + A^T A\right) x(u) = A^T b, \tag{2.11}$$

where $u(0, u_0) = u_0 \in U$.

For this system we obtain the following inequality

$$b^T \frac{du}{dt} = \|b - Ax(u)\|^2 + x^T(u)G(v(u))x(u) \geq 0.$$

Hence the objective function of the dual problem monotonically increases on a feasible set.

By applying the Euler numerical integration method we obtain the following iterative algorithm

$$u_{k+1} = u_k + \alpha_k(b - Ax_k), \quad v_{k+1} = v_k + \alpha_k G(v_k)x_k, \tag{2.12}$$

$$\left(G(v_k) + A^T A\right) x_k = A^T b + \tau \left(v_k + A^T u_k - c\right).$$

Similarly for the system (2.11) we have

$$u_{k+1} = u_k + \alpha_k \left(b - Ax_k\right), \quad \left(G(v_k), + A^T A\right) x_k = A^T b, \tag{2.13}$$

where $v_k = v(u_k)$. Both variants solve the primal and dual problems simultaneously.

Theorem 1 *Let x_* and u_* be unique nondegenerate solutions of Problems (2.1) and (2.2) respectively and let $v_* = c - A^T u_*$. Assume that the space transformation $\varphi(w)$ satisfies conditions C_1, C_2 and $\tau > 0$. Then the following statements are true:*

1. *The pair $[u_*, v_*]$ is an asymptotically stable equilibrium state of system (2.10).*

2. *The solutions $u(t, z_0), v(t, z_0)$ of system (2.10) converge locally to the pair $[u_*, v_*]$. The corresponding function $x(u(t, z_0), v(t, z_0))$ converges to the optimal solution x_* of the primal problem (2.1).*

3. *The point u_* is an asymptotically stable equilibrium state of system (2.11).*

4. *The solutions $u(t, u_0)$ of system (2.11) converge locally to the optimal solution u_* of the dual problem (2.2). The corresponding function $x(u(t, u_0))$ converges to the optimal solution x_* of the primal problem (2.1).*

5. *There exists an $\alpha_* > 0$ such that for any fixed $0 < \alpha_k < \alpha_*$ the sequence $\{u_k, v_k\}$ generated by (2.12) converges locally with a linear rate to $[u_*, v_*]$ while the corresponding sequence $\{x_k\}$ converges to x_*.*

6. *There exists an $\alpha_* > 0$ such that for any fixed $0 < \alpha_k < \alpha_*$ the sequence $\{u_k\}$ generated by (2.13) converges locally with a linear rate to u_* while the corresponding sequence $\{x_k\}$ converges to x_*.*

Proof. Let $\delta z^T = [\delta u^T, \delta v^T], \delta u = u(t, z_0) - u_*$ and $\delta v = v(t, z_0) - v_*$. We linearize system (2.10) in the neighborhood of the point $z_*^T = [u_*^T, v_*^T]$. Then we obtain the first approximation of (2.10) about point z_*:

$$\delta \dot{z} = -Q \delta z,$$

where

$$Q = \begin{bmatrix} \tau A \Phi^{-1} A^T & A \Phi^{-1} \left(\tau I_n - D(\dot{\theta}(v_*)) D(x_*) \right) \\ \tau G(v_*) \Phi^{-1} A^T & \left(I_n - G(v_*) \Phi^{-1} \right) D(\dot{\theta}(v_*)) D(x_*) + \tau G(v_*) \Phi^{-1} \end{bmatrix},$$

$\Phi = G(v_*) + A^T A$ and I_n is the $n \times n$ identity matrix.

Suppose that the first m columns of A are linearly independent and denote the $m \times m$ matrix determined by these columns as B. Assume that B is the optimal basis. With respect to this partition we can write

$$x_* = \begin{bmatrix} x_*^B \\ x_*^N \end{bmatrix}, \quad v_* = \begin{bmatrix} v_*^B \\ v_*^N \end{bmatrix}, \quad A = [B \mid N], \tag{2.14}$$

$$G(v_*) = \begin{bmatrix} 0_{mn} & 0_{md} \\ 0_{dm} & G_N \end{bmatrix}, \quad \Phi = \begin{bmatrix} B^T B & B^T N \\ N^T B & G_N + N^T N \end{bmatrix},$$

where $x_*^B > 0_m, v_*^B = 0_m, v_*^N > 0_d, d = n - m$ and $G_N = D(\theta(v_*^N))$ is the $d \times d$ matrix. Using the Frobenious formula we can find Φ^{-1} and obtain

$$\Phi^{-1} A^T = \begin{bmatrix} B^{-1} \\ 0_{dm} \end{bmatrix}, \quad Q = \begin{bmatrix} \tau I_m & Q_2 \\ 0_{nm} & Q_1 \end{bmatrix}, Q_1 = \begin{bmatrix} D(\dot{\theta}(v_*^B))D(x_*^B) & 0_{md} \\ Q_3 & \tau I_d \end{bmatrix},$$

where the matrices Q_2 and Q_3 are not essential.

It is obvious that the characteristic equation

$$\det \left(Q - \lambda I_{n+m} \right) = 0$$

has following roots: $\lambda_i = \dot{\theta}^i(0)x_*^i, \lambda_j = \tau, 1 \le i \le m, m+1 \le j \le n + n$. Since the transformation $\varphi(w)$ satisfies C_2, and since x_* is a nondegenerate optimal solution of the problem (2.1), all these roots are positive and the smallest root is

$$\lambda_* = \min \left[\tau, \min_{1 \le i \le m} \dot{\theta}^i(0)x_*^i \right] > 0.$$

Hence, according to Lyapunov's linearization principle, the equilibrium point z_* is asymptotically stable and the following estimate holds

$$\limsup_{t \to \infty} \frac{\ln \|z(t, z_0) - z_*\|}{t} < -\lambda_*.$$

58

Denote

$$\alpha_* = 2/\lambda^*, \quad \lambda^* = \max\left[\tau, \max_{1\leq i\leq m} \dot{\theta}^i(0)x_*^i\right].$$

If the stepsize $\alpha_k < \alpha_*$, then by Theorem 2.3.7 from [5] the linear convergence of the discrete versions (2.12) follows from the proof given above.

If $u_0 \in U$ then the solutions of (2.11) coincide with corresponding solutions of (2.10), if in (2.10) we take $v_0 = v(u_0)$. Therefore the local exponential convergence of (2.10) implies the local exponential convergence of (2.11). In the similar way the linear convergence of (2.12) implies the linear convergence of (2.13). $\square$

We proposed the dual method (2.11) in 1977. It was described in [7] where we also gave the following primal method

$$\frac{dx}{dt} = G(x)\left[c - A^T u(x)\right], \tag{2.15}$$

where $AG(x)A^T u(x) = AG(x)c, G(x) = D(x)$. Both methods are similar and both solve primal and dual problems simultaneously. The method (2.15) is very popular now. It was reinvented recently in [13], [15] and analyzed in the book [14].

3 OTHER VARIANTS OF DUAL METHODS

As before we assume that A has full rank, therefore the nullspace of A has dimension $d = n - m$. Let P be a full rank $d \times n$ matrix such that $AP^T = 0_{md}$. Therefore the columns of P^T are linearly independent and form a basis for the nullspace of A. We partition A as $A = [B, N]$, where the square matrix B is nonsingular. We can now write the matrix P as

$$P = \left[-N(B^T)^{-1} \mid I_d\right].$$

The definitions of the sets V and V_U can be rewritten as follows

$$V = \{v \in R^n : P(v - c) = 0_d\}, \quad V_U = \{v \in R_+^n : P(v - c) = 0_d\}.$$

Let $\bar{x} \in R^n$ be an arbitrary vector which satisfies the constraint $A\bar{x} = b$. Then

$$\max_{u\in U} b^T u = \max_{u\in U} \bar{x}^T A^T u = \max_{v\in V_U} \bar{x}^T(c - v) = \bar{x}^T c - \min_{v\in V_U} \bar{x}^T v.$$

Hence the solution of the dual problem (2.2) can be substituted by the following equivalent minimization problem

$$\min_{v\in V_U} \bar{x}^T v.$$

Applying the stable barrier-projection method [12] to this problem, we obtain

$$\frac{dv}{dt} = -G(v)\left(\bar{x} - P^T x(v)\right), \tag{3.1}$$

$$PG(v)P^T x(v) = PG(v)\bar{x} + \tau P(c - v). \tag{3.2}$$

If a point v is such that the matrix $PG(v)P^T$ is invertible, then we can solve the linear equation (3.2) and obtain

$$x(v) = \left(PG(v)P^T\right)^{-1}\left(PG(v)\bar{x} + \tau P(c - v)\right).$$

Let $H(v) = G^{1/2}(v)$ and introduce the pseudoinverse matrix $(PH)^+ = (PH)^T(PGP^T)^{-1}$ and the projection matrix $(PH)^\sharp = (PH)^+PH$. The system (3.1), (3.2) can be rewritten in the following projective form

$$\frac{dv}{dt} = H\left[\tau(PH)^+P(c - v) - \left(I_n - (PH)^\sharp\right)H\bar{x}\right]. \tag{3.3}$$

The first vector in the square brackets belongs to the null space of AH^{-1} and the second vector belongs to the row space of this matrix. Furthermore

$$P\frac{dv}{dt} = \tau P(c - v), \quad P(c - v(t, v_0)) = P(c - v_0)e^{-\tau t}.$$

Hence, the trajectories $v(t, v_0)$ approach the set V as $t \to \infty$.

If $v_0 \in V_U$ and $v_0 > 0$, then the entire trajectory does not leave the feasible set V_U, the objective function $\bar{x}^T v(t, v_0)$ is a monotonically decreasing function of t and (3.3) can be rewritten as follows

$$\frac{dv}{dt} = -G(v)\left(I_n - P^T\left(PG(v)P^T\right)^{-1}PG(v)\right)\bar{x}, \quad v_o \in riV_U. \tag{3.4}$$

Theorem 2 *Suppose that the conditions of Theorem 1 hold. Then:*

1. *the point v_* is an asymptotically stable equilibrium point of system (3.1);*

2. *the solutions $v(t, v_0)$ of (3.3) converge locally to v_* with an exponential rate of convergence;*

3. *there exists an $\alpha_* > 0$ such that for any fixed $0 < \alpha_k < \alpha_*$ the discrete version*

$$v_{k+1} = v_k - \alpha_k G(v_k)\left(\bar{x} - P^T x_k\right), \quad x_k = x(v_k) \tag{3.5}$$

converges locally with a linear rate to v_* while the corresponding sequence $\{x_k\}$ converges to x_*.

The proof is very similar to that of Theorem 1.

Since for system (3.4) $P\dot{v} = 0_d$, it follows that the vector $\dot{v}$ belongs to null-space of P which coincides with the row space of A. Therefore there exists a vector $\lambda \in R^m$ such that

$$\dot{v} = A^T\lambda. \tag{3.6}$$

If $v > 0_n$, then after left multiplying both sides of (3.6) with $AG^{-1}(v)$ and in view of (3.4) we obtain

$$\lambda = -\left(AG^{-1}(v)A^T\right)^{-1}A\bar{x} = -\left(AG^{-1}(v)A^T\right)^{-1}b.$$

Hence, on the set riV_U the method (3.4) takes the form

$$\frac{dv}{dt} = -A^T \left(AG^{-1}(v)A^T \right)^{-1} b, \quad v_0 \in riV_U.$$

In $u-$space this method can be written as

$$\frac{du}{dt} = \left(AG^{-1}(v(u))A^T \right)^{-1} b, \quad u_0 \in U_0.$$

If we use the quadratic and exponential space transformations (2.3), we obtain

$$\frac{du}{dt} = \left(AD^{-1}(v(u))A^T \right)^{-1} b, \quad u_0 \in U_0, \tag{3.7}$$

and

$$\frac{du}{dt} = \left(AD^{-2}(v(u))A^T \right)^{-1} b, \quad u_0 \in U_0, \tag{3.8}$$

respectively. The system (3.8) coincides with the continuous version of the dual affine scaling method proposed by I.Adler, N.Karmarkar, M.Resende and G.Veiga in 1989 (see [1]).

According to Theorem 2, the solution of (3.1) converges locally with an exponential rate to equilibrium point $v_* = v(u_*)$. Therefore the solutions of (3.7) also converge to the point u_* in the set U_0.

The discrete version of (3.7) consists of the iteration

$$u_{k+1} = u_k + \alpha_k \left(AD^{-1}(v_k)A^T \right)^{-1} b, \quad u_0 \in U_0, \tag{3.9}$$

where $v_k = v(u_k)$. Taking into account Theorem 2.3.7 from [5] we conclude that the exponential rate of convergence of (3.7) insures local linear convergence of the discrete variant (3.9) if the step length α_k is sufficient small.

4 NONLOCAL CONVERGENCE ANALYSIS

In this section we consider the global convergence of the dual barrier-projection method (3.9) on the set U. Suppose that the problem (2.2) is such that

$$A\bar{e} = 0_m, \tag{4.1}$$

where $\bar{e}$ is a vector of ones in R^n.

We assume that the dual problem (2.2) has a unique solution u_*. Let $v_* = v(u_*)$ and $J_*^N = \{1 \le i \le n : v_*^i > 0\}$. Then

$$0 < \sum_{i \in J_*^N} v_*^i = \bar{e}^T v_* = \bar{e}^T c = C.$$

Here we denoted $C = \sum_i^n c^i$ and showed that $C > 0$.

Condition (4.1) implies that along all trajectories of the system (3.7) the following property holds:

$$\sum_{j=1}^{n} v^j(u(t, u_0)) = const. \tag{4.2}$$

Introduce the Lyapunov function

$$F(u) = \sum_{i \in J_*^N} v_*^i \left(\ln v_*^i - \ln v^i(u) \right). \tag{4.3}$$

This function is well-defined and continuously differentiable everywhere on the set

$$U_1 = \left\{ u \in U : v^i(u) > 0_n, i \in J_*^N \right\}.$$

Moreover $F(u_*) = 0$ and $F(u) > 0$ for all $u \in U_1$ such that $u \neq u_*$. This follows from the well-known inequality

$$F(u) = -C \sum_{i \in J_*^N} \frac{v_*^i}{C} \ln \frac{v^i(u)}{v_*^i} = -C \ln \prod_{i \in J_*^N} \left(\frac{v^i(u)}{v_*^i} \right)^{v_*^i/C} > -C \ln \sum_{i \in J_*^N} \frac{v^i(u)}{C} = 0.$$

The derivative of the Lyapunov function (4.3) along the solutions of (3.7) is

$$\frac{dF(u)}{dt} = F_u^T \dot{u} = v_*^T D^{-1}(v(u)) A^T \left(A D^{-1}(v(u)) A^T \right)^{-1} b.$$

Let

$$p(u) = \left(A D^{-1}(v(u)) A^T \right)^{-1} b, \quad x(u) = D^{-1}(v(u)) A^T p(u).$$

These functions satisfy the following conditions

$$Ax(u) = b, \quad x^T(u)v(u) = \bar{e}^T A^T p(u) = 0.$$

Hence

$$\frac{dF(u)}{dt} = v_*^T x(u) = x^T(u) \left(c - A^T u_* \right) = b^T u - b^T u_* \leq 0, \tag{4.4}$$

where equality holds only if $u = u_*$.

For an arbitrary $u_0 \in U_0$ define a Lebesgue level set $Q = \{u \in U_1 : F(u) \leq F(u_0)\}$. In view of (4.2) the set V_U is compact. Hence, U and Q are also compact. The set Q does not contain any vertex from U other than u_*. The inequality (4.4) implies that $u(t, u_0) \in Q$ for all $t \geq 0$.

Let

$$K = \inf_{u \in Q} \frac{< b, u_* - u >}{F(u)}. \tag{4.5}$$

Here, $< \cdot, \cdot >$ stands for the standard scalar product in R^n. Using (4.4) and (4.5), we obtain

$$F(u(t, u_0)) \leq F(u_0)e^{-Kt}$$

for all $t \geq 0$.

Lemma 3 *Suppose that the dual problem (2.2) has a unique nondegenerate solution u_*. Then the following estimate holds*

$$K \geq \bar{K}(u_0) = \frac{1 - e^{-F(u_0)/C}}{F(u_0)} \min_{1 \leq j \leq m} s_j > 0, \tag{4.6}$$

where $s_j = b^T(u_* - u_j)$ and u_j is a vertex of U adjacent to u_*.

Proof. We introduce the variable $z = u - u_*$ and write $F(u)$ and K as

$$F(u_* + z) = \tilde{F}(z) = - \sum_{i \in J_*^N} v_*^i \ln \left(1 - \frac{< a_i, z >}{v_*^i} \right), \quad K = - \sup_{z \in Q_1} \frac{< b, z >}{\tilde{F}(z)},$$

where $Q_1 = \{z \in Z : \tilde{F}(z) \leq F(u_o)\}$, $Z = \{z \in R^m : A^T z \leq v_*\}$ and a_i is the i−th column of A.

The function $\tilde{F}(z)$ is convex on the set Q_1. Furthermore $\tilde{F}(0) = 0$, $\tilde{F}(z) > 0$ and $b^T z < 0$ for all $z \in Z, z \neq 0_m$. Thus, for any point $\bar{z} \in S = \{z \in Q_1 : \tilde{F}(z) = F(u_o)\}$ and any $0 \leq \alpha \leq 1$ the inequality $\tilde{F}(\alpha \bar{z}) \leq \alpha \tilde{F}(\bar{z})$ holds. Hence,

$$\frac{< b, \alpha \bar{z} >}{\tilde{F}(\alpha \bar{z})} \leq \frac{< b, \bar{z} >}{\tilde{F}(\bar{z})}, \quad K = - \frac{1}{F(u_0)} \max_{z \in S} < b, z > . \tag{4.7}$$

The point $z = 0$ is a vertex of the polytope Z. Let z_j be another adjacent vertex of this polytope and let β_j be a solution of the following equality

$$\sum_{i \in J_*^N} v_*^i \ln \left(1 - \beta_j q_{ij} \right) + F(u_0) = 0, \tag{4.8}$$

where $q_{ij} = a_i^T z_j / v_*^i$. Since $\tilde{F}(z_j) = +\infty$, we obtain that $0 < \beta_j < 1$ and

$$\max_{z \in S} < b, z >= \max_{1 \leq j \leq m} \beta_j < b, z_j >= \max_{1 \leq j \leq m} \beta_j s_j > 0. \tag{4.9}$$

The inequality $A^T z \leq v_*$ implies that $q_{ij} \leq 1$ for all $i \in J_*^N$. Moreover, for at least one i we have $q_{ij} = 1$. Therefore

$$\ln(1 - \beta_j q_{ij}) \geq \ln(1 - \beta_j).$$

Hence any β_j which satisfies (4.8) is such that $\beta_j \geq \bar{\beta}$, where $\bar{\beta}$ is a solution of the following equation

$$\ln(1 - \bar{\beta}) \sum_{i \in J_*^N} v_*^i + F(u_0) = 0.$$

We conclude that $\bar{\beta} = 1 - e^{-F(u_0)/C}$. Taking into account (4.7) and (4.9) we obtain the estimate (4.6). $\square$

Let $\mu(u) = \max_{1 \leq i \leq n} x^i(u)$. We note that for any $u \in U_0$ the inequality $\mu(u) > 0$ holds. By contradiction, assume that $\mu(u) \leq 0$. Then $x(u) \leq 0$, and also $x^i(u) < 0$ at least for one i. For any $\alpha > 0$ we have $\alpha x(u) \leq 0_n < \bar{e}$. Multiplying this inequality by $D(v(u))$ we obtain $\alpha A^T (AD^{-1}(v)A^T)^{-1}b \leq v(u)$ or equivalently

$$A^T \left(u + \alpha(AD^{-1}(v)A^T)^{-1}b \right) \leq c. \qquad (4.10)$$

Thus we must have that $u + \alpha(AD^{-1}(v)A^T)^{-1}b \in U$ for any $\alpha > 0$. This contradicts the compactness of the set U. From (4.10) it follows that the value $1/\mu(u)$ is the upper bound for α such that $u + \alpha x(u) \in U$.

Theorem 3 *Let a stepsize α_k in (3.9) be chosen such that*

$$0 < \alpha_k = \gamma/\mu(u_k), \quad 0 < \gamma < 1. \qquad (4.11)$$

Then for any $u_0 \in U_0$ there exists $\gamma(u_0)$ such that $0 < \gamma(u_0) < 1$ and for all $0 < \gamma \leq \gamma(u_0)$, $k \geq 0$ the following estimate holds

$$F(u_{k+1}) \leq F(u_k) \left(1 - \frac{\alpha_k K}{2} \right), \qquad (4.12)$$

where K is defined from (4.5).

Proof. Denote

$$W(u, \alpha) = \alpha^{-1} \sum_{i \in J_*^N} v_*^i \ln \left(1 - \alpha x(u) \right).$$

It follows from (3.9) that

$$F(u_{k+1}) = F(u_k) - \alpha_k W(u_k, \alpha_k). \qquad (4.13)$$

Using the Taylor series expansion, we obtain

$$W(u, \alpha) = -v_*^T x(u) - \frac{\alpha}{2} \sum_{i \in J_*^N} \frac{v_*^i (x^i(u))^2}{(1 - \alpha \zeta^i(u) x^i(u))^2},$$

where $0 \leq \zeta^i(u) \leq 1, i \in J_*^N$. The last equation and (4.4) imply that for any $\alpha \leq \gamma/\mu(u)$ we have

$$W(u, \alpha) \geq b^T(u_* - u) - \frac{\gamma}{2(1 - \gamma)^2 \mu(u)} \sum_{i \in J_*^N} v_*^i (x^i(u))^2. \qquad (4.14)$$

We introduce the function

$$r(u) = \mu(u) \frac{< b, u_* - u >}{\sum_{i \in J_*^N} v_*^i (x^i(u))^2}$$

and prove that

$$\bar{r} = \inf_{u \in Q} r(u) > 0. \qquad (4.15)$$

Consider a minimizing sequence $\{u_s\}$ such that all $u_s \in Q, \lim_{s \to \infty} u_s = \bar{u}$ and $\bar{r} = \lim_{s \to \infty} r(u_s)$. If $\bar{u} \neq u_*$, then $\bar{r} > 0$. We prove that if $\bar{u} = u_*$, then $\bar{r} > 0$. Suppose that the partition (2.14) holds, where $v_*^B = 0_m, v_*^N > 0_d$. The same partition will be used for vector

$v(u)$ and for matrix A. Denote $\Gamma^B(u) = BD^{-1}(v^B(u))B^T, \Gamma^N(u) = ND^{-1}(v^N(u))N^T$. Since the matrix B is nonsingular we have

$$\Gamma(u) = AD^{-1}(v(u))A^T = \Gamma^B(u) + \Gamma^N(u) = \Gamma^B(u)\left[I + (\Gamma^B(u))^{-1}\Gamma^N(u)\right],$$

$$\Gamma^{-1}(u) = (\Gamma^B(u))^{-1} + \Phi(u),$$

where $\|\Phi(u)\| = o(\|u - u_*\|)$. Hence we obtain

$$x^B(u) = D^{-1}(v^B)B^T(\Gamma^B(u))^{-1}b + D^{-1}(v^B)B^T\Phi(u)b = x_*^B + \phi_1(u),$$

$$x^N(u) = D^{-1}(v^N)N^T\Gamma^{-1}(u)b = \phi_2(u),$$

$$\mu(u) = \max_{1\leq i\leq n} x^i(u) = \max_{1\leq i\leq m} x_*^i(u) + \phi_3(u),$$

where $\|\phi_i(u)\| = O(\|u - u_*\|), i = 1, 2, 3$.

Assume the contrary, i.e. $\bar{r} = 0$, then $r(u_s) < 1$ for all s sufficiently large. It follows from $x^N(u) = O(\|u - u_*\|)$ that

$$\sum_{i\in J_*^N} v_*^i(x^i(u))^2 = o(\|u - u_*\|).$$

This means that the inequality $r(u_s) < 1$ does not hold for all s sufficiently large. Therefore, $\bar{r} > 0$. From (4.15) it follows that there exists sufficiently small $0 < \gamma(u_0) < 1$ such that for all $0 < \gamma < \gamma(u_0)$ and $u \in Q$ we have

$$\frac{\gamma}{(1-\gamma)^2\mu(u)}\sum_{i\in J_*^N} v_*^i(x^i(u))^2 \leq b^T(u_* - u).$$

Hence, for such γ, u and $\alpha \leq \gamma/\mu(u)$ we obtain from (4.14) that $W(u, \alpha) \geq b^T(u_* - u)/2$. From this inequality and in view of the inequality $b^T(u_* - u) \leq KF(u)$ and (4.13) we conclude that (4.12) holds for any $u_k \in Q$. $\square$

Let

$$S(u_0) = \max_{u\in Q} \max_{1\leq i\leq n} x^i(u).$$

If the stepsize α_k is such that $\alpha_k = \gamma/\mu(u_k)$, then $\alpha_k \geq \bar{\alpha}(u_0) = \gamma/S(u_0)$ for all $k \geq 0$. Hence we have

$$V(u_{k+1}) \leq V(u_k)\left[1 - \frac{\alpha K}{2}\right], \tag{4.16}$$

where $0 < \alpha \leq \bar{\alpha}(u_0)$.

Let ϵ be the tolerance for the Lyapunov function. Then it follows from (4.6) and (4.16) that the total number of iterations performed by algorithm (3.9), (4.11) is no greater than

$$\frac{2}{\bar{\alpha}(u_0)\bar{K}(u_0)}\ln\left(\frac{V(u_0)}{\epsilon}\right).$$

5   ACKNOLEGEMENTS

The authors would like to thank K.L.Teo and V.Rehbock for very useful corrections and comments. The work described in this paper is a part of research project sponsored by Russian Scientific fund.

References

[1] I.Adler, N.Karmarkar, M.G.C.Resende and G.Veiga. An Implementation of Karmarkar's Algorithm for Linear Programming, *Math. Programming* **44**, 297-335, (1989).

[2] A. Bacciotti, *Local stability of nonlinear control systems*, Series on Advances in Mathematics for Applied Sciences **8**, World Scientific Publishing Co. Ptc. Ltd., Singapore, (1992).

[3] I.I.Dikin. Iterative solution of problems of linear and quadratic programming, *Sov. Math. Dokl.* **8**, 674-675, (1967).

[4] Yu.G.Evtushenko. Two numerical methods of solving nonlinear programming problems, *Sov. Math. Dokl.* **15** (2), 420-423, (1974).

[5] Yu.G.Evtushenko. *Numerical Optimization Techniques. Optimization Software*, Inc. Publications Division, New York., (1985).

[6] Yu.G.Evtushenko and V.G.Zhadan. Numerical methods for solving some operations research problems, *U.S.S.R. Comput. Maths. Math. Phys.* **13** (3), 56-77, (1973).

[7] Yu.G.Evtushenko and V.G.Zhadan. A relaxation method for solving problems of nonlinear programming, *U.S.S.R. Comput. Maths. Math. Phys.* **17** (4), 73-87, (1977).

[8] Yu.G.Evtushenko and V.G.Zhadan. *Barrier-projective and barrier-Newton numerical methods in optimization (the nonlinear programming case)*, Computing Centre of the USSR Academy of Sciences, Reports in Comput. Math., (1991), (in Russian).

[9] Yu.G.Evtushenko and V.G.Zhadan. *Barrier-projective and barrier-Newton numerical methods in optimization (the linear programming case)*, Computing Centre of the USSR Academy of Sciences, Reports in Comput. Math., (1992), (in Russian).

[10] Yu.G.Evtushenko and V.G.Zhadan. Stable Barrier-Projection and Barrier-Newton Methods in Nonlinear Programming, *Optimization Methods and Software* **3** (1-3), 237-256, (1994).

[11] Yu.G.Evtushenko and V.G.Zhadan. Stable Barrier-Projection and Barrier-Newton Methods for Linear and Nonlinear Programming, In *Algorithms for Continuous Optimization*, (Edited by E.Spedicato), NATO ASI Series, 255-285, Kluwer Academic Publishers, (1994).

[12] Yu.G.Evtushenko and V.G.Zhadan. Stable Barrier-Projection and Barrier-Newton Methods in Linear Programming, *Computational Optimization and Applications* **3** (4), 289-303, (1994).

[13] L.E.Faybusovich. Hamiltonian structure of dynamical systems which solve linear programming problems, *Physica D* **53**, 217-232, (1991).

[14] U. Helmke and J. B. Moore. *Optimization and Dynamical Systems*, Springer-Verlag, (1994).

[15] S.Herzel, M.C.Recchioni and F.Zirilli. it A quadratically convergent method for linear programming, *Linear Algebra and its Applications* **152**, 255-289, (1991).

[16] K.Tanabe. A geometric method in nonlinear programming, *Journal of Optimization Theory and Applications* **30** (2), 181- 210, (1980).

WSSIAA 5 (1995) pp. 67–78

SOLVING PARAMETER IDENTIFICATION PROBLEMS
ON A PARALLEL-VECTOR COMPUTER

ILIO GALLIGANI and MARINA BERTAJA

Department of Mathematics, University of Bologna, Italy
I-40126 Bologna, Italy

ABSTRACT

In this paper we analyze a common procedure used for the identification of parameters in distributed systems in order to obtain an efficient solver of the parameter identification problem on a parallel-vector computer, such as the Cray Y-MP. The results of some numerical experiments are presented.

1. The Parameter-Identification Problem

Let t denote time, $t \in [0, t^*]$, and let Ω be an open, bounded and simply connected domain in $\Re^2$ with piecewise smooth boundary Γ. Let be given the following parabolic differential equation in $\Omega \times (0, t^*]$

$$\varepsilon(xy) \frac{\partial \phi(xyt)}{\partial t} = \frac{\partial}{\partial x} \left(\sigma(xy) \frac{\partial \phi(xyt)}{\partial x} \right) + \frac{\partial}{\partial y} \left(\sigma(xy) \frac{\partial \phi(xyt)}{\partial y} \right) + \theta(xyt) \tag{1}$$

with the initial-boundary conditions

$$\phi(xy0) = \phi_0(xy) \qquad \text{in } \Omega$$

$$\phi(xyt) = \phi_1(xyt) \qquad \text{on } \Gamma \times [0, t^*] \tag{2}$$

We assume that the coefficients $\varepsilon(xy)$ and $\sigma(xy)$ and the "source" term $\theta(xyt)$ of the equation (1) and the data $\phi_0(xy)$ and $\phi_1(xyt)$ appearing in (2) will be sufficiently smooth in order to have for the system (1)-(2) a unique regular solution $\phi(xyt)$ (For a problem of identifying discontinuous coefficients in parabolic equations, see [4])
This solution depends on the coefficients of (1) $\varepsilon = \varepsilon(xy)$ and $\sigma = \sigma(xy)$; in order to display this dependence explicitly, we write $\phi(xyt) = \phi(xyt; v)$, where $v = (\varepsilon, \sigma)$.

The *parameter-identification problem* associated with system (1)-(2) is that of minimizing the functional

$$J(v) = \int\limits_{0\Omega}\int (\phi(xyt;v) - \tilde{\phi}(xyt))^2 \, dx\,dy\,dt \tag{3}$$

where $\tilde{\phi}(xyt)$ is a "measurement" of $\phi(xyt)$. We assume that this "measurement" will be a sufficiently smooth map obtained, for example, by deterministic or stochastic interpolators.

The problem of identifying the coefficients $\varepsilon(xy)$ and $\sigma(xy)$ in the system (1)-(2) is *improperly posed*, because its solution can not be uniquely determined.

The following is a simple proof of this remarkable proposition.

Let be given two functions $f_1(t)$ and $f_2(t)$ sufficiently smooth in $[0, t^*]$. From equation (1) we obtain the two equations ($i=1,2$):

$$\frac{\partial\sigma}{\partial x}\int\limits_0^{t^*}\frac{\partial\phi}{\partial x}f_i\cdot dt + \frac{\partial\sigma}{\partial y}\int\limits_0^{t^*}\frac{\partial\phi}{\partial y}f_i\cdot dt + \sigma\int\limits_0^{t^*}\left(\frac{\partial^2\phi}{\partial x^2} + \frac{\partial^2\phi}{\partial y^2}\right)f_i\cdot dt$$

$$+ \int\limits_0^{t^*}\theta f_i\cdot dt = \varepsilon\int\limits_0^{t^*}\frac{\partial\phi}{\partial t}f_i\cdot dt$$

Then, combining linearly these two equations, we have

$$a(xy)\frac{\partial\sigma(xy)}{\partial x} + b(xy)\frac{\partial\sigma(xy)}{\partial y} + c(xy)\sigma(xy) = g(xy) \qquad \text{in } \Omega \tag{4}$$

where $\Delta\phi = \dfrac{\partial^2\phi}{\partial x^2} + \dfrac{\partial^2\phi}{\partial y^2}$ and

$$a(xy) = \int\limits_0^{t^*}\frac{\partial\phi}{\partial x}f_1\cdot dt \int\limits_0^{t^*}\frac{\partial\phi}{\partial t}f_2\cdot dt - \int\limits_0^{t^*}\frac{\partial\phi}{\partial x}f_2\cdot dt \int\limits_0^{t^*}\frac{\partial\phi}{\partial t}f_1\cdot dt$$

$$b(xy) = \int\limits_0^{t^*}\frac{\partial\phi}{\partial y}f_1\cdot dt \int\limits_0^{t^*}\frac{\partial\phi}{\partial t}f_2\cdot dt - \int\limits_0^{t^*}\frac{\partial\phi}{\partial y}f_2\cdot dt \int\limits_0^{t^*}\frac{\partial\phi}{\partial t}f_1\cdot dt$$

$$c(xy) = \int\limits_0^{t^*}\Delta\phi\, f_1\cdot dt \int\limits_0^{t^*}\frac{\partial\phi}{\partial t}f_2\cdot dt - \int\limits_0^{t^*}\Delta\phi\, f_2\cdot dt \int\limits_0^{t^*}\frac{\partial\phi}{\partial t}f_1\cdot dt$$

$$g(xy) = \int_0^{t^*} \theta f_1 \cdot dt \int_0^{t^*} \frac{\partial \phi}{\partial t} f_2 \cdot dt - \int_0^{t^*} \theta f_2 \cdot dt \int_0^{t^*} \frac{\partial \phi}{\partial t} f_1 \cdot dt$$

A *characteristic curve* $y = y(x)$ related to the hyperbolic equation (4) has the characteristic direction

$$\frac{dy}{dx} = \frac{b(xy)}{a(xy)} = \alpha(xy) \tag{5}$$

A part of the boundary Γ of Ω, called Γ_0, is characterized by the fact that the characteristic curves of (4) through each point of Γ_0 must cover completely the domain Ω. That is, if Σ_0 is the set of the characteristic curves starting from Γ_0 with characteristic direction (5), it holds $\Sigma_0 \cap \overline{\Omega} = \overline{\Omega}$, where $\overline{\Omega} = \Omega \cup \Gamma$.
Let's assume that $a(xy), b(xy), c(xy)$ and $g(xy)$ are functions of class $C^1(\overline{\Omega})$ with $a \neq 0$ and $b \neq 0$ in Ω and let's assume that $\sigma(xy)$ is known on Γ_0:

$$\sigma(xy) = d(xy) \qquad \text{on } \Gamma_0 \tag{6}$$

where the given function $d(xy)$ is of class $C^1(\Gamma_0)$.
From $P_0 = (x_0, y_0)$ on Γ_0 is starting a characteristic curve $y = y(x)$ which is the solution for the Cauchy problem

$$\frac{dy}{dx} = \alpha(xy) \qquad\qquad y(x_0) = y_0$$

for $x \in E_0$ where $E_0 = \{x \mid (x, y(x)) \in \Omega \cup \Gamma_0\}$.
Let's assume that this curve is open and has no multiple point.
On this curve the solution $\sigma(xy)$ of (4)-(6) is obtained by solving the Cauchy problem

$$\begin{cases} \dfrac{ds(x)}{dx} = \beta(xy(x)) s(x) + \gamma(xy(x)) \\[2mm] s(x_0) = \sigma(x_0 y(x_0)) \end{cases} \tag{7}$$

for $x \in E_0$, where

$$s(x) = \sigma(xy(x)), \qquad \beta(xy) = -\frac{c(xy)}{a(xy)} \qquad \text{and} \qquad \gamma(xy) = \frac{g(xy)}{a(xy)}$$

It is well known from the Theory of Partial Differential Equations that problem (7) has in $C^1(E_0)$ a unique solution. Thus, in order to obtain a *unique* solution $\sigma(xy)$ of (4) in

Ω, it is necessary to know the function $\sigma(xy)$ on Γ_0 (see condition (6)): if condition (6) is not satisfied, (4) will not necessarily have a solution.

Therefore, when the boundary condition (6) is lacking, the problem of identifying the parameter $\sigma(xy)$ is improperly posed and in general yields multiple solutions.

2. Numerical Solution of the Problem

One of the most widely used method for minimizing the functional (3) along a trajectory $\phi(xyt)$ of the system (1)-(2) is the *gradient method*[2].

Starting from an initial "approximation" $v^{(0)}$ of the local minimum u of the functional $J(v)$, the gradient method is defined by the iterative process

$$v^{(k+1)} = v^{(k)} - \rho_k \nabla J(v^{(k)}) \tag{8}$$

where $\nabla J(v^{(k)})$ is the "gradient" of $J(v)$ evaluated at $v^{(k)}$ and ρ_k is a positive "search parameter" determined is such a way that $J(v^{(k+1)}) < J(v^{(k)})$.

The parameter ρ_k can be determined with the (one-dimensional) "cubic method".[1] The gradient method stops when $|J(v^{(k+1)}) - J(v^{(k)})|$ is less then a given "convergence parameter".

Thus, at each stage k of the process (8) we have to evalutate $\nabla J(v)$ in $v^{(k)}$.

For the evaluation of the gradient of $J(v)$ in v we consider the "variation" δv of v that corresponds to the variation $\delta J(v)$ of (3) defined by

$$\delta J(v) = 2 \int_0^{t^*}\!\!\int_\Omega \left(\phi(xyt;v) - \tilde{\phi}(xyt)\right) \delta\phi \cdot dx\,dy\,dt =$$

$$= 2 \int_0^{t^*}\!\!\int_\Omega \left(\phi(xyt;v) - \tilde{\phi}(xyt)\right) \left(\phi(xyt;v+\delta v) - \phi(xyt;v)\right) \cdot dx\,dy\,dt \tag{9}$$

From equation (1) we obtain in $\Omega \times (0,t^*]$

$$\delta\varepsilon \frac{\partial\phi}{\partial t} + \varepsilon \frac{\partial\delta\phi}{\partial t} = \frac{\partial}{\partial x}\left(\delta\sigma \frac{\partial\phi}{\partial x} + \sigma \frac{\partial\delta\phi}{\partial x}\right) + \frac{\partial}{\partial y}\left(\delta\sigma \frac{\partial\phi}{\partial y} + \sigma \frac{\partial\delta\phi}{\partial y}\right)$$

From equation (2) we have

$$\delta\phi(xy0) = 0 \qquad\qquad \text{in } \Omega$$

$$\delta\phi(xyt) = 0 \qquad\qquad \text{on } \Gamma\times[0,t_*]$$

By introducing the "adjoint differential system" in $\Omega \times (0, t^*]$

$$-\varepsilon(xy) \frac{\partial \psi(xyt)}{\partial t} = \frac{\partial}{\partial x}\left(\sigma(xy) \frac{\partial \psi(xyt)}{\partial x}\right) + \frac{\partial}{\partial y}\left(\sigma(xy) \frac{\partial \psi(xyt)}{\partial y}\right) + \phi(xyt; v) - \tilde{\phi}(xyt) \quad (10)$$

$$\psi(xyt^*) = 0 \qquad \text{in } \Omega$$

$$\psi(xyt) = 0 \qquad \text{on } \Gamma \times [0, t*] \qquad (11)$$

we have

$$\int\limits_0^{t^*}\!\!\iint\limits_\Omega \delta\varepsilon \frac{\partial \phi}{\partial t} \psi \cdot dx\,dy\,dt + \int\limits_0^{t^*}\!\!\iint\limits_\Omega \varepsilon \frac{\partial \delta\phi}{\partial t} \psi \cdot dx\,dy\,dt = \iint\limits_0^{t^*}\limits_\Omega \left(\frac{\partial}{\partial x}\left(\sigma \frac{\partial \delta\phi}{\partial x}\right) + \right.$$

$$\left. + \frac{\partial}{\partial y}\left(\sigma \frac{\partial \delta\phi}{\partial y}\right)\right) \psi \cdot dx\,dy\,dt + \int\limits_0^{t^*}\!\!\iint\limits_\Omega \left(\frac{\partial}{\partial x}\left(\delta\sigma \frac{\partial \phi}{\partial x}\right) + \frac{\partial}{\partial y}\left(\delta\sigma \frac{\partial \phi}{\partial y}\right)\right) \psi \cdot dx\,dy\,dt \quad (12)$$

Integration by parts yields $(\delta\phi(xy0) = 0;\ \psi(xyt^*) = 0)$:

$$\int\limits_0^{t^*}\!\!\iint\limits_\Omega \varepsilon \frac{\partial \delta\phi}{\partial t} \psi \cdot dt\,dx\,dy = - \int\limits_0^{t^*}\!\!\iint\limits_\Omega \varepsilon \delta\phi \frac{\partial \psi}{\partial t} \cdot dt\,dx\,dy$$

Using Green's formula, we obtain (ψ and $\delta\phi$ are zero on $\Gamma \times [0, t^*]$)

$$\int\limits_0^{t^*}\!\!\iint\limits_\Omega \left(\frac{\partial}{\partial x}\left(\sigma \frac{\partial \delta\phi}{\partial x}\right) + \frac{\partial}{\partial y}\left(\sigma \frac{\partial \delta\phi}{\partial y}\right)\right) \psi \cdot dx\,dy\,dt =$$

$$= \int\limits_0^{t^*}\!\!\iint\limits_\Omega \delta\phi \left(\frac{\partial}{\partial x}\left(\sigma \frac{\partial \psi}{\partial x}\right) + \frac{\partial}{\partial y}\left(\sigma \frac{\partial \psi}{\partial y}\right)\right) dx\,dy\,dt$$

By the divergence theorem, we have ($\psi = 0$ on $\Gamma \times [0, t^*]$)

$$\int\limits_0^{t^*}\!\!\iint\limits_\Omega \left(\frac{\partial}{\partial x}\left(\delta\sigma \frac{\partial \phi}{\partial x}\right) + \frac{\partial}{\partial y}\left(\delta\sigma \frac{\partial \phi}{\partial y}\right)\right) \psi \cdot dx\,dy\,dt =$$

$$= - \int\limits_0^{t^*}\!\!\iint\limits_\Omega \delta\sigma \left(\frac{\partial \phi}{\partial x} \frac{\partial \psi}{\partial x} + \frac{\partial \phi}{\partial y} \frac{\partial \psi}{\partial y}\right) \cdot dx\,dy\,dt$$

72

Thus, taking into account (10), formula (12) becomes

$$\int\limits_0^{t^*}\!\!\int\limits_\Omega \delta\varepsilon\, \frac{\partial\phi}{\partial t}\, \psi \cdot dt\, dx\, dy + \int\limits_0^{t^*}\!\!\int\limits_\Omega \big(\phi(xyt;v) - \bar\phi(xyt)\big)\, \delta\phi \cdot dt\, dx\, dy =$$

$$= -\int\limits_0^{t^*}\!\!\int\limits_\Omega \delta\sigma \left(\frac{\partial\phi}{\partial x}\frac{\partial\psi}{\partial x} + \frac{\partial\phi}{\partial y}\frac{\partial\psi}{\partial y}\right) \cdot dt\, dx\, dy$$

Then, from (9) we have

$$\delta J(v) = -2 \int\limits_\Omega \delta\varepsilon \int\limits_0^{t^*} \frac{\partial\phi}{\partial t}\, \psi \cdot dt\, dx\, dy - 2 \int\limits_\Omega \delta\sigma \int\limits_0^{t^*} \left(\frac{\partial\phi}{\partial x}\frac{\partial\psi}{\partial x} + \frac{\partial\phi}{\partial y}\frac{\partial\psi}{\partial y}\right) \cdot dt\, dx\, dy$$

Therefore, the two components of the gradient of $J(v)$ in $v^{(k)}$ are

$$\frac{\partial J(v^{(k)})}{\partial\varepsilon} = -2 \int\limits_0^{t^*} \frac{\partial\phi(xyt;v^{(k)})}{\partial t}\, \psi(xyt;v^{(k)}) \cdot dt$$

$$(13)$$

$$\frac{\partial J(v^{(k)})}{\partial\sigma} = -2 \int\limits_0^{t^*} \left(\frac{\partial\phi(xyt;v^{(k)})}{\partial x}\frac{\partial\psi(xyt;v^{(k)})}{\partial x} + \frac{\partial\phi(xyt;v^{(k)})}{\partial y}\frac{\partial\psi(xyt;v^{(k)})}{\partial y}\right) \cdot dt$$

The evaluation of these components requires the determination of the solution $\phi(xyt;v^{(k)})$ of the system (1)-(2) for given approximations $\varepsilon^{(k)}$ and $\sigma^{(k)}$ of the coefficients ε and σ of (1) and the determination of the solution $\psi(xyt;v^{(k)})$ of the adjoint differential system (10)-(11) for $\phi(xyt;v) \equiv \phi(xyt;v^{(k)})$.

We shall henceforth assume that the domain Ω is a rectangle with the sides parallel to the coordinate axes and that the coefficients $\varepsilon(xy)$ and $\sigma(xy)$ of equation (1) are positive functions in Ω.

Then, a network Ω_Δ of horizontal and vertical lines is imposed on the domain Ω; we assume equal spacing Δy between the horizontal lines and equal spacing Δx between the vertical lines. At each interior mesh-point (i,j) of the network the following difference approximation to the differential equation (1) is obtained by substitution of difference ratios for derivatives:

$$\varepsilon_{ij}\, \frac{d\phi_{ij}(t)}{dt} = \nabla_x\left(\sigma\nabla_x\phi\right)\big|_{ij} + \nabla_y\left(\sigma\nabla_y\phi\right)\big|_{ij} + \theta_{ij}(t) \tag{14}$$

Taking into account the initial-boundary conditions (2), we obtain that the approximate solution $\Phi(t)$ of $\phi(xyt)$ on the mesh-points of the Ω_Δ satisfies the matrix equation

$$C\frac{d\Phi(t)}{dt} = -A\Phi(t) + q(t) \qquad (15)$$

with $\Phi(0) = \Phi_0$ given, where C is a diagonal matrix with positive diagonal elements and A is a block tridiagonal matrix; each square block on the diagonal of A is a non singular tridiagonal matrix and each block on the subdiagonal and on the superdiagonal of A is a diagonal matrix. The matrix A is symmetric with positive diagonal entries and non positive off-diagonal entries. Moreover, A is irreducibly diagonally dominant, so that A is positive definite. Then, the solution $\Phi(t)$ of (15) is asymptotically stable.
Similary, we obtain that the restriction on Ω_Δ of the approximation to the solution $\psi(xyt)$ of system (10)-(11) satisfies the ordinary matrix differential equation

$$-C\frac{d\psi(t)}{dt} = -A\psi(t) + \left(\Phi(t) - \tilde{\Phi}(t)\right) \qquad (16)$$

with $\psi(t^*) = 0$, where $\tilde{\Phi}(t)$ is the restriction on the network Ω_Δ of the given function $\tilde{\phi}(xyt)$. Having obtained the solutions of the systems (15) and (16), it is possible to obtain approximations to the integrals (13) by applying the "rectangular rule" to each cell of the network Ω_Δ.

3. Computational Complexity

The most expensive (with respect to the computer-time) steps of the gradient method are those related to the numerical integration of the system (1)-(2) and of the adjoint system (10)-(11). Therefore a deeper analysis of these steps is useful, expecially when we are interested to solve the above systems on a *parallel computer*.

In literature many methods have been proposed for solving the matrix equations (15) and (16) or the equivalent equations

$$\frac{d}{dt}\left(C^{1/2}\Phi(t)\right) = -C^{-1/2}AC^{-1/2}\left(C^{1/2}\Phi(t)\right) + C^{-1/2}q(t) \qquad (15')$$

$$-\frac{d}{dt}\left(C^{1/2}\psi(t)\right) = -C^{-1/2}AC^{-1/2}\left(C^{1/2}\psi(t)\right) + C^{-1/2}(\Phi(t) - \tilde{\Phi}(t)) \qquad (16')$$

written in a symmetrizable form
Most of these methods have been also re-structured for an implementation on parallel and parallel-vector computers.

74

An interesting result has been obtained[3]: the *Alternating-Direction Implicit (ADI)* method is a very efficient method for solving on parallel-vector computers those diffusion problems with time-dependent boundary value and time-dependent source term for which it is required to use a "medium-sized" time-step in the integration process. Thus, for many identification problems of practical interest the ADI method seems to be the most effective for solving systems (1)-(2) and (10)-(11).

The Alternating-Direction Implicit method associated with the matrix equation (15) is defined by $(0 \leq \rho \leq 1)$:

$$\left(C + \Delta t A_1\right) \Phi_{m+1} = \left(C - \Delta t A_2\right) \Phi_m + \Delta t \left((1-\rho) q(t_m) + \rho q(t_{m+1})\right) \qquad (17)$$

$$\left(C + \Delta t A_2\right) \Phi_{m+2} = \left(C - \Delta t A_1\right) \Phi_{m+1} + \Delta t \left((1-\rho) q(t_{m+2}) + \rho q(t_{m+1})\right) \qquad (18)$$

where A_1 and A_2 are two splittings of A, $A = A_1 + A_2$, corresponding to the discretization of equation (1) in the x-direction and in the y-direction, respectively. The vector Φ_m is the approximation to the solution $\Phi(t)$ of (15) at the time $t_m = m\Delta t$ for any time-level $m = 1, 2, \ldots$; Δt is the time-step.

The matrix A_1 is a direct sum of tridiagonal matrices. Thus, the system (17) can be solved "in parallel" with the usual cyclic reduction algorithm. Matrix A_2 is not a direct sum of tridiagonal matrices, but there exists a permutation matrix P such that $P A_2 P^T$ is a direct sum of tridiagonal matrices. Hence, the coefficient matrix of the system (18) is permutationally similar to a direct sum of tridiagonal matrices. Therefore, also the system (18) can be solved efficiently "in parallel".

The *stability* of the ADI method (17)-(18) is assured by the following general result.

Theorem – Let be given a matrix M of varying order n, having at most $r < n$ non zero elements on each row and column, r being constant for all n. Let M be expressed in the form

$$M = \left(H_1 + D_1\right) + \left(H_2 + D_2\right)$$

where the elements of H_i $(i=1,2)$ are inversely proportional to the square of the mesh-spacings Δx and Δy and the elements of D_i $(i=1,2)$ do not depend upon Δx and Δy. Assume that $H_1 + H_1^T$ and $H_2 + H_2^T$ are positive definite matrices and that $\Delta t / \Delta x^2$ and $\Delta t / \Delta y^2$ are constants as $\Delta x \to 0$, $\Delta y \to 0$ and $\Delta t \to 0$.

Then, the matrix $\left(T(\Delta t)\right)^m$, where

$$T(\Delta t) = \left(I + \Delta t (H_2 + D_2)\right)^{-1} \left(I - \Delta t (H_1 + D_1)\right) \cdot \left(I + \Delta t (H_1 + D_1)\right)^{-1} \left(I - \Delta t (H_2 + D_2)\right),$$

is ℓ_2-norm uniformly bounded for all n and all Δt: $0<\Delta t<\Delta t^*$, $0<m\Delta t\leq t^*$ $(m=1,2,...)$.

Thus, if we continually refine the time-step Δt, and hence also the spatial mesh-spacings Δx and Δy (in such a way that $\Delta t/_{\Delta x^2}$ and $\Delta t/_{\Delta y^2}$ are arbitrary positive constants), and require more and more steps to reach a fixed time-station $\bar{t}\leq t^*$, the computed solution (obtained with the ADI) is always bounded.

For the sake of completeness, we shall give a proof of this theorem.
The matrices D_i and $\Delta t H_i$ $(i=1,2)$ have ℓ_2-norm uniformly bounded for all n. Indeed, their elements are uniformly bounded because the elements of D_i are independent of the parameters Δt, Δx, Δy and the magnitude of the elements of $\Delta t H_i$ is bounded provided $\Delta t/_{\Delta x^2}$ and $\Delta t/_{\Delta y^2}$ are fixed as $\Delta t \to 0$. Now, let $\gamma>0$ be a bound on the absolute value of the elements of the matrices D_i and $\Delta t H_i$ with $i=1,2$. Then $(i=1,2)$:

$$\|\Delta t H_i\|_\infty \leq \gamma r, \quad \|\Delta t H_i\|_1 \leq \gamma r, \quad \|D_i\|_\infty \leq \gamma r \quad \text{and} \quad \|D_i\|_1 \leq \gamma r$$

The Cauchy-Schwarz inequality gives:

$$\|D_i\|_2^2 = \sup_{\|z\|_2=1} \sum_{\ell=1}^{r} \left| \sum_{j=1}^{r} d_{\ell j}^{(l)} z_j \right|^2 \leq \sup_{\|z\|_2=1} \sum_{\ell=1}^{r} \left(\sum_{j=1}^{r} \sqrt{|d_{\ell j}^{(l)}|} \cdot \sqrt{|d_{\ell j}^{(l)}|} \, |z_j| \right)^2 \leq$$

$$\leq \sup_{\|z\|_2=1} \sum_{\ell=1}^{r} \left(\left(\sum_{j=1}^{r} |d_{\ell j}^{(l)}| \right) \left(\sum_{j=1}^{r} |d_{\ell j}^{(l)}| \, |z_j|^2 \right) \right) \leq (\gamma r)^2 \sup_{\|z\|_2=1} \sum_{j=1}^{r} |z_j|^2 = (\gamma r)^2$$

Similarly, we have $\|\Delta t H_i\|_2 \leq \gamma r$.
Also the matrices $I-\Delta t H_1$ and $I-\Delta t H_2$ have bounded norms.
Now, we prove that the matrices $(I+\Delta t H_1)^{-1}$ and $(I+\Delta t H_2)^{-1}$ have bounded norms.
By definition we obtain $(i=1,2)$

$$\|(I+\Delta t H_i)^{-1}\|_2^2 = \sup_{z} \frac{z^T (I+\Delta t H_i^T)^{-1} \, (I+\Delta t H_i)^{-1} \, z}{z^T \cdot z} =$$

$$= \sup_{\xi} \frac{\xi^T \cdot \xi}{\xi^T (I+\Delta t H_i^T)(I+\Delta t H_i) \, \xi} = \sup_{\xi} \frac{\xi^T \cdot \xi}{\xi^T (I+\Delta t(H_i^T+H_i) + \Delta t^2 H_i^T H_i) \, \xi}$$

Since the matrix $H_i+H_i^T$ is positive definite, the last expression is less than one, and therefore the norm of $(I+\Delta t H_i)^{-1}$, $i=1,2$, is bounded less than one.
In similar way it is easy to prove that also the matrix $(I-\Delta t H_i)(I+\Delta t H_i)^{-1}$, $i=1,2$, has bounded norm less than one.
The next step in the proof of the theorem is to demonstrate that the matrix $T(\Delta t)$

may be written as $T(\Delta t) = T_1(\Delta t) + \Delta t\, T_2(\Delta t)$ where $\|(T_1(\Delta t)^m\|_2$ is uniformly bounded for $0 < \Delta t < \Delta t^*$, $0 < m\Delta t \le t^*$ and $T_2(\Delta t)$ is bounded.

Indeed, by direct computation, we have

$$T(\Delta t) = \left(I + \Delta t\,(I + \Delta t\,H_2)^{-1}\,D_2\right)^{-1}\,(I + \Delta t\,H_2)^{-1}\,(I - \Delta t\,(H_1 + D_1)) \cdot$$

$$\cdot \left(I + \Delta t\,(I + \Delta t\,H_1)^{-1}\,D_1\right)^{-1}\,(I + \Delta t\,H_1)^{-1}\,(I - \Delta t(H_2 + D_2)) =$$

$$= \left(I - \Delta t(I + \Delta t\,H_2)^{-1}\,D_2\right)(I + \Delta t\,H_2)^{-1}\,(I - \Delta t\,(H_1 + D_1)) \cdot$$

$$\cdot \left(I - \Delta t\,(I + \Delta t\,H_2)^{-1}\,D_1\right)(I + \Delta t\,H_1)^{-1}\,(I - \Delta t\,(H_2 + D_2)) + o\left(\Delta t^2\right)$$

The last result is obtained because $\|(I + \Delta t H_i)^{-1}\|_2 < 1$ and $\Delta t\,\|-D_i\|_2 < 1$ as $\Delta t \to 0$ $(i = 1,2)$. Therefore:

$$T(\Delta t) = (I + \Delta t H_2)^{-1}\,(I - \Delta t H_1)\,(I + \Delta t H_1)^{-1}\,(I - \Delta t H_2) + \Delta t\,T_2(\Delta t)$$

where the matrix $T_2(\Delta t)$ is bounded for $0 < \Delta t < \Delta t^*$. For the matrix

$$T_1(\Delta t) = (I + \Delta t H_2)^{-1}\,(I - \Delta t H_1)\,(I + \Delta t H_1)^{-1}\,(I - \Delta t H_2)$$

we have

$$\|(T_1(\Delta t))^m\|_2 = \| (I + \Delta t H_2))^{-1}\,(I - \Delta t H_1)\,(I + \Delta t H_1)^{-1}\,(I - \Delta t H_2) \cdot$$

$$\cdot (I + \Delta t H_2))^{-1}\,(I - \Delta t H_1)\,(I + \Delta t H_1)^{-1}\,(I - \Delta t H_2) \cdot$$

$$\ldots\ldots$$

$$(I + \Delta t H_2)^{-1}\,(I - \Delta t H_1))\,(I + \Delta t H_1)^{-1}\,(I - \Delta t H_2)\,\|_2 \qquad (m\ \text{times})$$

$$\le \|(I + \Delta t H_2)^{-1}\|_2\,\|(I - \Delta t H_1)\,(I + \Delta t H_1)^{-1}\|_2\,\|(I - \Delta t H_2)\,(I + \Delta t H_2)^{-1}\|_2$$

$$\cdot \|(I - \Delta t H_1)\,(I + \Delta t H_1)^{-1}\|_2 \ldots \|(I - \Delta t H_1)\,(I + \Delta t H_1)^{-1}\|_2\,\|(I - \Delta t H_2)\|_2 \le \|(I - \Delta t H_2)\|_2$$

Thus, the ℓ_2-norm $\|(T_1(\Delta t))^m\|_2$ is bounded for $0 < \Delta t < \Delta t^*$ and $0 < m\Delta t \le t^*$.

Now, by applying a theorem on perturbed operators[6], we have finally that the ℓ_2-norm $\|(T(\Delta t))^m\|_2$ is uniformly bounded for $0 < \Delta t < \Delta t^*$ and $0 < m\Delta t \le t^*$.#

The *effectiveness* of the gradient method has been evaluated by carrying out computational experiments on some "representative" problems for which exact results are known. A parallel implementation on the parallel-vector computer CRAY Y-MP of the gradient method, combined with the ADI method for solving the systems (1)-(2) and (10)-(11), has been developed. The multiprocessor Cray used can execute concurrently different tasks on four vector-processors with shared central memory. In each vector processor concurrency is provided by pipelining and independent instruction execution and the operations are performed with vector instructions in a vector register environment. A high-level parallelism among independent tasks is offered by the Cray-multitasking, using CMIC$ microtasking directives.

The first numerical study concerns the estimation of the parameters $\varepsilon(xy)$ and $\sigma(xy)$ in problems (1)-(2) with $\phi(xyt)$, $\theta(xyt)$, $\phi_0(xy)$, $\phi_1(xyt)$, $\varepsilon(xy)$ and $\sigma(xy)$ given exactly. The domain Ω is the rectangle $\Omega = [0,2] \times [0,1]$. The functions $\varepsilon(xy)$ and $\sigma(xy)$ are polynomials in x, y of degree varying from 1 to 3 and the function $\phi(xyt)$ has the form $\phi(xyt) = e^{-\alpha t}\phi_*(xyt)$ with $0 \le \alpha \le 1$, where $\phi_*(xyt)$ is a polynomial in x, y, t of degree varying from 1 to 3. The measurement $\tilde{\phi}(xyt)$ of $\phi(xyt)$ is exactly $\phi(xyt)$: $\tilde{\phi}(xyt) \equiv \phi(xyt)$ in Ω. In the gradient method the initial approximation $v^{(0)} = (\varepsilon^0(xy), \sigma^0(xy))$ to the parameters $\varepsilon(xy)$ and $\sigma(xy)$ is equal to two constants or is obtained by adding a random function to the exact functions $\varepsilon(xy)$ and $\sigma(xy)$. On Ω a rectangular network of 64×48 mesh-points is overlapped.

The second study is related to the numerical simulation of a three-years (15 days time-steps) pumping test conducted on a realistic underground aquifer with 48 wells distributed in a square of 12 km width. We suppose that the history of the production rate of the aquifer is known and, at a fixed time $t=0$, the piezometric, storage coefficient and transmissivity maps are available: it means that the functions $\theta(xyt)$, $\phi_0(xy)$, $\varepsilon(xy)$ and $\sigma(xy)$ are assigned. We put $\phi_1(xyt) = 0$. With these data we simulate the measures $\tilde{\phi}(xyt)$ of the piezometric head by integrating the system (1)-(2). Now, ignoring the exact values of $\varepsilon(xy)$ and $\sigma(xy)$ we solve the parameter identification problem with an initial guess $v^{(0)} = (\varepsilon^0(xy), \sigma^0(xy))$ equal to two costants or obtained by a random perturbation of the exact functions $\varepsilon(xy)$ and $\sigma(xy)$. On Ω a network of 55×55 mesh-points is overlapped.

From the experience we have gathered on this set of test-problems, the following conclusions can be drawn.

a) In both studies $\varepsilon(xy)$ and $\sigma(xy)$ are determined with satisfactory accuracy when the initial approximation $v^{(0)}$ is a random perturbation of the exact parameters $\varepsilon(xy)$ and $\sigma(xy)$.

b) When the initial approximation $\varepsilon^0(xy)$ and $\sigma^0(xy)$ are constants, the computed values of $\varepsilon(xy)$ and $\sigma(xy)$ are a little different from the initial guess values in the mesh-points near the boundary of Ω. The computed values and the exact values of $\varepsilon(xy)$ and $\sigma(xy)$ are in good agreement in sufficiently extended regions *inside* of Ω.

c) In case *a)* the convergence of the gradient method (10) is obtained after a number k of iterations varying from 45 to 70. In case *b)*, sometimes 100 iterations are not sufficient for the convergence. However, in both cases the value of the functional (3) $J(v^k)$

decreases with respect to the iterations k.

d) The computer-time necessary for performing a stage of the iterative process (8) is about 1.5 seconds, when the maximum number of time-levels m in the ADI method is 50.

e) In some experiments numerical instability has appeared, manifested by oscillations in the estimated ε and σ, the frequency and amplitude of which are inconsistent with the expected smoothness of the true $\varepsilon(xy)$ and $\sigma(xy)$.

In these cases the application of the regularization technique is strongly recommended[5].

f) When it is known that the partial derivative of $\phi(xyt)$ with respect to time is identically null at the initial time $t_0 = 0$, it is less costly to use a *two-stage gradient method* for solving the parameter identification problem (1)-(2).

In the first stage we estimate the parameter $\sigma(xy)$ by solving (with the gradient method) a parameter identification problem associated with the elliptic equation obtained from (1) with $t = t_0 = 0$. Having estimated $\sigma(xy)$, in the second stage, we determine (with the gradient method) the parameter $\varepsilon(xy)$ by minimizing (3) along a trajectory $\phi(xyt)$ of (1).

References

1. M. Avriel, *Nonlinear Programming. Analysis and Methods*, Prentice-Hall, Englewood Cliffs, N.J. 1976.

2. G. Chavent, P. Lemonnier, Identification de la non-linearité d'une equation parabolique quasilineaire, *Appl. Math. Optim.* **1** (1974), 121-162.

3. I. Galligani, V. Ruggiero, Solving large systems of linear ordinary differential equations on a vector computer, *Parallel Computing* **9** (1988/89), 359-366.

4. S. Gutman, Identification of discontinuous parameters in flow equations, *SIAM J. Control Optim.* **28**, **5** (1990), 1049-1060.

5. C. Kravaris, J.H. Seinfeld, Identification of parameters in distributed parameter systems by regularization, *SIAM J. Control Optim.* **23** (1985), 217-241.

6. R.D. Richtmyer, K.W. Morton, *Difference Methods for Initial-Value Problems*, Interscience Publ., New York, 1967.

WSSIAA 5 (1995) pp. 79–106
© World Scientific Publishing Company

IMAGES, SEPARATION OF SETS AND EXTREMUM PROBLEMS

FRANCO GIANNESSI

University of Pisa, Department of Mathematics

Via F. Buonarroti, 2 – 56127 Pisa, Italy

and

TAMÁS RAPCSÁK

Computer and Automation Institute,

Hungarian Academy of Sciences

Kende ut. 13–17, 1111 Budapest, Hungary

ABSTRACT

A constrained extremum problem is embedded in a generalized system and is associated with the space where the images of its functions run. In such a space, called image space, the image of the extremum problem is defined. This approach is shown to be effective for other kinds of problems, such as Variational Inequalites. In the last two decades there has been an increasing interest in analyzing the properties of the image of an extremum problem. Here we report on some results, concerned with optimality conditions, duality, penalty methods, which have been obtained through this approach. This is extended to other fields, including Variational Inequalities. Further research in the field is discussed.

1. Introduction

The concept of image of a constrained extremum problem has been shown to be a powerful tool for analyzing most topics, as for instance optimality conditions, duality, penalty methods [10,4]. Traces of the idea of studying the images of the functions involved in an extremum problem go back to the work of Carathéodory [4,19]. In the last two decades some authors, independently from each other, have introduced explicitly the

80

concept of image of a constrained extremum problem. Within this approach separation of sets plays the role of the root for developing the theory. Recently such an approach has been extended to other fields, like Complementarity Systems and Variational Inequalities [32,33,9]. Here we report on the main results obtained through this approach. To this end we start with a general format, namely a generalized system, which embodies all the above problems.

Definition 1.1. Assume that ν is a positive integer, $\mathcal{H} \subseteq \mathbf{R}^\nu$ a cone, H a real Hilbert space, $K \subseteq H$ a nonempty set, Y a parameter set and $F : K \times Y \to \mathbf{R}^\nu$ a real–valued mapping. Then,

$$(1.1) \qquad F(x;y) \in \mathcal{H} \subseteq \mathbf{R}^\nu, \quad x \in K \subseteq H, \quad y \in Y,$$

is a *generalized system* (in the variable x).

Theoretically, the definition of a generalized system may be more general, e.g., a set might be considered instead of a cone; however, the case of a cone embodies all the applications we know. A major question related to (1.1) consists in finding values of y such that the generalized system be impossible, and in finding methods which show the impossibility of such a system. Separation schemes seem to be one of the most important tools for studying the impossibility of system (1.1); to this end we will use the concept of image of a set.

Definition 1.2. Assume that ν is a positive integer, H a Hilbert space, $K \subseteq H$ a nonempty set, Y a parameter set and $F : K \times Y \to \mathbf{R}^\nu$ a mapping; then $\mathcal{K}_y := F(K;y)$, $y \in Y$, is the *image* of the set K (through the map F) at a given value of the parameter y.

The impossibility of a generalized system means that the intersection of the image $\mathcal{K}_y$ and the given cone $\mathcal{H}$ is empty, namely $\mathcal{H} \cap \mathcal{K}_y = \emptyset$.

A first important case, which can be reduced to the study of the impossibility of (1.1), is that of constrained extremum problems having finite–dimensional image. Let ν be a positive integer, $f : K \to \mathbf{R}$, $g : K \to \mathbf{R}^\nu$ be real–valued functions, and $\mathcal{C} \subset \mathbf{R}^\nu$ be a convex and closed cone with apex at the origin; the problem is:

$$(1.2) \qquad \min f(x), \quad \text{subject to} \quad x \in R := \{x \in K : g(x) \in \mathcal{C}\}.$$

Given the integers m and p, with $0 \le p \le m$ and m positive, a particularly important specialization of (1.2) is the so–called nonlinear programming problem (in short, NLP), where $K \subseteq \mathbf{R}^n$ is open, $\mathcal{C} = O_p \times \mathbf{R}_+^{m-p}$, $f, g_i \in C^0(K)$, so that (1.2) becomes:

$$(1.2)' \qquad \min f(x), \quad \text{subject to} \quad g_i(x) = 0, \;\; i \in I^0, \;\; g_i(x) \geq 0, \;\; i \in I^+, \;\; x \in K,$$

where $I^0 := \{1, \ldots, p\}$, $I^+ := \{p+1, \ldots, m\}$; $p = 0 \Rightarrow I^0 = \emptyset$; $p = m \Rightarrow I^+ = \emptyset$; and $I := I^0 \cup I^+$.

The optimality conditions for (1.2) can be investigated in the image space. A feasible point y is a minimum point of problem (1.2) iff the system

$$(1.3) \qquad \phi_y(x) := f(y) - f(x) > 0, \;\; g(x) \in \mathcal{C}, \;\; x \in K,$$

has no solution. In the case of a NLP, (1.3) becomes:

$$(1.3)' \qquad \phi_y(x) > 0, \;\; g_i(x) = 0, \;\; i \in I^0, \;\; g_i(x) \geq 0, \;\; i \in I^+, \;\; x \in K.$$

The impossibility of $(1.3)'$ can be written as $\mathcal{K}_y \cap \mathcal{H} = \emptyset$, where

$$(1.4) \qquad \begin{aligned} \mathcal{K}_y &:= \{(u,v) \in \mathbf{R}^{1+m} : u = \phi_y(x), \; v = g(x), \; x \in K\} \\ \mathcal{H}_y &:= \{(u, v^0, v^+) \in \mathbf{R}^{1+m} : u > 0, \; v_i = 0, \; i \in I^0, \; v_j \geq 0, \; j \in I^+\}, \end{aligned}$$

where $v^0 := (v_i, i \in I^0)$, $v^+ := (v_i, i \in I^+)$, $v := (v^0, v^+)$. Hence, in order to state optimality for (1.2) we are led to prove disjunction of sets $\mathcal{H}$ and $\mathcal{K}_y$. Let $F(K;y) := \{(\phi_y(x), g_i(x), g_j(x)), \;\; i \in I^0, \;\; j \in I^+ : x \in K\}$; then the image of $K \subseteq \mathbf{R}^n$ under F with parameter y is the set $\mathcal{K}_y := F(K;y)$, i.e., the image of problem (1.2). By means of the present definition of F, and setting $\nu = 1 + m$, the optimality of y for (1.2) and $(1.2)'$ is reduced to the impossibility of (1.1). Let us observe that local versions of the results are obviously obtained by replacing K with a neighbourhood of a given point and $\mathcal{K}_y$ with the image of this neighbourhood under F. We remark that optimization problems can be considered in a Hilbert space instead of a Euclidean one in this framework.

To prove directly whether or not $\mathcal{K}_y \cap \mathcal{H} = \emptyset$ is generally impracticable, therefore, in order to show such a disjunction, it should be proved that the two sets lie in two disjoint halfspaces or, more generally, in two disjoint level sets, respectively. This separation approach is exactly equivalent to finding a theorem of the alternative related to (1.3) (see, e.g. [8]); only the mathematical languages are different. A separation scheme enjoys mainly geometric motivations, while in studying alternatives algebraic characters are dominating. (This equivalence is valid in mathematics, because the meaning of the alternative may be different in decision support systems, logical propositions, etc.).

82

Let us observe that also some infinite–dimensional problems can be considered within such a scheme.

Example 1.1. Let $K = C^1([a, b])$, with $a, b \in \mathbf{R}$, where K is equipped with the norm $\|z\|_\infty := \max\limits_{t \in [a,b]} |z(t)|$; and where

$$f(x) = \int_a^b \psi_0(t, x(t), x'(t))dt, \quad g_i(x) = \int_a^b \psi_i(t, x(t), x'(t))dt, \quad i \in I,$$

the functions $\psi_i : \mathbf{R}^3 \to \mathbf{R}$, $i \in \{0\} \cup I$, with $I := \{1, \ldots, m\}$ are given. Fixed–endpoint conditions can be included in the definition of K. Of course, only formal changes occur if x is a vector of functions.

The optimization problems which can be reduced to scheme (1.2) share the characteristic of having a finite–dimensional image even if $\dim K = \infty$. Nevertheless, certain problems, for instance those of geodesic–type, elude (1.2). In [10] we show how the present approach can be modified by means of multifunction theory, in order to handle problems having an infinite–dimensional image.

Several real problems (e.g., from Mathematical Physics, Structural Mechanics, Network Equilibrium flows) often lead to the solution of a Variational Inequality (in short, VI), of a Quasi–Variational Inequality (in short, QVI), or of a Complementarity System (in short, CS). First, VI and QVI will be briefly recalled. Let K be a nonempty closed and convex set in a real Hilbert space H (finite or infinite–dimensional) and let $A : H \to H$ or $A : K \to H$ be a single–valued mapping that may be linear or nonlinear. The VI for K and A is to determine an element $y \in K$ satisfying the conditions:

$$(1.5) \qquad \langle A(y), x - y \rangle \geq 0, \quad \forall x \in K \subseteq H.$$

Here $\langle \cdot, \cdot \rangle$ denotes the inner product in H. If K equals H and b is an element of H such that the mapping is given in the form $A(y) - b$, then (1.5) reduces to the equation $A(y) = b$. If there is a differentiable function $f(y)$ on H with the gradient mapping satisfying

$$(1.6) \qquad A(y) = \nabla f(y), \quad y \in H,$$

then the VI expresses a first–order necessary condition for y to be a local solution to the NLP:

$$(1.7) \qquad \min f(y), \quad y \in K.$$

Not every VI, however, even with a monotone mapping $A(y)$, can be interpreted in this way; for example, if $A(y)$ is linear but not symmetric, a gradient representation (1.6) is impossible. It is shown in [5] that VI, where the mapping A is monotone but is not the gradient of any functional, can nevertheless give the optimality conditions for some minimum problems of convex type when Lagrange multipliers are used. It seems to be an open question whether by some extension of these ideas, all monotone and generalized monotone VI could be interpreted as arising from optimization problems. But it is clear that monotone mappings not of gradient type do arise in a number of ways, particularly in certain physical problems involving friction.

If we consider VI (1.5), where it is assumed that the closed and convex set K is bounded, then – as is well known – the existence of a solution to the VI can be proved by Brouwer fixed–point theorem. In general, the problem does not always admit a solution. For example, if $K = \mathbf{R}$, the VI becomes $f(y)(x - y) \geq 0$, $\forall x \in \mathbf{R}$ and has no solution for $f(y) = e^y$.

In certain infinite–dimensional problems of VI type arising in the applications, such as those related to boundary–value problems for partial differential operators, the formulation can often lead to a more general VI, namely the QVI originally introduced by Bensoussan and Lions in 1973. A QVI consists in determining an element $y \in K(y) \neq \emptyset$ such that:

$$(1.8) \qquad \langle A(y), x - y \rangle \geq 0, \quad \forall x \in K(y) := \{x \in X(y) \subseteq H : g(y; x) \in \mathcal{C}\},$$

where $g : K(y) \times K(y) \to \mathbf{R}^m$. This format can be interpreted as the search, among the fixed–points of the point–to–set map $K(y)$, for one satisfying (1.8). It obviously includes a classic form of QVI, when, $\forall x \in K(y)$, the condition $g(y; x) \in \mathcal{C}$ is identically true. When $K(y)$ is independent of y, (1.8) is equivalent to a VI.

In [9] it has been proposed to associate an image space to QVI. This approach starts with the obvious remark that $y \in K(y)$ is a solution of (1.8) iff the system (depending on the variable x):

$$(1.9) \qquad u = \langle A(y), y - x \rangle > 0, \quad v = g(y; x) \in \mathcal{C}, \quad x \in K(y)$$

is infeasible. With an obvious position of F, the impossibility of (1.9) is a particular case of that of (1.1). The space where (u, v) runs is the image space associated to (1.8) and the set

$$\mathcal{K}_y := \{(u, v) \in \mathbf{R} \times \mathbf{R}^m : u = \langle A(y), y - x \rangle, \quad v = g(y; x), \quad x \in K(y)\}$$

84

is the image of (1.8). To system (1.9) we associate the set

$$\mathcal{H} := \{(u, v) \in \mathbf{R}^{1+m} : u > 0, \quad v \in \mathcal{C} \subseteq \mathbf{R}^m\},$$

which depends on the cone $\mathcal{C}$ only. Another obvious remark is that the impossibility of (1.9) is equivalent to $\mathcal{H} \cap \mathcal{K}_y = \emptyset$. To show this disjunction a separation scheme is proposed in [9] which leads to a general class of gap functions having a Lagrangian aspect, in the sense that they allow us to free ourselves from the constraints $g(y; x) \in \mathcal{C}$.

Consider a nonlinear complementarity system (in short, NLCS) as follows:

$$(1.10) \qquad \langle x, g(x) \rangle = 0, \quad g(x) \geq 0, \quad x \geq 0, \quad x \in \mathbf{R}^n,$$

and, more generally,

$$(1.11) \qquad \langle \hat{g}(x), g(x) \rangle = 0, \quad \hat{g}(x) \geq 0, \quad g(x) \geq 0, \quad x \in \mathbf{R}^n,$$

where $\hat{g}, g : \mathbf{R}^n \to \mathbf{R}^n$. It is well known that, if VI (1.5) is defined on $\mathbf{R}^n_+$ (or, more generally, on a convex cone), then VI is equivalent to CS (1.10), i.e., the point y is a solution to (1.10) iff the point y is a solution to VI:

$$(1.12) \qquad \langle g(y), x - y \rangle \geq 0, \quad \forall x \in \mathbf{R}^n_+,$$

where $\mathbf{R}^n_+$ is the non–negative orthant of $\mathbf{R}^n$. The connection between QVI and NLCS (1.11) has been investigated and the equivalence has been proved in locally convex spaces. The image representations of VI and QVI provide also the image representations of NLCS by the equivalence theorems.

Finally, a classical notion, the Lebesgue integral based on the images of the subsets in a decomposition and the measures of these subsets demonstrate the usefulness of the image in a different field of mathematics, since sets become measurable for which the Riemannian measure does not exist.

Some advantages of the image approach consist in a deeper knowledge of a general problem class and in a more transparent view about them providing us with a geometric interpretation of the properties. Moreover, the image approach can be valid for both finite–dimensional and infinite–dimensional problems overcoming certain irregularities which may happen in the given space and the proofs may be obtained more easily. A disadvantage is that properties related to the given space (e.g., to find an optimal solution of an optimization problem) require more sophisticated investigations.

2. Separation in the image space

In this section, a separation theorem (or a theorem of the alternative) is stated for generalized systems. It will be shown later how to deduce from such a theorem known and new optimality conditions, like saddle–point ones, weak and strong duality, penalty. The asymptotic weak and strong separation theorems give the theoretical basis of different methods for solving generalized systems, e.g., exterior, interior penalty function methods and augmented Lagrangian methods, etc.

Assume that m be a positive integer, Y and Ω parameter sets, and $w : Y \times \mathbf{R}^\nu \times \Omega \to \mathbf{R}$, a function. Let us introduce the following notation:

$$(2.1) \qquad \mathcal{H}^w := \mathrm{lev}_{>0}\, w := \{h \in \mathbf{R}^\nu :\ w(y; h; \omega) > 0\},$$

where lev marks level set.

Definition 2.1. Assume that m is a positive integer, Y and Ω are parameter sets, $w : Y \times \mathbf{R}^\nu \times \Omega \to \mathbf{R}$ and $s : Y \times \mathbf{R}^\nu \times \Omega \to \mathbf{R}$ are functions. Then, w is a *weak separation function* iff

$$(2.2) \qquad \mathcal{H}^w = \mathrm{lev}_{>0}\, w \supseteq \mathcal{H}, \quad \forall \omega \in \Omega , \quad \forall y \in Y;$$

and s is a *strong separation function* iff

$$(2.3) \qquad \mathcal{H}^s := \mathrm{lev}_{>0}\, s \subseteq \mathcal{H}, \quad \forall \omega \in \Omega, \ \forall y \in Y.$$

The image problem (1.1) can be characterized by the following theorem:

Theorem 2.1. (i) If a generalized system and a weak separation function are given, then for each $\omega \in \Omega$ and for each $y \in Y$ the system:

$$(2.4) \qquad \exists x \in K \subseteq H, \quad \text{such that} \quad F(x; y) \in \mathcal{H} \subseteq \mathbf{R}^m,$$

and the system:

$$(2.5) \qquad w(y; F(x; y); \omega) \le 0, \quad \forall x \in K \subseteq H,$$

are not simultaneously possible.

(ii) If a generalized system and a strong separation function are given, then, for each $\omega \in \Omega$ and for each $y \in Y$, (2.4) and the system:

$$(2.6) \qquad s(y; F(x; y); \omega) \le 0, \quad \forall x \in K \subseteq H,$$

are not simultaneously impossible.

Proof. (i) If (2.4) is possible, i.e., there exists $(x,y) \in K \times Y$ such that $F(x;y) \in \mathcal{H}$, then $w(y; F(x;y); \omega) > 0$, $\forall \omega \in \Omega$, so that (2.5) is false. (ii) If (2.4) is impossible, i.e., $F(x;y) \notin \mathcal{H}$, $\forall (x,y) \in K \times Y$, then

$$s(y; F(x;y); \omega) \le 0, \quad \forall (x,y,\omega) \in K \times Y \times \Omega,$$

so that (2.6) is true. □

An important question consists in finding conditions on F and K under which a weak (or strong) separation function exists.

If a weak separation function is given and (2.5) is impossible, then Theorem 2.1 does not state anything about (2.4). However, the concept of weak separation can be enlarged. Consider the function $w^* : Y \times \Omega \to \mathbf{R}$ given by

$$(2.7) \qquad w^*(y; \omega) = \sup_{x \in K} w(y; F(x;y); \omega),$$

and assume that $+\infty$ belongs to Ω. We will study the limit points of $w^*(y; \omega)$ for $\omega \to +\infty$ (if ω has several components, then every component tends to $+\infty$).

Theorem 2.2. Let w be such that $w(y; z; \omega) > 0$ implies $\liminf_{\omega \to +\infty} w(y; z; \omega) > 0$. If

$$(2.8) \qquad \liminf_{\omega \to +\infty} w^*(y; \omega) \le 0,$$

then the generalized system (2.4) is impossible.

Proof. If (x,y) is a solution of (2.4), then $w(y; F(x;y); \omega) > 0$ for all $\omega \in \Omega$, because w is a weak separation function. If $\omega \to +\infty$, then

$$\liminf_{\omega \to +\infty} w^*(y; \omega) > 0,$$

from which the statement follows. □

Condition (2.8) can be weakened if a continuity assumption is done on w. The weak and strong separations (or alternatives) introduce a partitioning in the study of (1.1) and in all problems which are reduced to it. The weak side corresponds to sufficient conditions of saddle–point type, to exterior penalty, to classic duality, and so on. The strong side corresponds to necessary (not in **Lagrangian** sense) optimality conditions, to interior penalty, to a new (with respect to the literature) type of duality (called strong with a different meaning with respect to the literature), and so on.

3. Saddle–point type conditions

Every time a weak separation function w is given, a sufficient condition can be established for a given $y \in K$ to be a solution of (1.1); such conditions generalize the classic saddle–point conditions. In the context of extremum problems with unilateral constraints, a class of w has been discussed in [31]. Now let us consider a particular subclass, say W, of w and apply it to (1.2) and (1.8). When we consider (1.2), the subfix y of $\mathcal{K}_y$ will be deleted since in this case the dependence of F in (1.1) on the parameter is merely a translation. W is now the set of functions:

$$(3.1) \qquad w(u, v; \omega) := u + \gamma(v; \omega) \quad , \quad \omega \in \Omega,$$

where Ω must be such that:

$$(3.2a) \qquad \mathrm{lev}_{\geq 0}\, \gamma(v; \omega) \supseteq C \quad , \quad \forall \omega \in \Omega;$$

$$(3.2b) \qquad \bigcap_{\omega \in \Omega} \mathrm{lev}_{\geq 0}\, \gamma(v; \omega) = C.$$

W is the set of (3.1) which fulfils (3.2). Hence it is easy to see that W is unbounded: if γ fulfils (3.2), then $\alpha\gamma$ does whatever the positive real α may be. It is easy to show that $w \in W \Rightarrow (2.2)$.

An example of (3.1) for the case $C = O_p \times \mathbf{R}_+^{m-p}$ is given by:

$$(3.3) \qquad w(u, v; \lambda, \mu) := u + \sum_{i \in I^0}(\lambda_i v_i + \mu_i v_i^2) + \sum_{i \in I^+} \lambda_i v_i \exp\left(-\mu_i v_i\right)$$

$$\text{with} \quad \lambda_i \geq 0,\ i \in I^+,\ \mu_i \geq 0,\ i \in I,$$

which contains the linear one at $\mu = 0$. For the general case an example is given by:

$$(3.4) \qquad w(u, v; \lambda) := u + \langle \lambda, v \rangle \quad , \quad \lambda \in C^*,$$

88

where C^* is the (positive) polar of C.

Proposition 3.1. Assume that the following conditions are satisfied:
(i) $y \in R$; (ii) there exists $\bar{\omega} \in \Omega$ such that:

$$(3.5) \qquad w(\phi(x),\ g(x); \bar{\omega}) \le 0, \quad \forall x \in K.$$

Then, y is a global minimum point of (1.2).

Proof. $(3.5) \Rightarrow \mathcal{K} \subseteq \mathrm{lev}_{\le 0}\, w \Rightarrow \mathcal{H} \cap \mathcal{K} = \emptyset.$ $\qquad\qquad\qquad\square$

Now introduce the (generalized) Lagrangian function:

$$\mathcal{L}(x; \omega) := f(x) - \gamma(g(x); \omega),$$

and let us prove the following theorem.

Theorem 3.2. Conditions (i) and (ii) of Proposition 3.1 are equivalent to the following one: there exist $y \in K$ and $\bar{\omega} \in \Omega$ such that:

$$(3.6) \qquad \mathcal{L}(y; \omega) \le \mathcal{L}(y; \bar{\omega}) \le \mathcal{L}(x; \bar{\omega}) \ , \quad \forall x \in K, \quad \forall \omega \in \Omega.$$

Proof. (i)–(ii) of Proposition (3.1) imply (3.6). In fact, $y \in R$ and $\bar{\omega} \in \Omega$ imply $\gamma(g(y); \bar{\omega}) \ge 0$; at $x = y$ (3.5) implies $\gamma(g(y); \bar{\omega}) \le 0$; it follows that $\gamma(g(y); \bar{\omega}) = 0$. Hence, (3.5) is equivalent to the 2–nd of inequalities (3.6). Because of (3.2), $g(x) \in C \Leftrightarrow \gamma(g(x); \omega) \ge 0 \ \forall \omega \in \Omega$; it follows that $\gamma(g(y); \bar{\omega}) \le \gamma(g(y); \omega)$, $\forall \omega \in \Omega$, which is equivalent to the 1–st part of (3.6). Now, let us show that (3.6) implies (i)–(ii) of Proposition 3.1. We have just noted that the 1–st part of (3.6) is equivalent to $\gamma(g(y); \bar{\omega}) \le \gamma(g(y); \omega)$, $\forall \omega \in \Omega$, which implies $g(y) \in C$, and hence $\gamma(g(y); \bar{\omega}) \ge 0$. Assume that $\gamma(g(y); \bar{\omega}) > 0$. Then, because of (3.2), the 1–st part of (3.6) is contradicted. Thus, $\gamma(g(y); \bar{\omega}) = 0$ is achieved. Now it is easy to show that the 2–nd part of (3.6) implies (3.5). $\qquad\qquad\square$

When (3.5) or (3.6) holds, from the proof of Theorem 3.2 we find that $(y, \bar{\omega})$ fulfils the *generalized complementarity* condition:

$$\gamma(g(x); \omega) = 0,$$

which collapses to the well–known ordinary one [33], when $K \subseteq \mathbf{R}^n$, $p = 0$, $C = \mathbf{R}_+^m$, γ is linear and hence $\Omega = \mathbf{R}_+^m$; in this case Theorem 3.2 becomes the well–known *saddle–point sufficient condition* for y to be (global) minimum point for $(1.2)'$.

Now let us consider (1.8). Instead of Proposition 3.1 we find now:

Proposition 3.3. Assume that the following conditions are satisfied: (i) $y \in K(y)$; (ii) there exists $\bar{\omega} \in \Omega$ such that:

$$(3.7) \qquad w(\langle A(y), y - x \rangle \, , \; g(y; x); \bar{\omega}) \leq 0 \, , \quad \forall x \in X(y).$$

Then, y is a solution of the QVI (1.8).

Proof. $(3.7) \Rightarrow \mathcal{K}_y \subseteq \mathrm{lev}_{\leq 0} \, w \Rightarrow \mathcal{H} \cap \mathcal{K}_y = \emptyset.$ $\qquad\qquad\qquad\qquad\qquad \square$

Note that $\mathcal{K}_y, w$ and Ω depend on y (as in the case (1.2) where such a dependence has been understood); however such a dependence is not merely a translation.

Now, the generalized Lagrangian is defined by:

$$\mathcal{L}(y; x; \omega) := \langle A(y), x \rangle - \gamma(y; g(y; x); \omega).$$

Theorem 3.4. Conditions (i) and (ii) of Proposition 3.3 are equivalent to the following one: there exist $y \in K(y)$ and $\bar{\omega} \in \Omega$ such that:

$$(3.8) \qquad \mathcal{L}(y; y; \omega) \leq \mathcal{L}(y; y; \bar{\omega}) \leq \mathcal{L}(y; x; \bar{\omega}), \quad \forall x \in K(y), \quad \forall \omega \in \Omega.$$

Proof. (i)–(ii) of Proposition 3.3 imply (3.8). In fact, because of (3.2a), $y \in K(y) \Rightarrow g(y; y) \in C \Rightarrow \gamma(y; g(y; y); \bar{\omega}) \geq 0$; at $x = y$ (3.7) $\Rightarrow \gamma(y; g(y; y); \bar{\omega}) \leq 0$; it follows that $\gamma(y; g(y; y); \bar{\omega}) = 0$. Hence, (3.7) turns out to be equivalent to the 2–nd of (3.8). Because of (3.2), $g(y; x) \in C \Leftrightarrow \gamma(y; g(y; x); \omega) \geq 0 \;\; \forall \omega \in \Omega$; it follows that $\gamma(y; g(y; y); \bar{\omega}) \leq \gamma(y; g(y; y); \omega) \; \forall \omega \in \Omega$, which is equivalent to the 1–st part of (3.8). Now, let us show that (3.8) implies (i)–(ii) of Proposition 3.3. We have just noted that the 1–st part of (3.8) is equivalent to $\gamma(y; g(y; y); \bar{\omega}) \leq \gamma(y; g(y; y); \omega)$, $\forall \omega \in \Omega$; this inequality means that γ is bounded on Ω and then implies $g(y; y) \in C$, since $g(y; y) \notin C \Rightarrow \exists \hat{\omega} \in \Omega$ such that $\gamma(y; g(y; y); \hat{\omega}) < 0$ and, taking into account that $\alpha\gamma$ fulfils (3.2) if γ does, the boundedness of γ would be contradicted. It follows $\gamma(y; g(y; y); \bar{\omega}) \geq 0$. Assume that $\gamma(y; g(y; y); \bar{\omega}) > 0$. This and the 1–st of (3.8) become:

$$0 < \gamma(y; g(y; y); \bar{\omega}) \leq \gamma(y; g(y; y); \omega), \quad \forall \omega \in \Omega,$$

which is absurd since $\alpha\gamma \in \Omega$ with $\alpha > 0$ if γ does, and α can be as small as desired. Thus, $\gamma(y; g(y; y); \bar{\omega}) = 0$ is achieved. Now it is easy to show that the 2–nd part of (3.8) implies (3.7). $\qquad\qquad\square$

Theorem 3.4 represents a *saddle–point sufficient condition* for $y \in X(y)$ to be a solution of QVI (1.8). Condition (3.6) can be specialized to a complementarity system, and (3.8) to implicit complementarity system [33].

In presence of a solution y to (1.2) or (1.8), the corresponding saddle–point conditions (3.6) or (3.8) cannot be satisfied at a certain – for instance, linear – function w; then, remaining in the class Ω, we can search for another element of W such that the sufficient conditions be fulfilled.

4. Duality

The duality theory for constrained extremum problems can be derived from the separation approach [8]. Since we have split the separation into weak and strong ones, we must expect two kinds of duality. We will call them weak and strong duality; hence, these terms are here used with a meaning different from that traditionally used in the literature. We are conscious that the introduction of new meanings is not suitable; however, we must note that the traditional terminology has a superficial motivation: strong (weak) stands for an inequality which is (not necessarily) verified as equality. While here the two sides, weak and strong ones, have a logical correspondence with weak and strong alternatives, weak and strong separation, sufficient and necessary conditions, exterior and interior penalty, respectively.

The separation approach allows us to associate to system (1.1) two extremum problems. Let w and s be a weak and a strong separation function, respectively. The problems

$$(4.1) \qquad\qquad \inf_{\omega \in \Omega} \sup_{x \in K} w(y; F(x; y); \omega) \quad , \quad y \in Y,$$

and

$$(4.2) \qquad\qquad \inf_{\omega \in \Omega} \sup_{x \in K} s(y; F(x; y); \omega) \quad , \quad y \in Y,$$

will be called, respectively, *weak and strong dual problems* associated with system (1.1). They depend on the parameter y to the degree that (1.1) does.

The above way of introducing duality shows that it can be viewed as a by–product of separation and that it depends heavily on the tools we analyse (1.1).

There are two levels for our analysis: to study the properties of (4.1)–(4.2); and, when (1.1) comes from a particular problem, to specialize (4.1)–(4.2). Here we will give a short sample of the latter level.

Consider the case where (1.1) comes from (1.3), and adopt a weak separation function of type (3.1). Then (4.1) can be put in the form:

$$(4.3) \qquad \sup_{\omega \in \Omega} \Phi(\omega),$$

where

$$\Phi(\omega) := \inf_{x \in K} [f(x) - \gamma(g(x); \omega) := \Lambda(x; \omega)].$$

When γ is linear, (4.3) is the traditional dual problem of (1.2); γ being not necessarily linear, (4.3) can be called *generalized weak dual problem* of (1.2); this is motivated by the following:

Theorem 4.1. We have:

$$(4.4) \qquad \sup_{\omega \in \Omega} \Phi(\omega) \leq \inf_{x \in R} f(x) \ ,$$

whatever the weak separation function w may be.

Proof. W is a weak separation function of type (3.1), so that γ satisfies (3.2). Hence, both systems (2.4) and (2.5), which here become respectively the system:

$$(4.5) \qquad f(y) - f(x) > 0, \quad g(x) \geq 0, \quad x \in K,$$

and the system:

$$(4.6) \qquad f(y) - f(x) + \gamma(g(x); \omega) \leq 0, \ \forall x \in K, \ \omega \in \Omega$$

or, equivalently, the system:

$$(4.6)' \qquad \Phi(\omega) \geq f(y) \ , \quad \omega \in \Omega,$$

cannot be possible simultaneously. This statement does not depend on the value of $f(y)$; in fact, whatever y may be, the possibility of (4.6) implies the impossibility of (4.5); moreover, the statement is still valid if $f(y)$ is replaced with any real α. Now, Lemma 1 of [25] implies (4.4). $\qquad \square$

When we can ensure that a stationary point of the problem $(1.2)'$ is also a global minimum point (this happens, for instance, when K and Λ are convex), then problem (4.3) can be equivalently written as:

$$(4.7) \qquad \sup \Lambda(x; \omega) \ , \quad \text{s.t.} \quad \liminf_{y \to x} \frac{\Lambda(y; \omega) - \Lambda(x; \omega)}{\|y - x\|} \geq 0.$$

When $K = \mathbf{R}^n$, $p = 0$, f and $-g$ are convex and differentiable, so that we can set $\Lambda := \omega$, $\gamma(v; \omega) = \langle \lambda, v \rangle$, then (4.7) takes the more familiar form [22]:

$$(4.8) \qquad \max L(x; \lambda) \ , \quad \text{s.t.} \quad \nabla_x L(x; \lambda) = 0, \quad \lambda \geq 0 \ ,$$

where $L(x; \lambda) := f(x) - \langle \lambda, g(x) \rangle$ (namely Λ at γ linear) is the ordinary Lagrangian function, and we assume (as traditionally in this field) that the supremum is achieved.

The difference between the right–hand and the left–hand sides of (4.4) is called the *weak duality gap*; it is zero, when (1.3) is convex (as above) and (4.3) becomes (4.8); however, note that, in spite of the convexity of (1.2), in general (4.8) is not concave.

Now let us consider again (1.2) and adopt a strong separation function of type:

$$(4.9) \qquad s(u, v; \omega) := u - \delta(v; \omega) \ , \quad \omega \in \Omega,$$

where Ω is the domain of parameter ω.

It is easy to see that, in place of (4.3), we find now:

$$(4.10) \qquad \inf_{\omega \in \Omega} \Psi(\omega),$$

where

$$\Psi(\omega) := \inf_{x \in K} [f(x) + \delta(g(x); \omega)].$$

Problem (4.10) can be called *generalized strong dual problem* of (1.2); this is motivated by the following theorem [8]:

Theorem 4.2. We have:

$$(4.11) \qquad \inf_{\omega \in \Omega} \Psi(\omega) \geq \inf_{x \in R} f(x) \ ,$$

whatever the strong separation function s may be.

The difference between the left–hand and right–hand sides is the *strong duality gap*. Note that the strong dual problem looks for the same kind of extremum as the primal one, unlike what happens in the weak side.

Now let us consider (1.8). By adopting the following weak separation function:

$$(4.12) \qquad w(y; u, v; \lambda, \omega) := u + \langle \lambda, G(y; v; \omega) \rangle \,, u \in \mathbf{R}, \; v \in \mathbf{R}^m, \; \lambda \in \mathcal{C}^*, \; \omega \in \Omega,$$

where $\mathcal{C}^* := \{\lambda \in \mathbf{R}^m : \lambda_i \geq 0, \; i = p+1, \ldots, m\}$ is the (positive) polar of $\mathcal{C}$, and

$$G(y; v; \omega) := (G_i(y; v_i; \omega_i), \; i \in I) \,,$$

$$G_i : H \times \mathbf{R} \times \Omega_i \to \mathbf{R}, \quad \omega = (\omega_i, \; i \in I) \,, \quad \omega_i \in \Omega_i \,, \quad \Omega = \prod_{i=1}^{m} \Omega_i;$$

according to (3.2), G and Ω must be such that, $\forall y \in X(y)$,

$$\mathrm{lev}_{>0} w \supset \mathcal{H} \quad ; \quad \bigcap_{\omega \in \Omega} \mathrm{lev}_{>0} w = \mathrm{cl}\, \mathcal{H},$$

where the level sets are considered with respect to (u, v) only. Under these conditions (4.12) is a weak separation function.

Each G_i may be considered as a transformation of g_i; for this reason, $\forall y \in X(y)$, and $\forall \omega_i \in \Omega_i$, G_i must be such that:

$$(G_i(y; g_i(y; x); \omega_i), \; i \in I) \in \mathcal{C} \iff (g_i(y; x), \; i \in I) \in \mathcal{C}.$$

In the case of a VI, if $p = 0$, so that $\mathcal{C} = \mathbf{R}_+^m$, examples of G_i are:

$$G(y; v_i; \omega_i) = v_i \exp(-\omega_i v_i), \quad \omega_i \in \Omega_i := \mathbf{R}_+,$$

$$G(y; v_i; \omega_i) = 1 - \exp(-\omega_i v_i), \quad \omega_i \in \Omega_i := \mathbf{R}_+ .$$

The explicit dependence of w on y is motivated by the fact that, in spite of what happens for constrained extremum problems, the change of y does not merely imply a shift of $\mathcal{K}(y)$.

The above comments lead us to consider the *transformed image* of (1.8):

$$\mathcal{K}(y; \omega) := \{(u, v) \in \mathbf{R} \times \mathbf{R}^m : u = \langle A(y), y - x \rangle, \quad v = G(y; g(y; x); \omega), \; x \in X(y)\}$$

and its *conic extension*:

$$\mathcal{E}(y; \omega) := \mathcal{K}(y; \omega) - \mathrm{cl}\, \mathcal{H} \,,$$

where cl denotes closure.

94

Definition 4.1. A function $\psi : K^0 \to \mathbf{R}$ with $K^0 := \{y \in H : y \in K(y)\}$ is called *gap function* iff $\psi(y) \geq 0\ \forall y \in K^0$ and $\psi(y) = 0$ iff y is a solution of (1.8); K^0 is the set of fixed–points of the point–to–set map K.

Since we will set up a gap function as a by–product of the separation scheme in the image space, it is natural to expect to find two classes of gap functions, corresponding to weak and strong separation functions [9]. The preceding definitions and notations correspond to the weak separation; hence the function (where the dependence on ω is understood):

$$(4.13) \qquad \psi(y) := \min_{\lambda \in \mathcal{C}^*} \max_{x \in X(y)} [\langle A(y), y - x \rangle + \langle \lambda, G(y; g(y; x); \omega) \rangle],$$

can be shown to be a gap function for (1.8), and will be called *weak gap function*, since it comes from (4.12). Note that (4.13) is no more than the weak dual problem (4.1) associated to (1.9), which is (1.1) specialized to the present case.

In the definition of ψ there is a maximization; this is performed on $X(y)$, i.e. on the relaxation of the domain of the QVI, and not on $K(y)$, unlike what happens (in the case of a VI) to the existing gap functions.

Note that, when $K(y)$ is a closed and convex cone with apex at the origin and $K^*(y)$ denotes its positive polar, then we easily find:

$$\min_{y \in K(y)} \psi(y) = \min_{\substack{y \in K(y) \\ A(y) \in K^*(y)}} \langle A(y), y \rangle,$$

which shows a gap function for a Quasi–Complementarity System. It is possible to prove the following theorem [9].

Theorem 4.3. Let $y \in K^0$. Assume that the extrema in (4.13) exist and for each $y \in K(y)$ there exists $\omega(y) \in \Omega$, such that

(i) $\mathcal{E}(y; \omega(y))$ is convex;

(ii) $U := \{(u, v) \in \mathcal{H} : v = 0\} \not\subseteq T(\mathcal{E}(y; \omega(y)))$, where T denotes Bouligand tangent cone at $(u = 0,\ v = g(y; y))$.

Then y is a solution of (1.8) iff $\psi(y) = 0$.

A duality scheme for (1.8) is one which shows a Quasi–Variational Inequality, whose argument runs in the dual of space H. The gap function (4.13), namely the dual problem (4.1), is a starting point. It consists in showing that the minimization of (4.13) on $K(y)$ can be expressed in terms of the difference of two constrained extremum problems; the

1-st order optimality condition of the latter can be taken, by definition, as dual of (1.8). For a particular case this has been done in [1,5,9].

Another way consists in transforming the minimization of (4.13) on $K(y)$ – which requires preliminarily two extremizations – equivalently into the minimization of a function which does not depend on other extremizations. In a particular case this has been done in [1,5,9].

5. Penalty

We have already announced that exterior and interior penalty approaches can be derived from weak and strong separations, respectively. This will be outlined here. To this end consider again problem $(1.2)'$, and introduce a sequence of continuous functions $p_\omega : K \to \mathbf{R}$, $\omega = 1, 2, \ldots$, such that

$$(5.1a) \qquad p_\omega(v) \begin{cases} = 0 , & \text{if } v \in C, \\ > 0 , & \text{if } v \notin C, \end{cases}$$

$$(5.1b) \qquad p_{\omega+1}(v) > p_\omega(v) , \quad \omega = 2, 3, \ldots,$$

$$(5.1c) \qquad \lim_{\omega \to +\infty} p_\omega(v) = +\infty , \quad \forall v \notin C.$$

It is easy to see that the function ($\mathbf{Z}_+$ denotes the set of positive integers):

$$w(y; u, v; \omega) := u + p_\omega(v) , \quad y \in K , \quad \omega \in \Omega := \mathbf{Z}_+,$$

is a weak separation function. Hence (i) of Theorem 2.1 can be applied and (2.5) becomes:

$$(5.2) \qquad f(y) - f(x) - p_\omega(g(x)) \leq 0 , \quad \forall x \in K,$$

and is a sufficient condition for the feasible y to be optimal. Such a condition can be weakened by applying Theorem 2.2, which leads to the sufficient condition:

$$(5.3) \qquad \limsup_{\omega \to +\infty} \inf_{x \in K} [f(x) + p_\omega(g(x))] \geq f(y),$$

which is weaker than (5.2). Denote by Φ_ω the infimum in (5.3). From (5.1) we deduce that:

$$(5.4) \qquad \Phi_1 \leq \Phi_2 \leq \ldots \leq \Phi := \inf_{x \in R} f(x).$$

Assume that $\exists \bar{\omega}$ such that $\Phi_{\bar{\omega}} > -\infty$, and that $\exists x_{\bar{\omega}} \in K$ such that $f(x_{\bar{\omega}}) + p_{\bar{\omega}}(g(x_{\bar{\omega}})) = \Phi_{\bar{\omega}}$, $\forall \omega \geq \bar{\omega}$. If x^* is any limit point of sequence $\{x_\omega\}_{\omega=1}^{\infty}$, then condition (5.3) is fulfilled and Theorem 2.2 gives the optimality of x^*. The construction of sequence $\{x_\omega\}_{\omega=1}^{\infty}$ by solving the infimum problems in (5.3) is the well known *exterior penalty method* and p_ω is called a penalty function; if the above convergence can be ensured after a finite number of steps, i.e. if $\exists \bar{\omega} \in Z_+$ such that (5.2) is fulfilled at $\omega = \bar{\omega}$, then p_ω is called *exact penalty function*. Hence, the conditions for a penalty function to be exact can be regarded as conditions which ensure (5.2) instead of (5.3). It follows that the exterior penalty approach can be formulated in terms of **weak** separation. Among other things, this fact enables us to extend the penalty approach to solve systems (1.1). This allows us to conceive a penalty approach for Variational and Quasi–Variational Inequalities.

Now it is easy to see that the interior penalty approach can be formulated in terms of strong separation, namely that any interior penalty function produces a strong separation function.

6. Topological properties of the image

In order to deepen and extend the analysis of the structure of the image, the different separation schemes in the image space (e.g., proper and strong separation), their topological characteristics may be important. In this section, we use the standard topological terminology and give a brief summary of some important classical topological properties of the image.

Let K, K_1 be arbitrary sets. A mapping $F : K \to K_1$ associates an element of K_1 to each element $x \in K$. This mapping is defined by $F(x)$ which is the image of x under F. A mapping $F : K \to K_1$ is injective iff all $x_1, x_2 \in K$ with $x_1 \neq x_2$ we have $F(x_1) \neq F(x_2)$. The mapping F is surjective iff $F(K) = K_1$, i.e., iff the image of F covers all K_1. The mapping F is bijective iff it is both injective and surjective. If K, K_1 are topological spaces, then a mapping $F : K \to K_1$ is continuous iff the inverse image of an open set in K_1 is open in K, i.e., iff U is open in K_1, then $F^{-1}(U)$ is open in K. Equivalently, a mapping is continuous iff the inverse image of a closed set is closed. A mapping $F : K \to K_1$ is continuous at a point $x \in K$ iff for a given neighbourhood V of $F(x)$ there exists a neighbourhood U of x such that $F(U) \subseteq V$. It follows that F is continuous on K iff F is continuous at every point of K.

A topological space is connected iff it is not a union of two disjoint non-empty open or closed sets. If K is a topological space and K' is a connected subset, then the closure of K', denoted by $\bar{K}'$, is connected. The connected component of a point is closed. If the topological spaces $K_i, i \in I$ are connected, then the product $\prod_{i \in I} K_i$ is connected.

Theorem 6.1. If $F : K \to K_1$ is continuous and K is connected, then the image $F(K)$ is connected.

A mapping is closed iff it maps closed sets onto closed sets. A continuous mapping needs not be either closed or open. For instance, the graph of the tangent function is closed in the plane, but the projection map on the x-axis maps it on an open interval. In Optimization Theory, the closed mappings have a central position, because the existence of a solution for a large class of problems can be guaranteed by them. The basic assumption requests a compact set which should be the feasible region in Optimization Theory.

Theorem 6.2. A continuous image of a compact set is compact.

A direct consequence of this fundamental theorem is that if K is a compact set and $F : K \to \mathbf{R}$ a continuous function on K, then F has the minimum (and maximum) because the image $F(K)$ is compact, thus closed, bounded and the greatest lower (and least upper) bound of $F(K)$ lies in $F(K)$. In the finite-dimensional case, this is the well known Weierstrass theorem. In [37], by exploiting the image space analysis, a necessary and sufficient condition for the existence of the minimum is given and such a theorem is recovered as a special case. It follows from the theorem that the compactness of the image provided by a continuous mapping does not imply the compactness of the feasible domain in the given space (see Example 6.1).

Example 6.1. Consider the optimization problem $\min \sin(x)$ $s.t.$ $\cos(x) \geq 0$, $x \in \mathbf{R}$, the image of which at x^* is equal to $\mathcal{K}_{x^*} = \{(u, v) \in \mathbf{R}^2 : u = \sin(x^*) - \sin x; v = \cos(x)\}$. The explicit form of the image at $x^* = 3/2\pi$ is $\mathcal{K}_{x^*} = \{(u, v) \in \mathbf{R}^2 : (u+1)^2 = -v^2 + 1\}$ which is a compact set.

$\Lambda_m := \{\lambda \in \mathbf{R}^m : \lambda_i \geq 0, \sum_{i=1}^{m} \lambda_i = 1\}$ is the $(m-1)$-dimensional standard simplex.

Theorem 6.3. (Pshenichnyi [30]). Let K be a nonempty convex subset of a real vector space, $g_i : K \to \mathbf{R}$, $i = 1, \ldots, m$, concave functions and suppose that the set

$F(K) = (g_1(x), \ldots, g_m(x), x \in K)$ is compact in $\mathbf{R}^m$. Then, the inequality system

$$(6.1) \qquad g_i(x) \geq 0, \quad i = 1, \ldots, m,$$

is compatible iff

$$(6.2) \qquad \sup_{x \in K} \sum_{i=1}^{m} \lambda_i g_i(x) \geq 0, \quad \forall \lambda \in \Lambda.$$

Thus, the image approach and the compactness can provide some additional structural properties of optimization problems. Let H and H_1 denote real Hausdorff topological vector spaces and $\Phi : H \to H_1$ a multifunction (set-valued mapping), i.e., $\Phi(x)$ is a nonvoid subset in H_1 for each $x \in H$. A multifunction $\Phi : H \to H_1$ is said to be closed iff the graph of Φ, given by $\{(x, y) : x \in H, y \in \Phi(x)\}$, is closed in $H \times H_1$. A closedness criterion for the image of a convex closed locally compact set under a convex multifunction is proved in [12] and applications are given to the solvability of linear systems over cones, the existence of generalized spline functions and a duality theory with zero duality gap for abstract convex optimization problems containing inequality constraints. In the latter case, the crucial assumption permitting the application of the strong separation theorem is the closedness of the conic extension of the image. Naturally, it comes into question whether the image of a closed convex set, by a linear continuous mapping, is closed [28]. Theorems of the alternative and separation schemes have been stated for generalized systems of multifunction and it was shown how to apply them to extremum problems [10]. The cone approximation of the image seems to be a successful application domain of multifunctions.

A mapping is open iff it maps open sets onto open sets. A continuous image of an open set is not necessarily open. If a map is continuous and bijective, then a necessary and sufficient condition for being a homeomorphism is that it should be open. In Banach spaces, a characterization of open mappings is as follows:

Theorem 6.4. Let K, K_1 be Banach spaces and $F : K \to K_1$ a continuous, surjective and linear mapping. Then F is open.

A direct consequence is that a continuous, bijective and linear mapping is invertible. The following well known statement is another corollary:

Closed graph theorem. Let K, K_1 be Banach spaces, $F : K \to K_1$ a linear mapping and assume that the graph is closed. Then F is continuous.

Let K be an open set in a Banach space H, $F : K \to H_1$ a C_1 mapping into a Banach space H_1 and $x_0 \in H$. F is locally open in a neighbourhood of x_0 iff there exists an open neighbourhood $U \subseteq K$ of x_0 such that for all $x \in U$ and open ball $B_x \subseteq U$ centered at x, the image $F(B_x)$ contains an open neighbourhood of $F(x)$.

Surjective mapping theorem (Graves). Let K be an open set in a Banach space H, $F : K \to H_1$ a C_1 map into a Banach space H_1 and $x_0 \in H$. If $F'(x_0)$ is surjective, then F is locally open in a neighbourhood of x_0.

The surjective or interior mapping theorems ensure that $F(x_0)$ belong to the interior of the image. Carathéodory [4] was the first to understand the relevance of surjective mapping theorems for equality constrained optimization in Euclidean spaces and to prove the Lagrange multiplier rule in the image space. The main idea of the proof is based on a necessary condition of the optimality at a given point, showing that its image is not an interior point, which follows from the impossibility of the corresponding generalized system. By the surjective mapping theorem, the derivative of the image mapping F' is not surjective at the given point, i.e., the gradients of the equality constraints are linearly dependent. The surjective mapping theorems can be extended to the case of generalized differentiability (e.g., strong differentiability) and inequality constraints obtaining even stronger results [29]. The same idea can be used to characterize the first-order conditions of a general NLP, containing inequality constraints and vector extremum problems defined in a Banach space, by multifunctions based on the conic extension of the image and its first-order approximation [35]. In the theory of differentiable manifolds, open mapping or domain invariance theorems play an important role.

Open mapping theorem in $\mathbf{R}^n$. Let $U \subseteq \mathbf{R}^n$ be open, $F : U \to \mathbf{R}^n$ continuous and one-to-one, then $F(U) \subseteq \mathbf{R}^n$ is open.

7. Convexity in the image space

The impossibility of a generalized system (1.1) means that the intersection of the given cone $\mathcal{H}$ and the image $\mathcal{K}_y$ is empty, namely $\mathcal{H} \cap \mathcal{K}_y = \emptyset$. One of the simplest cases of generalized systems is the convex optimization where this disjunction should be proved by using a separation theorem. Since the image of a convex optimization problem is not convex in general (e.g., $\min x^2$ s.t. $x \geq 0$, $x \in \mathbf{R}$, the image of which is equal to $K_0 = \{(u,v) \in \mathbf{R}^2 : u = -v^2\}$), its conic extension with respect to the cone $cl\mathcal{H}$, denoted by $\mathcal{E}_y$, is introduced as $\mathcal{E}_y = F(K) - cl\mathcal{H}$ from which it follows that $\mathcal{K}_y \subseteq \mathcal{E}_y$. The importance of the conic extension for the image of generalized systems is enforced by the following statement [8]:

Theorem 7.1 System (1.1) is impossible iff $\mathcal{H} \cap \mathcal{E}_y = \emptyset$.

Hence, proving optimality is equivalent to showing disjunction between $\mathcal{H}$ and $\mathcal{K}_y$ or between $\mathcal{H}$ and $\mathcal{E}_y$. In certain cases, it is easier to prove the latter instead of the former, because the conic extension may have some advantageous properties that $\mathcal{K}_y$ has not. In the case of convex optimization, the conic extension is a convex set so that the weak separation theorem of nonempty disjoint convex sets in $\mathbf{R}^n$ (Minkowski, 1911) could be used. Ky Fan introduced the concept of convexlike functions [7]. Recalling that $F : K \to \mathbf{R}^m$ is convexlike iff $\forall x', x'' \in K$, and $\forall \alpha \in [0,1]$, there exists a point $z \in K$ such that

$$(7.1) \qquad F(z) \leq (1 - \alpha)F(x') + \alpha F(x''),$$

we formulate the following statement [36]:

Theorem 7.2. Let K be convex. Then $\mathcal{E}_y$ is convex iff F is convexlike.

While the convexity of $\mathcal{K}_y$ implies the convexity of $\mathcal{E}_y$ in an obvious way from its definition, Theorem 7.2 leads us to find some examples in which $\mathcal{E}_y$ is convex while $\mathcal{K}_y$ is not. Thus, a class of nonconvex optimization problems is defined by the convexity of the conic extension of the image set, providing a linear separation in the image space between $\mathcal{H}$ and $\mathcal{E}_y$. The convex functions are convexlike and so is any real function on any part of the real line. A consequence of the linear separation is that, considering a convexlike nonlinear optimization problem with inequality constraints, the duality theorem with zero duality gap can be proved [6,15]. The results are based on the following statement [15] by replacing the Farkas theorem in the image space:

Theorem 7.3. Let K be a nonempty set and K_1 an ordered linear topological space with a convex cone K' whose interior is nonempty. If $F : K \to K_1$ is a convexlike mapping, then either (i) or (ii) holds:
 (i) there is $x_0 \in K$ such that $F(x_0) < 0$;
 (ii) there is $x' \in K'$ such that $x' \neq 0$ and $< F(x), x' > \geq 0$, $\forall x \in K$.
 The two alternatives (i) and (ii) exclude each others.

Some generalizations of the concept of convexlike functions based on the generalizations of Theorem 7.3, can be found in [13,19]. From the viewpoint of image approach, it seems to be natural to define the nonconvex class of functions by the convexity of the sets $\mathcal{K}_y - \mathrm{int}\,\mathcal{H}$ and $\mathrm{cl}(\mathcal{K}_y - \mathrm{cl}\,\mathcal{H})$, respectively [3,17]. These classes embody obviously the convexlike functions, $\mathcal{K}_y - \mathrm{int}\,\mathcal{H}$ the König convex functions [17,21], and $\mathrm{cl}(\mathcal{K}_y - \mathrm{cl}\,\mathcal{H})$ is equivalent to the almost convexlike functions [13], where $F : K \to \mathbf{R}^m$ is almost convexlike iff $\forall x', x'' \in K$, $\forall \alpha \in [0,1]$, and $\forall \varepsilon > 0$, there exists a point $z \in K$ such that

$$(7.2) \qquad F(z) \le (1-\alpha)F(x') + \alpha F(x'') + \varepsilon.$$

In [13,20], the concepts of the almost convexlike and the more general weakly convexlike were used to characterize the solvability of special generalized systems consisting of nonlinear inequalities. By using the statement that, if $\mathcal{K}_y$ is a subset of a linear topological space and $\mathrm{cl}\,\mathcal{H}$ a convex cone with nonempty interior in it [36], then

$$(7.3) \qquad \mathcal{K}_y - \mathrm{int}\,\mathcal{H} = \mathrm{int}(\mathcal{K}_y - \mathcal{H}),$$

we obtain a clearer view on this generalization. If $\mathcal{K}_y$ is a compact subset and $\mathrm{cl}\,\mathcal{H}$ a subset of a metrizable topological linear space with a translation-invariant metric on it, then

$$(7.4) \qquad \mathrm{cl}(\mathcal{K}_y - \mathrm{cl}\,\mathcal{H}) = \mathcal{K}_y - \mathrm{cl}\,\mathcal{H};$$

hence this case is equivalent to the convexlike one [36]. Explicit characterizations for convexlike and generalized convexlike functions are still open questions.

8. Geodesic convexity in the image approach

In Optimization Theory, the concept of the geodesic convexity has been recently introduced on a Riemannian manifold instead of a linear vector space in order to generalize the local-global property (every local optimum is global) of a smooth nonlinear optimization problem with equality constraints [32]. The advantage of this approach is the recognition of the importance of the geometrical structure of optimization problems.

When geodesic convexity has been proved, it is concluded that a stationary point is a global optimum point and, consequently, every algorithm which gives a stationary point gives a global minimum point, too. Note that convex optimization is a special case of geodesic convex optimization. Moreover, a great number of generalized convex function classes can be defined depending on the different Riemannian metrics, and infinitely many coordinate representations of the Riemannian manifold may be possible in every metric class [34].

The geodesic convexity with respect to the Euclidean metric plays an important role in Nonlinear Optimization, e.g., optimality criteria [32], global optimization, linear and nonlinear complementarity systems [33], variable metric methods [34] and interior point methods. The introduction of this notion makes the differential geometric analysis of the structure of the smooth image problem possible. First, the image representations will be recalled [34].

In the case of NLP (1.2)', the image representations based on the image mapping are given in the form of

$$(8.1) \quad (u, v_1, v_2)^T = (\phi(x), g(x))^T, \quad x \in K \subseteq \mathbf{R}^n, \quad (u, v_1, v_2) \in \mathbf{R} \times \mathbf{R}^p \times \mathbf{R}^{m-p},$$

where K is an open set. The image mapping represents an $(m+1)$-dimensional manifold in $\mathbf{R} \times \mathbf{R}^m \times \mathbf{R}^n$, because its Jacobian matrix is nonsingular on K. Therefore, the image set can be represented in the neighbourhoods of the $(m+1)$-dimensional manifold. We distinguish three cases.

(1) $m = n$; the rank of the Jacobian matrix of the image mapping $(\phi(x), g(x))^T$ is equal to n.

This case corresponds to the classical surface definition based on the parameters x; this is equivalent to an Euler-Monge form. Thus, the image can be given by a level set of a function with variables (u, v_1, v_2), because a suitable Jacobian submatrix of the image mapping is nonsingular.

(2) $n > m$; the rank of the Jacobian matrix of $(\phi(x), g(x))^T$ is equal to m.

This case corresponds to an m-dimensional surface definition with $n - m$ additional parameters; this is equivalent to an Euler-Monge form depending on $n - m$ parameters, i.e., the image can be given by an equality level set of a function with variables (u, v_1, v_2) and $n - m$ parameters.

(3) $m > n$; the rank of a Jacobian submatrix of $(\phi(x), g(x))^T$ is equal to n.

A system of equations will be obtained for the image by using only the variables of the image space by eliminating variables x from n equalities with a nonsingular Jacobian matrix.

The following problem can be associated to (1.2)':

$$(8.2) \quad \max u \quad \text{s.t.} \quad (u, v_1, v_2) \in \mathcal{K}_y \subseteq \mathbf{R} \times \mathbf{R}^p \times \mathbf{R}^{m-p}, \quad v_1 = 0, \quad v_2 \geq 0.$$

A direct consequence of [36] is a relation between the optimal solutions of (1.2)' and (8.2).

Theorem 8.1 The following conditions are equivalent:

(i) problem (1.2)' has a global minimum point;

(ii) problem (8.2) has a global minimum point.

An image representation subject to the conditions $v_1 = 0$, $v_1 \in \mathbf{R}^p$ is regular if it is a differentiable manifold in the image space (by fixing the parameter values in parametrical cases). Now, problem (8.2) is defined on a differentiable manifold, thus by proving geodesic convexity of u and the components of v_2, $v_2 \in \mathbf{R}^{m-p}$ with respect to a suitable Riemannian metric, a sufficient condition of the global optimality for a wider class of problems [33], and by applying variable metric methods along geodesics [34], convergent methods can be obtained in the image space.

References

[1] ANTONI C., "On a separation approach to Variational Inequalities". In "Variational inequalities and network equilibrium problems", F.Giannessi and A. Maugeri (Ed.s), Plenum, N.Y. 1995, pp.1-7.

[2] BEONI C., "A generalization of Fenchel duality theory". Tech. Report N. 107, Dept. of Mathem., Optim. Group, Univ. of Pisa, 1983, pp. 1-31. Published in Jou. Optim. Th. Appl., Vol. 49, N. 3, 1986, pp. 375-387.

[3] BLAGA L. and KOLUMBÁN J., "Optimization on closely convex sets". In: "Generalized convexity", S. Komlósi, T. Rapcsák and S. Schaible, (Eds.), Springer-Verlag, 1994, pp. 1934.

[4] CARATHÉODORY C., "Variationsrechnung und Partielle Differential Gleichungen Erster Ordnung". B. G. Teuber, Berlin, 1935.

[5] CASTELLANI M. and G. MASTROENI, "On the duality theory for finite dimensional Variational Inequalities". In "Variational inequalities and network equilibrium problems", F.Giannessi and A. Maugeri (Ed.s), Plenum, N.Y. 1995, pp.21-31.

[6] ELSTER K. H. and NEHSE R., "Optimality conditions for some nonconvex problems". Springer-Verlag, New York, New York, 1980.

[7] FAN K., "Minimax theorems". Proc. of the National Academy of Sciences of the United States of America, Vol. 39, 1953, pp. 42-47.

[8] GIANNESSI F., "Theorems of the alternative and optimality conditions". Jou. Optim. Th. Appl., Vol. 42, N. 3, 1984, pp. 331–365.

[9] GIANNESSI F., "Separation of sets and gap functions for Quasi–Variational Inequalities". In "Variational inequalities and network equilibrium problems", F.Giannessi and A. Maugeri (Ed.s), Plenum, N.Yo. 1995, pp.101–121.

[10] GIANNESSI F., "Theorems of the alternative for multifunctions with applications to optimization: general results". Tech. Report N. 127, Dept. of Mathem., Optim. Group, Univ. of Pisa, 1985, pp. 1–40. Published in Jou. Optim. Th. Appl., Vol. 55, N. 2, 1987, pp. 233–256.

[11] GIANNESSI F., "General Optimality Conditions via a separation scheme". In "Algorithms for continuous optimization", E. Spedicato (Ed.), Kluwer Acad. Publishers, 1994, pp. 1–23.

[12] GWINNER J., "Closed images of convex multivalued mappings in linear topological spaces with applications". Jou. of Mathem. Analysis and Appl., Vol. 60, 1977, pp. 75–86.

[13] GWINNER J. and OETTLI W., "Theorems of the alternative and duality for inf-sup problems". Mathematics of Operations Research, Vol. 19, 1994, pp. 238-256.

[14] HALKIN H., "Necessary Conditions for Optimal Control problems with differentiable or nondifferential data." Lecture Notes in Mathematics, N. 680, W.A. Cappel (ed.), Springer–Verlag, Berlin, 1978, pp. 77–118.

[15] HAYASHI M. and KOMIYA H., "Perfect duality for convexlike program". Jou. of Optimization Th. and Appl., Vol. 38, 1982, pp. 179-189.

[16] HESTENES M.R., "Calculus of Variations and optimal control theory". J. Wiley, New York, 1966.

[17] ILLÉS T. and KASSAY G., "Farkas type theorems for generalized convexities". PU.M.A. Vol. 5, 1994. (in print)

[18] IOFFE A.D., "Necessary Conditions in Nonsmooth Optimization", Mathematics of

Operations Research, Vol. 9, 1984, pp. 159–189.

[19] JEYAKUMAR V., "Convexlike alternative theorems and mathematical programming". Optimization, Vol. 16, 1985, pp. 643–652.

[20] KASSAY G. and KOLUMBÁN J., "On a generalized sup-inf problem". Jou. of Optimization Th. and Appl. (to appear)

[21] KÖNIG H., "Über das von Neumannsche minimax theorem". Archives of Mathematics, Vol. 19, 1968, pp. 482-487.

[22] MANGASARIAN O.L., "Nonlinear Programming". SIAM, Philadelphia, 1994.

[23] MASTROENI G., PAPPALARDO M. and YEN N.D., "Image of a Parametric Optimization Problem and Continuity of the perturbation Function", Jou. of Optim. Th. Appl., Vol. 81, N. 1, 1994, pp. 193–202.

[24] MAUGERI A., "Variational and Quasi–Variational Inequalites in network models. Recent developments on theory and algorithms", In "Variational inequalities and network equilibrium problems", F.Giannessi and A. Maugeri (Ed.s), Plenum, N.Y. 1995, pp.195–211.

[25] McLINDEN L., "Duality theorems and theorems of the alternative". Proceedings of Annals of Mathematical; Society, Vol. 53, 1975, pp. 172–175.

[26] PAPPALARDO M., "Image space approach to penalty methods", Jou. Optim. Th. Appl., Vol. 64, No. 1, 1990, pp. 141–152.

[27] PELLEGRINI L., "Some remarks on semistationarity and optimality conditions". In "Nonsmooth Optimization Methods and Applications", Giannessi (ed.), Gordon and Breach, 1992, pp. 295–302.

[28] POMEROL J.-Ch., "Is the image of a closed convex set by a continuous linear mapping closed?". 5° Symposium on Operations Research, Heidelberg, 1976, Verlag Hain.

[29] POURCIAU B.H., "Multipliers rules". Am. Math. Monthly, Vol. 87, 1980, pp. 443–452.

[30] PSHENICHNYI V. N., "Necessary conditions for an extremum" (Russian), Nauka,

Moscow, 1969, (English ed.: Dekker, New York, 1971).

[31] QUANG P.H., "Lagrangian multiplier rules via image space analysis". In "Nonsmooth Optimization Methods and Applications", F. Giannessi (ed.), Gordon and Breach, 1992, pp. 354–365.

[32] RAPCSÁK T., "Geodesic convexity in nonlinear optimization", Jou. of Optimization Th. and Appl., Vol. 69, 1991, pp. 169-183.

[33] RAPCSÁK T., "On the connectedness of the solution set to nonlinear complementarity systems", Jou. of Optim. Th. and Appl., Vol. 81, N. 3, pp. 619–631.

[34] RAPCSÁK T., " On nonlinear coordinate representations of nonsmooth optimization problems", Jou. of Optim. Th. and Appl., Vol. 86, N.2, 1995 (in print).

[35] ROBINSON S.M., "First order conditions for general nonlinear optimization". SIAM Jou. on Appl. Mathem., Vol. 30, N. 4, 1976, pp. 597–607.

[36] TARDELLA F., "On the image of a constrained extremum problem and some applications to the existence of a minimum". Jou. Optim. Th. Appl., Vol. 60, N. 1, 1989, pp. 93–104.

[37] WARGA J., "Controllability and Necessary Conditions in Unilateral problems without differentiability assumptions". SIAM Jou. on Control and Optimiz., Vol. 14, 1976, pp. 546–573.

WSSIAA 5 (1995) pp. 107–121

PARAMETER ESTIMATION UNDER DISTRIBUTED CONSTRAINTS: TWO ALGORITHMS FOR CONVEX SEMI-INFINITE QUADRATIC PROGRAMMING PROBLEMS

GUIDO O. GUARDABASSI
Department of Electronic and Information Engineering, Politecnico di Milano
20133 Milano, Italy

JOSE' C. GEROMEL
FEE-UNICAMP
13081 Campinas S.P., Brazil

XIAO JIAN
Department of Electrical Engineering, Southwest Jaotong University
Chengdu, China

ABSTRACT

The convex semi-infinite quadratic program dealt with in this paper originates from a least-squares parameter estimation problem for systems in linear regression form subject to distributed constraints. A batch algorithm and a recursive algorithm for the considered problem are presented, and their convergence proved. The former is a proper extension of Kelley's convex cutting plane algorithm, the latter a suitably adjusted projection algorithm.

1. Introduction

Given a time series $\{d(t); t=1,2,...,N\}$, $d(t) = |y(t)'\ v(t)'|' \in \mathbf{R}^{r+s}$ (prime denotes transpose) and a prediction model $M(\vartheta)$ in linear regression form, i.e.:

$$M(\vartheta) : \quad \begin{cases} \hat{y}(t+1) = \varphi(t)\,\vartheta & , \quad t = 0, 1, 2, \ldots, N\text{-}1 \quad , \\[2ex] \varphi(t) = g(z(t), d(t)) & , \quad z(t+1) = f(z(t), d(t)) \quad , \end{cases} \tag{1}$$

where $\vartheta \in \mathbf{R}^n$ is an unknown parameter vector and the linear regressor $\varphi(t) \in \mathbf{R}^{r \times n}$ is the output of a known dynamical system $(f,g;z(0))$ driven by d, consider the problem of estimating the parameter vector ϑ, subject to a *distributed constraint* of the form

$$A(u)\,\vartheta + b(u) \geq 0 \quad , \quad \forall\, u \in \mathbf{U} \quad ,$$

where $\mathbf{U}$ is a given compact subset of $\mathbf{R}^p$ while $A: \mathbf{R}^p \to \mathbf{R}^{m \times n}$ and $b: \mathbf{R}^p \to \mathbf{R}^m$ are given matrix (vector) functions of u continuous on $\mathbf{U}$.

108

As is well known, the Minimum Least Squares estimate $\hat{\vartheta}$ of ϑ is obtained from the data $\{d(t);\ t=1,2,...,N\}$ by minimizing:

$$J := \sum_{t=1}^{N} \|\hat{y}(t) - y(t)\|^2 = \|\Phi\,\vartheta - y\|^2$$

where

$$\Phi := \begin{bmatrix} \varphi(1) \\ \varphi(2) \\ ... \\ \varphi(N) \end{bmatrix} \quad , \quad y := \begin{bmatrix} y(1) \\ y(2) \\ ... \\ y(N) \end{bmatrix} \quad ,$$

subject to the constraint

$$A(u)\,\vartheta + b(u) \geq 0 \quad , \quad \forall\, u \in \mathbf{U} \quad .$$

The problem to solve is therefore a *convex semi-infinite quadratic program P* of the form

$$P: \qquad \hat{\vartheta} = \mathrm{argmin}\ \vartheta'\,Q\,\vartheta + q\,\vartheta \qquad \text{s.t.} \quad A(u)\,\vartheta + b(u) \geq 0 \quad , \quad \forall\, u \in \mathbf{U} \quad , \tag{2}$$

where

$$Q := \Phi'\,\Phi \quad , \quad q := -2\,y'\,\Phi \quad .$$

If Φ is full rank (identifiability condition), then $Q > 0$. If, in addition, the feasible set

$$\Theta := \{\vartheta:\ A(u)\,\vartheta + b(u) \geq 0\,,\ \forall\, u \in \mathbf{U}\} = \bigcap_{u\,\in\,\mathbf{U}} \{\vartheta:\ A(u)\,\vartheta + b(u) \geq 0\}$$

is nonempty (viability condition), then problem P is *well-posed* (the solution exists and is unique). It should also be noted that $b(u) \geq 0$, $\forall\, u \in \mathbf{U}$, implies viability ($\vartheta = 0$ is an obvious feasible solution). Finally, note that Θ is convex; as a matter of fact, the intersection of an infinity of convex sets is convex.

A special instance of P, where:

i) $\quad m = p \quad ,$ \hfill (3)

ii) $\quad A(u) = \begin{bmatrix} A_1(u_1) \\ A_2(u_2) \\ \cdots \\ A_p(u_p) \end{bmatrix} \qquad , \qquad b(u) = c \in (\mathbf{R}^+)^p \qquad ,$ (4)

and each A_i, $i = 1, 2, \ldots, p$, is a row vector function mapping $\mathbf{R}$ into $\mathbf{R}^n$,

iii) $\quad u_i \in U_i \subset \mathbf{R} \qquad , \qquad i = 1, 2, \ldots, p \qquad , \qquad U = U_1 x U_2 x \ldots x U_p \qquad ,$ (5)

is discussed in detail in Guardabassi and Jian[1]. A totally different (harmonic balance) approach to a strongly related nonlinear identification problem equally leads[2] to a convex semi-infinite quadratic program of the form (2)-(5).

2. Batch algorithms

In this section, a batch two-levels algorithm for the convex semi-infinite quadratic program P introduced in Section 1 (Eq.2) is described, and its convergence proved. The algorithm, henceforth referred to as Algorithm 1, is a proper extension of Kelley's cutting plane algorithm[3] (see also Luenberger[4], Chapter 13) and can be modified in several ways in order to deal with interesting variants of the basic problem. Some of these variants are shortly discussed at the end of the present section.

Letting, for any $(u, \vartheta, i) \in U x \mathbf{R}^n \, x \{1, 2, \ldots, m\}$:

$$V_i(u, \vartheta) := A_i(u) \, \vartheta + b_i(u)$$

where $A_i(u)$ is the i-th row of $A(u)$ and $b_i(u)$ is the i-th element of $b(u)$, the basic algorithm can be described as follows.

Algorithm 1

Step 0 Set: $\quad W = 0 \, , \; w = 0 \, .$

Step 1 Find: $\quad \hat{\vartheta} := \underset{W \vartheta + w \geq 0}{\mathrm{argmin}} \; \vartheta' Q \, \vartheta + q \, \vartheta$

Step 2 Find: $\quad \hat{u}(i) := \underset{u \,\in\, U}{\mathrm{argmin}} \; V_i(u, \hat{\vartheta}) \quad , \quad \hat{V}_i := V_i(\hat{u}(i), \hat{\vartheta}) \quad , \quad \forall \, i \in \{1, 2, \ldots, m\} \; ,$

and: $\quad j := \underset{i \in \{1, 2, \ldots, m\}}{\mathrm{argmin}} \; \hat{V}_i \; , \qquad \hat{V} := \hat{V}_j \quad .$

110

Step 3   If $\hat{V} \geq 0$, stop; else set:

$$W = \begin{bmatrix} W \\ A_j(\hat{u}(j)) \end{bmatrix} \quad , \quad w = \begin{bmatrix} w \\ b_j(\hat{u}(j)) \end{bmatrix} \quad ,$$

and return to Step 1.

$\square$

In order to prove convergence of Algorithm 1, it is convenient to fix a few preliminary points, stated below as lemmas.

Lemma 1

Let $\Theta \subset \mathbf{R}^n$ be the feasible parameter set and $\hat{\vartheta}$ an unfeasible value of ϑ; i.e. $\hat{\vartheta} \notin \Theta$. Furthermore, let $\hat{u}(i)$ and j be defined as in Step 2 of Algorithm 1. Then,

$$h := \{\vartheta \in \mathbf{R}^n : V_j(\hat{u}(j),\vartheta) := A_j(\hat{u}(j))\, \vartheta + b_j(\hat{u}(j)) = 0\}$$

is a separating hyperplane (relative to $\hat{\vartheta}$ and Θ).

Proof. First, note that, by definition, $V_j(\hat{u}(j),\vartheta) = 0$, $\forall \vartheta \in h$. Since, by assumption, $\hat{\vartheta}$ is not feasible, there must exist $i \in \{1, 2, ..., m\}$ and $u \in \mathbf{U}$ such that $V_i(u,\hat{\vartheta}) < 0$. Hence, in keeping with the notation used in Step 2,

$$V_j(\hat{u}(j),\hat{\vartheta}) := \hat{V}_j \leq \hat{V}_i := V_i(\hat{u}(i),\hat{\vartheta}) \leq V_i(u,\hat{\vartheta}) < 0 \quad .$$

On the other hand, for any $\vartheta \in \Theta$, one has: $V_i(u,\vartheta) \geq 0$, $\forall (u,i) \in \mathbf{U} \times \{1, 2, ..., m\}$. Therefore,

$$V_j(\hat{u}(j),\vartheta) \geq 0 \quad , \quad \forall\, \vartheta \in \Theta .$$

Thus, the proof is complete.

Lemma 2

Denote $\{\vartheta(k)\}$ the sequence of points generated by Algorithm 1 at Step 1, $\{\hat{u}(i,k)\}$ and $\{j(k)\}$ the sequences generated at Step 2, $\{W(k)\}$ and $\{w(k)\}$ those generated at Step 3; furthermore, let $\Theta(k) := \{\vartheta \in \mathbf{R}^n : W(k)\, \vartheta + w(k) \geq 0\}$ so that: $\vartheta(k) \in \Theta(k)$, $\forall k$. Then:

i) $\Theta \subseteq \Theta(k+1) \subseteq \Theta(k)$

ii) $V_{j(k)}(\hat{u}(j(k),k),\vartheta) \geq 0$, $\quad \forall\, \vartheta \in \Theta(k+1)$.

Proof. In view of Step 3, one may write:

$$\Theta(k+1) := \{\vartheta \in \Theta(k) : A_{j(k)}(\hat{u}(j(k),k))\, \vartheta + b_{j(k)}(\hat{u}(j(k),k)) \geq 0\} \quad .$$

Therefore: $\Theta(k+1) \subseteq \Theta(k)$ and $V_{j(k)}(\hat{u}(j(k),k),\vartheta) \geq 0$, $\forall \vartheta \in \Theta(k+1)$. To complete the proof, one has to show that $\Theta \subseteq \Theta(k+1)$. As a first step, note that, in view of Lemma 1, $\Theta \subseteq \Theta(k)$ implies $\Theta \subseteq \Theta(k+1)$; as a matter of fact, $\{\vartheta \in \mathbf{R}^n : V_{j(k)}(\hat{u}(j(k),k),\vartheta)=0\}$ is a separating hyperplane (relative to $\vartheta(k)$ and Θ), whereby $V_{j(k)}(\hat{u}(j(k),k),\vartheta) \geq 0$, $\forall \vartheta \in \Theta$. Finally, since $\Theta \subseteq \Theta(0) = \mathbf{R}^n$, the proof trivially follows by induction.

Lemma 3

For any $\hat{\vartheta} \notin \Theta$ and $i \in \{1, 2, ..., m\}$, let:

$$\hat{u}(i) := \underset{u \in \mathbf{U}}{\mathrm{argmin}}\ V_i(u,\hat{\vartheta}) \qquad , \qquad \hat{V}_i := V_i(\hat{u}(i),\hat{\vartheta})$$

then

$$V_i(\hat{u}(i),\vartheta) = \hat{V}_i + A_j(\hat{u}(i))\,(\vartheta - \hat{\vartheta}) \quad , \quad \forall\, \vartheta \in \mathbf{R}^n .$$

Proof. It is immediate to check that, $\forall \vartheta \in \mathbf{R}^n$:

$$V_i(\hat{u}(i),\vartheta) := A_i(\hat{u}(i))\,\vartheta + b_i(\hat{u}(i)) = A_i(\hat{u}(i))\,\hat{\vartheta} + b_i(\hat{u}(i)) + A_i(\hat{u}(i))\,(\vartheta - \hat{\vartheta}) =$$

$$= \hat{V}_i + A_i(\hat{u}(i))\,(\vartheta - \hat{\vartheta}) .$$

Theorem 1

Any limit point of a sequence $\{\vartheta(k)\}$ generated by Algorithm 1 is a solution to Problem P.

Proof. Suppose $\{\vartheta(k),\ k \in K\}$ is a subsequence of $\{\vartheta(k)\}$ converging to a limit point ϑ°. Then the proof is in two steps; first, ϑ° is shown to be feasible; second, to solve P.

For any integer $k \geq 0$ and $i \in \{1, 2, ..., m\}$, denote $\{\hat{u}(i,k)\}$, $\{\hat{V}_i(k)\}$ and $\{j(k)\}$ the sequences generated by Algorithm 1 at Step 2, $\{W(k)\}$ and $\{w(k)\}$ those generated at Step 3; furthermore, let $\Theta(k) := \{\vartheta \in \mathbf{R}^n : W(k)\,\vartheta + w(k) \geq 0\}$ so that: $\vartheta(k) \in \Theta(k)$. For any $k \in K$, one has (Lemma 2):

$$V_{j(k)}(\hat{u}(j(k),k),\vartheta) \geq 0 \quad , \quad \forall\, \vartheta \in \Theta(k+1) \quad ,$$

and (Lemma 3)

$$V_{j(k)}(\hat{u}(j(k),k),\vartheta) = \hat{V}_{j(k)}(k) + A_{j(k)}(\hat{u}(j(k),k))\,(\vartheta - \vartheta(k)) \quad , \quad \forall\, \vartheta \in \mathbf{R}^n \quad ,$$

therefore:

$$\hat{V}_{j(k)}(k) + A_{j(k)}(\hat{u}(j(k),k))\,(\vartheta - \vartheta(k)) \geq 0 \quad , \quad \forall\, \vartheta \in \Theta(k+1) \quad .$$

112

Since, still in view of Lemma 2, $\vartheta(\tilde{k}) \in \Theta(\tilde{k}) \subseteq \Theta(k+1)$, $\forall\, \tilde{k} > k$, it follows that:

$$\hat{V}_{j(k)}(k) + A_{j(k)}(\hat{u}(j(k),k))\,(\vartheta(\tilde{k}) - \vartheta(k)) \geq 0 \quad , \quad \forall\, k,\tilde{k} \in K \ , \ \tilde{k} > k \quad .$$

Compactness of **U** and continuity of A over **U** imply that there exists a positive $S < \infty$ such that:

$$\|A_i(u)\| \leq S \quad , \quad \forall\, (u,i) \in \mathbf{U} \times \{1, 2, ..., m\} \quad ;$$

then one has:

$$\hat{V}_{j(k)}(k) \geq - A_{j(k)}(\hat{u}(j(k),k))\,(\vartheta(\tilde{k}) - \vartheta(k)) \geq - \|A_{j(k)}(\hat{u}(j(k),k))\|\,\|\vartheta(\tilde{k}) - \vartheta(k)\| \geq$$

$$\geq - S\,\|\vartheta(\tilde{k}) - \vartheta(k)\| \quad , \quad \forall\, k,\tilde{k} \in K \ , \ \tilde{k} > k \quad .$$

Since, by definition:

$$V_i(u,\vartheta(k)) \geq \hat{V}_i(\hat{u}(i,k),\vartheta(k)) = \hat{V}_i(k) \geq \hat{V}_{j(k)}(k) \quad , \quad \forall\, (u,i) \in \mathbf{U} \times \{1, 2, ..., m\} \quad ,$$

it follows that

$$V_i(u,\vartheta(k)) \geq - S\,\|\vartheta(\tilde{k}) - \vartheta(k)\| \quad , \quad \forall\, (u,i) \in \mathbf{U} \times \{1, 2, ..., m\} \quad \text{and} \quad \forall\, k,\tilde{k} \in K \ , \ \tilde{k} > k \quad .$$

But, if ϑ° is a limit point of $\{\vartheta(k)\}$, $\|\vartheta(\tilde{k}) - \vartheta(k)\|$ goes to zero as k and $\tilde{k} > k$ go to infinity (within K); hence, ϑ° is feasible.

Finally, let: $J(k) := \vartheta'(k)\,Q\,\vartheta(k) + q\,\vartheta(k)$, and denote by J^* the optimal value of Problem P. Since, by Lemma 2,

$$\Theta \subseteq \Theta(k+1) \subseteq \Theta(k) \quad , \quad \forall\, k \quad ,$$

one has:

$$J(k) \leq J(k+1) \leq J^* \quad , \quad \forall\, k \quad ;$$

thus, $J^\circ := \vartheta^{\circ\prime}\,Q\,\vartheta^\circ + q\,\vartheta^\circ \leq J^*$ and hence ϑ° is an optimal solution to P.

$\square$

In the interesting special case of Eqs.3-5, Algorithm 1 may be slightly modified so as to take advantage of an effective decomposition of Step 2. The first line of Step 2 splits in fact into p scalar subproblems and produces as many cutting hyperplanes (relative to $\hat{\vartheta}$ and Θ) as are the violated constraints (when $\vartheta = \hat{\vartheta}$). Precisely, Step 2 and Step 3 may be modified as follows.

Step 2'   Find: $\hat{u}_i := \underset{u_i \in \mathbf{U}_i}{\mathrm{argmin}} \; V_i(u_i, \hat{\vartheta})$, $\hat{V}_i := V_i(\hat{u}_i, \hat{\vartheta})$, $\forall \, i \in \{1, 2, ..., p\}$

and: $\hat{V} := \underset{i \in \{1, 2, ..., p\}}{\min} \hat{V}_i$.

If $\hat{V} \geq 0$, set: $\vartheta^\circ = \hat{\vartheta}$ and stop; else, set i=1 and go to Step 3'.

Step 3'   If $\hat{V}_i < 0$, set:

$$W = \begin{bmatrix} W \\ A_i(\hat{u}_i) \end{bmatrix} \quad , \quad w = \begin{bmatrix} w \\ b_i(\hat{u}_i) \end{bmatrix} \quad ,$$

else, if $i<p$, set $i=i+1$ and repeat Step 3'; otherwise ($i=p$), return to Step 1.

□

If, in addition to distributed constraints (Eq.2), one has also to comply with standard inequalities of the form:

$$A_0 \, \vartheta + b_0 \geq 0 \quad ,$$

where, for some positive integer m_0, the matrix $A_0 \in \mathbf{R}^{m_0 \times n}$ and the vector $b_0 \in \mathbf{R}^{m_0}$ are given, then Algorithm 1 keeps on solving Problem P provided Step 0 is substituted by

Step 0'   Set: $W = A_0$, $w = b_0$.

In the presently available MATLAB implementation of Algorithm 1, Step 1 consists of a suitable implementation of the Goldfarb-Idnani[5] quadratic programming algorithm. As for Step 2, which in general may indeed be very demanding from a computational point of view (global optimization), substantial simplifications occur when A_i and b_i, $i \in \{1, 2, ... , m\}$, either are (approximated by) polynomial functions[6] of (the components of) u on $\mathbf{U}$, or depend on the i-th component of u only (Eqs.3-5); in the latter case, in fact, even simple brute-force methods may prove satisfactory.

3. Recursive algorithms

In this section, attention is focused on the so called normalized least-mean-squares (NLMS) or projection[7] algorithm. For the sake of simplicity, reference is herein made to the special case $r = 1$; namely, to the case of a single output variable y. Thanks to its *recursive* nature, the projection algorithm plays a fundamental role in solving real-time parameter estimation problems as those encountered, for instance, in adaptive system design. Of course, in the standard situation, where no distributed constraints act on the parameter vector, other recursive algorithms can be used that in many practical situations are preferable to the projection one because of their generally higher rate of convergence; to the best of the authors'

114

knowledge, though, none of them could be extended yet to accomodate the general case dealt with in this section.

The standard form of the projection algorithm, for single-output systems ($r = 1$) with unconstrained parameters ($\vartheta \in \Theta = \mathbf{R}^n$), is as follows.

Projection Algorithm

$$\hat{\vartheta}(t) = \hat{\vartheta}(t\text{-}1) + \frac{a\,\varphi(t\text{-}1)'}{c + \|\varphi(t\text{-}1)\|^2}\,[y(t) - \varphi(t\text{-}1)\,\hat{\vartheta}(t\text{-}1)] \quad ; \quad a \in (0,2) \quad , \quad c > 0 \ .$$

$\square$

Setting $a=1$ and $c=0$ would result in the *Basic Projection Algorithm*, the rationale of which is as follows. Given $y(t)$ and $\varphi(t\text{-}1)$, all possible values of ϑ satisfying at time t the model equation:

$$y(t) = \varphi(t\text{-}1)\,\vartheta$$

belong to the hyperplane $h_t := \{\vartheta \in \mathbf{R}^n :\ y(t) = \varphi(t\text{-}1)\,\vartheta\}$. Given the estimate $\hat{\vartheta}(t\text{-}1)$ of ϑ at time $t\text{-}1$, the element of h_t closest to $\hat{\vartheta}(t\text{-}1)$ is conceivably taken as subsequent estimate of ϑ; i.e.:

$$\hat{\vartheta}(t) = \operatorname*{argmin}_{\vartheta \in h_k} \|\vartheta - \hat{\vartheta}(t\text{-}1)\| \ .$$

This immediately leads to the basic algorithm. The "modifications" $a \in (0,2)$ and $c > 0$ are introduced to add flexibility and avoid the (remote) risk of division by zero. The main properties of the Projection Algorithm are summarized in the following statement.

Theorem 2

For any sequence $\{y(t)\}$ generated by $M(\vartheta°)$, $\vartheta° \in \Theta = \mathbf{R}^n$, let:

$e(t) := y(t) - \varphi(t\text{-}1)\,\hat{\vartheta}(t\text{-}1)$ be the *output prediction error* at time t, and

$\eta(t) := \hat{\vartheta}(t) - \vartheta°$ be the *parameter estimation error* at time t .

Then, the sequences produced by the Projection Algorithm are such that:

1) $\|\eta(t)\| \le \|\eta(t\text{-}1)\|$, $\forall\, t \ge 1$,

2) $\displaystyle \lim_{N \to \infty} \sum_{t=1}^{N} \frac{e(t)^2}{c + \|\varphi(t\text{-}1)\|^2} < \infty$

       and this implies:

a) $\quad \lim\limits_{t \to \infty} \dfrac{e(t)}{[c + \|\varphi(t-1)\|^2]^{1/2}} = 0 \quad$;

b) $\quad \lim\limits_{t \to \infty} e(t) = 0 \quad$, if $\varphi(\cdot)$ is bounded ;

c) $\quad \lim\limits_{N \to \infty} \sum\limits_{t=1}^{N} \|\hat{\vartheta}(t) - \hat{\vartheta}(t\text{-}1)\|^2 < \infty \quad$;

d) $\quad \lim\limits_{t \to \infty} \|\hat{\vartheta}(t) - \hat{\vartheta}(t\text{-}i)\| = 0 \quad$, for any finite integer i .

Furthermore, if there exists a positive ε such that:

$$\sum_{i=1}^{N} \frac{\varphi(t+i)' \, \varphi(t+i)}{\|\varphi(t+i)\|^2} \geq \varepsilon I$$

for all t and some fixed $N > 0$, then $\hat{\vartheta}(t)$ is globally exponentially convergent to $\vartheta°$.

Proof. The proof of Theorem 2 can be found in Goodwin et al.[7,8]. It is omitted here for the sake of brevity.

Remark 1

Global exponential convergence of $\hat{\vartheta}(t)$ to $\vartheta°$ (Theorem 2) is proved by showing that, under the given assumptions, the Lyapunov function:

$$L(\eta) := \|\eta\|^2 \quad , \quad \eta(t) := \hat{\vartheta}(t) - \vartheta°$$

is indeed decreasing along any trajectory of the Projection Algorithm. Therefore, any *Extended Projection Algorithm* of the form:

$$\tilde{\vartheta}(t) = \hat{\vartheta}(t\text{-}1) + \frac{a \, \varphi(t-1)'}{c + \|\varphi(t-1)\|^2} [y(t) - \varphi(t-1) \, \hat{\vartheta}(t\text{-}1)] \quad ; \quad a \in (0,2) \quad , \quad c > 0 \quad ,$$

$$\hat{\vartheta}(t) = F[\tilde{\vartheta}(t)] \quad ,$$

where F is any mapping such that $F(\vartheta) \in \Theta$ and

$$\|\eta\| := \|F(\tilde{\vartheta}) - \vartheta°\| \leq \|\tilde{\vartheta} - \vartheta°\| := \|\tilde{\eta}\|$$

retains (and possibly improves) the convergence properties of the Projection Algorithm.

$\square$

Before moving to constrained parameter estimation problems, a basic property of orthogonal projections onto separating hyperplanes is here pointed out.

Lemma 4

Let h be any hyperplane in $\mathbf{E}^n$, H one of the two halfspaces in which h splits $\mathbf{E}^n$, $\tilde{\vartheta}$ any point of $\mathbf{E}^n$ not belonging to the closure $\bar{H}$ of H, and ϑ^+ the orthogonal projection of $\tilde{\vartheta}$ onto h; then

$$\|\vartheta^+ - \vartheta\| < \|\tilde{\vartheta} - \vartheta\| \quad , \quad \forall\, \vartheta \in \bar{H}\,.$$

Proof. For any $\vartheta \in \bar{H}$, let ϑ^* be the orthogonal projection of ϑ onto the straight line traversing $\tilde{\vartheta}$ and ϑ^+. Then,

$$\|\vartheta^+ - \vartheta^*\| < \|\tilde{\vartheta} - \vartheta^*\|$$

immediately implies $\|\vartheta^+ - \vartheta\| < \|\tilde{\vartheta} - \vartheta\|$.

$\square$

Assume now that the parameter vector ϑ is subject to a standard inequality constraint of the form:

$$A_0\, \vartheta + b_0 \geq 0 \quad ,$$

where, for some positive integer m_0, $A_0 \in \mathbf{R}^{m_0 \times n}$ and $b_0 \in \mathbf{R}^{m_0}$ are a given constant matrix and a given constant vector, respectively. For any $\tilde{\vartheta} \in \mathbf{R}^n$, define:

$$F_1(\tilde{\vartheta}) := \begin{cases} \tilde{\vartheta} & , \quad \text{if } \tilde{\vartheta} \in \Theta \\[2mm] \vartheta^+ & , \quad \text{otherwise} \end{cases}$$

where $\Theta := \{\vartheta \in \mathbf{R}^n : A_0\, \vartheta + b_0 \geq 0\}$ and ϑ^+ is the orthogonal projection of $\tilde{\vartheta} \notin \Theta$ onto Θ. Hence, $F_1(\tilde{\vartheta}) \in \Theta$. Furthermore, since the closed polytope Θ is obviously convex,

$$h := \{\vartheta \in \mathbf{E}^n : (\vartheta - \vartheta^+)'\, (\tilde{\vartheta} - \vartheta^+) = 0\}$$

is a supporting hyperplane separating Θ and $\tilde{\vartheta}$. Then, by Lemma 4,

$$\|F_1(\tilde{\vartheta}) - \vartheta^\circ\| \leq \|\tilde{\vartheta} - \vartheta^\circ\| \quad , \quad \forall\, \vartheta^\circ \in \Theta\,.$$

Thus, for any initial estimate $\vartheta_0 \in \mathbf{R}^n$, the extended projection algorithm of Remark 1, with $F = F_1$ and $\hat{\vartheta}(0) = F_1(\vartheta_0)$, solves recursively the problem of estimating ϑ subject to $\vartheta \in \Theta$. Of course, such an algorithm, henceforth referred to as *Algorithm* **2.1**, applies unchanged to any constrained estimation problem where the feasible parameter set Θ is closed and convex[7].

Remark 2

Algorithm 2.1 calls, at each step, for the solution of:

$$\vartheta^+ = \underset{\vartheta \in \Theta}{\text{argmin}} \; \|\tilde{\vartheta} - \vartheta\| \; .$$

This may be a very easy job in some important special cases (e.g. when each element of ϑ is constrained to an independent interval, whereby Θ is the hyperparallelepipedon resulting from the cartesian product of n independent feasibility intervals) but, in general, it is nearly as demanding as solving the overall constrained estimation problem. It should be apparent, then, that Algorithm 2.1 looses most of its interest in the generic case.

$\square$

Finally consider the case in which the parameter ϑ is subject to a distributed constraint of the form:

$$A(u)\, \vartheta + b(u) \geq 0 \quad , \quad \forall\, u \in \mathbf{U} \quad ,$$

where $\mathbf{U}$ is a given compact subset of $\mathbf{R}^p$ while $A\colon \mathbf{R}^p \to \mathbf{R}^{m \times n}$ and $b\colon \mathbf{R}^p \to \mathbf{R}^m$ are given matrix (vector) functions of u continuous on $\mathbf{U}$. Precisely, let:

$$\Theta := \{\vartheta \in \mathbf{R}^n : A_0\, \vartheta + b_0 \geq 0 \, ; \; A(u)\, \vartheta + b(u) \geq 0 \, , \, \forall\, u \in \mathbf{U}\} \; ;$$

furthermore, denoting by a subscript h or i the h-th or i-th row of any matrix, define:

$$V_0(h,\vartheta) := A_{0h}\, \vartheta + b_{0h} \quad , \quad V(i,u,\vartheta) := A_i(u)\, \vartheta + b_i(u) \; ,$$

for any $(h,i,u,\vartheta) \in \{1,2,...,m_0\} \times \{1, 2, ...,m\} \times \mathbf{U} \times \mathbf{R}^n$.

In this case, the ineffective Algorithm 2.1 can be substituted by an extended projection algorithm (Remark 1), henceforth referred to as *Algorithm* **2.2**, defined by setting $F = F_2$ and $\hat{\vartheta}(0) = F_2(\vartheta_0)$, $\vartheta_0 \in \mathbf{R}^n$, where the mapping F_2 is computed, for any $\tilde{\vartheta} \in \mathbf{R}^n$, by the following

Algorithm $\mathcal{F}_2$

Step 1   Find:

$$j_0 := \underset{h \in \{1,2,...,m_0\}}{\text{argmin}} \{V_0(h,\tilde{\vartheta})\} \; , \quad \hat{V}_0 := V_0(j_0,\tilde{\vartheta})$$

$$\hat{u}(i) := \underset{u \in \mathbf{U}}{\text{argmin}} \; V(i,u,\tilde{\vartheta}) \; , \quad \hat{V}(i) := V(i,\hat{u}(i),\tilde{\vartheta}) \quad , \quad \forall\, i \in \{1, 2, ..., m\} \; ,$$

$$j := \underset{i \in \{1, 2, ..., m\}}{\text{argmin}} \; \hat{V}(i) \; , \quad \hat{V} := \hat{V}_j \quad .$$

118

Step 2   If $\hat{V}_0 \geq 0$ and $\hat{V} \geq 0$, set: $F_2(\tilde{\vartheta}) = \tilde{\vartheta}$ and stop. Else,

$$\text{if } \hat{V}_0 \leq \hat{V}, \text{ set: } \vartheta^+ = \tilde{\vartheta} - \frac{A_{0j_0}\tilde{\vartheta} + b_{0j_0}}{\|A_{0j_0}\|^2} A'_{0j_0}$$

$$\text{else, set: } \vartheta^+ = \tilde{\vartheta} - \frac{A_j(\hat{u}(j))\,\tilde{\vartheta} + b_j(\hat{u}(j))}{\|A_j(\hat{u}(j))\|^2} A'_j(\hat{u}(j))$$

end if

end if.

Set: $\tilde{\vartheta} = \vartheta^+$ and go to Step 1.

$\square$

In order to ensure convergence to $\vartheta°$ of any sequence $\{\hat{\vartheta}(t)\}$ produced by of Algorithm 2.2 when the time series $\{d(t)\}$ is generated by $M(\vartheta°)$, the mapping computed by Algorithm $\mathcal{J}_2$ at any unfeasible $\tilde{\vartheta} \in \mathbf{R}^n$, has to return a $\tilde{\vartheta}^* \in \Theta$, not more distant from $\vartheta°$ than $\tilde{\vartheta}$ was (Remark 1). Before proving that Algorithm $\mathcal{J}_2$ has in fact this property, it is convenient to point out a preliminary lemma.

Lemma 5

Referring to Algorithm $\mathcal{J}_2$, assume $\tilde{\vartheta} \notin \Theta$, and let

$$a' := \begin{cases} A_{0j_0} & , \text{ if } \hat{V}_0 \leq \hat{V} \\ A_j(\hat{u}) & , \text{ if } \hat{V} < \hat{V}_0 \end{cases} \quad , \quad b := \begin{cases} b_{0j_0} & , \text{ if } \hat{V}_0 \leq \hat{V} \\ b_j(\hat{u}) & , \text{ if } \hat{V} < \hat{V}_0 \end{cases} \quad ;$$

furthermore, for any $\vartheta \notin \Theta$, denote by $D(\vartheta)$ the distance of ϑ from Θ, i.e.:

$$D(\vartheta) := \min_{\theta \in \Theta} \| \vartheta - \theta \|.$$

Then:

1) $h := \{\vartheta \in \mathbf{R}^n : a'\vartheta + b = 0\}$ is a separating hyperplane of $\tilde{\vartheta}$ and Θ,

2) ϑ^+ is the orthogonal projection of $\tilde{\vartheta}$ onto h,

3) $D(\vartheta^+) < D(\tilde{\vartheta})$,

4) $\|\vartheta - \vartheta^+\| \leq D(\tilde{\vartheta})$.

Proof. 1) Note that $a'\mathfrak{I}+b=\min(\hat{V}_0, \hat{V})$, while $\mathfrak{I}\notin\Theta$ implies $\min(\hat{V}_0, \hat{V})<0$; then: $a'\mathfrak{I}+b<0$. On the other hand, for any $\mathfrak{I}\in\Theta$, one has:

$$A_{0i}\,\mathfrak{I} + b_{0i} \geq 0 \qquad , \qquad \forall\ i\in\{1, 2, ..., m_0\}$$

$$A_i(u)\,\mathfrak{I} + b_i(u) \geq 0 \quad , \quad \forall\ (i,u) \in \{1, 2, ..., m\}\mathrm{x}\mathbf{U}$$

whereby

$$a'\,\mathfrak{I} + b \geq 0 \quad , \quad \forall\ \mathfrak{I}\in\Theta.$$

Thus, **h** is a separating hyperplane.

2) The orthogonal projection of $\mathfrak{I}$ onto **h** is in fact given by $\mathfrak{I} - \|a\|^{-2}\,(a'\,\mathfrak{I} + b)\,a = \mathfrak{I}^+$.

3) In view of (1) and Lemma 4, one has: $\|\mathfrak{I}^+ - \mathfrak{I}\| < \|\tilde{\mathfrak{I}} - \mathfrak{I}\|$, $\forall\mathfrak{I}\in\Theta$. Then, denoting by θ^+ and $\tilde{\theta}$ the orthogonal projections of $\mathfrak{I}^+$ and $\tilde{\mathfrak{I}}$, respectively, onto Θ, it follows that:

$$\|\mathfrak{I}^+ - \tilde{\theta}\| < \|\tilde{\mathfrak{I}} - \tilde{\theta}\| .$$

On the other hand, since:

$$\theta^+ = \operatorname*{argmin}_{\mathfrak{I}\in\Theta} \|\theta^+ - \mathfrak{I}\| ,$$

one also has:

$$\|\theta^+ - \mathfrak{I}^+\| \leq \|\mathfrak{I}^+ - \tilde{\theta}\| ;$$

therefore

$$D(\mathfrak{I}^+) = \|\theta^+ - \mathfrak{I}^+\| < \|\tilde{\mathfrak{I}} - \tilde{\theta}\| = D(\tilde{\mathfrak{I}}) .$$

4) Finally, denote by θ_h the intersection of **h** with the straight line running through $\mathfrak{I}$ and $\tilde{\theta}$, namely through $\mathfrak{I}$ and its orthogonal projection onto Θ. It should then be obvious that

$$\|\mathfrak{I} - \mathfrak{I}^+\| \leq \|\mathfrak{I} - \theta_h\|$$

and

$$\|\mathfrak{I} - \theta_h\| \leq \|\mathfrak{I} - \tilde{\theta}\| = D(\mathfrak{I})$$

whereby $\|\mathfrak{I} - \mathfrak{I}^+\| \leq D(\mathfrak{I})$. Thus, the proof of Lemma 5 is complete.

Theorem 3

If Θ is nonempty, the sequence $\{\mathfrak{S}(k)\}$ generated by Algorithm $\mathcal{I}_2$, starting from $\mathfrak{S}=\mathfrak{S}(0)\in\mathbf{R}^n$, either stops after a finite number of steps or converges to a limit point. Denote by $\mathfrak{S}^*:=F_2(\mathfrak{S})$ the limit (or end) point of $\{\mathfrak{S}(k)\}$. Then, $\mathfrak{S}^*\in\Theta$ and, for any $\vartheta^\circ\in\Theta$, $\|\mathfrak{S}^*-\vartheta^\circ\|\le\|\mathfrak{S}-\vartheta^\circ\|$.

Proof. If $\{\mathfrak{S}(k)\}$ is finite and $\mathfrak{S}^*$ is its end point, then by Step 2 of Algorithm $\mathcal{I}_2$ it is $\mathfrak{S}^*\in\Theta$. Furthermore, if $\mathfrak{S}\in\Theta$, $\mathfrak{S}^*=\mathfrak{S}$ and the theorem is trivially true; otherwise ($\mathfrak{S}\notin\Theta$), by Lemma 5, (1) and (2), and by Lemma 4 the theorem immediately follows.

Suppose now that the sequence $\{\mathfrak{S}(k)\}$ is of infinite length, namely $\mathfrak{S}(k)\notin\Theta$ for all finite k. Then $\mathfrak{S}(k+1)=\mathfrak{S}^+(k)$ and, in view of Lemma 5, point (3), $D(\mathfrak{S}(k+1))<D(\mathfrak{S}(k))$ for all k. Since $D(\mathfrak{S})$ is apparently positive definite on $\mathbf{R}^n-\Theta$, $D(\mathfrak{S}(k))$ goes to zero as k goes to infinity, what amounts to saying that the feasible set Θ is a domain of attraction for $\{\mathfrak{S}(k)\}$. For any positive real ε and any positive integer N, let $\bar{k}$ be such that: $D(\mathfrak{S}(\bar{k}))<\varepsilon\,N^{-1}$; whence:

$$D(\mathfrak{S}(\bar{k}+i))<D(\mathfrak{S}(k))<\varepsilon\,N^{-1}\quad,\quad\forall\,i\ge 0\,.$$

Then, for any $k\ge\bar{k}$, one has:

$$\|\mathfrak{S}(k)-\mathfrak{S}(k+N)\|=\|\mathfrak{S}(k)-\mathfrak{S}(k+1)+\mathfrak{S}(k+1)-\mathfrak{S}(k+2)+\ldots+\mathfrak{S}(k+N-1)-\mathfrak{S}(k+N)\|\le$$

$$\le\|\mathfrak{S}(k)-\mathfrak{S}(k+1)\|+\|\mathfrak{S}(k+1)-\mathfrak{S}(k+2)\|+\ldots+\|\mathfrak{S}(k+N-1)-\mathfrak{S}(k+N)\|\le$$

$$\le\sum_{i=0}^{N}D(\mathfrak{S}(k+i))<N\varepsilon\,N^{-1}=\varepsilon\,.$$

Hence, by the Cauchy's Convergence Criterion, there exists a unique limit point $\mathfrak{S}^*$ of $\{\mathfrak{S}(k)\}$ on the boundary of Θ. Finally, by Lemma 4, one has:

$$\|\mathfrak{S}(k+1)-\vartheta^\circ\|\le\|\mathfrak{S}(k)-\vartheta^\circ\|\quad,\quad\forall\,\vartheta^\circ\in\Theta$$

then, for any $k\ge 1$,

$$\|\mathfrak{S}(k)-\vartheta^\circ\|\le\|\mathfrak{S}(0)-\vartheta^\circ\|\quad,\quad\forall\,\vartheta^\circ\in\Theta$$

whereby $\|\mathfrak{S}^*-\vartheta^\circ\|\le\|\mathfrak{S}-\vartheta^\circ\|$, $\forall\vartheta^\circ\in\Theta$. Thus, the proof of Theorem 3 is complete.

4. Concluding Remarks

A general constrained parameter estimation problem has been considered in this paper. Different algorithms to solve it have been presented, and their convergence proved. Though the main emphasis was on the newer aspect, namely on the distributed nature of certain constraints, it may be worth pointing out that, even in the standard case, Algorithm 2.2 presented in Section 3 may compare favorably, from a real-time computational viewpoint, with Goodwin et al.'s "double projection" algorithm[7] because of its inherent capability to decompose into much simpler substeps the computation needed to properly recover, whenever necessary, a feasible current estimation of the unknown parameter vector. Research on several extensions of the results presented in this paper is underway.

Acknowledgments. The present work has been supported by MPI, and by CNR - Centro di Teoria dei Sistemi.

REFERENCES

1. G. O. Guardabassi and X. Jian, Constrained least-squares identification of deterministic nicely-nonlinear models, *Proc. IFAC World Congress*, Sidney 1993.

2. D. Abastanotti, P. Colaneri, J. C. Geromel, R. Ghidoni and G. O. Guardabassi, Nicely nonlinear modelling: A DF approach with application to 'Linearized Process Control', *Proc. IFAC Symp. on Nonlinear Control Systems Design* (A. Isidori, Ed.). Pergamon Press, Oxford (1989).

3. J. E. Kelley, The cutting plane algorithm for solving convex programs, *J. SIAM*, **8** (1960), n.4, 703-712.

4. D. G. Luenberger, *Introduction to Linear and Nonlinear Programming*, Addison Wesley, 1973.

5. D. Goldfarb and A. Idnani, A numerically stable dual method for solving strictly convex quadratic programs, *Math. Programming*, **27** (1983), 1-33.

6. S. Malan, M. Milanese, M. Taragna and J. Garloff, B^3 Algorithm for robust performance analysis in presence of mixed parametric and dynamic perturbations, *31st IEEE Decision and Control Conference*, 1992.

7. G. C. Goodwin and K. S. Sin, *Adaptive Filtering, Prediction, and Control*, Prentice-Hall, 1984.

8. G. C. Goodwin, P. J. Ramadge, and P. E. Caines, Discrete-time multivariable adaptive control, Tech. Rep. Harvard University (1978), and *IEEE Trans. on Automatic Control*, **AC-25** (1980), 449-456.

WSSIAA 5 (1995) pp. 123–143
© World Scientific Publishing Company

APPELL'S EQUATIONS OF MOTION

AND

THE GENERALIZED INVERSE FORM

C. ITIKI [1] , R. E. KALABA[2] , F. E. UDWADIA [3]

Abstract. The actual accelerations of constrained mechanical systems are characterized implicitly by Appell's equations. However, the generalized inverse form provides an explicit characterization. It is shown that Appell's equations and the generalized inverse form minimize the same function. Both methods are then applied to a non-holonomic system, and provide exactly the same actual accelerations.

Keywords. Appell's equations, generalized inverse form, proof of equivalence, minimization

1. INTRODUCTION

The purpose of this paper is to show that the generalized inverse form is an explicit characterization of the actual acceleration of constrained mechanical systems, which is also characterized implicitly by the augmented set of Appell's equations. Both forms may be thought of as arising from a fundamental problem of constrained optimization in mechanics.

---

[1] University of Southern California, Biomedical Engineering Department.

[2] University of Southern California, Economics and Biomedical Engineering Departments.

[3] University of Southern California, Mechanical Engineering, Civil Engineering and Decision Systems

124

Consider a system of n particles, without constraints. External forces are applied to the system so that the free acceleration of the k-th particle is given by the components (ax_k, ay_k, az_k), in an inertial frame of reference.

The inclusion of constraints generates a distinction between actual and free accelerations. The actual acceleration vector of the k-th particle $(\tilde{\ddot{x}}_k, \tilde{\ddot{y}}_k, \tilde{\ddot{z}}_k)$ is such that

$$m_k\, \tilde{\ddot{x}}_k \;=\; m_k\, ax_k \;+\; Fx_k^C \,, \tag{1.a}$$

$$m_k\, \tilde{\ddot{y}}_k \;=\; m_k\, ay_k \;+\; Fy_k^C \,, \tag{1.b}$$

$$m_k\, \tilde{\ddot{z}}_k \;=\; m_k\, az_k \;+\; Fz_k^C \;; \quad \text{for } k=1,2,\ldots,n\,; \tag{1.c}$$

where (Fx_k^C, Fy_k^C, Fz_k^C) are the k-th particle's force components generated by the constraints, and m_k is the k-th particle's mass.

Suppose the constraint equations may be differentiated. One differentiation of non-holonomic constraint equations results in linear relationships between the acceleration components. For holonomic constraint equations, two differentiations are necessary. The resulting equations will be called differentiated constraint equations, and they are of the form

$$A_{k,1}\, \ddot{x}_1 + A_{k,2}\, \ddot{y}_1 + A_{k,3}\, \ddot{z}_1 + \cdots + A_{k,3n-2}\, \ddot{x}_n + A_{k,3n-1}\, \ddot{y}_n + A_{k,3n}\, \ddot{z}_n \;=\; b_k \,,$$

$$\text{for } k = 1, 2, \ldots, m\,. \tag{2}$$

The matrix notation for this set of differentiated constraint equations is

$$A\, \ddot{X} = b \,, \tag{3}$$

where
$$A = \begin{bmatrix} A_{1,1} & A_{1,2} & \cdots & A_{1,3n} \\ A_{2,1} & A_{2,2} & \cdots & A_{2,3n} \\ & \vdots & & \\ A_{m,1} & A_{m,2} & \cdots & A_{m,3n} \end{bmatrix}, \quad b = \begin{bmatrix} b_1 \\ b_2 \\ \vdots \\ b_m \end{bmatrix}, \quad \text{and} \quad \ddot{X} = \begin{bmatrix} \ddot{x}_1 \\ \ddot{y}_1 \\ \ddot{z}_1 \\ \vdots \\ \ddot{x}_n \\ \ddot{y}_n \\ \ddot{z}_n \end{bmatrix}.$$

2. APPELL'S EQUATIONS

Let us assume m independent differentiable non-holonomic constraints, as described before. Because of the differentiated constraint equations (2), m of the acceleration components will be dependent on the others. The $3n-m$ independent acceleration components will be denoted by ($\ddot{p}_1$, $\ddot{p}_2$, $\cdots$, $\ddot{p}_{3n-m}$).

Let us define function S as the Gibbs' function [1]

$$S = \frac{1}{2} \sum_{k=1}^{n} m_k (\ddot{x}_k^2 + \ddot{y}_k^2 + \ddot{z}_k^2) \ , \tag{4}$$

where all acceleration components $(\ddot{x}_1, \ddot{y}_1, \ddot{z}_1 ..., \ddot{x}_n, \ddot{y}_n, \ddot{z}_n)$ are functions of the independent accelerations $(\ddot{p}_1, \ddot{p}_2, ..., \ddot{p}_{3n-m})$, velocities $(\dot{x}_1, \dot{y}_1, \dot{z}_1 ..., \dot{x}_n, \dot{y}_n, \dot{z}_n)$ and positions $(x_1, y_1, z_1 ..., x_n, y_n, z_n)$ *.

---

*   The quantities (p_1, p_2, ..., p_{3n-m}) are the coordinates associated with the independent accelerations. The cordinates (x_1, y_1, z_1, ..., x_n, y_n, z_n) are viewed as unknown functions of (p_1, p_2, ..., p_{3n-m}) and time.

126

Function P_r is defined as [3]

$$P_r = \sum_{k=1}^{n} m_k \left(\ddot{x}_k \frac{\partial x_k}{\partial p_r} + \ddot{y}_k \frac{\partial y_k}{\partial p_r} + \ddot{z}_k \frac{\partial z_k}{\partial p_r} \right) , \quad \text{for } r=1,2,...3n-m; \tag{5}$$

where the work done by the external forces is given by

$$\sum_{r=1}^{3n-m} P_r \, \delta \, p_r = \sum_{r=1}^{3n-m} \sum_{k=1}^{n} m_k \left(\ddot{x}_k \frac{\partial x_k}{\partial p_r} + \ddot{y}_k \frac{\partial y_k}{\partial p_r} + \ddot{z}_k \frac{\partial z_k}{\partial p_r} \right) \delta \, p_r . \tag{6}$$

Appell's equations [2], which characterize the equations of motion, are given by

$$\frac{\partial S}{\partial \ddot{p}_r} = P_r \, , \quad \text{for } r = 1, 2, ... 3n\text{-}m; \tag{7}$$

where functions P_r and S are defined by equations (4) and (5) respectively.

The independent acceleration components $(\ddot{p}_1, \ddot{p}_2, ..., \ddot{p}_{3n-m})$ are determined by solving equations (7). However, the determination of the actual acceleration components $(\tilde{\ddot{x}}_1, \tilde{\ddot{y}}_1, \tilde{\ddot{z}}_1 ..., \tilde{\ddot{x}}_n, \tilde{\ddot{y}}_n, \tilde{\ddot{z}}_n)$ demands the application of the m differentiated constraint equations in addition to the $3n\text{-}m$ Appell's equations.

The complete set of differential equations, which will be called augmented set of Appell's equations, involves Appell's equations (7), the differentiated constraint equations (2), and relationships between all acceleration components $(\ddot{x}_1, \ddot{y}_1, \ddot{z}_1 ..., \ddot{x}_n, \ddot{y}_n, \ddot{z}_n)$ and the independent ones $(\ddot{p}_1, \ddot{p}_2, ..., \ddot{p}_{3n-m})$.

3. THE GENERALIZED INVERSE FORM

Let us define the free acceleration vector of the unconstrained system as $a = [ax_1 \ ay_1 \ az_1 \cdots ax_n \ ay_n \ az_n]^T$. The mass matrix M is a positive definite diagonal matrix with elements $[m_1 \; m_1 \; m_1 \; m_2 \; m_2 \; m_2 \cdots m_n \; m_n \; m_n]$ on its main diagonal. All possible candidates for the actual acceleration vector are represented by $\ddot{X} = [\ddot{x}_1 \quad \ddot{y}_1 \quad \ddot{z}_1 \cdots \ddot{x}_n \quad \ddot{y}_n \quad \ddot{z}_n]^T$.

Gauss' function is given by

$$G = (\ddot{X} - a)^T M (\ddot{X} - a) \quad . \tag{8}$$

According to Gauss' principle [4], the actual acceleration is the one that minimizes function G, subject to the constraints.

In our approach, constraints are represented by differentiated constrained equations in matrix notation (3). According to Udwadia and Kalaba, the minimization of function G is obtained by the generalized inverse form of the actual acceleration [5,6]

$$\tilde{\tilde{X}} \;=\; a \;+\; M^{-1/2} \, (A \, M^{-1/2})^+ \, (b - A \, a) \quad ; \tag{9}$$

where $(A \, M^{-1/2})^+$ is the Moore-Penrose generalized inverse matrix of the product $(A \, M^{-1/2})$.

It will next be shown that the generalized inverse form provides an explicit characterization of the actual acceleration, which is characterized implicitly by the augmented set of Appell's equations.

4. PROOF OF EQUIVALENCE

Let us define function G_A as

$$G_A = \frac{1}{2} \sum_{k=1}^{n} m_k (\ddot{x}_k^2 + \ddot{y}_k^2 + \ddot{z}_k^2) - \sum_{k=1}^{n} m_k (ax_k \, \ddot{x}_k + ay_k \, \ddot{y}_k + az_k \, \ddot{z}_k) . \tag{10}$$

Even though there are $3n$ acceleration components $(\ddot{x}_1, \ddot{y}_1, \ddot{z}_1 ..., \ddot{x}_n, \ddot{y}_n, \ddot{z}_n)$, they may be considered functions of the $3n$-m independent ones $(\ddot{p}_1, \ddot{p}_2, ..., \ddot{p}_{3n-m})$, according to the differentiated constraint equations (2). Consequently, G_A is expressed in terms of the independent accelerations.

We may also define a function G_G, which has the same algebraic form as G_A

$$G_G = \frac{1}{2} \sum_{k=1}^{n} m_k (\ddot{x}_k^2 + \ddot{y}_k^2 + \ddot{z}_k^2) - \sum_{k=1}^{n} m_k (ax_k \, \ddot{x}_k + ay_k \, \ddot{y}_k + az_k \, \ddot{z}_k). \tag{11}$$

However, G_G is seen as a function of all $3n$ acceleration components $(\ddot{x}_1, \ddot{y}_1, \ddot{z}_1 ..., \ddot{x}_n, \ddot{y}_n, \ddot{z}_n)$.

We now state that the equations of motion are derived by minimizing either function G_G or G_A. The minimization of function G_A is an unconstrained one, for the constraint equations have been already used to eliminate m of the acceleration components. While, the minimization of G_G is subject to the m differentiated constraint equations (2).

Appell's equations are obtained by minimizing G_A. The minimization of the function G_A is achieved by making its partial derivatives, regarding the $3n$-m independent acceleration components, equal to zero

$$\frac{\partial G_A}{\partial \ddot{p}_r} = 0, \quad \text{for } r=1,2,...,3n\text{-}m. \tag{12}$$

Substituting equation (10) into equation (12), we obtain

$$\frac{\partial}{\partial \ddot{p}_r} \left[\frac{1}{2} \sum_{k=1}^{n} m_k (\ddot{x}_k^2 + \ddot{y}_k^2 + \ddot{z}_k^2) - \sum_{k=1}^{n} m_k (ax_k \, \ddot{x}_k + ay_k \, \ddot{y}_k + az_k \, \ddot{z}_k) \right] = 0. \tag{13}$$

$$\text{for } r=1,2,...,3n\text{-}m.$$

Using the definition (4) of function S, we obtain

$$\frac{\partial S}{\partial \ddot{p}_r} = \frac{\partial}{\partial \ddot{p}_r} \left[\sum_{k=1}^{n} m_k \left(ax_k\, \ddot{x}_k + ay_k\, \ddot{y}_k + az_k\, \ddot{z}_k \right) \right], \quad \text{for } r=1,2,...,3n\text{-}m. \quad (14)$$

Since masses m_k and free accelerations (ax_k, ay_k, az_k) are known, and they do not depend on the independent acceleration components $\ddot{p}_r$, we obtain

$$\frac{\partial S}{\partial \ddot{p}_r} = \sum_{k=1}^{n} m_k \left(ax_k\, \frac{\partial \ddot{x}_k}{\partial \ddot{p}_r} + ay_k\, \frac{\partial \ddot{y}_k}{\partial \ddot{p}_r} + az_k\, \frac{\partial \ddot{z}_k}{\partial \ddot{p}_r} \right), \quad \text{for } r=1,2,...,3n\text{-}m. \quad (15)$$

The partial derivatives of the coordinates x_k with respect to the independent ones $(p_1, p_2,..., p_{3n-m})$ are called π_r [3] and defined as

$$\pi_r = \frac{\partial x_k}{\partial p_r}, \quad \text{for } r=1,2,...,3n\text{-}m; \text{ and } k=1,2,...,n; \quad (16)$$

where π_r are functions of the coordinates. The velocities are given by

$$\frac{dx_k}{dt} = \sum_{r=1}^{3n-m} \frac{\partial x_k}{\partial p_r} \frac{dp_r}{dt} + \frac{\partial x_k}{\partial t}, \quad \text{for } k=1,2,...,n. \quad (17)$$

Substituting equation (16) into equation (17), we obtain the following expression for the velocities [3]

$$\dot{x}_k = \sum_{r=1}^{3n-m} \pi_r\, \dot{p}_r + \alpha_k, \quad \text{for } k=1,2,...,n; \quad (18)$$

where α_k are functions of the coordinates. Differentiating the above equation, we obtain

$$\ddot{x}_k = \sum_{r=1}^{3n-m} \pi_r \ddot{p}_r + \sum_{r=1}^{3n-m} \frac{d\pi_r}{dt} \dot{p}_r + \frac{d\alpha_k}{dt}, \quad \text{for } k=1,2,...,n. \quad (19)$$

130

From equation (19), one may see that the partial derivatives of the acceleration components, regarding the independent ones, are given by

$$\frac{\partial \ddot{x}_k}{\partial \ddot{p}_r} = \pi_r \quad , \qquad \text{for } r=1,2,...,3n\text{-}m; \text{ and } k=1,2,...,n. \tag{20}$$

Applying the same procedure to the coordinates y_k, and z_k, we obtain

$$\frac{\partial x_k}{\partial p_r} = \frac{\partial \ddot{x}_k}{\partial \ddot{p}_r} \quad , \tag{21.a}$$

$$\frac{\partial y_k}{\partial p_r} = \frac{\partial \ddot{y}_k}{\partial \ddot{p}_r} \quad , \text{ and} \tag{21.b}$$

$$\frac{\partial z_k}{\partial p_r} = \frac{\partial \ddot{z}_k}{\partial \ddot{p}_r} \quad , \qquad \text{for } r = 1, 2, ..., 3n\text{-}m; \text{ and } k = 1, 2, ..., n. \tag{21.c}$$

Substituting equations (21) into equations (15), we obtain

$$\frac{\partial S}{\partial \ddot{p}_r} = \sum_{k=1}^{n} m_k \left(ax_k \frac{\partial x_k}{\partial p_r} + ay_k \frac{\partial y_k}{\partial p_r} + az_k \frac{\partial z_k}{\partial p_r} \right), \qquad \text{for } r=1,2,...,3n\text{-}m; \tag{22}$$

where ($\delta p_1, \delta p_2, \cdots, \delta p_{3n-m}$) are small independent displacements. The application of equations (1) would cause an expansion of equation (22) to

$$\frac{\partial S}{\partial \ddot{p}_r} = \sum_{k=1}^{n} m_k \left(\ddot{x}_k \frac{\partial x_k}{\partial p_r} + \ddot{y}_k \frac{\partial y_k}{\partial p_r} + \ddot{z}_k \frac{\partial z_k}{\partial p_r} \right) - \sum_{k=1}^{n} \left(Fx_k^C \frac{\partial x_k}{\partial p_r} + Fy_k^C \frac{\partial y_k}{\partial p_r} + Fz_k^C \frac{\partial z_k}{\partial p_r} \right),$$

$$\text{for } r=1,2,...,3n\text{-}m. \tag{23}$$

According to the principle of virtual work, constraint forces cannot do any work in these virtual displacements [3]

$$\sum_{r=1}^{3n-m} \sum_{k=1}^{n} \left(Fx_k^C \frac{\partial x_k}{\partial p_r} + Fy_k^C \frac{\partial y_k}{\partial p_r} + Fz_k^C \frac{\partial z_k}{\partial p_r} \right) \delta p_r = 0 \quad . \tag{24}$$

The displacements ($\delta p_1, \delta p_2, \cdots, \delta p_{3n-m}$) are independent; therefore,

$$\sum_{k=1}^{n} \left(Fx_k^C \frac{\partial x_k}{\partial p_r} + Fy_k^C \frac{\partial y_k}{\partial p_r} + Fz_k^C \frac{\partial z_k}{\partial p_r} \right) = 0, \quad \text{for } r=1,2,\ldots,3n\text{-}m. \tag{25}$$

According to equation (25), the last term of equation (23) is zero, which would result in

$$\frac{\partial S}{\partial \ddot{p}_r} = \sum_{k=1}^{n} m_k \left(\ddot{x}_k \frac{\partial x_k}{\partial p_r} + \ddot{y}_k \frac{\partial y_k}{\partial p_r} + \ddot{z}_k \frac{\partial z_k}{\partial p_r} \right), \quad \text{for } r=1,2,\ldots,3n\text{-}m. \tag{26}$$

Using equation (5), we finally obtain Appell's equations of motion

$$\frac{\partial S}{\partial \ddot{p}_r} = P_r, \quad \text{for } r = 1, 2, \ldots 3n\text{-}m. \tag{27}$$

We obtained Appell's equations by minimizing function G_A regarding $(\ddot{p}_1, \ddot{p}_2, \ldots, \ddot{p}_{3n-m})$ as the independent acceleration components.

The accelerations, which are characterized implicitly by the augmented Appell's equations, are also characterized explicitly by the generalized inverse form.

It will be shown now that the minimization of function G_G results on the generalized inverse form. Using matricial notation, equation (11) becomes

$$G_G = \frac{1}{2} \ddot{X}^T M \ddot{X} - a^T M \ddot{X}, \tag{28}$$

where $\ddot{X} = [\ddot{x}_1 \quad \ddot{y}_1 \quad \ddot{z}_1 \cdots \ddot{x}_n \quad \ddot{y}_n \quad \ddot{z}_n]^T$ is any possible candidate for the actual acceleration vector of the constrained system; $a = [ax_1 \quad ay_1 \quad az_1 \cdots ax_n \quad ay_n \quad az_n]^T$ is the free acceleration vector of the unconstrained system; and M is the mass matrix, a positive definite diagonal matrix with elements $[m_1 \; m_1 \; m_1 \; m_2 \; m_2 \; m_2 \cdots m_n \; m_n \; m_n]$ on its main diagonal.

Similarly, the set of differentiated constraint equations are represented by their matricial form (3).

Equations (28) and (3) may be rewritten as

$$G_G = \frac{1}{2}(M^{1/2}\ddot{X})^T (M^{1/2}\ddot{X}) - (M^{1/2}a)^T (M^{1/2}\ddot{X}), \text{ and} \qquad (29)$$

$$(A\,M^{-1/2})(M^{1/2}\ddot{X}) = b\,. \qquad (30)$$

Since M is a positive definite matrix, G_G should be minimized, subject to the constraint equation (30).

For simplification purposes, we will define the following matrices $C = (A\,M^{-1/2})$, $Y = (M^{1/2}\ddot{X})$, and $g = (M^{1/2}a)$. The expression for G_G becomes

$$G_G = \frac{1}{2}Y^T Y - g^T Y\,, \qquad (31)$$

while the differentiated constraint equations reduce to

$$C\,Y = b\,. \qquad (32)$$

The general solution to equation (32) is given by [7]

$$Y = C^+ b + (I - C^+C)\,w\,, \qquad (33)$$

where w is any vector of the appropriate dimension.

Substituting Y into equation (31), we obtain

$$G_G = \frac{1}{2}[C^+b + (I - C^+C)\,w]^T [C^+b + (I - C^+C)\,w] - g^T [C^+b + (I - C^+C)\,w].$$

$$(34)$$

Expanding the above equation, we obtain

$$G_G = \frac{1}{2}(C^+b)^T(C^+b) + \frac{1}{2}(C^+b)^T(I-C^+C)w + \frac{1}{2}[(I-C^+C)w]^T(C^+b) + \dots$$

$$+ \frac{1}{2}[(I-C^+C)\,w]^T[(I-C^+C)\,w] - g^T C^+b - g^T(I-C^+C)\,w \quad .$$

$$(35)$$

Since $(I-C^+C)$ is a symmetric idempotent matrix, and

$$C^{+^T}(I-C^+C) = C^{+^T}(I-C^+C)^T = C^{+^T}(I-C^TC^{+^T}) = C^{+^T} - C^{+^T}C^TC^{+^T} = 0 \quad , \qquad \text{we}$$

obtain

$$G_G = \frac{1}{2}(C^+b)^T(C^+b) + \frac{1}{2}\,w^T(I-C^+C)\,w - g^TC^+b - g^T(I-C^+C)\,w. \qquad (36)$$

Since w is the only unknown variable in the above equation, we should determine $\tilde{w}$ that minimizes G_G, where $\tilde{w} = \begin{bmatrix} w_1 & w_2 & \cdots & w_{3n} \end{bmatrix}^T$. The minimization is obtained by

$$\frac{\partial G_G}{\partial w_k} = 0 \quad , \qquad \text{for } k=1, 2, \dots, 3n. \qquad (37)$$

Substituting equation (36) into equation (37), and considering only the components that depend on $\tilde{w}$, we obtain

$$\frac{\partial\left[\dfrac{1}{2}\tilde{w}^T(I-C^+C)\,\tilde{w} - g^T(I-C^+C)\,\tilde{w}\right]}{\partial w_k} = 0, \quad \text{for } k=1,2,\dots,3n\text{-}m. \qquad (38)$$

134

If we consider $g^T = [g_1 \quad g_2 \quad \cdots \quad g_{3n}]$, and

$$(I - C^+C) = \begin{bmatrix} d_{1,1} & d_{1,2} & \cdots & d_{1,3n} \\ d_{2,1} & d_{2,2} & & d_{2,3n} \\ \vdots & & & \\ d_{3n,1} & d_{3n,2} & \cdots & d_{3n,3n} \end{bmatrix}$$,we obtain the following relationships

$$\tilde{w}^T (I - C^+C) \, \tilde{w} = \sum_i \sum_j d_{i,j} \, w_i \, w_j \; , \tag{39}$$

$$g^T (I - C^+C) \, \tilde{w} = \sum_i \sum_j d_{i,j} \, g_i \, w_j \; . \tag{40}$$

According to equations (38), (39) and (40), the partial derivatives of G_G regarding the components w_k are given by

$$\frac{\partial \, G_G}{\partial \, w_k} = \frac{1}{2} \, (2 \sum_i d_{i,k} \, w_i) - \sum_i d_{i,k} \, g_i = 0, \qquad \text{for } k = 1, 2, ..., 3n. \tag{41}$$

The matricial representation of equation (41) is given by

$$(I - C^+C) \, \tilde{w} = (I - C^+C) \, g \; . \tag{42}$$

Substituting equation (42) into equation (33), we obtain

$$Y = C^+ b + (I - C^+C) \, g \; . \tag{43}$$

The resultant expression for the actual acceleration $\tilde{\tilde{X}}$ is

$$\tilde{\tilde{X}} = M^{-1/2} g + M^{-1/2} C^+ (b - C g) \; ; \tag{44}$$

which is equivalent to the generalized inverse form [6]

$$\tilde{\tilde{X}} = a + M^{-1/2} (A \, M^{-1/2})^+ (b - A \, a) . \tag{45}$$

It has been shown that the generalized inverse form minimizes function G_G, subject to the constraint equations. There are several vectors $\tilde{w}$ that are solutions to equation (42). However, the actual acceleration (45) is unique.

Both Appell's equations and the generalized inverse form are obtained from the minimization of the same algebraic function. The generalized inverse form provides an explicit characterization to the actual acceleration, while the augmented set of Appell's equations are an implicit characterization.

5. EXAMPLE

The generalized inverse method is useful in obtaining explicitly the actual acceleration of constrained systems. A comparison with the characterization given by the augmented set of Appell's equations will be shown through an illustrative example.

The example consists of a particle constrained to move according to a non-holonomic constraint equation

$$\dot{y} = z\,\dot{x}\,. \tag{46}$$

No external forces are applied to the particle, and the particle's mass is given by m.

One differentiation of the non-holonomic constraint equation provides a linear relationship on the acceleration components, as in equations (2)

$$\ddot{y} = \dot{x}\,\dot{z} + z\,\ddot{x}\ \,. \tag{47}$$

APPELL'S EQUATIONS

According to Appell's procedure, dependent acceleration components should be expressed in terms of the independent ones. The dependent acceleration component is given by equation (47).

The function S may be expressed in terms of the independent acceleration components $(\ddot{p}_1 = \ddot{x},\ \ddot{p}_2 = \ddot{z})$, velocities $(\dot{x}, \dot{y}, \dot{z})$ and positions (x, y, z)

$$S = \frac{1}{2}\, m \left\{ \ddot{p}_1^2 + (\dot{x}\dot{z} + z\,\ddot{p}_1)^2 + \ddot{p}_2^2 \right\}. \tag{48}$$

Applying equations (7), we obtain Appell's equations

$$\frac{1}{2}\, m\ \{2\,\ddot{p}_1 + 2(\dot{x}\dot{z} + z\ddot{p}_1)z\} \ = \ 0 \ , \tag{49.a}$$

$$\frac{1}{2}\, m\ \{2\,\ddot{p}_2\} \ = \ 0 \ . \tag{49.b}$$

The augmented set of Appell's equations is composed by equations (49), the relationships $(\ddot{p}_1 = \ddot{x},\ \ddot{p}_2 = \ddot{z})$, and equation (47). This system of equations provides an implicit characterization to the accelerations of the constrained system.

GENERALIZED INVERSE FORM

The generalized inverse form provides an explicit characterization to the actual acceleration of constrained mechanical systems.

The linear relationships (47) on the acceleration components may be represented in matrix notation $A\ \ddot{X}\ =\ b$. Matrix A, and vectors $\ddot{X}$ and b are given by

$$A = \begin{bmatrix} -z & 1 & 0 \end{bmatrix}, \tag{50}$$

$$\ddot{X} = [\ddot{x} \quad \ddot{y} \quad \ddot{z}]^T, \tag{51}$$

$$b = \dot{x}\,\dot{z}. \tag{52}$$

The free acceleration vector is the null vector, since no external force is applied to the particle. The generalized inverse of the product $(A\, M^{-1/2})$ is given by

$$(A\, M^{-1/2})^+ = \frac{m^{1/2}}{1+z^2} \begin{bmatrix} -z \\ 1 \\ 0 \end{bmatrix}. \tag{53}$$

Applying equation (45), we obtain

$$\ddot{X} = \begin{bmatrix} 0 \\ 0 \\ 0 \end{bmatrix} + \begin{bmatrix} m^{-1/2} & 0 & 0 \\ 0 & m^{-1/2} & 0 \\ 0 & 0 & m^{-1/2} \end{bmatrix} \begin{bmatrix} -\dfrac{m^{1/2}}{(1+z^2)}z \\[2mm] \dfrac{m^{1/2}}{(1+z^2)} \\[2mm] 0 \end{bmatrix} \left(\dot{x}\,\dot{z} - \begin{bmatrix} -z & 1 & 0 \end{bmatrix} \begin{bmatrix} 0 \\ 0 \\ 0 \end{bmatrix} \right). \tag{54}$$

The actual acceleration vector is given explicitly by the generalized inverse form

$$\ddot{X} = \begin{bmatrix} \ddot{x} \\ \ddot{y} \\ \ddot{z} \end{bmatrix} = \dot{x}\,\dot{z} \begin{bmatrix} -z/(1+z^2) \\ 1/(1+z^2) \\ 0 \end{bmatrix}. \tag{55}$$

Let us find the actual accelerations, given by the augmented set of Appell's equations, and compare them with the explicit characterization given by equation 55. Solving equations (49) and substituting $\ddot{p}_1$ by $\ddot{x}$, and $\ddot{p}_2$ by $\ddot{z}$, we obtain the actual acceleration components

138

$$\ddot{x} = -\frac{z}{1+z^2}\,\dot{x}\dot{z}\,, \tag{56.a}$$

$$\ddot{z} = 0\,. \tag{56.b}$$

Substituting equation (56.a) into equation (47), we obtain the other acceleration component

$$\ddot{y} = \frac{1}{1+z^2}\,\dot{x}\dot{z}\,. \tag{56.c}$$

This example shows that the generalized inverse form provides an explicit characterization to the actual acceleration of mechanical constrained systems, while the augmented set of Appell's equations provides an implicit characterization of the acceleration. Indeed, the explicit characterization (55) of the actual acceleration, given by the generalized inverse form, is exactly the same as the one (56) obtained by solving the augmented set of Appell's equations.

NUMERICAL INTEGRATION

We will next do a numerical integration of the actual acceleration given by the generalized inverse form.

Consider a particle of mass 1 kg, with no external forces applied to it. Matrix A and vector b are defined as in equations (50) and (52). The actual acceleration $\ddot{X}$ is calculated by equation (45). MATLAB [8] performs sums, subtractions and multiplications of matrices. Furthermore, the calculation of the generalized inverse matrix is done by a built-in function called *pinv(.)*.

Using the fourth-order Runge Kutta method for numerical integration, we may obtain positions, velocities and accelerations as functions of time. The integration is performed in MATLAB for 10 seconds, with an integration step size of 0.005 second, and initial conditions $(x_0, y_0, z_0, \dot{x}_0, \dot{y}_0, \dot{z}_0) = (-3.0, 2.1, -2.0, 0.5, -1.0, 0.4)$. Observe that initial conditions are consistent with the non-holonomic constraint equation $\dot{y} = z\dot{x}$.

Figure 1.a shows the coordinate (x,y,z). The coordinate z is a linear function of time. Figure 1.b shows the corresponding velocities, while figure 1.c shows the accelerations. One may observe that the velocity corresponding to the coordinate z is constant and equal to 0.4 meters/second^2; and the respective acceleration is equal to zero, as in equations (55) and (56.b).

This numerical integration of the generalized inverse form did not present any difficulty.

6. DISCUSSION AND CONCLUSION

Even though we assumed a non-holonomic system, Appell's equations are valid for holonomic systems as well. In this case, holonomic constraints should be differentiated twice in order to generate linear relationships on the acceleration components as in equation (2).

One may also use generalized coordinates instead of rectangular coordinates. In this case, the differentiated constraint equations should be written in terms of generalized coordinates, velocities and accelerations [3].

Even though the generalized inverse form (45) looks simple, it has deep implications. The use of Calculus of Variations for non-holonomic systems, as suggested incorrectly by Bucy [9], would result in wrong solutions. In fact, Pars [10] points out that only someone with a superficial understanding of Mechanics would make such a mistake.

It has been shown that the generalized inverse provides an explicit characterization to the actual acceleration of constrained mechanical systems, while the augmented set of Appell's equations provides an implicit characterization. As shown from the derivation, the generalized inverse form is in all senses equivalent to Appell's equations, yet the two set of equations are not the same. Clearly, when we use the generalized inverse method, we do not need to evaluate Gibbs' function. Whenever Appell's equations are valid, the generalized inverse form is valid. Any shortcomings [9] of the generalized inverse form will reflect on Appell's equations, as well.

We should be careful with the use of the generalized inverse form, since there is an interplay between physics and mathematics. As indicated by Bucy [9], anyone may create pathological examples, such as changes in rank of the matrix $(A\, M^{-1/2})$, in order to defy the validity of the generalized inverse form. However, these pathological cases were not presented as an invalidation of Lagrange and Appell's works. In fact, both Lagrange and Appell assumed that there was a set of m independent constraint equations. In the same way they did not consider the case when the number of independent equations varies from one time instant to another, we did not dedicate our study to exceptional cases with changes in rank. The generalized inverse form is highly applicable to most mechanical systems, to the same extent that Appell's equations are.

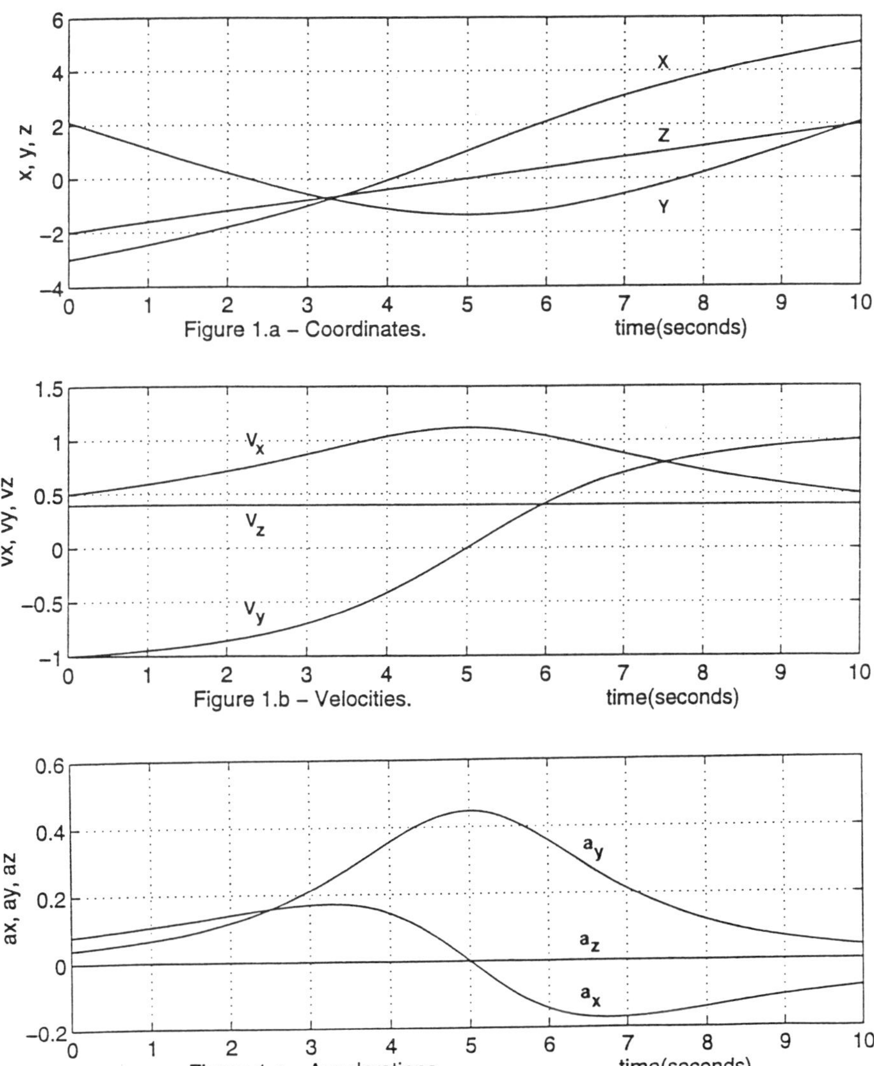

Figure 1.a – Coordinates.

Figure 1.b – Velocities.

Figure 1.c – Accelerations.

ACKNOWLEDGMENT

This work was partially funded by scholarship #200904/91.4 from CNPq-Brasilia/Brasil.

REFERENCES

[1] GIBBS, J. W.   On the fundamental formulae of dynamics. *American Journal of Mathematics*, v.2, pp.49-64, 1879.

[2] APPELL, P. Sur une forme générale des équations de la dynamique. *Mémorial des Sciences Mathématique*. Gauthier-Villars, Paris, 1925.

[3] WHITTAKER, E. *A Treatise on the Analytical Dynamics of Particles and Rigid Bodies*. pp. 258-259, Cambridge, Cambridge Press, 1904.

[4] GAUSS, C. F. Uber ein Neues Allgemeines Grundgesatz der Mechanik. *J. Reine Angewandte Mathematik*. v. 4, pp. 232-235, 1829.

[5] UDWADIA, F. E. and KALABA, R. E. A new perspective on constrained motion. *Proc. R. Soc. Lond.* **A**. v. 439, pp. 407-410, 1992.

[6] KALABA, R. E. and UDWADIA, F. E. On constrained motion. *Applied Mathematics and Computation*. v. 51,   pp. 85-86, 1992.

[7] GRAYBILL, F. A. *Matrices with Applications in Statistics*. 2nd. ed., p.114, Wadsworth, Pacific Grove, 1983.

[8] *MATLAB Reference Guide.* The MathWorks, Inc. August 1992.

[9] BUCY, R.S. Comments on a paper by Udwadia & Kalaba. *Proc. R. Soc. Lond.* **A.**, v.444, pp.253-255, 1994.

[10] PARS, L.A. *A Treatise on Analytical Dynamics.* 2nd. ed. Ox Bow Press, Woodbridge, 1979.

WSSIAA 5 (1995) pp. 145–169
© World Scientific Publishing Company

145

PLANT OPTIMIZATION AND PERFORMANCE MONITORING

MICHAEL A. JOHNSON
Industrial Control Centre University of Strathclyde
Glasgow G1 1QE, Scotland, UK.

And

M. REZA KATEBI
Industrial Control Centre University of Strathclyde
Glasgow G1 1QE, Scotland, UK.

ABSTRACT

World-wide competitive trends require large-scale industrial processes to be fully optimised across all levels of the process hierarchy. This paper reports a methodological framework for the problems to be solved. Industrial examples are used to motivate the use of generic performance monitoring indices for different types of industrial process control problems. Proposals to extend the statistical process control paradigm to incorporate new tools are also discussed. The potential industrial benefits for adopting a holistic plant optimization strategy close the paper.

1. The Control of Large-Scale Industrial Processes

Large-scale industrial processes range from the widely geographically dispersed national electric power network to the relatively compact single site conglomeration comprising, say, an oil refinery or a steel works. Whatever the geographical scale of such plant, their operation tends to be characterised by high energy consumption, feedstock and material usage. Utility support in terms of power and materials are often very high.The operation of these important industrial processes is always complex and demanding. Large-scale distributed computer systems are now very common in providing the global control, synchronisation and co-ordination needed to operate these processes. Newly installed plant will always be controlled using a significant input of new computer technology whilst older plant is often refurbished with the installation of distributed computer control systems. Such new online computer technology offers the potential to progress beyond simple control and co-ordination and address the problems of global plant optimisation.

1.1 *The Process Control Hierarchy*

The standard process control hierarchy[1] , as shown in Fig. 1, is an extremely convenient generic framework for discussing many aspects of large-scale industrial processes. It can be used to describe and conceptualise process control technology,

146

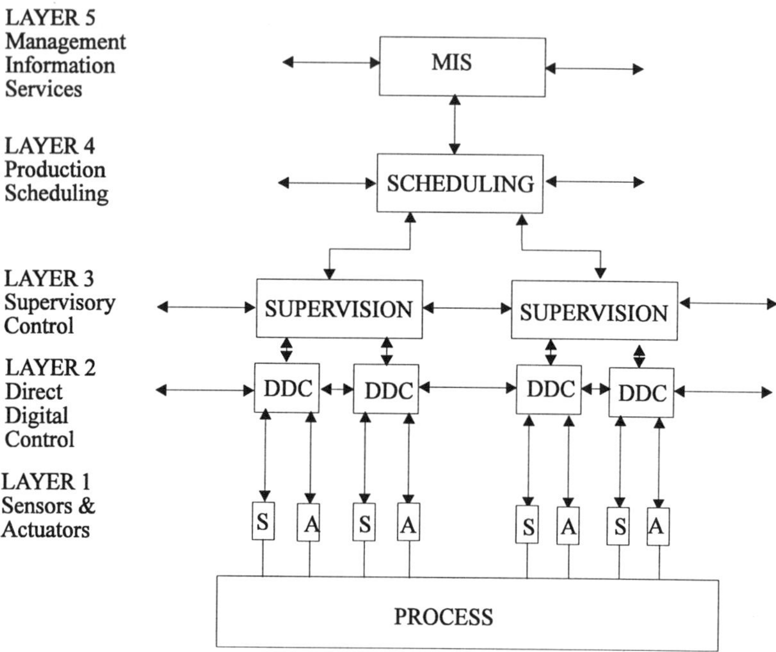

Fig. 1 : The Process Control Hierarchy

plant control strategies and information flows within the large-scale industrial process.

1.2   *Process Control Technology*

The technological aspect of the process control hierarchy is straightforward in the lower layers of the hierarchy where direct digital control (DDC), sensors and actuators interface with the industrial process. In the supervisory, scheduling and management layers it is less obvious that the technological framework is one of interconnected computer networks. These provide the information links between the enterprise management (commercial/business divisions) and the plant management (production divisions). The industrial communications networks thus follow a hierarchy: field bus at the DDC level, system bus at supervisory level and Local Area Network (LAN)/Long Distance Bus at scheduling and management levels. Typical communications update times are shown in Table 1. One implication of enhanced network technology is the potential to transfer more and more data to the higher levels of the operational hierarchy. However, for rational decision taking it is necessary to both decrease the volume of data transfer and enhance its quality.   Performance indices are one method of achieving this.

Table 1 : **Industrial Communication Network Speeds**[2]

Network speed	Predicable Updates	Application
High	1-20 msec	Motion control, Drive-coordinate
Medium	20-200 msec	Machine sequencing, alarms, supervisor parameters, Limited data collection
Slow	200 msec-2 sec	Operation interfaces, data collection/archiving

1.3 *Process Control Strategy*

The strategy of global plant control usually has two components, one geographical, where the plant is subdivided into operational units and one of control where the plant control and command framework is devised. Fig. 2 shows an example for a hydrocarbon processing plant[3].

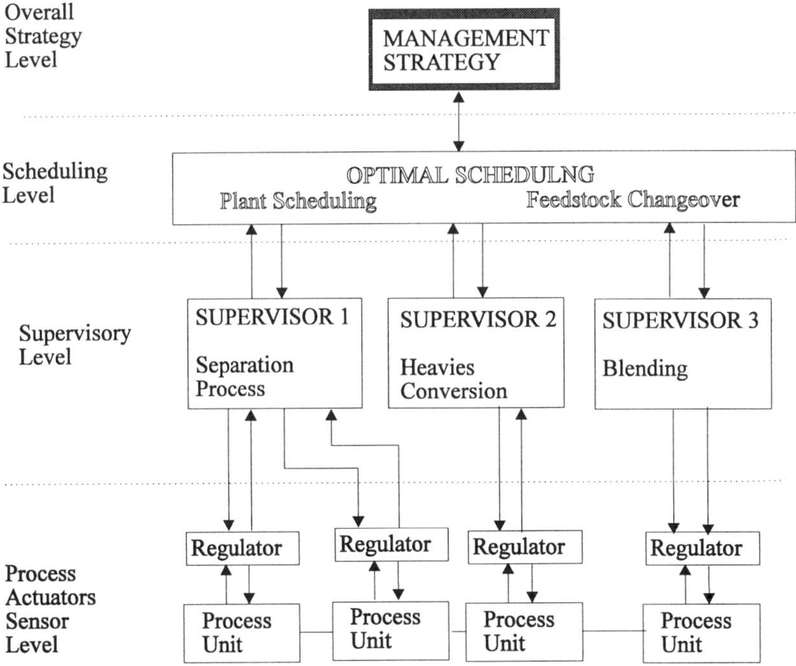

Fig. 2 : **Control Hierarchy for Hydro-Carbon Processing Plant**

It is interesting to note how a global plant optimisation strategy would encompass every layer of the control hierarchy:

(i) *Distributed Digital Control Level* : Control loops should be optimised and correctly tuned, with constant vigilance to re-tune loops if operational conditions change.

(ii) *Supervisory Level* : Here the control centres on the process unit, comprising a number of pieces of plant equipment, and a supervisor. In the case of feedstock changeover or new local conditions, setpoints supplied from the scheduling layer have to be implemented. This requires routines and algorithms, firstly to automate the transition and secondly to optimise the changeover. Such operational setpoint manoeuvres are usually subject to actuator and process output constraints, consequently constrained optimization and constrained control algorithms characterise the solution methods for this problem.[4]

(iii) *Scheduling Level* : The response time of the processes through the hierarchy generally becomes slower at higher levels, so that there is a longer time frame in which to provide control and commands to the lower layers. This feature is one reason why the hierarchical structure is able to function. At the scheduling level, plant wide monitoring, co-ordination and planning occurs. The problems at this level generally involve steady-state optimisation algorithms concerned with longer term plant changes and plans. Production strategies to accommodate unplanned outages, planned maintenance and repair periods originate from this layer. In the hierarchy of plant monitoring, unit performance measures will be collated at this level to give global plant performance figures. This topic is considered in more detail in the next section.

1.4 *Hierarchical Information Structure*

Information, real-time operating information of an appropriate quality, is the key to global plant optimisation. A diagram which uses the standard process control hierarchy to structure typical large-scale system information flows is shown in Fig. 3. It is in this structure that three types of performance indices or diagnostics can be identified:

(i) *Control performance indices.* These are associated with the direct digital control and the supervisory layer. They are the mathematical cost indices on which many control design methods are based.

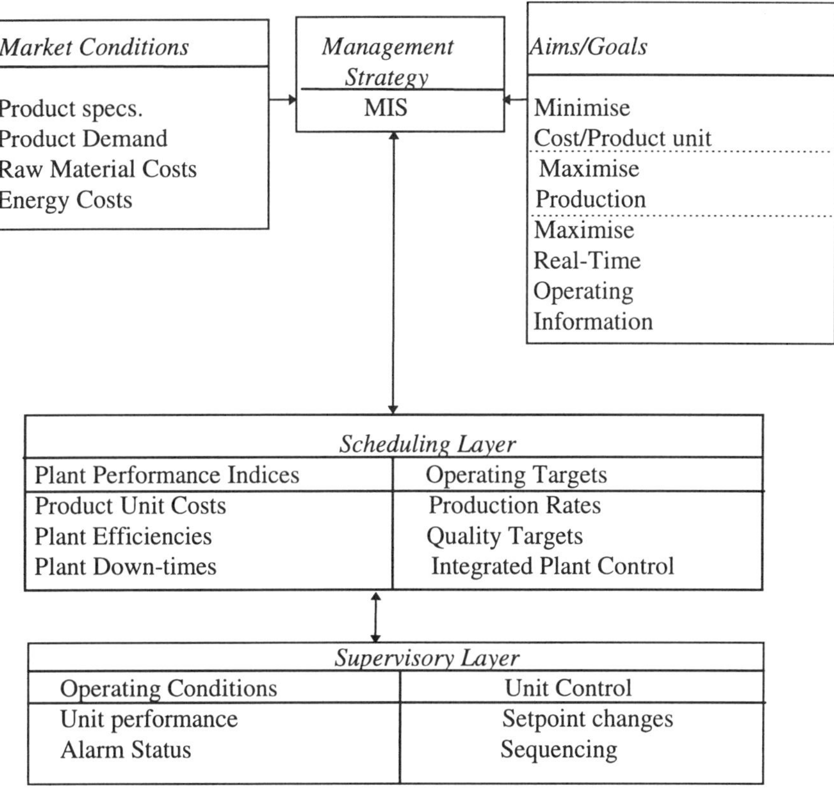

Fig. 3 : **Hierarchical Information Structure**[5]

(ii) *Plant Performance Indices.* These are invariably plant or process specific and are devised to capture some cost aspect of plant/process/unit performance. Often these indices have a direct monetary interpretation. Consequently, they are usually associated with the scheduling and management layers of the operational hierarchy.

(iii) *Fault Detection Indices.* In the sensor/actuator layer, the direct digital control layer and the supervisory layer will be indices or diagnostics designed to detect and identify fault conditions. The outcome of these (often simple) algorithms will be operator alarms and subsequent action.

The optimisation of control performance indices has a long history as exemplified by Kalman's linear quadratic performance index. Of interest in this paper is the use of such indices in supervisory control design, and for assessing controller performance. Given the availability of plant performance indices, it is surprising that more use is not made of this approach to global optimization. This approach is discussed in the sequel. Although, fault detection indices are not involved in plant optimisation, these techniques are part of the portfolio which make up the techniques of modern supervisory control. Hence, it is useful to include them in the preliminary delineation of the subject.

2.   An Overview of Modern Supervisory Control

Supervisory control is rarely perceived as a coherent discipline. There are only a few textbooks which attempt to describe the full depth of the subject.[5,6]   Further, it is only recently that an identifiable group of techniques have emerged which might justifiably comprise part of the supervisory control discipline, for example, statistical process control,[7] or model based fault detection[8] routines.

Modern supervisory control comprises at least three distinct topics:

(i)   *Supervisory command and control design*
(ii   *Performance monitoring*
(iii)   *Fault detection*

The pertinent definitions are:

Definition 1 : *Supervisory command and control design*
Supervisory command and control design is concerned with the unconstrained, constrained, static and dynamic algorithms used to schedule and optimise the control actions throughout the process control hierarchy.

Definition 2 : *Performance monitoring techniques*
The group of methods used to provide an online assessment of the quality or status of process outcomes are termed performance monitoring techniques.

Definition 3 : *Fault detection methods*
Techniques used to automate the recognition and diagnosis of system malfunctions are collectively termed fault detection methods.

2.1   *Supervisory command and control design*

The basic principle behind the classical process control hierarchy is a time-scale decomposition between the scheduling and optimization levels of the control hierarchy and the control loop levels of the process. Thus constrained, static optimisation algorithms are used to solve the allocation and optimised setpoint problems at the scheduling level.[9] Independent of this, many design algorithms are available for the control problems of the loop level. These range from straightforward PID controllers through to more advanced multivariable H_∞ and H_2 optimal control design methods. The connection between the two control layers has often been engineered by *ad hoc* solutions based on past experience.[5]   Industrially motivated solutions to this problem has lead to the use of model based predictive control algorithms to automate and optimise the manoeuvres of plant between different operation conditions and scenarios.[4,10]   The use of a mathematical performance index in a constrained optimization framework is common to most of these methods.

2.2   *Performance monitoring techniques.*

These online methods provide a means of quantifying and analysing process output quality, plant efficiencies and process unit performance. Once such an analysis is available, it is possible to consider whether a process operation needs to be optimised, and how this might be achieved. Although the detection of equipment malfunction is part of the process monitoring exercise, these techniques are more concerned with identifying non-optimal process performance and rectifying this situation whatever the cause.

In a large number of industries the basic online tool for performance monitoring and optimization is statistical process control or procedures based on this *philosophy* via the *magnificent seven* (the histogram, the check sheet, the Pareto chart, the cause and effect diagram, the defect concentration diagram, the scatter diagram and the control chart). [11]

Over recent years, some new methodologies have been devised, and these include performance quality indices,[12] function curve diagnostics[13] and controller performance indices.[14,15]   As many of the names indicate, a performance index or diagnostic often plays an important role in these methods.

2.3   *Fault detection methods*

In contrast to the plant performance monitoring techniques are the fault detection methods concerned with system malfunctions and failures. These methods generally involve a *diagnostic* (index, or rule-based test), a fault *detection* mechanism and a fault *isolation* method as shown in Fig. 4. The outcome of the fault detection system would usually be registered at supervisory level for operator action.

The traditional fault detection method is the simple threshold or limit test as applied to selected process variables and is generally based on a one-to-one measurement-to-fault relationship. The use of a fault dictionary adds more sophistication to the threshold method. The fault dictionary (rather like a cause and effect tree), is a rule-based library of anticipated fault conditions which often originated as a repository of operational experience with recognised fault conditions.

A group of related techniques which also involve performance monitoring but with more emphasis on fault detection, are based on the methods of signal spectrum

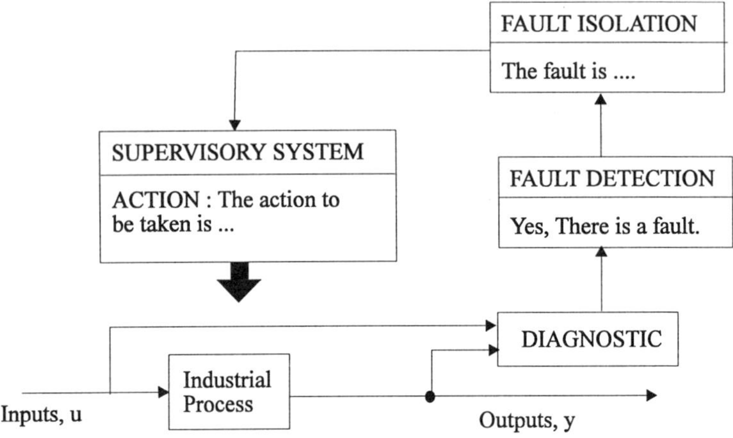

Fig. 4 : **A Fault Detection Architecture**

analysis. For example, the methods of conditioning monitoring involve the use of signal measurements and signal processing to measure performance attributes and predict the failure of machines. This class of methods is not currently developed as generic performance (quality) monitoring procedures with the potential for application to process optimisation. Hence, these methods are classed as fault-detection procedures for the purposes of this overview.

2.4 *Summary*

The three topics of process command and control, performance monitoring and fault detection are viewed as being key topics in a modern supervisory control discipline. The full depth of the supervisory control design paradigm has not been detailed here since this is an on-going development. In this paper the aspect of interest is the use of performance indices of different types to achieve local and global process optimization. For this reason the main exposition concentrates on command and control, and performance monitoring, so that fault detection is not considered further.

3. Supervisory Command and Control Design

The optimised control and operation of large-scale industrial processes can yield significant economical benefits and savings. The industrial techniques used to pursue these benefits are still evolving since past design and operational experience is usually the main guide for current industrial practice. There is considerable industrial interest in discrete event and continuous simulation tools like *SIMPLE++, WITNESS* or *SIMFACTORY* as an inexpensive means for testing alternative operational or production strategies. However, the canon of optimization tools, methods and algorithms for these problems has plenty of opportunity for growth. The command and control structure of supervisory systems is perhaps a more well established area of research and development. However, although industrial experience and practise has been extensive, there has not been so much exchange between industry and the research community on these topics. For this reason, it is sometimes difficult to be precise about the problems faced by industry.

3.1   *The Supervisory System Command Structure*

Motivated by a need to define precisely the supervisory command structure for large-scale combined cycle and combined heat and power generation plant, a study produced the structure shown in Fig. 5.

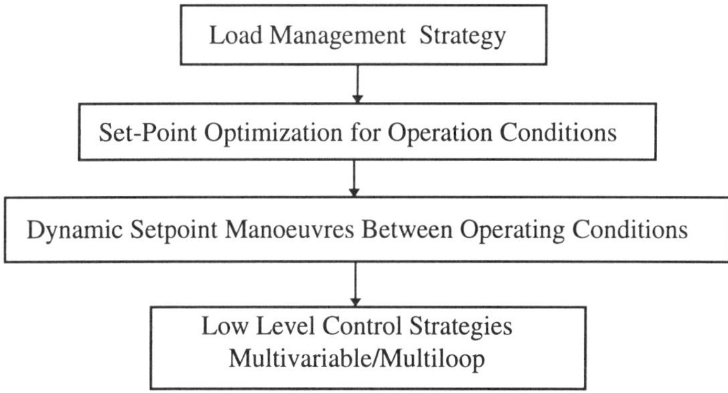

Fig. 5 : **Power Plant Supervisory Concepts**

The findings of the study were found to be generic for large-scale processes[16] and may be discussed from the bottom layer of the supervisory structure upward.

Low Level Control Strategies

The technology at the unit level tends to be supplied with all the local control loops supplied. In large scale industrial plant fine-tuning the local loops is not such a high priority, it is the global integration and optimization which are of more importance. The actual low level control design strategies were found to be remarkable classical, being primarily multiloop, cascade loop three term controllers with simple controller scheduling for adaptation to changing process conditions. One important implication here is that in designing algorithms for integrating and co-ordinating the process units, the low level units can for all essential purposes be regarded as interacting stable (closed loop) systems.

154

Dynamic Setpoint Manoeuvres

To effect feedstock changeover, or respond to new external load conditions, the process operation usually has to be transferred to a new operating condition. However, it was often found that the changeover or setpoint manoeuvre has to occur in the presence of process operational constraints. For example, actuator limits and actuator rate limits might have to be observed or certain outputs were constrained to lie within tight limits of their nominal value. Standard industrial practice is to ramp setpoints over a pre-selected time period so that constraints are not violated.[5] In some cases there have been successful attempts to provide an optimization framework to automate this constrained changeover of operating point.[10] More recently the techniques of constrained model based control have been used to solve this problem.[4,17] It is useful to observe that for these supervisory problems, the selection and use of a quadratic performance index is justifiable as well as practically convenient. A typical formulation[17] comprises:

System
$$x(k+1) = Ax(k) + Bu(k)$$

$$y(k) = Cx(k)$$

Cost function
$$J(U) = \sum_{j=k}^{\infty} \left\{ \left(y(j) - y_s \right)^T Q_c \left(y(j) - y_s \right) \right.$$
$$\left. + \left(u(j) - u_s \right)^T R_c \left(u(j) - u_s \right) + \Delta u(j)^T S_c \Delta u(j) \right\}$$

Operational Constraints

Control limits : $u_{min} \leq u(j) \leq u_{max}$

Rate limits: $\Delta u_{min} \leq \Delta u(j) \leq \Delta u_{max}$

Output limits: $y_{min} \leq Cx(j) \leq y_{max}$

The dynamic setpoint problem is to minimise the performance index $J(U)$ in the presence of constraints, and for a given future setpoint trajectory. The solution (with suitable assumptions) involves sequentially solving constrained Quadratic Programming problems and implementing only the currently calculated control input.

Setpoint Optimisation and Load Management Strategy

At this point in the hierarchy, the new application of constrained model based predictive control algorithms meets the more established procedures of setpoint selection and optimisation unit allocation, and plant scheduling.[9] Many of these procedures and strategies will have their source in the original designs studies and flowsheet exercises that were conducted prior to plant construction. For this reason, there is often a close link between the supervisory control structure, the equipment sizing and the original design objectives for the plant. Post plant construction process optimization is almost exclusively a constrained optimisation problem for this very reason. As has been already mentioned, the increasingly slow system response times that occur in the higher reaches of the hierarchy lead to a predominance of static or steady-state optimization problems at these levels.

4. Performance Monitoring Techniques

Motivated by the success of the online methods of Statistical Process Control[7,11], performance monitoring techniques are methods which seek to extend these principles to create entirely new routines. After a brief introduction to the SPC method, the techniques of process and controller performance indices are introduced and described.

4.1 *Statistical Process Control*

Statistical Process Control (SPC) has its origin in the 1920's when Dr. W.A. Shewhart of the Bell Telephone Laboratories developed the control chart. The key result from statistics which is being exploited is the Central Limit Theorem. This indicates that if $\bar{x}_n$ is the sample mean of a random sample of size n from any distribution with finite variance σ^2 and mean, μ, then $\bar{x}_n$ is approximately distributed as a normal variate, mean μ and variance σ^2/n. Consequently, it is possible to assign a probability to a sample mean occurring in a specific range, for example:

$$pr\{|\bar{x}_n - \mu| < \sigma / \sqrt{n}\} = 0.682 \quad or \quad pr\{|\bar{x}_n - \mu| > \sigma / \sqrt{n}\} = 0.318$$

$$pr\{|\bar{x}_n - \mu| < 2\sigma / \sqrt{n}\} = 0.9546 \quad or \quad pr\{|\bar{x}_n - \mu| > 2\sigma / \sqrt{n}\} = 0.0454$$

$$pr\{|\bar{x}_n - \mu| < 3\sigma / \sqrt{n}\} = 0.9973 \quad or \quad pr\{|\bar{x}_n - \mu| > 3\sigma / \sqrt{n}\} = 0.0027$$

Thus if a series of sample means occurs outside the $+2\sigma/\sqrt{n}$ range, it is highly likely that the underlying process variable mean has changed. Shewhart's innovation was to translate this into a control chart where sample means are plotted in real-time in the presence of upper and lower control limits, UCL and LCL, respectively:

$$UCL = \mu_v + k\sigma_v$$

$$Centre\ Line = \mu_v$$

$$LCL = \mu_v - k\sigma_v$$

156

with the process variable mean μ_v, the process variable standard deviation σ_v and a distance constant k related to the number of standard deviations for the control limits ($2\sigma_v$ and <5% limits ; $3\sigma_v$ and <1%limits). This theory can be developed for both attributes and for continuous variables ; the latter being more appropriate for process and industrial control applications. A typical chart might be as in Fig. 6.

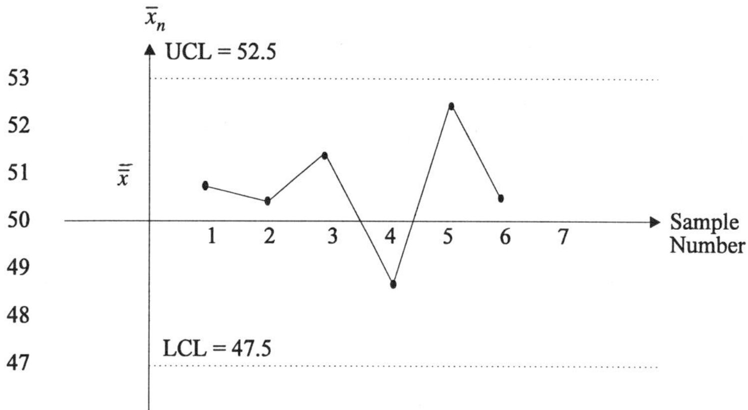

Fig. 6 : **A Control Chart**

Abnormal patterns on control charts then form the means by which deteriorating process performance is identified. For example:

(a)   A sequence of (eight) successive points in one direction or a trend is often indicative of component wear, a slow deterioration of a process component or even operator fatigue.

(b)   A sequence of (seven) successive points on one side of the process mean indicates a shift in process variable levels. This may be due to a systematic change in feedstock or raw material quality, a change in shift or a change in machine components.

(c)   A cyclic pattern can indicate systematic environmental changes, rotation of process operators or component cyclically affecting the variable.

The performance monitoring and diagnostic capability of statistical process control is extensive. The various types of charts are supported by the use of online statistical process control alarms and cause and effect diagnostic software.[7,11] In addition, off-line studies using pareto analysis, scatter diagrams and so on ensure adverse process effects are eradicated. A further component of the statistical process control philosophy is the involvement of the workforce in the need to achieve process quality objectives. The importance of this for process optimization cannot be stressed too highly. For it is only after the process is in control with repeatable behaviour can online optimization and fine tuning of process performance begin. The types of adverse process effects identifiable by statistical process control includes:

Material variability
Variations in the quality of raw materials and utility supplies.
Personnel Problems
Some operators are more vigilant than others ; some operators may be overworked by the excessive demands arising.
Equipment Problems
Maintenance or servicing of equipment required ; excessive wear identified ; sensors inaccurate, improperly calibrated, or broken ; actuators undersized and working at limit ; leaks or loss of materials identified.
Process Operational Problems
Disturbances due to load changes disrupting the process ; process interactions causing instability ; control loops incorrectly tuned ; quality specifications too demanding for the installed process or production strategy being followed.
Environmental Factors
Ambient conditions affecting the process

4.2   *Performance Quality Indices*

The methods of statistical process control usually apply to selected variables of the process, for example, temperature, pressure, or a physical property like viscosity. These are primarily the steady-state variable values of the process and will be highly dependent on the reference setpoints applied to the process. Optimization of performance quality unusually requires more than the attainment of selected setpoint values since online optimization exploits the degrees of freedom remaining in the process operation. Following the mien of the statistical process control philosophy, a method using performance quality indices[12] is proposed:

(i)   Determine the performance objectives for each process unit in the global context of process separation.
(ii)   Translate these objectives into a set of performance indices and related constraints.

Typically these indices might quantify energy and material usage, emissions and effluent produced, quality (tolerance) costs and constraints. More precisely two types of indices are common:

A Cost Index, I_c

$$I_c = \frac{\text{Total cost (monetary) of process inputs and outputs}}{\text{Process product yield}}$$

158

A Performance Index, I_p

$$I_p = \frac{\text{Actual performance achieved by process}}{\text{Design performance specified for process}}$$

Sometimes multiple performance indices and constraints arise for a process unit. Whilst some standard mathematical norms are used, indices are usually process specific and reflect some special performance objectives.

Example 1 : *Boiler-Turbine System*

Dieck-Assad, et al[18] chose a nonlinear performance index which quantified the energy input requirement for a boiler-turbine system:

$$J_E = \int_0^t c_f^{-1} \left\{ HHV.W_{fuel} + W_{fe}(h_{po} - h_{pi}) \right\} d\tau$$

where *HHV* is the high heating value for fuel, W_{fuel} is the fuel flow, W_{fe} is the feedwater flow and h_{po}, h_{pi} are the outlet and inlet enthalpy of the boiler feedwater ; c_f is a simple correction factor to normalise the time step. This performance quality index was to be minimised subject to a set of process constraints involving heat rates and temperature fluctuations. After parameterization of the reference inputs, the problem was solved using a constrained finite dimensional non-linear optimization algorithm.

Example 2 : *Extractive Distillation Process*

An example which illustrates the online utility of performance indices is that due to *Bhandari, et al*[12]. From the process industries, this concerned the optimised performance of the process towers to extract high purity 1,3-butadiene product from a mixture of C_4 hydrocarbons. The three towers involved each had a different function as seen in Fig. 7 and quantifiable performance objectives.

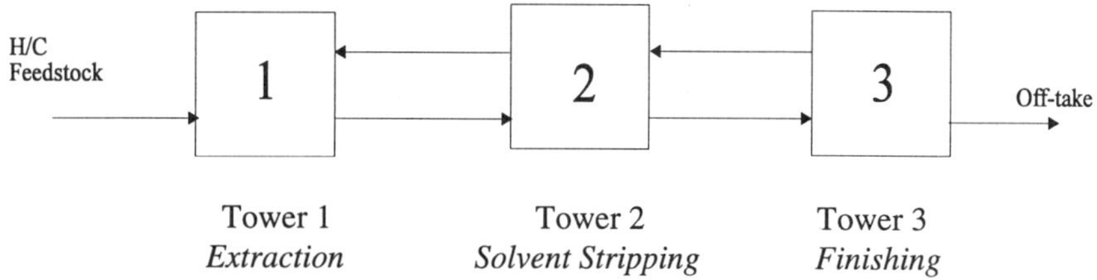

Fig. 7 : **Extractive Distillation of 1,3-Butadiene**

Tower 1 : In the extractive distillation tower, separation is the operational objective, thus the cost index is defined as:

$$J_E = S_E / Q_E$$

where S_E is the separation factor achieved and Q_E is the total energy consumed to attain the separation level.

Tower 2: The second tower is a recycle tower used to separate the butadiene from the solvent used in the extractive process. For this tower, a performance index is used which compares the actual energy consumption Q_A to the design energy consumption Q_D, viz. $J_S = Q_A / Q_D$. Optimal operation of the solvent stripping occurs when $J_S = 1$.

Tower 3 : The third tower in the process is a finishing tower with an objective of achieving the high purity product required. A cost index was chosen for this tower as $J_F = S_F / Q_F$ where S_F is the finishing tower separation factor and Q_F the energy consumption required to achieve the separation.

Since optimal operation of the three towers does not coincide with each tower achieving a minimal performance index, a global cost performance index was constructed:

$$J_{Total} = \frac{Q_T F_c}{P_f}$$

where Q_T is the total energy consumption of the three towers, P_f is the product flow rate fraction and F_c is the monetary cost per energy unit consumed. Thus, index J_{total} is the monetary equivalent of the energy cost per unit of finished product.

The three cost indices, J_E, J_F and J_T, and the performance index J_S were evaluated and plotted online as though they were just another measurement. Online optimisation to minimise the production costs and operate the extraction unit at its design performance was a trial and error process utilizing the plant operators experience and establishing empirical cause and effect relationships. A systematic investigation into how to optimise these process performance monitoring indices brought real monetary savings and operational benefits.[12]

Example 3 : *Hot-Strip Mill Optimization*
The steel industry has some of the most substantial and energy intensive processes in its plant,. Many of the process lines from the production of molten steel through to the rolling and coating of strip are linear and sequential. In a recent study,[19] the global optimization of a hot-strip mill was undertaken using performance quality indices in a discrete event simulation created in a SIMPLE++ environment. The optimization study was based on the real hot-mill layout of a Korean installation.

160

Fig. 8 shows the block diagram for the hot mill process at the point where hot slabs for rolling are produced. Slabs for rolling are produced in a continuous sequence by the Casting Shop.

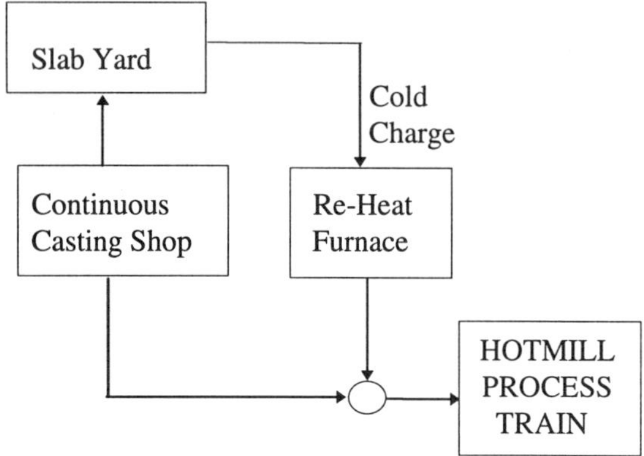

Fig. 8 : **Slab Production for Hot Strip Rolling Mill**

These slabs are either required immediately and transported at about 1200°C to the hotmill process train or are kept in reserve and despatched to the Slab Yard where they are allowed to cool to ambient temperature. Due to a requirement for special slabs, or problems in the Casting Shop, cold slabs are drawn from the Slab Yard and reheated to approximately 1200°C by pusher mechanised reheat furnaces. Clearly the reheat furnaces are high energy consumers and hence it is necessary to optimise and balance the flow of slabs coming directly from the Casting Shop and those requiring reheating and processing by the reheat furnaces. Operational delays in the hotmill or in the slabbing operations can produce significant energy wastage, and economic loss. This particular example shows how performance quality indices can be used in an off-line manner to achieve optimized process operation. The components in a study to optimise process performance are as follows:

(i)   *Simple process models*
(ii)   *Performance quality indices for energy and material consumption costs.*
(iii)   *Constraints on process conditions and operation*
(iv)   *Mixed discrete and continuous process simulation.*
(v)   *Optimized operational strategy obtained by testing different scenarios to optimise the performance quality indices.*

Performance quality indices are viewed as one technique for extending the methods of statistical process control into the field of optimising process performance. Such indices are not the usual mathematical norms used in numerical optimisation procedures but are designed to capture the objectives of good process operation. Online or offline they can be used as though they were an additional measurement. They are usually simple and are therefore easy to display (plot) and track. Performance quality indices have three useful features:

(i)   *For online optimisation.* In complex processes, such indices can be used by operators and engineering staff to nudge the process into a more economical operating condition.

(ii)   *As early-warning indicators.* As in the control charts of SPC, performance quality indices might be used to detect the onset of abnormal conditions producing uneconomical process operation.

(iii)   *As online and offline management analysis tools.* Since these indices are often closely related to monetary operating costs, they can be used by management as a real-time indicator of the economic performance of process units and operational procedures.

4.3   *Controller Performance Indices*

Statistical process control and performance quality indices are just two methods in the portfolio of process performance monitoring techniques. These methods tend to concentrate on the quality of process performance although they can be used to indicate unsatisfactory controller performance. However, there are a small number of methods in development with the objective of specifically quantifying good or bad controller performance leading to a recommendation as to whether controller retuning is necessary.

4.3.1   *Åström's Knowledge-Base Approach*

Following the success of the *autotune* method[20,] Åström has devoted some publications to the concepts of *expert control*. One contribution to this research has been to initiate a knowledge-based approach for the achievable performance obtainable with PID controllers on single loops.[21] The idea is that armed with different levels of information about the process dynamics, a knowledge-base can be used to assess the potential achievable performance from different controller structures (for example, P, PI, PD, PID). This information can be compared with actual performance to see if re-tuning or re-configuring the controller is required. Consequently, the approach is investigative as shown in Fig. 9 with a long term objective of producing intelligent PID control.

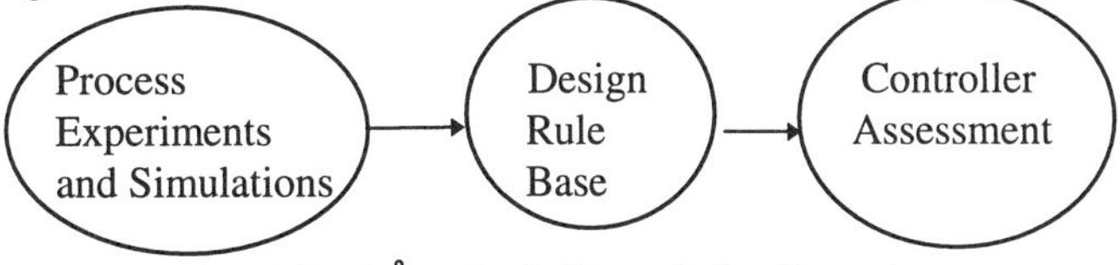

Fig. 9 **Åström's Knowledge-Base Approach**

162

4.3.2 *A Real-Time Controller Monitoring Procedure*

Ortega, et al[15] have devised an automated procedure to switch on the adaptation of a parameterized controller when this will improve the controller performance. This procedure thus not only attempts to determine when control parameter update is desirable but also initiates the update automatically. A schematic is shown in Fig. 10. The main components of the procedure are:

(i)   A performance index $J(t,\gamma)=\dfrac{1}{2}\int_0^t e^2(\tau)d\tau$ where the error

$e(t) = r(t) - y(t)$, and $\gamma(t)$ is an adaptation switch ; $\gamma(t)>0$ implies adaptation is active and inactive for $\gamma(t)\leq 0$.

(ii)   To determine whether the adaptation should be active, the sensitivity of $J(t,\gamma)$ with respect to γ is evaluated. *Ortega, et al* devised an explicit

expression for this sensitivity $S_\gamma = \left.\dfrac{\partial J}{\partial \gamma}\right|_{\gamma=0}$. Thus if $S_\gamma < 0$ then the

performance index would decrease if adaptation were initiated.

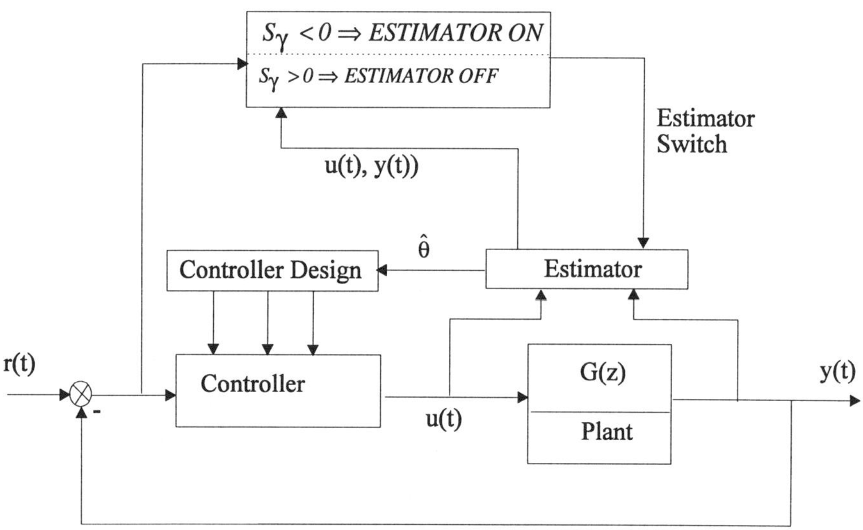

Fig. 10 Schematic for a Controller Monitoring Procedure

The explicit expression for performance index sensitivity enabled the procedure to be implemented in real-time. An interesting set of experimental results is given for the PI control of laboratory scale heat exchanger[15]. Despite some practical problems this represents a significant contribution to the concept of controlled performance monitoring using a performance quality index.

4.3.3   *Controller Assessment Benchmarks*

The use of an optimisation framework based on the deterministic Linear Quadratic (LQ) and stochastic Linear Quadratic Gaussian (LQG) optimal control problem for various types of controller and filter assessments has been long standing.[22,23]   The key concept which enables these optimal control methods to be used is that of maximal accuracy. For example, the LQ optimal control problem is defined as:

$$Minimise J(\rho) = \frac{1}{2}\int_{o}^{\infty}\left\{y^{T}(t)Qy(t) + \rho u^{T}(t)Ru(t)\right\}dt$$

$$\text{subject to } S: \begin{cases} \dot{x} = Ax + Bu \quad x(o) = x_{o} \\ y = Cx \end{cases}$$

Maximal accuracy is then an analysis of the behaviour of the optimal control cost as the control weighting ρ becomes vanishingly small. In a practical sense this limit is interpreted as representing the best possible process performance as the controller action is permitted to become unrealistically ideal. Whilst these benchmark concepts retain their theoretical interest for some researchers,[24,25], little progress has been made in turning these results into online monitoring tools.

This situation was substantially changed by the recent contributions of *Harris, et al.*[14,26]   In this work the benchmark concept used the framework of minimum variance control. This was enjoined with the development of a performance monitoring procedure for *online* implementation. That this continues the maximal accuracy lineage is clear because minimum variance control can be conceptually related to the methods originating from the general cost function:

$$J = E\{y^2(t+k|t) + \rho u^2(t)\}, \text{ where } \rho = 0.$$

The *Harris, et al* procedure starts from the basic minimum variance control problem and its solution:

Given system description:

$$A(z^{-1})y(t) = z^{-k}B(z^{-1})u(t) + C(z^{-1})e(t)$$

where z^{-1} represents a one step delay operator.

164

Introduce polynomials:

$$F(x) = 1 + f_1 x + \ldots + f_{k-1} x^{(k-1)}$$

$$G(x) = g_o + g_1 x + \ldots + g_{ng} x^{ng}$$

with $n_g = max(n_a\text{-}1, n_c\text{-}k)$; $n_a = deg(A)$; $n_c = deg(C)$.

Introduce Diophantine equation:

$$C(x) = A(x)F(x) + x^k G(x)$$

The minimum variance control strategy is devised to minimise the mean square output performance cost index:

$$J_{MV} = E\{y^2(t+k|t)\}$$

The analysis is as follows:

(i)   Advance the output equation:

$$y(t+k) = \frac{B(z^{-1})}{A(z^{-1})} u(t) + \frac{C(z^{-1})}{A(z^{-1})} e(t+k)$$

(ii)   Use Diophantine equation and re-arrange to obtain:

$$y(t+k) = \left[\frac{B(z^{-1})F(z^{-1})}{C(z^{-1})} u(t) + \frac{G(z^{-1})}{C(z^{-1})} y(t) \right] + F(z^{-1})e(t+k)$$

(iii)   Introduce k-step ahead predictor:

$$\hat{y}(t+k|t) = \frac{B(z^{-1})F(z^{-1})}{C(z^{-1})} u(t) + \frac{G(z^{-1})}{C(z^{-1})} y(t)$$

(iv)   Pursue a cost function analysis:

$$J_{MV} = E\{y^2(t+k|t)\}$$

$$= E\{(\hat{y}(t+k|t) + F(z^{-1})e(t+k))^2\}$$

$$= E\{\hat{y}^2(t+k|t)\} + (1 + f_1^2 + \cdots + f_k^2)\sigma_e^2$$

The cost J_{MV} is minimised if $\hat{y}^2(t+k|t) = 0$ and hence:

Optimal control is: $u^o(t) = -\left[\dfrac{G(z^{-1})}{F(z^{-1})B(z^{-1})}\right]y(t)$

Optimal cost is: $\quad J_{MV}^0 = (1 + f_1^2 + \cdots + f_{k-1}^2)s_e^2$

The use of a minimum variance approach to a benchmark requires the pre-condition that the system is minimum phase, since the inverse $B(\cdot)$ polynomial appears in the control law (see above).

However, the benchmark value is given by:

$$J_{MV}^0 = (1 + f_1^2 + \cdots + f_{k-1}^2)\sigma_e^2$$

If a controller is able to reduce the mean square output error variance close to the value of J_{MV}^0, then the implemented controller is considered adequate and the cause of poor process performance must be sought elsewhere. In a later work, *Desborough and Harris*[26] introduced two more innovations:

(i) The minimum variance optimal cost was used in a normalised controller performance index,

$$0 \le \eta(k) = \frac{mse(y(t)) - J_{MV}^0}{mse(y(t))} \le 1$$

  where *mse(y(t))* is the mean square error for the process output, *y(t)*. Thus if η*(k)* is near zero then the applied controller is operating *close* to a minimum variance controller, and performance improvements arising from adjusting the controller will be minimal. However, if η*(k)* is nearer unity, then adjustment or re-tuning of the implemented controller should bring improved performance.

(ii) The second innovation was to devise a *recursive* scheme for estimating η*(k)* online. This enabled the method to be used as a performance monitoring procedure which has a very specific controller assessment interpretation. It is this potential of the method which makes it such a promising monitoring routine.

5.   SUMMARY

This paper reports work devoted to establishing a discipline of modern supervisory control for large scale industrial systems and processes. The basic concept of the hierarchy was used to structure the presentation of technology, control and information flows. This was followed by an overview section which delineated the topics of supervisory command and control, performance monitoring and fault detection. The latter was not considered again in the paper.

A slightly more detailed discussion of supervisory command and control followed. In this, the basic supervisory command structure was outlined and defined. The use of constrained model based predictive control algorithms was identified as a recent contribution to advance the methods used in dynamic setpoint transitions. This utilises a convenient constrained quadratic programming formulation to automate the setpoint changeover:

The main exposition in the paper concerned the techniques of performance monitoring. In contrast to the methods of fault detection, performance monitoring methods are a set of procedures which can be used to *detect* and *analyse* both *adverse* and *beneficial* process conditions. The philosophy of statistical process control was described. This method forms a framework and guide for the development of new online techniques.

The method of performance quality indices was proposed and examples given for online and offline use. These indices tend to be either a cost index or a performance index. They are designed to capture desirable process properties and are to be used to infer departures from non-optimal conditions. Sometimes they are used in numerical optimization algorithms to give optimised process conditions. In this case the performance quality index has the role of a non-linear cost function. In other arrangements *ad hoc* online optimization by process operators and engineers occurs.

Another set of performance monitoring techniques were described to assess controller performance. The objective here was to distinguish when retuning the controller used on a loop could give improved performance benefits. Three different approaches were describe : an investigative knowledge-based approach due to *Åström*, an automated controller monitoring procedure due to *Ortega et al* and a controller benchmark framework with emphasis on recent seminal work by *Harris et al*.

In conclusion, performance monitoring techniques like Statistical Process Control, the Performance Quality Indices and the new Control Monitoring Procedures are important and valuable modern supervisory tools. These techniques have the following uses:

- *Identifying good process conditions*
- *Detecting poor process performance*
- *Some fault detection capability*
- *On-line (nudge) optimization method*
- *Controller tuning diagnostic applications*
- *Management analysis tool*

6. Acknowledgements

The authors wish to thank the EPSRC for their financial support of award GR/H/44431 : Robust procedures of the operation and optimization of combined cycle of power generation. The authors are also extremely grateful to Mrs. Ann Frood for typing the manuscript of this paper with such skill, accuracy and speed.

7. References

1. E.O. Doebelin, *Control System Principles and Design*, John Wiley, New York, 1985 (ISBN 0-471-08815-3).

2. K.M. Middaugh, A comparison of industrial communications networks, *IEEE Trans. Ind. Applics,* **29**,. No. 5, pp. 846-853, Sept/Oct, 1993.

3. S. Pirie, M.A. Johnson, C.D. Grant, B.E. Postlethwaite, Simulation and Control of a Hydrocarbon Processing Plant, *Workshop CIM in the Process Industry*, Athens, 1991.

4. M.R. Katebi and M.A. Johnson, Predictive Control Design for Large-Scale Systems, *IFAC Conference on Integrated Systems Engineering*, Baden-Baden, Germany, 1994.

5. T.H., Tsai, J.W. Lane and C.S. Lin, *Modern Control Techniques for the Processing Industries*, Marcel Dekker, New York, 1990, (ISBN 0-8247-7549-X).

6. D. Popovic and V.P.Bhatkar, *Distributed Computer Control for Industrial Automation,* Marcel Dekker, New York, 1990, (ISBN 0-8247-8118-X).

7. C.L. Mamzic and T.W. Tucker, Incorporating statistical process control within a distributed control systems, *43rd Annual Symposium : Instrumentation for the Process Industries*, Tech. Paper 3912, Texas, 1988.

8.   J. Gertler, Analytical redundancy methods in fault detection and isolation, *IFAC Conference on Fault Detection, Supervision and Safety for Technical Processes*, Baden-Baden, Germany, 1991.

9.   D.E. Seborg, T.F. Edgar and D.A. Mellichamp, *Process Dynamics and Control*, John Wiley and Sons, New York, 1989 (ISBN 0-471-86389-0)

10.   M. Uchida, N. Nakamura and K. Kawai, Application of linear programming to thermal power plant control, *IFAC 8th Triennial World Congress*, Vol. XX, Paper 97.12, Kyoto, Japan.

11.   D.C. Montgomery, *Statistical Quality Control* , John Wiley and Sons Ltd., New York, 1991, (ISBN 0-471-52993-1)

12.   V.A. Bhandari, R. Paradis and A.C. Saxena, Using performance indices for better control, *source unknown*, ca 1990

13.   S.B.Dolins and J.D. Reese, A curve interpretation and diagnostic technique for industrial processes, *IEEE Trans. Ind. Applics*, **28**, No. 1, 261-267, Feb, 1992.

14.   T.J. Harris, Assessment of control loop performance *Can. Journal Chem. Engrg*, **67**, 856-861, Oct., 1989.

15.   R. Ortega, G. Escobar, and F. Garcia. To tune or not to tune? : A monitoring procedure to decide, *Automatica,* **28**, No. 1, 179-184, 1992.

16.   M.A. Johnson, M.R. Katebi, C. Cloughley and R. Farnham, Integrated Control of Power Generation Plant : New Research Directions, *2nd IEEE Mediterranean Symposium on New Directions in Control and Automation, Crete*, June 1994.

17.   J.B. Rawlings and K.R. Muske, The stability of constrained receding horizon control, *IEEE Trans. Automatic Control*, **38,** No. 10, 1512-1516, Oct., 1993.

18.   G. Dieck-Assad, G.Y. Masada,. and R.H. Flake, Optimal setpoint scheduling in a boiler-turbine system, *IEEE Trans. Energy Conversion*, **2**, No.3, 388-295, 1987.

19.   Y-W Yoon, *Energy optimization of hot rolling process using discrete event simulation,* MPhil Thesis (In Preparation), Industrial Control Centre, University of Strathclyde, Glasgow, Scotland, U.K., 1995.

20.   T. Hägglund and K.J. Åström, U.S. Patent 4,549,123 ; Method and an apparatus in tuning a PID regulator, October 22nd, 1985.

21. K.J. Åström, Assessment of achievable performance of simple feedback loops, *Int. Journal of Adaptive Control and Signal Processing*, **5**, No. 1, 3-20, Jan-Feb, 1991.

22. M.A. Johnson and M.J. Grimble, On the maximal accuracy of linear optimal systems, *IMA Journal of Mathematical Control and Information*, **1**, 95-106, 1984.

23. M.A. Johnson and M.J. Grimble, On measurement structures and the maximal accuracy of linear filters, *21st IEEE Conference on Decision and Control*, Session TP5, 297-204, Orlando, Florida, USA, 1982.

24. M.A. Johnson, Bounds for an optimization problem with a linear programming structure, *J. Inst. Maths. Applics.*, **124**, 295-410, 1979.

25. R.J. Veillette, Reliable linear-quadratic state feedback control, *Automatica,* **31**, No. 1, 137-143, 1995.

26. L. Desborough and T. Harris, Performance assessment measured for univariate feedback control, *The Canadian Journal of Chemical Engineering*, **70**, 1186-1197, 1992.

WSSIAA 5 (1995) pp. 171–175
© World Scientific Publishing Company

PLANNING FOR THE "BIG ONE"

ROBERT KALABA

*Departments of Economics and Biomedical Engineering
University of Southern California, Los Angeles, CA 90089*

and

RONG XU

*School of Urban and Regional Planning
University of Southern California, Los Angeles, CA 90089*

ABSTRACT

In this note, the tool of the calculus of variations is used to determine optimal schedules in preparation for a major earthquake. The simple model presented here is one of the many applications of the dynamic optimization techniques to planning and management problems.

1. Introduction

Seismic forecasters have made a prediction that there will be a great earthquake in California during the next quarter century with probability near unity. A central point in preparing for this catastrophe is to select the appropriate mode of spending in preparation. This note is devoted to a simplified model that aims at minimizing the expected cost of this earthquake. It incorporates seismic knowledge in the form of a probability function $P = P(t)$ which gives the probability that the earthquake will occur by time t or sooner. It also relates, crudely, the rate of spending at time t on the total of medical services, fire protection, and police protection to ameliorating the earthquake's physical effects. Also considered is the worst possible form of the probability function, since the form of the function $P(t)$, $0 \leq t \leq T$, may not be accurately known.

2. The Model [1,2]

Suppose it is definite that a large earthquake will occur by time T, say, a quarter century. Let $P(t)$ be the probability of the earthquake by time t with $P(0) = 0$ and $P(T) = 1$. Accordingly, $\dot{P}(t) dt$ is the probability that the temblor will occur in the interval $(t, t + dt)$.

Assume that in preparing for the earthquake, the amount of all expenses spent by time t ($0 \leq t \leq T$) is $x(t)$. These expenses include, among others, the expenses spent on medical services, fire protection, and police protection. Then $\dot{x}(t)$ is the spending rate at time t. Generally speaking, the greater the rate of spending on preparation at time t, the less the physical damage will be if the earthquake really occurs at that time. That is to say, the amount of damage done ought to be in inverse proportion to the spending rate $\dot{x}(t)$. For simplicity, let it take the form $1/\dot{x}(t)$. Consequently, the total cost by time t if the earthquake occurs at time t is $x(t)+1/\dot{x}(t)$. The expected total cost from time 0 to T is thus given by the integral

$$\int_0^T [x(t)+1/\dot{x}(t)]\dot{P}(t)dt. \tag{1}$$

We wish to choose the optimal spending function $x = x(t)$, so that the expected total cost from time 0 to T is minimal. In other words, we want to choose $x(t)$ $(0 \le t \le T)$ to minimize the above integral, subject to the boundary conditions

$$x(0) = 0, \tag{2}$$

and $x(T)$ is free.

3. Analysis [3]

Essentially, the model under consideration is a problem in the calculus of variations with a free end condition. To solve it, we may use the necessary conditions for the function $x(t)$ to be optimal. These conditions include the Euler equation, the initial condition, and the transversality condition.

The Euler equation in this case is

$$\dot{P}(t) - \frac{d}{dt}[-\dot{P}(t)/\dot{x}^2(t)] = 0, \tag{4}$$

or

$$P(t) + \dot{P}(t)/\dot{x}^2(t) = k, \tag{5}$$

where k is an arbitrary constant. The transversality condition is

$$\dot{P}(T)/\dot{x}^2(T) = 0, \tag{6}$$

Assuming $\dot{P}(T) \ne 0$, this condition implies that $\dot{x}(T) = \infty$. Eq. 6, together with the knowledge that $P(T) = 1$, tells us that the value of k in Eq. 5 is actually

$$k = P(T) + \dot{P}(T)/\dot{x}^2(T) = 1+0 = 1. \tag{7}$$

Hence, Eq. 5 turns out to be

$$P(t) + \dot{P}(t)/\dot{x}^2(t) = 1. \tag{8}$$

We thus see that

$$\dot{x}(t) = \sqrt{\dot{P}(t)/[1-P(t)]}. \tag{9}$$

which is the optimal rate of spending on earthquake preparation. Notice that $\dot{x}(t)$ is a function of $P(t)$ and its first-order derivative $\dot{P}(t)$. In casual terms, the radicand is the probability of immediate occurrence at time t given no earthquake up to then.

Substituting Eq. 9 into Eq. 2, we have

$$\begin{aligned}
C &= \int_0^T [x(t)+1/\dot{x}(t)]\dot{P}(t)dt = \int_0^T [x+\sqrt{(1-P)/\dot{P}}]\dot{P}\,dt \\
&= \int_0^T x\dot{P}\,dt + \int_0^T \dot{P}\sqrt{(1-P)/\dot{P}}\,dt \\
&= \int_0^T x\dot{P}\,dt + \int_0^T \sqrt{\dot{P}(1-P)}\,dt. \tag{10}
\end{aligned}$$

By using integration by parts, the first term in Eq. 10 becomes

$$\int_0^T x\dot{P}\,dt = xP\big|_0^T - \int_0^T P\dot{x}\,dt = x(T) - \int_0^T P\dot{x}\,dt = \int_0^T \dot{x}\,dt - \int_0^T P\dot{x}\,dt$$

$$= \int_0^T (1-P)\dot{x}\,dt = \int_0^T (1-P)\sqrt{\dot{P}/(1-P)}\,dt$$

$$= \int_0^T \sqrt{(1-P)\dot{P}}\,dt. \tag{11}$$

Substituting Eq. 11 back into Eq. 10 then gives

$$C = \int_0^T \sqrt{(1-P)\dot{P}}\,dt + \int_0^T \sqrt{\dot{P}(1-P)}\,dt$$

$$= 2\int_0^T \sqrt{\dot{P}(1-P)}\,dt, \tag{12}$$

which is the least expected total cost given the probability function $P(t)$.

4. The Worst Probability Function and The Largest Value of The Minimum Expected Total Cost

Consider now the worst possible case, in which Nature chooses the probability function of the earthquake, $P(t)$, to maximize the expected total cost given by Eq. 12. Put in another way, $P(t)$ is to be chosen so as to maximize the integral

$$\int_0^T \sqrt{\dot{P}(1-P)}\,dt, \tag{13}$$

subject to the initial condition

$$P(0) = 0 \tag{14}$$

and the terminal condition

$$P(T) = 1. \tag{15}$$

This is a simplest problem in the calculus of variations. Let $F = \sqrt{(1-P)\dot{P}}$. Then, the Euler equation in this case can be written as

$$F - \dot{P}F_{\dot{P}} = \text{constant}, \quad 0 \le t \le T, \tag{16}$$

or

$$\sqrt{(1-P)\dot{P}} - \dot{P}\frac{1-P}{2\sqrt{(1-P)\dot{P}}} = k_1, \tag{17}$$

where k_1 is an arbitrary constant. Multiplying both sides of Eq. 17 by $2\sqrt{(1-P)\dot{P}}$ gives

$$2(1-P)\dot{P} - \dot{P}(1-P) = 2k_1\sqrt{(1-P)\dot{P}}, \tag{18}$$

which can be reduced to

$$(1-P)\dot{P} = 2k_1\sqrt{(1-P)\dot{P}}, \tag{19}$$

or

$$\sqrt{(1-P)\dot{P}} = n, \tag{20}$$

where $n = k_1$. Squaring both sides of Eq. 20 produces

$$(1-P)dP/dt = n^2. \tag{21}$$

Separating variables in Eq. 21 then gives

174

$$(1-P)dP = n^2 dt. \tag{22}$$

Integrating both sides of Eq. 22, we obtain

$$\int (1-P)dP = \int n^2 dt, \tag{23}$$

or

$$P - P^2/2 = n^2 t + m, \tag{24}$$

where m is an arbitrary constant. Applying the boundary conditions $P(0) = 0$ and $P(T) = 1$ to Eq. 24 provides

$$m = 0, \tag{25}$$

and

$$n^2 = \frac{1}{2T}. \tag{26}$$

Finally, we substitute Eqs. 25 and 26 into Eq. 24 and obtain

$$P - \frac{P^2}{2} = \frac{t}{2T}, \tag{27}$$

or

$$P^2 - 2P + t/T = 0. \tag{28}$$

The roots of the quadratic Eq. 28 are

$$P = \frac{2 \pm \sqrt{4 - 4t/T}}{2} = 1 \pm \sqrt{1 - t/T}, \qquad 0 \le t \le T. \tag{29}$$

Given the knowledge that $0 \le P \le 1$, the only possible solution is

$$P(t) = 1 - \sqrt{1 - t/T}, \qquad 0 \le t \le T, \tag{30}$$

which is the worst probability function of the earthquake that Nature can choose (See Figure 1). This worst probability function leads to the largest value of the minimum expected total cost

$$C_{largest} = 2\int_0^T \sqrt{\dot{P}(1-P)}\, dt = 2\int_0^T \sqrt{\left(-\frac{-1/T}{2\sqrt{1-t/T}}\right)\sqrt{1-t/T}}\, dt$$

$$= \int_0^T \sqrt{2/T}\, dt = \sqrt{2T}. \tag{31}$$

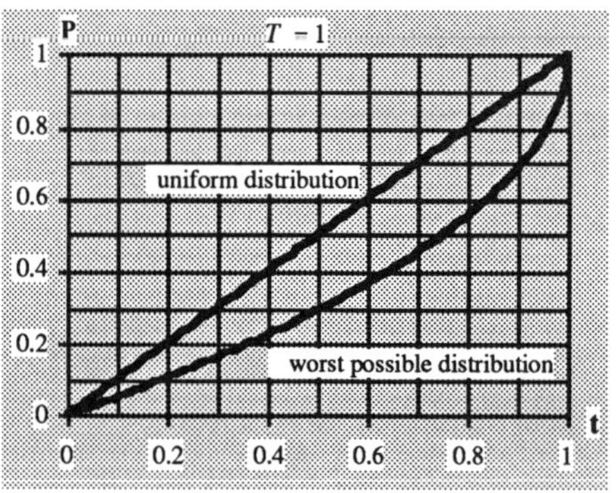

Figure 1. Cumulative Probability that the Earthquake Will Occur by Time t or Earlier

For comparison, let us consider the case where the probability that the earthquake will occur is uniformly distributed over time. That is, let

$$P(t) = t/T, \quad 0 \le t \le T. \tag{32}$$

Then, according to Eq. 12, the minimum expected total cost will be

$$C_{uniform} = 2\int_0^T \sqrt{\dot{P}(1-P)}\,dt = 2\int_0^T \sqrt{(1/T)(1-t/T)}\,dt = 2\int_0^T \sqrt{(T-t)/T^2}\,dt = \frac{4}{3}\sqrt{T}. \tag{33}$$

Comparing Eq. 33 with Eq. 31, we see that there really is a lowering of the cost when Nature departs from the total cost given in Eq. 31. It is noteworthy that these minimal costs depend upon the square root of T.

5. Concluding Remarks

The model of earthquake preparation developed in this note is extremely simple. Much can be done to improve the realism. In particular, the aggregate spending stream considered can be subdivided into three expenditures, one for medical service, one for fire and one for police protection at the time of the event. More realistic assumptions on the form of the damage function can also be chosen. Automation of the solution of such problems in the calculus of variations is considered in Kagiwada et al. [4] and Taylor et al. [5] Extensions, including discounting, will be discussed in future communications.

6. References

1. M. I. Kamien and N. L. Schwartz, *Dynamic Optimization*, North-Holland, New York, 1985, 50-52
2. J. B. Keller, Optimum Checking Schedules for Systems Subject to Random Failure, *Management Science* **2** (1974), 256-260.
3. R. Kalaba and K. Spingarn, *Control, Identification and Input Optimization*.
4. H. Kagiwada, R. Kalaba, N. Rasakhoo and K. Spingarn, *Numerical Derivatives and Nonlinear Analysis*, Plenum Press, New York, 1986.
5. G. Taylor, S. Luk and R. Kalaba, *CALVAR: Automatic Solution of Variational Problems -- with Economic Applications*, Forthcoming.

WSSIAA 5 (1995) pp. 177–192
© World Scientific Publishing Company

177

Dynamic Nonlinear Optimisation Using the Gate Function Approach

By

M R Katebi[1], J C Kalkkuhl and M J Grimble

Industrial Control Centre, Strathclyde University, Glasgow, G1 1QE, UK.

Abstract

The Wiener model of a nonlinear process is an optimal input/output representation of the system when the input is a Gaussian process. If the control input is not Gaussian, the Wiener model parameters are difficult to calculate. To circumvent this problem a set of gate functions, whose orthognality does not depend on the statistical properties of the input, is used. This paper is concerned with the application of the gate function approach to the identification and control of nonlinear systems. The model parameters are estimated by minimising the mean-square error between the measurement and the model output. An interpolation scheme is employed to update the model output and a predictive type controller is used to design a nonlinear compensator. The novel control design approach is applicable to general nonlinear systems and no a priori information is required. Simulation results suggest that the gate function approach is superior to the Volterra series based methods, neural network and fuzzy logic techniques for some applications. The proposed method is applied to pH control problems.

1. Introduction

The structure of a generalised Wiener model for a SISO nonlinear system with finite memory is shown in Fig. 1. The model consists of a SIMO linear model in series with a MISO nonlinear element. The linear part represents the memory of total system which is nonlinear. A set of orthogonal transfer functions such as a Laguerre network is usually used to represent this part of the model. The orthognality of the signals $v_1(t),...,v_m(t)$ is necessary to ensure that the gains in the nonlinear part are calculated independently. The nonlinear part consists of a network representing the static nonlinearities and a set of coefficients are used to weight the outputs of the network. A finite dimensional network of nonlinear elements is used to represent the static nonlinearity. A set of polynomials of order N is used in the original Wiener model and the gain vector of the parameters is found by minimising the mean-square error or other possible cost functions.

---

[1] *Author for correspondence*

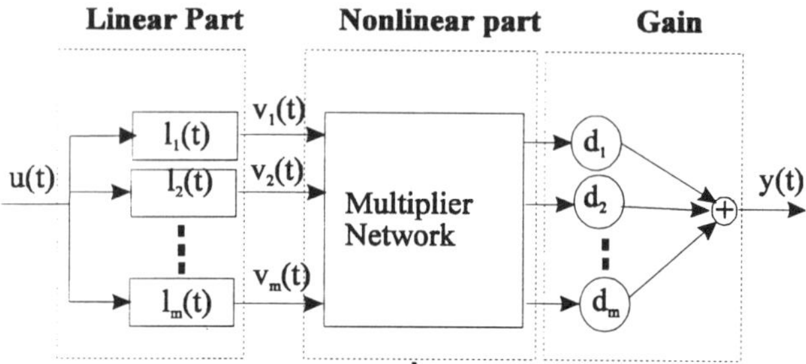

Fig 1 The generalised Wiener Model

This model is an optimal representation of the system if the input u(t) is Gaussian and the model parameters are found by minimising the mean-square error. In practice, the control input is not Gaussian and the Wiener model parameters are difficult to calculate. To circumvent this problem a set of gate functions, whose orthognality does not depend on the statistical properties of the input is usually used. The gate function model was developed by Bose (1954) as an alternative to the original Wiener white noise approach. A comparison of the gate function model with the Wiener model was given by Schetzen (1986). The need to estimate a large number of parameters to obtain a reasonable accuracy in the model output and the high computational load have been the main obstacles in the application of the gate function model to nonlinear control problems. Kalkkuhl (1993) has proposed the use of interpolation schemes (from the theory of the numerical solution of partial differential equations) to improve the output accuracy and to reduce the number of parameters. This technique is extended below to design nonlinear controllers based on the gate function model.

The results below are concerned with nonlinear control design using the gate function model. The controller is developed using model-based predictive control techniques. The special form of the output relationship and the structure of the gate function model using the interpolation scheme proposed by Kalkkuhl (1993) enables the derivative term $\partial y / \partial v$ to be easily calculated. This derivative can then be used to develop a time-varying linear model of the nonlinear system. A time-varying Kalman filter is then used to estimate the state variables. The prediction of the state variables and the output are based on the current estimates from the Kalman filter. A standard predictive control cost criterion is then minimised to find the optimal control signal. The design method is used to solve a pH control problem. Simulation results are presented and the advantages and disadvantages of the gate function approach are discussed.

2. The Gate Function Model

Consider the set of one-dimensional gate functions R_n and a bounded scalar input, the time functions $R_n[v(t)]$ are defined as[2] :

$$R_0[v(t)] = \begin{cases} 1 \text{ for } -\infty < v(t) < V_1; \\ 0 \qquad \text{otherwise} \end{cases}$$

$$R_i[v(t)] = \begin{cases} 1 \text{ for } V_i \le v(t) < V_{i+1}; \ i = 1,\dots N-1 \\ 0 \qquad \text{otherwise} \end{cases} \tag{1}$$

$$R_N[v(t)] = \begin{cases} 1 \text{ for } v_N \le v(t) < \infty; \\ 0 \qquad \text{otherwise} \end{cases}$$

where $V_i \in [V_1, V_N]$

It is clear that only one of the functions $R_n[v(t)]$ has the value of one at any instant of time and all other functions are zero. If $R_n[v(t)]$ operates on v(t), the signal v(t) may now be approximated as a staircase signal:

$$v(t) = \sum_{i=0}^{N} a_i R_i[v(t)] \tag{2}$$

where a_i's are selected such that an error criterion is minimised.

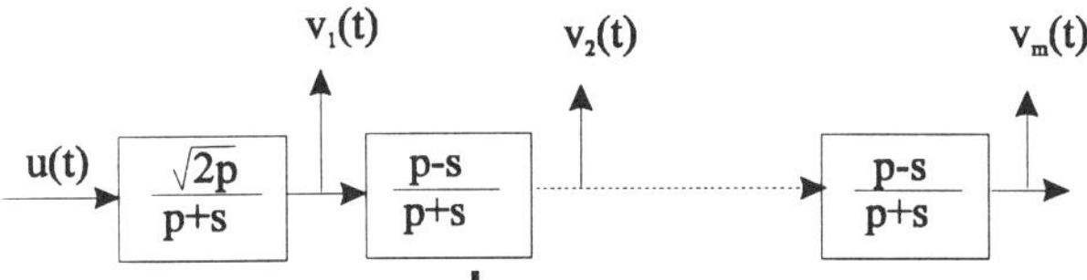

Figure 2 A Laguerre filter network with m elements

Consider now the case where the input of the nonlinear system, u(t), is passed through a network of Laguerre filters of dimension m to model the memory of the nonlinear process as shown in Fig. 2. If the linear section consists of a linear system of order m, then the output of the Wiener model can be approximated by:

180

$$y(t) \approx \sum_{k=1}^{\infty} \sum_{i_1=1}^{m} \ldots \sum_{i_k=1}^{m} A_{i_1 \ldots i_k} v_{i_1} \ldots v_{i_k} \tag{3}$$

Inserting Eq. 2 into Eq. 3, any step approximation to y(t) can be expressed as:

$$y(t) \approx \sum_{k=1}^{m} \sum_{i_1=0}^{N} \ldots \sum_{i_k=0}^{N} C_{i_1 \ldots i_k} R_{i_1}[v_{i_1}] \ldots R_{i_k}[v_{i_k}] \tag{4}$$

Using the relationship $\sum_{n=0}^{N} R_n[v_i(t)] = 1$, it can be shown that:

$$\sum_{i_m+1=0}^{N} \ldots \sum_{i_m=0}^{N} R_{i_m+1}[v_{i_1}] \ldots R_{i_m}[v_{i_m}] = 1 \tag{5}$$

Eq. 4 can then be simplified to obtain:

$$y(t) \approx \sum_{i_1=0}^{N} \ldots \sum_{i_m=0}^{N} D_{i_1 \ldots i_m} R_{i_1}[v_{i_1}] \ldots R_{i_m}[v_{i_m}] \tag{6}$$

Now, define the m-dimensional gate function as:

$$S_\alpha[v] = R_{i_1}[v_1(t)] \ldots R_{i_m}[v_m(t)]; \alpha = (i_1, \ldots, i_n), v = [v_1(t), \ldots, v_m(t)] \tag{7}$$

Then, the gate function model can be written as:

$$\begin{aligned} v(k+1) &= A_s v(k) + B_s u(k) \\ y(t) &= \sum_\alpha D_\alpha S_\alpha[v] \end{aligned} \tag{8}$$

3. Model Identification

The parameters of the gate function model, D_α can be determined by minimising the mean value of an arbitrary error function[2]:

$$J[e(t)] = F[z(t) - \sum_\alpha D_\alpha S_\alpha[v]] \tag{9}$$

where z(t) is the measurement and the error e(t) is defined as:

$$e(t) = z(t) - \sum_{\alpha} D_{\alpha} S_{\alpha} [v]$$ (10)

Assuming the mean-square error is to be minimised, the identification criterion can be written as:

$$F[e(t)] = [z(t) - \sum_{\alpha} D_{\alpha} S_{\alpha} [v]]^2 = [z(t) \sum_{\alpha} S_{\alpha} [v] - \sum_{\alpha} D_{\alpha} S_{\alpha} [v]]^2$$
$$= \{\sum_{\alpha} [z(t) - D_{\alpha}] S_{\alpha} [v]\}^2 = \{\sum_{\alpha} [z(t) - D_{\alpha}]^2 S_{\alpha} [v]\}$$ (11)

where the relation $\sum_{\alpha} S_{\alpha} [v] = 1$ is used to simplify the expression. The mean square error

$J = \overline{e^2}$ can now be written as:

$$J = \overline{F[e(t)]} = \overline{\{\sum_{\alpha} [z(t) - D_{\alpha}]^2 S_{\alpha} [v]\}}$$ (12)

Since J is a convex function, the sufficient condition for minimising J with respect to the model parameters D_{α} is to make the derivative $\partial J / \partial D_{\alpha} = 0$. This implies that:

$$\overline{[z(t) - D_{\alpha}] S_{\alpha} [v]\}} = 0 \implies D_{\alpha} = \frac{\overline{z(t) S_{\alpha} [v]}}{\overline{S_{\alpha} [v]}}$$ (13)

The optimal coefficient D_{α} can be interpreted as the average value of the desired output z under the condition that the filter output v is within the gate S_{α}. The parameters can therefore be calculated from the input-output data by an elementary filtering and averaging operation.

The accuracy of the gate function model depends on the number of the gates $(N + 1)^m$ and the order of the linear model, m. The number of gates increases much faster than the order of the linear system. The approach is therefore suitable for slow highly nonlinear systems. In order to improve the model accuracy, the interpolation techniques used for the solution of the partial differential equation is employed here to calculate the value of the output within the gates.

3.2 The Interpolation

To improve the model accuracy, mesh points are defined at the centre of each gate S_{α} and the corresponding parameter values S_{α} are assigned to these meshpoints as shown in Fig. 3.a. For the gates at the boundaries, the grid is extended equidistantly to the outside.

182

A gate function model with an m-dimensional Laguerre filter and N+1 gates in each dimension requires a grid of $(N+1)^m$ meshpoints. Once the grid is defined, an interpolation technique may be used to calculate the value of the output within the meshpoints. The m-linear interpolation functions are used here, since this type of interpolation is easy to implement for any m. Within the grid, groups of 2^m adjacent meshpoints form an m-linear C^0-interpolation element depicted in Fig 3.b for the two-dimensional case. For $(N+1)^m$ meshpoints, N^m elements are numbered consecutively by $n = 1,...,N^m$.

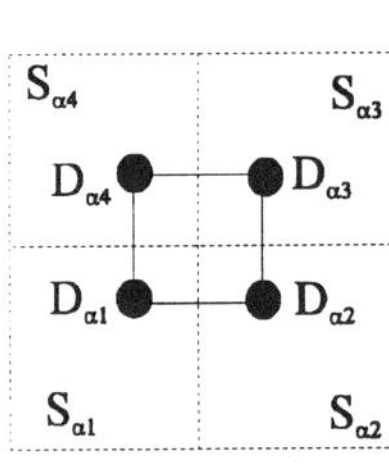
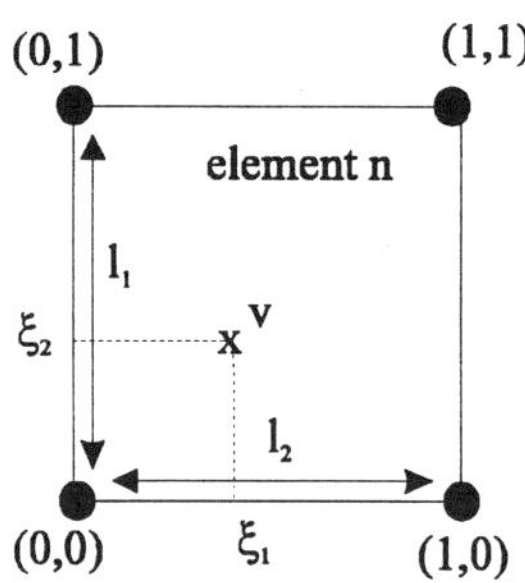

Figure 3.a A two dimensional grid.　　　Figure 3.b Bilinear C^0 element.

Each element is uniquely determined by the coordinate $(x_1,...,x_m)$ relative to a reference point P and the height, length and breadth from point P. The meshpoint with the lowest coordinate value is selected as the reference point and the geometrical dimension is defined by the lengths l_i, $i = 1,...,m$.

Each meshpoint of an element is given an index, $k = 1,...,2^m$. A local coordinate $(\xi_1,...,\xi_m)$ is assigned to each meshpoint such that the reference point has the local coordinate $(0,...,0)$ and a unit-hypercube is formed with the other meshpoints. This makes the local coordinate independent of the size of the elements and their position within the grid. Each Laguerre filter output vector $[v_1(t),...,v_m(t)]$ corresponds uniquely to one of the elements of the grid and can therefore be expressed in the local coordinate of this element using the relationship:

$$\xi_i = (v_i - x_i)/l_i; \quad i = 1,...,m. \tag{14}$$

For interpolating within an element, a 2^m-vector, λ of barycentric coordinate with the elements given for a bilinear C^0-element is formulated. The vector for m=2 is defined as:

$$\lambda = [(1-\xi_1)(1-\xi_2), \xi_1(1-\xi_2), \xi_1\xi_2, (1-\xi_1)\xi_2]^T \tag{15}$$

The scheme for generating the λ-coordinates can be easily carried over to higher dimensions[5], since there is a unique relationship between the local coordinate of a meshpoint and the pattern of the m-linear terms expressing the corresponding λ-coordinate. At a meshpoint k only λ_k assumes the value of 1 whereas all the other coordinates assume the value zero. Note, that

$$\lambda_k \geq 0, \quad \sum_{k=1}^{2^m} \lambda_k = 1 \tag{16}$$

Now if a Laguerre output vector is within an element n and has the local coordinates ξ_i then the corresponding output value y is calculated as:

$$y = \sum_{k=1}^{2^m} \lambda_k D_k \tag{17}$$

If the Laguerre filter output matches exactly a meshpoint k then y will assume exactly the value D_k. Otherwise the output y will be interpolated between the parameter values of 2^m adjacent gates.

4. Control Design

Using the interpolation scheme described in the previous section, y(t) may be defined as a polynomial in the local coordinate for any point v(t). The derivative of y with respect to v will then exist and may be found as follows:

$$\frac{\partial y}{\partial v} = \sum_{k=1}^{2^m} \frac{\partial \lambda_k}{\partial v} D_k \tag{18}$$

where

$$\frac{\partial \lambda_k}{\partial v_i} = \frac{\partial \lambda_k}{\partial \xi_i} \frac{\partial \xi_i}{\partial v_i} \tag{19}$$

Assuming the estimate of the state variables are available at time t, the output relationship of the gate function model given by Eq. 8 can be linearised around v(t) as follows:

$$y(t+1) = y(t) + \frac{\partial y}{\partial v}\bigg|_{v(t)} [v(t+1) - v(t)] \tag{20}$$

Using Eq. 8, this relationship can be simplified to:

$$y(t+1) = y(t) + [d(t)A_v - I]v(t) + d(t)B_s u(t) \tag{21}$$

where $d(t) = \dfrac{\partial y}{\partial v}\bigg|_{v(t)}$

To ensure zero steady state error in tracking control problems, the integral of the error can be introduced as a state variable in the state space model, i.e.:

$$s(t+1) = s(t) + k_i[r(t) - y(t)] \tag{22}$$

where r(t) is the reference signal generated by the reference model:

$$r(t+1) = A_r r(t) + B_r u_r(t)$$

A low pass filter can also introduced in the linear model to smooth the derivative term d(t), i.e.,

$$p(t+1) = A_f p(t) + B_f d(t) \tag{23}$$

The measurement model is defined as:

$$z(t)=y(t)+v(t) \tag{24}$$

where v(t) is white noise of zero mean value and covariance R.

Combining Eqs. 8, 10, 11 and 12, the overall time varying linear model can be written as:

$$x(t+1) = A(t)x(t) + B(t)u(t) + D\rho(t) + w(t)$$
$$z(t) = Cx(t) + v(t) \tag{25}$$

where $x = [v', y, s, r, p]$; $\rho(t) = [0, 0, u_r(t), d(t)]'$, w(t) is zero mean white noise of covariance Q and the system matrices are defined by:

$$A(t) = \begin{bmatrix} A_s & 0 & 0 & 0 \\ 0 & p(t)A_s - I & 0 & 0 \\ 0 & 0 & A_r & 0 \\ 0 & 0 & 0 & A_f \end{bmatrix}; B(t) = \begin{bmatrix} B_s \\ p(t)B_s \\ 0 \\ 0 \end{bmatrix}; C = \begin{bmatrix} 0 \\ 1 \\ 0 \\ 0 \end{bmatrix}'; D = \begin{bmatrix} 0 \\ 0 \\ B_r \\ B_f \end{bmatrix}'$$

Using the model in (14) and the measurement z(t), the following time-varying Kalman filter can be designed to estimate the states:

$$\hat{x}(t+1|t) = A(t)\hat{x}(t) + B(t)u(t) + D\rho(t) \tag{26}$$

$$\hat{x}(t+1) = \hat{x}(t+1|t) + K(t)[z(t) - C\hat{x}(t+1|t)] \tag{27}$$

$$K(t+1) = P(t+1)C^T R^{-1} \tag{28}$$

$$P(t+1|t) = A(t)P(t)A^T(t) + Q \tag{29}$$

$$P(t+1) = P(t+1|t)[C^T R^{-1} CP(t+1|t) + Q]^{-1} \tag{30}$$

The performance objective to be minimised is in standard quadratic form:

$$J = \sum_{j=t_1}^{t_2} (r_i - y_i)^T (r_i - y_i) + \lambda \sum_{j=1}^{t_u} u^2 \tag{31}$$

where $(t_2 - t_1) > 1$ is the output horizon, t_u is the control horizon, l is the weighting on the control input and n is the number of subprocesses. It is assumed that the control inputs outside the control horizon are zero and the future set points are known or set to a constant level.

Using the filter estimate at time t, the states and the outputs at any time $j > t$ may be predicted using the following relationships:

$$\hat{x}(t+j|t) = A^j \hat{x}(t) + \sum_{i=0}^{j-1} A^{(j-i-1)} Bu(t+i) \tag{32}$$

$$\hat{y}(t+j|t) = C\hat{x}(t+j|t)$$

For $j = t_1$ to t_2 the output prediction model can be written in the following matrix form:

$$Y = f + GU \tag{33}$$

where

$$Y_t' = [y'(t+n_1),..., y'(t+n_2)],$$

$$f_t' = \left\{ [CA^{n_1}\hat{x}(t)]',....,[\hat{x}(t)CA^{n_2}]' \right\},$$

$$U = [u_1(1),...,u_n(1),......,u_1(t_u),...,u_n(t_u)]'$$

$$G = [g_{ij}]; i = n_1,...,n_2; j = 1,...,n_u \quad \text{where } g_{ij} = \begin{bmatrix} CA^{j-i}B & j \geq i \\ 0 & j < i \end{bmatrix} \tag{34}$$

The constrained GPC minimises the performance objective (28) subject to constraints on the each subprocess input and output amplitudes and rates:

$$AU \le b; i = 1,...,n \tag{35}$$

By inserting the predictor model into the GPC cost index, the following performance index can be derived.

$$\min J_t = U'PU + qU + e'e \tag{36}$$

where $P = (G'G + \Lambda); q = -2e_i'G$

subject to: $A_i^c U_i \le b_i; i = 1,...,n$

This optimisation problem can be solved using linear Quadratic Programming (QP). A number of fast and efficient QP algorithms are available in the literature[1] . A method which is widely used is the active set method. In this method, the inequality constraints are divided into two sets, namely, active set and inactive set. The active set constraints are those which are violated or not satisfied. An equality constraint problem is then repeatedly solved until the optimal solution is found. This method is, however, found to be too complicated for real time applications. Another class of algorithms is based on ellipsoid of decreasing volume. These algorithms are simple to code and require only the function evaluation and the direction of at least one sub-gradient. Efficient stopping criteria are also available to find the optimal solution to any degree of accuracy[3,8] .

5. Simulation Studies

The process of neutralising a strong acid in a stirred tank can be described by the nonlinear differential equation:

$$\frac{d}{dt}x = [F_2C_2 - F_1C_1 + (F_1 + F_2 + F_w)(10^{-x} - k_w10^x)] / [V \ln 10(10^{-x} + k_w10^x)] \tag{37}$$

where x is the pH value at the outlet of the tank, F_1 is the acid in-flow (1/sec), F_2 is the caustic in-flow (control input) (1/sec), F_w is the water in-flow (1/sec), C_1, C_2 are the acid influent and the caustic influent concentrations (gmol/lit), V is the total volume of the tank and k_w is the water constant.

The pH problem is known to be a difficult nonlinear control problem[4] . Different linear control design techniques have been applied unsuccessfully to control the pH process[9] . A Volterra representation has been found to work well in the range 8<pH<10. The gate function model can be identified for a much wider range of pH values as demonstrated by the simulation studies which follow.

In the simulation studies, a sampling period of 20 seconds is used. The Laguerre filter parameter is tuned to be p=0.9656. Fig. 4 shows a comparison between the actual and the gate function model open loop responses for a sinusoidal input. The two responses are very close despite the severe nonlinearity at pH=7.

The closed-loop responses using the control scheme proposed in the previous section is shown in Fig. 5. This response demonstrates the high accuracy of the gate function model for a wide range of input and output variations. The set point change from pH=9 to pH=6 is extremely difficult to handle with existing controllers. In fact, for large changes in pH, the system is often controlled manually.

The control input is shown in Fig. 6. In these simulation, there is no constraint on the input or output signals, but these can be easily introduced using the proposed scheme. A comparison of the derivative d(t) and its filtered value is shown in Fig.7. The filtering action on the derivative term is commonly used in PID control and its main purpose is to reduce the effect of the noise which may be generated by the differentiation process.

6. Conclusions

A new approach was proposed for the identification and control of slow highly-nonlinear systems based on the gate function approach. Due to the special structure of the gate function model, a linear time-varying model could be generated at each sample time. Using this model and a quadratic cost function, the control input was found using the generalised predictive control approach. The control design problem is therefore linear at each sample time. Simulation results shows that the controller based on the gate function approach may be used to control nonlinear processes for a wide range of operating conditions.

7. References

1. M S Bazaraa, H D Sherali and C M Shetty, Nonlinear programming : *Theory and Algorithms*, Wiley-Interscience Series in Discrete math and optim., John Wiley & Sons., 1993.
2. G. Bose, *A theory of nonlinear systems*, Res. Lab. for Electronics, Tech Rep. 309, Cambridge, Massachusetts, (1954).
3. P. Boyd and C Barratt, *Linear controller design: Limits of performance*, Prentice Hall Inc., 1991.
4. G Du Mont, Y Fu and G Lu, *Nonlinear adaptive GPC and applications*, Advances in model-based predictive control, **Vol.** 2, Oxford, (1993), pp 189-213.

5. Ch. Grobmann and H G Roos, *Numerik partieller Differentialgleichungen*, B G Teubner., (1992)
6. C. Kalkkuhl, *Identification of Nonlinear systems using the gate function approach*, Strathclyde Uni, ICC Rep No. 62, (1993).
7. Schetzen, *The Volterra and Wiener theories of nonlinear systems*, John Wiley & sons, New York., 1980.
8. A Schrijver, *Theory of linear and integer programming*, Wiley-Interscience Series in Discrete math and optim., John Wiley & Sons., 1986.
9. Williams, R S Rhinehart and J Riggs, *In-line process-model-based control of wastewater pH using dual base injection*, Ind. Eng. Chem. Res.,29, (1990), pp 12565-1259.

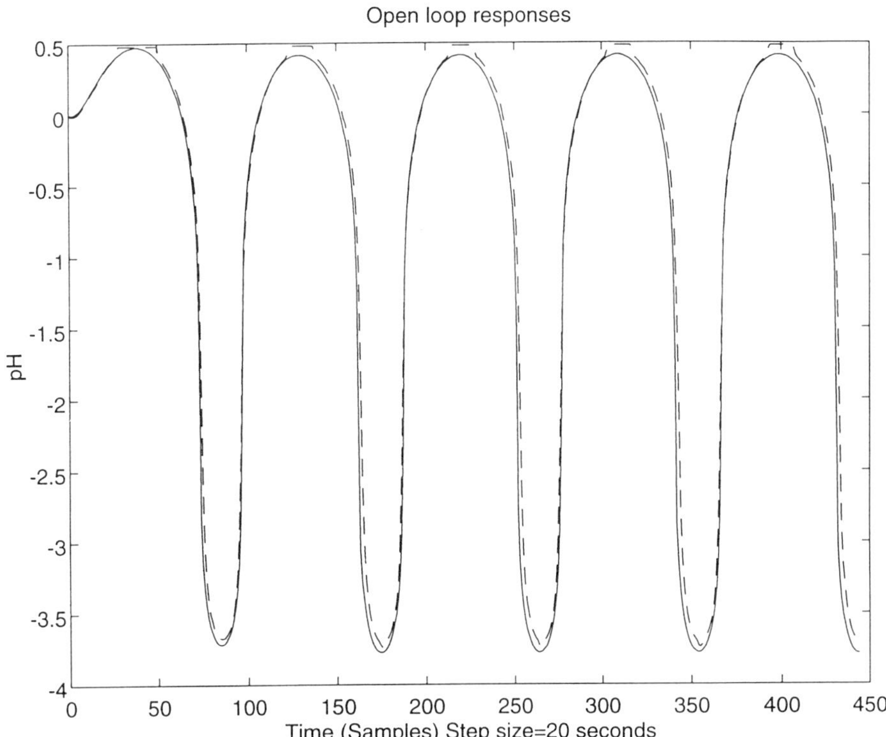

Figure 4 A comparison of the open loop process and the model outputs for a sinusoidal input.

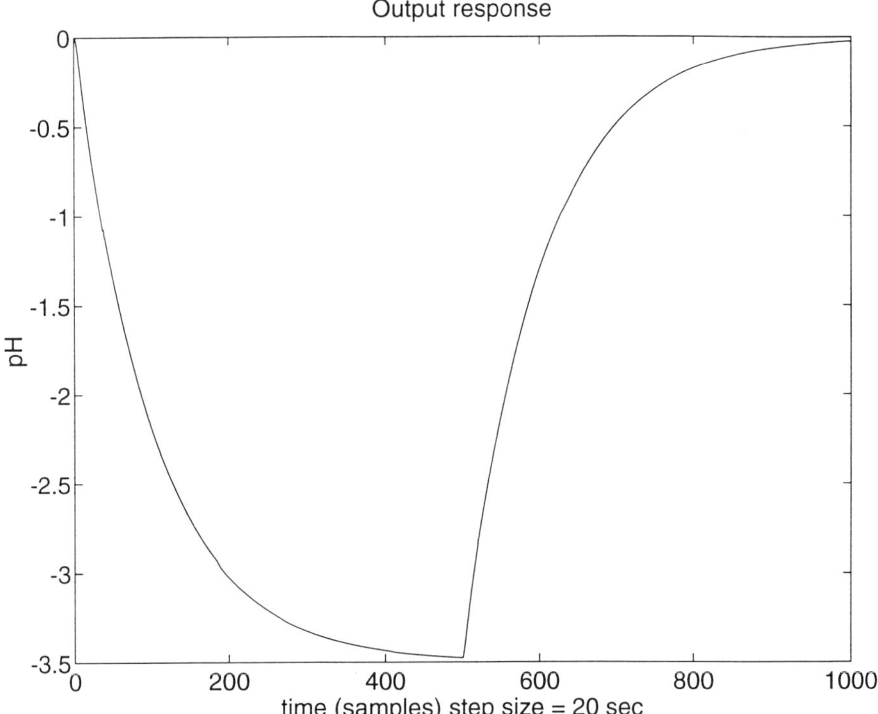

Figure 5 A set point change of magnitude -3 in the pH. The initial pH is 9.

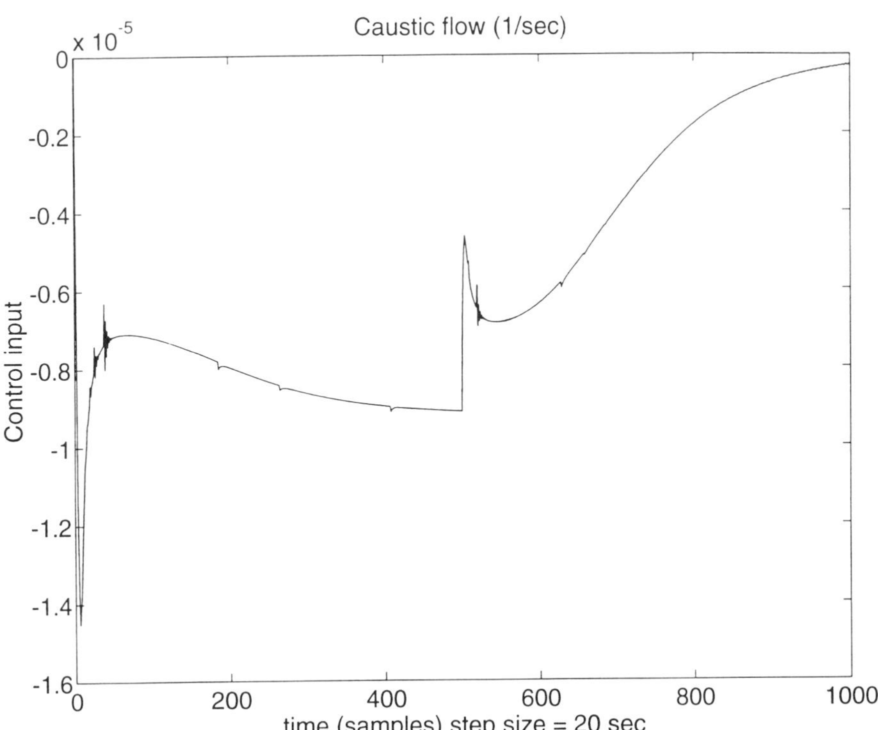

Figure 6 The control input for a set point change of pH=-3.

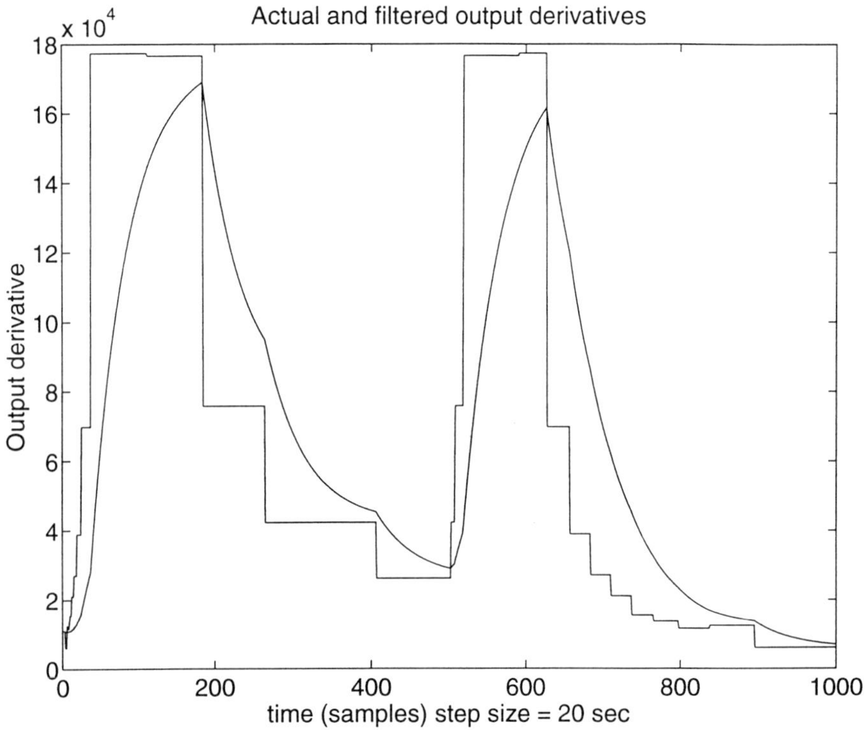

Figure 7 A comparison of the actual and filtered derivative term, d(t).

WSSIAA 5 (1995) pp. 193–206

OPTIMAL CONTROL DESIGN USING ERROR COMPENSATION

THEODORE J. KIM

Aided Navigation and Remote Sensing Department
Sandia National Laboratories
Albuquerque, NM 87185, USA

and

DAVID G. HULL

Department of Aerospace Engineering and Engineering Mechanics
The University of Texas
Austin, TX 78712, USA

ABSTRACT

A technique is presented which enables the optimal control designer
to compensate for the approximations usually made to obtain an analytical optimal control. Instead of discarding the approximation terms, they are
treated as bounded controls (called **error compensation** controls), and the
bounds are replaced by penalty terms in the performance index. The resulting optimal control, expected to be analytical, contains the penalty weights
as parameters. The error compensation control is then used in the exact
optimal control problem, and optimal weights are computed by nonlinear
programming. What results is an improved approximate analytical optimal control. An optimal guidance example shows that error compensation
improves the performance of the approximate optimal control.

1. Introduction

One approach to guidance is to repeatedly use an optimal control from the current
state to the desired final state in a sample and hold fashion. For such a guidance
law to be accepted, it must be analytical (explicit). To obtain an analytical optimal
control, it is usually necessary to make approximations in the mathematical model
defining the optimization problem. In general, small terms are discarded, and the
approximate optimal control is derived. However, it is possible to treat the small
terms as bounded controls and obtain an improved optimal control. This process is
called error compensation.

This work is motivated by a recent paper where uncertainties in the model are
treated as bounded controls and the control bounds are replaced by penalty terms in
the performance index[1]. If the regular controls are used to minimize the performance
index, the uncertainty controls are used to maximize it (worst case scenario). The
effect of the uncertainty controls is to make the optimal control a function of the
penalty weights. While these constants are supposed to be chosen so that the mag-
nitudes of the uncertainties are kept within bounds, they are more effectively used

to tune the optimal control to get the best performance when used with the exact model. This procedure, however, requires that a model of the expected uncertainties be created. Hence, why not abandon the worst case format and just determine the uncertainty control constants which give the best performance relative to the expected uncertainties?

This is the concept behind error compensation for deriving improved approximate optimal controls. The error compensation controls are used to extremize the performance index, and the penalty weights are chosen to get the best performance from the control law when used with the exact model. The worst case technique requires the weights to have particular signs, whereas the error compensation procedure allows the weights to be positive or negative.

In this paper, the error compensation technique is derived and applied to the development of an optimal guidance law for a hypersonic glider descending between two altitudes.

2. Error Compensation Controls

The general optimal control problem is to find the control $\boldsymbol{u}(t)$ that minimizes the performance index

$$J = \phi(t_f, \boldsymbol{x}_f) + \int_{t_0}^{t_f} \mathcal{L}(t, \boldsymbol{x}, \boldsymbol{u}) dt \tag{1}$$

subject to the state differential equations

$$\dot{\boldsymbol{x}} = \boldsymbol{f}(t, \boldsymbol{x}, \boldsymbol{u}), \tag{2}$$

the initial conditions

$$t_0, \boldsymbol{x}_0 \text{ specified}, \tag{3}$$

and the final constraint manifold

$$\boldsymbol{\psi}(t_f, \boldsymbol{x}_f) = 0. \tag{4}$$

Additional constraints on the controls and states can be included as necessary.

To obtain an analytical solution, it is usually necessary to introduce approximations and changes of variables into the state equations. Suppose that instead of simply eliminating terms from the differential equations, the equations of motion are rewritten as

$$\dot{\boldsymbol{x}} = \bar{\boldsymbol{f}}(t, \boldsymbol{x}, \boldsymbol{u}) + \boldsymbol{g}(t, \boldsymbol{x}, \boldsymbol{u}). \tag{5}$$

The vector $\bar{\boldsymbol{f}}$ approximates the true state derivative vector $\boldsymbol{f}$ and each element of $\boldsymbol{g}$ is small compared with the corresponding element of $\bar{\boldsymbol{f}}$. The vector $\boldsymbol{g}$ contains the terms that would normally be eliminated to obtain the model equations of motion, which lead to an approximate analytic optimal control.

The structure of the vector $\boldsymbol{g}(t, \boldsymbol{x}, \boldsymbol{u})$ prevents the analytical solution from occurring. Therefore, this vector is replaced with the unstructured vector $\boldsymbol{e}(t)$, which will be called the *error compensation* (EC) control. (In this paper, the "normal" control vector $\boldsymbol{u}$ is called the aerodynamic control vector.) The model equations of motion, then, are

$$\dot{\boldsymbol{x}} = \bar{\boldsymbol{f}}(t, \boldsymbol{x}, \boldsymbol{u}) + \boldsymbol{e}(t). \tag{6}$$

The EC controls are unknown functions of time only, but they are bounded. They embody the differences between the model equations and the true dynamics.

3. Performance Index

The EC controls are determined by hypothesizing that they *extremize* (either maximize or minimize) the same performance index that the aerodynamic controls are minimizing. In other words, the EC controls have their largest possible effect on the optimal control $\boldsymbol{u}$ in order to reduce the differences between the model and true differential equations.

The optimal control problem is restated as

$$\mathop{\text{ext}}_{\boldsymbol{e}} \mathop{\min}_{\boldsymbol{u}} \ J \tag{7}$$

subject to the model differential equations, Eq. 6, and the boundary conditions, Eqs. 3 and 4. The nonstandard notation "ext" emphasizes that the EC controls are *ext*remizing the performance index. As stated above, the EC controls are bounded. Here, the bounds are replaced by penalty terms in the performance index so that J is given by

$$J = \phi(t_f, \boldsymbol{x}_f) + \int_{t_0}^{t_f} \left[\mathcal{L}(t, \boldsymbol{x}, \boldsymbol{u}) + \frac{1}{2} \boldsymbol{e}^T \boldsymbol{M}^{-1} \boldsymbol{e} \right] dt. \tag{8}$$

The matrix $\boldsymbol{M}$ is composed of constants and is used to keep the $\boldsymbol{e}$'s from becoming too large. This matrix is assumed to be diagonal for simplicity so each element is the tuning parameter (or weight) associated with the corresponding error compensation control. The inverse of $\boldsymbol{M}$ is used only to simplify the subsequent tuning procedure.

4. Optimal Controls

The optimal control problem is defined by Eqs. 6, 7, 8, and the boundary conditions Eqs. 3 and 4. The differential equations are separable in $\boldsymbol{u}$ and $\boldsymbol{e}$ by construction. If the performance index, Eq. 8, is separable in $\boldsymbol{u}$ and $\boldsymbol{e}$ (i.e., if $\phi(t_f, \boldsymbol{x}_f)$ is sufficiently simple), the control solution is independent of the order of optimization[2]. The Bolza function and Hamiltonian for this problem are

$$G = \phi + \boldsymbol{\nu}^T \boldsymbol{\psi} \tag{9}$$

$$H = \mathcal{L}(t, \boldsymbol{x}, \boldsymbol{u}) + \frac{1}{2} \boldsymbol{e}^T \boldsymbol{M}^{-1} \boldsymbol{e} + \boldsymbol{\lambda}^T \left[\bar{\boldsymbol{f}}(t, \boldsymbol{x}, \boldsymbol{u}) + \boldsymbol{e} \right] \tag{10}$$

where ν and λ are Lagrange multiplier vectors. The Hamiltonian is also separable in u and e.

The first variation conditions for an extremal path are written in terms of G and H as

$$\frac{d\boldsymbol{x}}{dt} = \bar{\boldsymbol{f}}(t, \boldsymbol{x}, \boldsymbol{u}) + \boldsymbol{e} \tag{11}$$

$$\frac{d\boldsymbol{\lambda}}{dt} = -H_{\boldsymbol{x}}^T(t, \boldsymbol{x}, \boldsymbol{u}, \boldsymbol{e}, \boldsymbol{\lambda}) \tag{12}$$

$$0 = H_{\boldsymbol{u}}^T(t, \boldsymbol{x}, \boldsymbol{u}, \boldsymbol{\lambda}) \tag{13}$$

$$0 = H_{\boldsymbol{e}}^T(t, \boldsymbol{x}, \boldsymbol{e}, \boldsymbol{\lambda}) \tag{14}$$

subject to the boundary conditions

$$t_0, \ \boldsymbol{x}_0 \text{ specified} \tag{15}$$

$$\boldsymbol{\psi}(t_f, \boldsymbol{x}_f) = 0 \tag{16}$$

$$H(t_f, \boldsymbol{x}_f, \boldsymbol{u}_f, \boldsymbol{e}_f, \boldsymbol{\lambda}_f) = -G_{t_f}(t_f, \boldsymbol{x}_f, \boldsymbol{\nu}) \tag{17}$$

$$\boldsymbol{\lambda}_f = G_{\boldsymbol{x}_f}^T(t_f, \boldsymbol{x}_f, \boldsymbol{\nu}). \tag{18}$$

This solution is valid regardless of whether $\boldsymbol{e}$ minimizes or maximizes the performance index.

If the Hamiltonian does not depend explicitly on the time, the first integral holds along an extremal path. This condition is stated mathematically as

$$H(\boldsymbol{x}, \boldsymbol{u}, \boldsymbol{e}, \boldsymbol{\lambda}) = \text{constant}. \tag{19}$$

The first integral is of limited value since it can only be used to replace one of the above equations.

A necessary condition for the optimality of the extremal controls defined by Eqs. 13 and 14 is the Legendre-Clebsch condition[3]. If the aerodynamic controls are to minimize the performance index, this condition requires the second derivative of the Hamiltonian with respect to u to be positive semi-definite, i.e.

$$H_{\boldsymbol{uu}}(t, \boldsymbol{x}, \boldsymbol{u}, \boldsymbol{\lambda}) \geq 0. \tag{20}$$

Both $\bar{\boldsymbol{f}}$ and $\mathcal{L}$ must be defined for the specific problem before this inequality can be verified.

A general result for the EC controls can be obtained for any problem because of the formulation of the model equations. From Eq. 10, the second derivative of the Hamiltonian with respect to e is

$$H_{\boldsymbol{ee}} = \boldsymbol{M}^{-1}. \tag{21}$$

The sign of the elements of $H_{\boldsymbol{ee}}$ depend only on the sign of the individual tuning parameters. A positive tuning parameter corresponds to a minimization of the performance index (similar to the aerodynamic controls) and a negative value corresponds to a maximization. However, this information is not critical to the optimal control solution and does not affect the tuning procedure. In addition, the EC controls are independent so M^{-1} can be comprised of both positive and negative constants.

5. Hypersonic Glider Problem

The above technique is now applied to maximize the final velocity of a hypersonic glider that is descending between two altitudes. The differential equations that describe a nonthrusting vehicle moving in a vertical plane over a spherical, non-rotating Earth and modeled as a point mass are[4]

$$\frac{dX}{dt} = V\frac{r_s}{r_s + h}\cos\gamma \tag{22}$$

$$\frac{dh}{dt} = V\sin\gamma \tag{23}$$

$$\frac{dV}{dt} = -\frac{D}{m_0} - g_s\left(\frac{r_s}{r_s + h}\right)^2\sin\gamma \tag{24}$$

$$\frac{d\gamma}{dt} = \frac{L}{m_0 V} + \left[\frac{V}{r_s + h} - \frac{g_s}{V}\left(\frac{r_s}{r_s + h}\right)^2\right]\cos\gamma. \tag{25}$$

In the above equations, t denotes the time and vehicle states (X, h, V, γ) are the downrange position measured along a great circle arc, the altitude above the Earth's surface, the velocity of the glider, and the flight path angle. The mean radius of the Earth and gravity at the Earth's surface have the numerical values of

$$r_s = 20,925,672 \text{ ft}, \quad g_s = 32.172 \text{ ft/sec}^2. \tag{26}$$

The glider has a mass and reference surface area of

$$m_0 = 15.52 \text{ slugs}, \quad S_R = 1.5 \text{ ft}^2. \tag{27}$$

The aerodynamic lift and drag forces are written in terms of nondimensional aerodynamic coefficients as

$$L = \frac{1}{2}\rho V^2 S_R C_L \tag{28}$$

$$D = \frac{1}{2}\rho V^2 S_R C_D \tag{29}$$

where ρ denotes the density of the air, which is determined from interpolated data of the 1976 standard atmosphere[5]. The lift and drag coefficients are determined by linear interpolation of wind tunnel data that is stored as a function of angle of attack, Mach number, and Reynolds number.

5.1 Appoximate Model

The above equations of motion define the simulation environment (or truth model) of the glider. Since they are highly nonlinear, a reduction to a set of model equations must be performed in order to get an analytic solution to the optimal control problem. The following approximations are considered to be valid for this problem:

- **Low Altitude Flight**
  The altitude of the glider is considered to be small compared with the radius of the Earth. Therefore, $r_s + h \cong r_s$.

- **Parabolic Drag Polar**
  The glider is assumed to be flying hypersonically throughout the trajectory and the nondimensional drag coefficient is approximated by

$$C_D \cong C_{D_0}(1 + \bar{C}_L^2) \tag{30}$$

  where $\bar{C}_L \overset{\Delta}{=} C_L/C_L^*$ is the scaled lift coefficient and is the aerodynamic control for this problem. The drag polar constants are

$$C_{D_0} = 0.043, \quad C_L^* = 0.2888 \tag{31}$$

  and the maximum lift-to-drag ratio is $E^* = 3.281$.

- **Loh's Approximation**
  Loh's term is defined as[6]

$$\mathcal{M} \overset{\Delta}{=} \frac{2m_o}{C_L^* S_R} \left[1 - \frac{g_s r_s}{V^2}\right] \frac{\cos\gamma}{\rho r_s}. \tag{32}$$

  Loh shows that $\mathcal{M}$ is constant over γ integration along re-entry trajectories. It is assumed to remain constant along each trajectory for this problem, but its value is recomputed at each sample point.

- **Exponential Density**
  An analytic approximation for the air density is

$$\rho \cong \rho_s e^{-h/h_R} \tag{33}$$

  where $\rho_s = 0.0023769$ slug/ft^3 is the density at sea level and $h_R = 23,800$ ft is the density scale height chosen by fitting an exponential curve to a plot of density vs. altitude.

- **Linearized Flight Path Angle**
  The flight path angle is linearized about γ_0 to simplify the trigonometric functions to

$$\sin\gamma \ \cong \ a_0 + a_1\gamma \tag{34}$$

$$\cos\gamma \ \cong \ b_0 + b_1\gamma. \tag{35}$$

The linearization coefficients are the Taylor series expansion coefficients, i.e.

$$a_0 = \sin\gamma_0 - \gamma_0\cos\gamma_0, \quad a_1 = \cos\gamma_0 \tag{36}$$

$$b_0 = \cos\gamma_0 + \gamma_0\sin\gamma_0, \quad b_1 = -\sin\gamma_0. \tag{37}$$

The accuracy of these angular approximations depends on the range of values of the flight path angle compared with γ_0.

The dimensional states (X, h, V) are replaced with the following nondimensional variables:

$$\xi \triangleq \frac{X}{h_R} \tag{38}$$

$$w \triangleq \frac{C_L^* S_R h_R}{2m_0}\rho(h) \tag{39}$$

$$v \triangleq \ln\left(\frac{V^2}{g_s r_s}\right), \tag{40}$$

and the differential equation for the nondimensional density replaces Eq. 23. In addition, a new integration variable (to replace time) is introduced such that its time derivative is[7]

$$\frac{dz}{dt} \triangleq \frac{Vw}{h_R}. \tag{41}$$

The value of z always increases along any trajectory (unlike γ, which is often used[8]) since its time derivative is always positive. The initial value of z is chosen to be equal to 0 for convenience.

The exact state differential equations are now rewritten by adding and subtracting the approximations to obtain

$$\frac{d\xi}{dz} = \frac{(b_0 + b_1\gamma)}{w} + \left\{\frac{r_s}{r_s + h}\frac{\cos\gamma}{w} - \frac{(b_0 + b_1\gamma)}{w}\right\} \tag{42}$$

$$\frac{dw}{dz} = -(a_0 + a_1\gamma) + \left\{\frac{h_R}{\rho}\frac{d\rho}{dh}\sin\gamma + (a_0 + a_1\gamma)\right\} \tag{43}$$

$$\frac{dv}{dz} = -\frac{(1 + \bar{C}_L^2)}{E^*} + \left\{\frac{(1 + \bar{C}_L^2) - C_D/C_{D_0}}{E^*} - \frac{2h_R g_s}{V^2 w}\left(\frac{r_s}{r_s + h}\right)^2\sin\gamma\right\} \tag{44}$$

$$\frac{d\gamma}{dz} = \bar{C}_L + \mathcal{M}_0 + \left\{\frac{r_s}{r_s + h}\mathcal{M} - \mathcal{M}_0\right\} \tag{45}$$

where $\mathcal{M}_0$ is Loh's constant evaluated at the initial conditions. Each of the terms in braces is small compared with the corresponding non-braced terms. Therefore, the four error compensation controls $e_\xi(t)$, $e_w(t)$, $e_v(t)$, and $e_\gamma(t)$ are defined to replace those braced terms. The model equations of motion, then, are

$$\frac{d\xi}{dz} = \frac{(b_0 + b_1\gamma)}{w} + e_\xi \tag{46}$$

$$\frac{dw}{dz} = -(a_0 + a_1\gamma) + e_w \tag{47}$$

$$\frac{dv}{dz} = -\frac{(1 + \bar{C}_L^2)}{E^*} + e_v \tag{48}$$

$$\frac{d\gamma}{dz} = \bar{C}_L + \mathcal{M}_0 + e_\gamma. \tag{49}$$

5.2 Optimal Controls

The optimal control problem is to find the two sets of controls $\boldsymbol{u}$ and $\boldsymbol{e}$ that respectively minimize and extremize the performance index

$$J = -v_f + \frac{1}{2}\int_o^{z_f} \boldsymbol{e}^T \boldsymbol{M}^{-1}\boldsymbol{e}\, dz \tag{50}$$

subject to the model differential equations, Eqs. 46 - 49, and the boundary conditions

$$z_0 = 0, \quad \boldsymbol{x}_0 \text{ specified} \tag{51}$$

$$\boldsymbol{\psi}(z_f, \boldsymbol{x}_f) = w_f - w_D = 0 \tag{52}$$

where the subscript D denotes the desired final location. The aerodynamic control vector $\boldsymbol{u}$ minimizes the performance index and therefore *maximizes* the final velocity of the glider.

The performance index is separable for this problem since $dv/dz = f_1(\boldsymbol{u}) + f_2(\boldsymbol{e})$ and integration is a linear operator. The Bolza function and the Hamiltonian for this problem are

$$G = -v_f + \nu(w_f - w_D) \tag{53}$$

$$H = \frac{1}{2}\left[\frac{e_\xi^2}{M_\xi} + \frac{e_w^2}{M_w} + \frac{e_v^2}{M_v} + \frac{e_\gamma^2}{M_\gamma}\right] + \lambda_\xi\left[\frac{(b_0 + b_1\gamma)}{w} + e_\xi\right]$$
$$+ \lambda_w\left[-(a_0 + a_1\gamma) + e_w\right] + \lambda_v\left[-\frac{(1 + \bar{C}_L^2)}{E^*} + e_v\right] + \lambda_\gamma\left[\bar{C}_L + \mathcal{M}_0 + e_\gamma\right] \tag{54}$$

Since the downrange is not constrained, the Lagrange multiplier associated with that differential equation is zero, i.e. $\lambda_\xi = 0$. From Eqs. 12 and 18, the other multipliers are

$$\lambda_w = \text{unknown constant} \tag{55}$$

$$\lambda_v = -1 \tag{56}$$

$$\lambda_\gamma = a_1\lambda_w(z - z_f). \tag{57}$$

The optimal controls are determined from Eqs. 13 and 14 to be

$$\bar{C}_L = -\frac{E^*}{2}a_1\lambda_w(z - z_f) \tag{58}$$

$$e_\xi = 0 \tag{59}$$

$$e_w = -\lambda_w M_w \tag{60}$$

$$e_v = M_v \tag{61}$$

$$e_\gamma = -a_1\lambda_w M_\gamma(z - z_f) \tag{62}$$

using the above expression for λ_γ. The aerodynamic control satisfies Eq. 20 since

$$H_{\bar{C}_L\bar{C}_L} = -\frac{2\lambda_v}{E^*} > 0. \tag{63}$$

The only remaining unknown quantities are λ_w and z_f. These values are determined by analytically integrating Eq. 49, then integrating Eq. 47, and then applying the first integral. The details of this derivation are presented in detail elsewhere[9] and will not be covered here. The expression for λ_w in terms of z_f is

$$\lambda_w = \frac{d_1 + \sqrt{d_1^2 - 2K_v d_2}}{d_2} \tag{64}$$

where

$$d_1 \triangleq a_0 + a_1\gamma_0 + a_1\mathcal{M}_0 z_f \tag{65}$$

$$d_2 \triangleq -M_w - a_1^2 K_\gamma z_f^2 \tag{66}$$

and the constants K_v and K_γ are defined as

$$K_v \triangleq \frac{1}{E^*} - \frac{M_v}{2} \tag{67}$$

$$K_\gamma \triangleq M_\gamma + \frac{E^*}{2}. \tag{68}$$

The value of z_f is a root of the sixth-order polynomial

$$p_6 z_f^6 + p_4 z_f^4 + p_3 z_f^3 + p_2 z_f^2 + p_0 = 0 \tag{69}$$

whose coefficients are

$$p_6 = a_1^4\left[8K_v K_\gamma^2 + 3K_\gamma\mathcal{M}_0^2\right] \tag{70}$$

$$p_4 = a_1^2 M_w\left[48K_v K_\gamma + 27\mathcal{M}_0^2\right]$$
$$\qquad -12K_\gamma\left[a_1^3\mathcal{M}_0\Delta w + a_1^4\gamma_0^2 + a_0^2 a_1^2 + 2a_0 a_1^3\gamma_0\right] \tag{71}$$

$$p_3 = 72\mathcal{M}_0 M_w\left[a_1^2\gamma_0 + a_0 a_1\right] - 48K_\gamma\Delta w\left[a_1^3\gamma_0 + a_0 a_1^2\right] \tag{72}$$

$$p_2 = 72K_v M_w^2 - 36a_1^2 K_\gamma\Delta w^2$$
$$\qquad +36M_w\left[a_1\mathcal{M}_0\Delta w + a_1^2\gamma_0^2 + 2a_0 a_1\gamma_0 + a_0^2\right] \tag{73}$$

$$p_0 = -36\Delta w^2 M_w, \tag{74}$$

where $\Delta w \triangleq w_f - w_0$. Simulation work shows that the correct value of z_f is the second positive root of Eq. 69. This value must be determined numerically since there is no known analytical solution to a sixth-order polynomial.

5.3 Numerical Results

The derived optimal aerodynamic control listed in Eq. 58 is tested using a computer simulation of the glider. The commanded lift coefficient is assumed to be instantaneously achievable so the vehicle rotational dynamics are neglected. During the simulation, the commanded lift coefficient is recomputed every 0.1 seconds and held constant until the next control computation. In addition, the linearization coefficients and $\mathcal{M}_0$ are recomputed at every guidance step using the current values of the states.

The EC controls do not enter directly in the simulation environment since they do not appear explicitly in the simulation equations of motion. However, they are included implicitly throughout the problem, as evidenced by the presence of the tuning parameters M_w, M_v, and M_γ in the equations for λ_w and the polynomial coefficients $p_0, p_2, \ldots, p_6$.

The glider is initially flying at a constant altitude of $100,000$ ft with a flight path angle of 0 deg. The initial velocity of the vehicle is $11,000$ ft/sec and both the initial time and downrange are set to zero. The desired final altitude is 0 ft.

Figure 1 shows the effect of increasing the magnitude of the tuning parameter M_v on the final velocity of the vehicle. (The graphs for M_w and M_γ are similar and are not included for brevity.) From Eq. 61, a increase in the size of M_v increases the magnitude

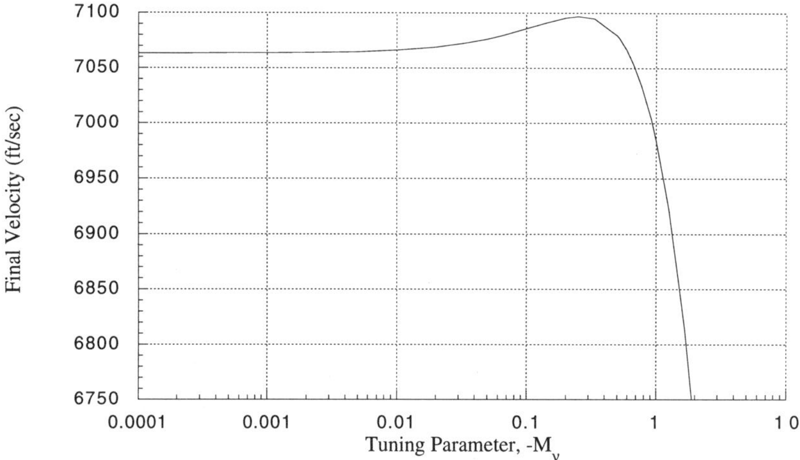

Figure 1: Final velocities for different values of M_v.

of the EC control e_v. Any increase in the EC control should monotonically change the performance index. However, the final velocity initially increases and then decreases. This initial increase of the final velocity is due to the mismatch between the model and simulation environments and is exploited to achieve the highest performance. (The initial increase in V_f does **not** occur when the model equations of motion are used as the simulation environment. Instead, the final velocity decreases monotonically as M_v gets more negative.)

The results obtained using three different control laws are compared in Figs. 2 and 3. The first two controllers are of the form derived in this paper but use two different sets of tuning parameters. The third controller is a seven-node, piecewise-linear control law that is determined by a numerical parameter optimization code[10] that uses the true vehicle dynamics. The final states of the glider using the three controllers are compared in Table 1.

The first controller, which will be called the *untuned* control law, has all of the tuning parameters set to 0. From Eqs. 59 - 62, the EC controls are therefore 0. The *tuned* control law uses parameter values of $M_w = -1/20$, $M_v = -2/9$, and $M_\gamma = -1/800$, which are chosen by testing a variety of combinations. A better set of parameters can be found if a more sophisticated search technique is used. The use of M^{-1} in the problem setup allows the elements of M to change signs during the tuning procedure without any numerical problems. All of the tuning parameters for this problem are negative so the EC controls maximize the performance index.

As shown in Fig. 2, the tuning procedure modifies the control to behave more like the numerically-determined controller. The resulting trajectory using the tuned con-

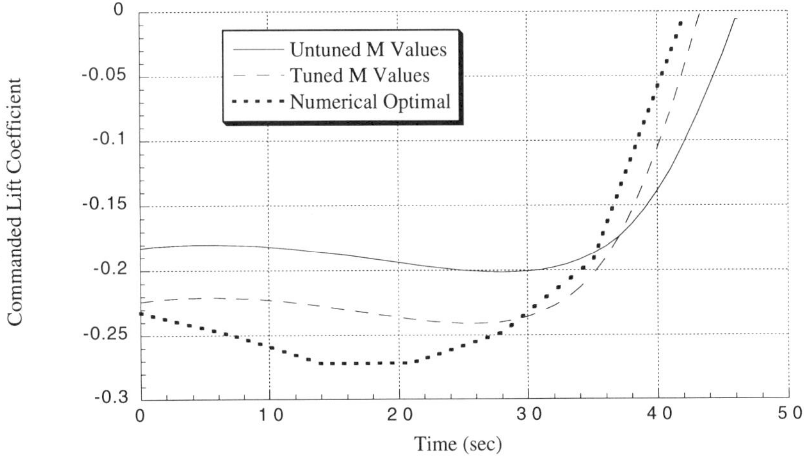

Figure 2: Comparison of the control histories.

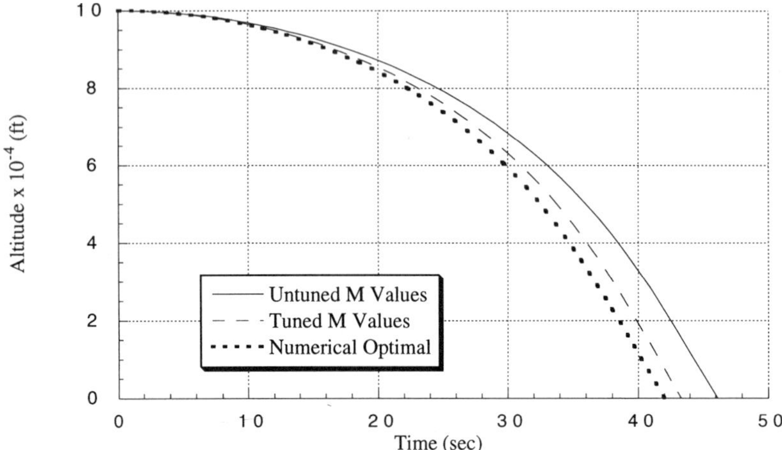

Figure 3: Trajectories for the three controllers.

trol law is therefore more like the trajectory followed by the piecewise-linear optimal control.

The tuned controller produces a final velocity that is higher than the velocity obtained using the untuned control law and is very close to the final velocity obtained using the numerically-determined control. However, the final velocity of the glider does not change tremendously for the three control laws. Since the control histories and the trajectories are considerably different, this specific problem most likely has a "flat" optimal point.

The derived controllers are further tested by varying the simulation air density by $\pm 15\%$ of the standard air density profile. These changes in the density model simulate changes in the ambient temperature (representing a hot day or a cold day). Even though the environment changes, the tuning parameters are fixed at the values that were selected using the nominal density profile.

As shown in Fig. 4, the performance of the untuned control law degrades more for the heavier density profiles. This degradation occurs since the exponential atmosphere approximation, Eq. 33, underestimates the actual nominal density. However, the

Table 1: Final glider states for the three controllers.

	X_f (nm)	V_f (ft/sec)	γ_f
Untuned Controller	73.4	7063.3	-47.1°
Tuned Controller	68.2	7096.0	-51.1°
Numerical Controller	65.8	7101.9	-50.5°

tuned control law restores the performance to the level obtained with the numerically-determined control law for the entire range of air densities even though the tuning parameters were chosen using the nominal density profile.

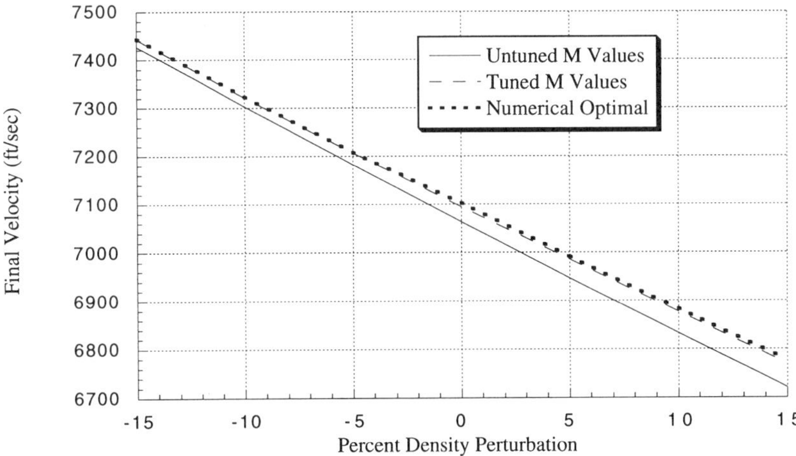

Figure 4: Final velocities for different density profiles.

6. Conclusions

A technique called error compensation has been developed to account for the approximations that are applied to the differential equation of motion describing a physical system in order to obtain an analytical optimal control. Instead of discarding the approximation terms, they are replaced by bounded controls where the bounds are imposed as penalty terms in the performance index. The derived controls contain the penalty weights as parameters, and their values are chosen (tuned) to achieve the best performance when the optimal control is applied to the exact or true physical system. What results is an improved analytical optimal control. Simulations of a descending glider have shown that the tuning parameters can be adjusted to improve the performance of the approximate (untuned) optimal control. The tuned control produces a final velocity higher than that of the untuned control and close to the final velocity obtained using a numerically-determined piecewise-linear optimal control.

7. Acknowledgment

This work was supported in part by the Department of Energy under Contract DE-AC04-94AL-85000.

8. References

1. M. R. Ilgen, J. L. Speyer, and C. T. Leondes, Robust Approximate Optimal Guidance Strategies for Aeroassisted Plane Change Missions: A Game Theoretic Approach, *Control and Dynamic Systems: Advances in Theory and Applications. Vol. 52: Integrated Technology Methods and Applications in Aerospace Systems Design,* C. T. Leondes, ed. Academic Press, Inc., 1992.

2. R. Isaacs, *Differential Games,* Robert E. Krieger Publishing Co., 1975.

3. A. E. Bryson and Y. C. Ho, *Applied Optimal Control,* Hemisphere Publishing Corporation, 1975.

4. T. S. Feeley and J. L. Speyer, A Real-Time Approximate Optimal Guidance Law for Flight in a Plane, *Proceedings of the 1990 American Control Conference,* San Diego, California, May 1990.

5. United States Committee on Extension to the Standard Atmosphere, *U. S. Standard Atmosphere,* Washington, D.C., 1976.

6. W. H. Loh, *Re-entry and Planetary Entry Physics and Technology,* Springer-Verlag, 1968.

7. J. L. Speyer and E. Z. Crues, Approximate Optimal Atmospheric Guidance Law for Aeroassisted Plane-Change Maneuvers, *Journal of Guidance, Control, and Dynamics* **13** (1990), 792-802.

8. G. R. Eisler and D. G. Hull, Guidance Law for Hypersonic Descent to a Point, *Journal of Guidance, Control, and Dynamics* **17** (1994), 649-654.

9. T. J. Kim, *A Mixed Nonlinear/Linear Control Design for Optimal Descent Trajectories.* Ph.D. Thesis, The University of Texas, 1994.

10. L. S. Lasdon and A. D. Waren, *GRG2 User's Guide,* 1989.

WSSIAA 5 (1995) pp. 207–220
© World Scientific Publishing Company

System Identification in Fuzzy Logic Control

T. KOBYLARZ

USAF

M. PACHTER AND C.H. HOUPIS

Air Force Institute of Technology, USA

Abstract: Two System Identification paradigms using Fuzzy Logic modeling are explored. The first entails the representation of an unknown input/output mapping by a fuzzy inference engine, using the input/output data. The second addresses the class of problems where the prior information about the system is imbedded in the structure of a fuzzy logic model and where one is concerned with the identification of the underlying fuzzy rules from input/output data. Examples, using polynomial models and a logical XOR device, respectively, illustrate the two fuzzy logic modeling/identification paradigms.

1 Introduction

Fuzzy logic control (FLC) problems may arise in one of two scenarios, depending on the level of detail of the available a priori information. First, one may have a good model (e.g., differential equations) describing the plant's behavior, and a FLC design is desired. Hence, a model based FLC design is to be undertaken. It has the advantage of using all the available information encapsulated in the detailed mathematical model of the plant along with the benefits of FLC concepts: Specifically, by relaxing some unnecessary control accuracy specifications in the output space, and using relatively rough but easily obtained measurements, a significant robustness benefit can be attained. In the second scenario, due to the lack of confidence in a mathematical model of the *plant* but the availability of empirical operational data, one may attempt fuzzy logic identification based solely upon the input/output data pairs. The *plant* referred to is either the actual controlled plant or a controller which needs to be designed to capture an experienced operator's modus operandi. That is, one wants to either obtain the fuzzy logic model of the plant based upon the (fuzzy) observations about the plant's behavior for subsequent design of a compatible fuzzy logic controller; or directly model the controller (by observing the operator's action, viz., inputs and outputs). The second scenario revolves around the lack of a good mathematical model of the plant and is prevalent in the FLC literature. This paper explores the implications for the second scenario of fuzzy logic identification. Thus, the problem of FLC modeling is addressed.

Indeed, Fuzzy Logic is often advertised as capable of providing control solutions for "difficult" to model and/or control plants, where other control design methods presumably fail. A fuzzy logic controller, it is claimed, does not require a mathematical model of the plant but instead is able to capture the expertise of a human operator [3]. Hence, in this approach to FLC, the control engineer encodes the operator's rules governing his actions into a fuzzy inference engine controlling the *plant*. Indeed, it's conceivable that in "simple systems", where the operating rules are easily articulated, this straight forward Expert System-like approach works well, and a suitable controller can be designed after some tuning of the inference engine during simulations [3]. In the case of more complex control systems, the fuzzy rules may not be so easy to articulate. Thus, there is a need for formal system (rule) identification as an integral part of the FLC based design process. In particular, when the resultant controller description is to be in the form of rules in a fuzzy inference engine, the need for fuzzy identification using available input/output data becomes obvious. One is then interested in the identification of either a functional relationship, or the underlying fuzzy rules. These two fuzzy logic modeling paradigms are investigated in the present paper.

In this paper a basic knowledge of FLC viz., familiarity with the concepts of universe of discourse, fuzzy variable, membership function, and defuzzification is assumed. For a good reference, see Kosko's book [6]. In Section 2 the Fuzzy Logic identification (ID) paradigm is presented. Section 3 provides the mathematical formulation of the proposed approach. In Sections 4 and 5 the important effects of measurement noise and persistent excitation in fuzzy logic ID are addressed. Illustrative examples concerning the identification of nonlinear polynomial mappings and a logical XOR gate, respectively are discussed. Concluding remarks are made in Section 6.

2 Current "Intelligent" ID Paradigms

Much of the work in this area assumes total ignorance of the plant and attempts identification based purely on plant input/output data. Thus, a data driven approach is pursued. This situation can be handled by using Neural Networks which are being trained to respond similarly to the actual plant, and therefore may be used as the implicit repository of the underlying rule base [1], [4]. An alternative, Fuzzy Logic, approach assumes a generic form of parametrically represented membership functions and performs an optimization on these parameters to minimize the output error, thus obtaining a "good" fit to the input/output data. The result provides a fuzzy inference engine which best fits the available data, given the form of the membership functions [5], [7]. Unfortunately, the deleterious effects of measurement noise and questions of persistent excitation are not addressed. The later play a critical role in conventional system ID, and obviously also are of concern, and in the present paper are being addressed, in fuzzy logic ID.

Furthermore, in this work a class of systems is considered in which prior information on the plant's operation is available. That is, much is known about the plant in question, but mathematical equations are not available due to either unmodeled dynamics or parameter uncertainties. This rich class of problems includes those in which experienced operators can control the system but can not articulate or verbalize the <u>rule set</u>. In particular, problems are discussed for which sufficient prior information exists, and it is convenient from a modeling viewpoint to formulate a FLC problem and stipulate the following:

1. The universe of discourse

2. The number of fuzzy input/output variables required

3. Fuzzy variable values (i.e. small, medium or large)

4. Appropriate membership functions

To complete the hypotheses, fuzzification (min/max, product, etc.) and defuzzification (e.g., centroid) algorithms must be specified.

Finally, the proposed approach allows for the "plant" being modeled to either represent an actual physical plant or a human operator controlling the physical plant. In the case of modeling the plant's controller, one is trying to emulate in the FLC the experienced operator's rules.

3 Identification Concept

R is defined as the finite set of all feasible rule sets, which relate the output fuzzy variables (FV) to the input FVs, and are consistent with the hypotheses. One can perform an exhaustive search to find the optimal rule set element $R^* \in$ R which best conforms to the available input and output data. In line with the classical system ID paradigm, the mean-squared output error metric of the rule's action, $R(x_i)$, on the input data set, versus actual output data, y_i, is used. Hence, in the scalar output case with input/output data (x, y), the optimal rule set R^* satisfies:

$$R^* = \arg \min_{R \in \mathsf{R}} \ \frac{1}{n} \sum_{i=1}^{n} (R(x_i) - y_i)^2 \tag{3.1}$$

Obviously, analysis cannot be brought to bear on the solution of the above finite optimization problem. As the number of fuzzy variables increases, minimization by exhaustive search over the set of all possible rules R suffers from a combinatorial explosion. However, it allows for a well posed problem in which an optimum exists. Suboptimal solutions are provided by heuristic optimization methods, e.g. Genetic Algorithms [2], [5], or other discrete/combinatorial optimization methods.

In this paper, the proposed Fuzzy Logic system ID paradigms are illustrated for two cases: 1) The identification of noise corrupted polynomial input/output mappings, and 2) The identification, in the presence of measurement noise, of a logic XOR gate.

4   Polynomial I/O Mappings

A static system with the following Fuzzy Logic hypotheses is considered. Thus, our prior information about the system is encapsulated in the FLC problem formulation.

The universe of discourse is scaled to the interval [0,1]. Five input fuzzy variables and three output FVs are used. Their linguistic labels are 1-5 and 1-3, respectively, where triangular membership functions represent these fuzzy variables. These are illustrated in Figures 1a and 1b. Furthermore, the fuzzification and defuzzification algorithms are the min/max and centroid methods, respectively.

A rule, e.g.,

<u>Rule $\mathcal{R}$:</u>   **IF** *input* is 2,   **THEN** *output* is 1

maps an input to an output. A rule set is a set of rules mapping each input FV to a unique output FV. So, for n input FVs and m output FVs, the set R contains m^n possible rule sets (combinations of rules).

Results for data generated from noise free 1st and 2nd order polynomials are shown in Figures 2a and 2b respectively. The identification using Eq (3.1) was performed in MATLAB® [8]. In the first experiment, R contains 243 possible rule sets with five rules each; hence, the second order fit requires $5^5 = 3125$ different possibilities to be investigated. In Figure 2a the linear truth model $y = x$ is compared to the action of the best rule set as applied to the universe of discourse [0,1]. In Figure 2b, the true values are the discrete points, while the curve is the output of the best fuzzy rule set. In both figures the fuzzification is performed by the min/max operators. The exhaustive search may be greatly reduced by first eliminating any of the possible rules that are never activated by a point in the data set.

In the above cases, the underlying rule set in FLC is identified. The truth models are

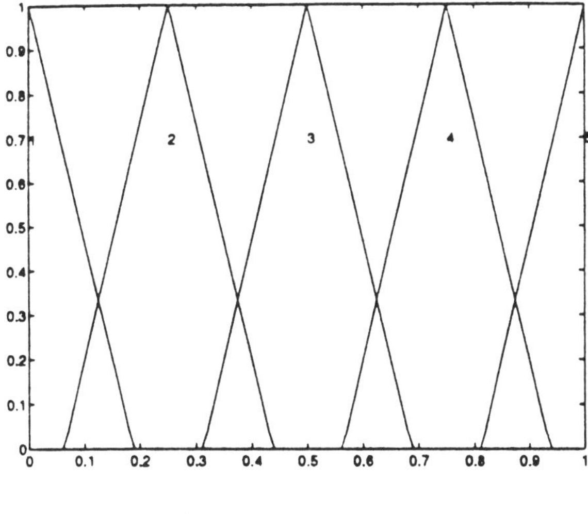

a) 5 Fuzzy Inputs

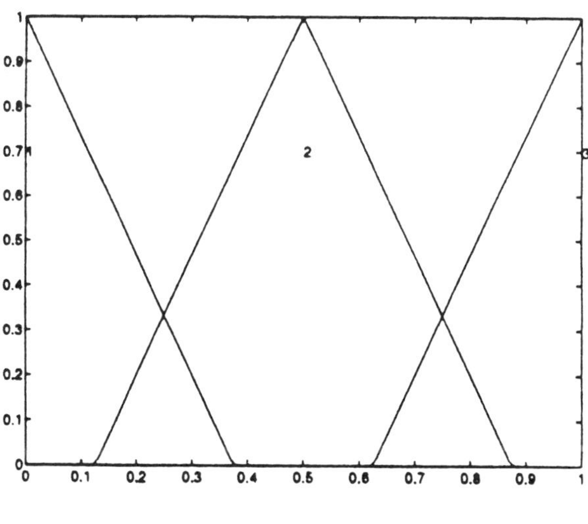

b) 3 Fuzzy Outputs

Figure 1: Triangular Fuzzy Variables for Input/Output

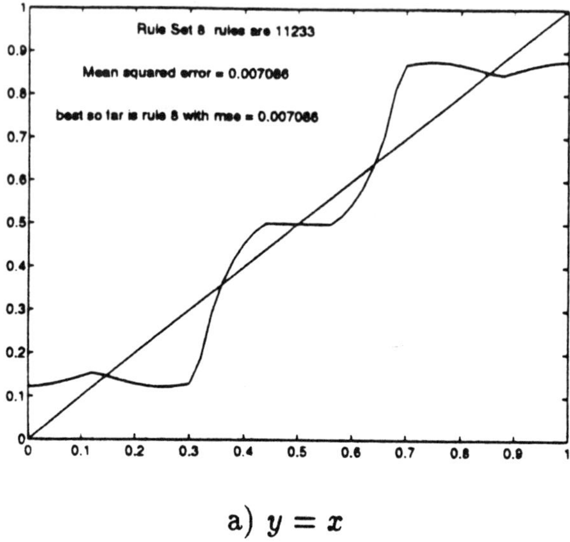

a) $y = x$

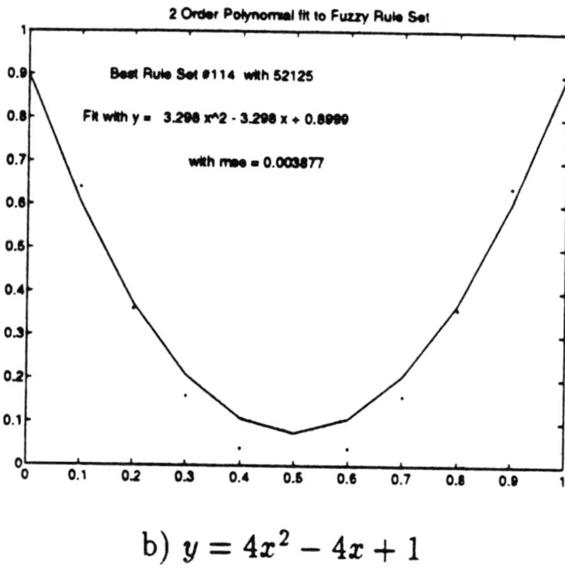

b) $y = 4x^2 - 4x + 1$

Figure 2: Fuzzy Fit of Polynomials

polynomial functions and therefore the performance of the FLC ID algorithm is gauged by comparing the ensuing input/output mapping with the underlying known polynomial relations.

Of course, there are many ways to fit polynomials, but this validates the usefulness of the above outlined approach. Furthermore, Wang and Mendel have shown that fuzzy inference engines as outlined above, are dense in the space of continuous functions on a compact universe of discourse [9].

A great advantage of a fuzzy modeled system becomes evident when the finite, and hence incomplete, data set appears to be discontinuous, while one has every reason to believe that the underlying true system is continuous. An example of such a data set is illustrated in Figure 3a, along with two attempts at fitting the data. To avoid the dangers of over-fitting data with high-order polynomials, the solid lines represent 1st order fits. The one using the entire domain [0,2] is continuous, but yields a poor fit to the data. When the interval is broken up into two subintervals, [0,1] and [1,2], excellent fits are obtained, but this results in a discontinuous mapping. However, if one uses overlapping membership functions, a continuous transition is obtained along the entire [0,2] interval. Figure 3b depicts this for several values of membership function overlap from 5% to 40%. Again, the fuzzification uses the min/max algorithms. A smooth blending of the two separate fits is obtained when a "product rule" is used for FV combination, along with analytic membership functions.

5 XOR Gate Plant

A celebrated example in Neural Networks is considered and the issues of measurement noise and persistent excitation in FLC ID are investigated. A mapping with two inputs and one output which is believed to represent a noise corrupted binary device is identified. Naturally, it is thus decided to model the inputs and the output using two FVs, "small" and "big". This establishes a truth table which shows that there are only 16 possible rule sets, of 4 rules each, to search. Since it is known that nearly all input and output measurements are in the interval [-1,2], this interval will be used as the universe of discourse. This time, Gaussian membership functions μ centered at 0 and 1 are chosen to represent the FVs "small" and "big", respectively. The fuzzy AND operator is implemented with the product rule.

$$\mu(x) = \mu_s(x)\mu_b(x) = e^{-\frac{x^2}{2\sigma_s^2}} e^{-\frac{(x-1)^2}{2\sigma_b^2}}$$

Defuzzification is again accomplished by the center of area method.

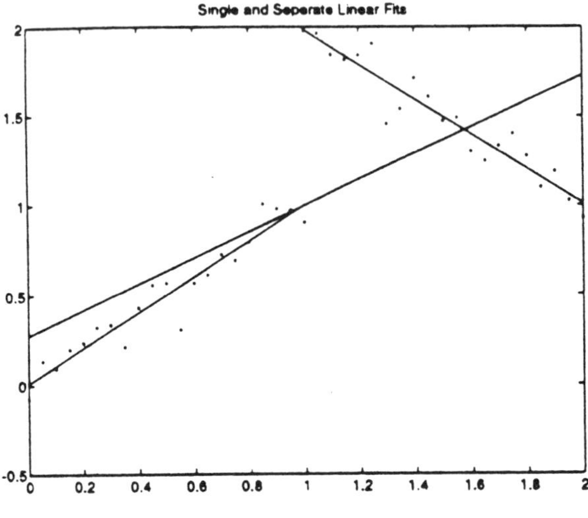

a) Single vs. Separate Fit

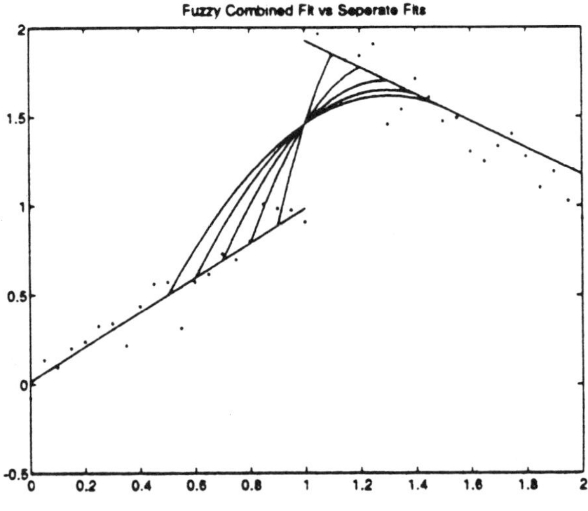

b) Fuzzy vs. Discontinuous

Figure 3: Fuzzy Blending as a Function of Overlap

Inputs		16 Possible Outputs															
1	2	1	2	3	4	5	6	**7**	8	9	10	11	12	13	14	15	16
S	S	S	S	S	S	S	S	**S**	S	B	B	B	B	B	B	B	B
S	B	S	S	S	S	B	B	**B**	B	S	S	S	S	B	B	B	B
B	S	S	S	B	B	S	S	**B**	B	S	S	B	B	S	S	B	B
B	B	S	B	S	B	S	B	**S**	B	S	B	S	B	S	B	S	B

Table 1: All Possible Rule sets for 2-input 1-output Binary Device

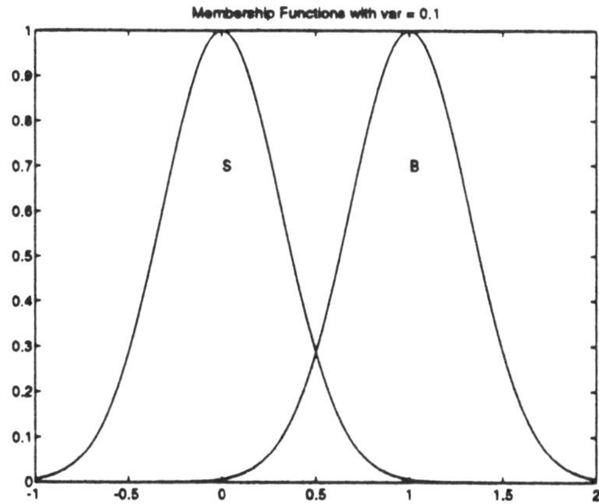

Figure 4: Gaussian Fuzzy Variables for all Channels

The input/output data set for identification is generated using a fuzzy XOR gate as the underlying truth model (rule 7 in Table 1) with membership functions $\mu_s(x)$ and $\mu_b(x)$ with variance parameter $\sigma_s^2 = \sigma_b^2 = \sigma^2 = 0.1$. These membership functions are shown in Figure 4. The inputs are fed to every possible rule set and the output is compared to that obtained from the truth model (the original XOR). During an actual search one would not know the variance either, so that the problem entails both structural identification (rule set identification) and parameter identification (σ).

An exhaustive search with fixed σ is made to find the minimum output error, and hence the best fit. However, σ is unknown. The problem is not convex with respect to the rule sets in general, so finding the global minimum is no longer assured by a non-exhaustive (finite) search.

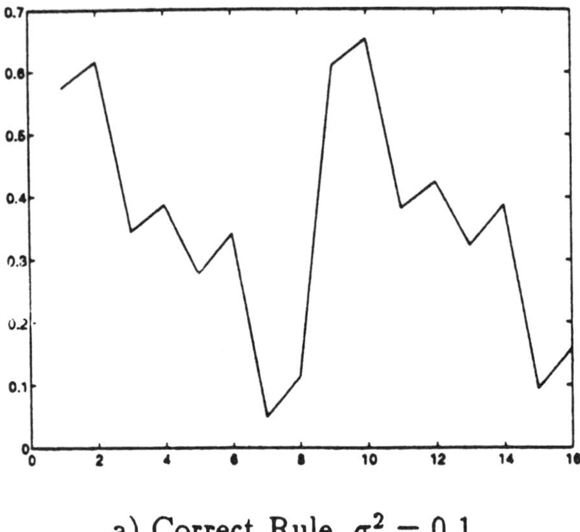

a) Correct Rule, $\sigma^2 = 0.1$

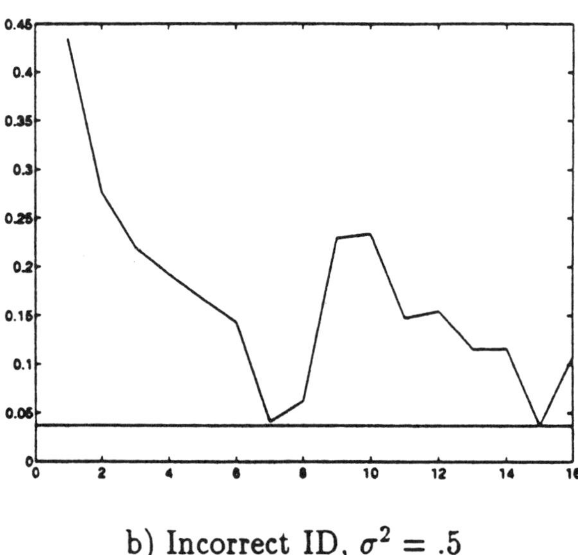

b) Incorrect ID, $\sigma^2 = .5$

Figure 5: Identification Results: Cost vs. Rule Set

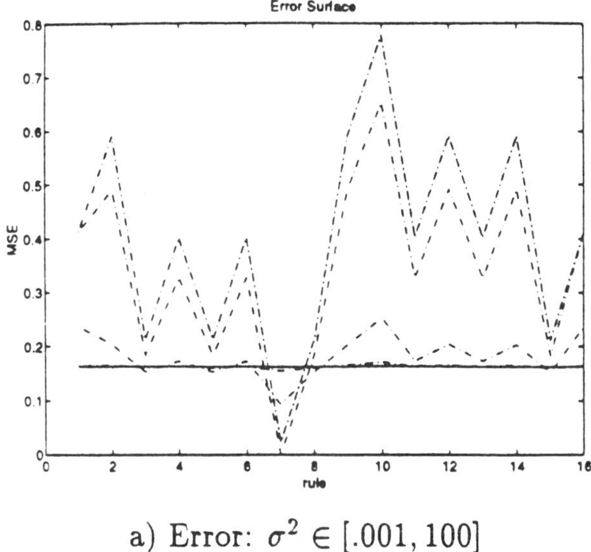

a) Error: $\sigma^2 \in [.001, 100]$

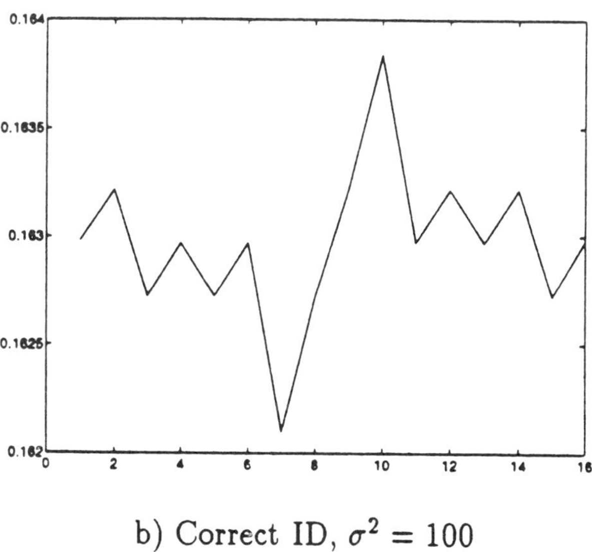

b) Correct ID, $\sigma^2 = 100$

Figure 6: Performance for Spanning Input/Output Data

If there were a guarantee that for any fixed σ one could identify the correct rule set the process would indeed be straight forward. One would fix σ and identify the true rule set (truth table), and then one would merely minimize with respect to σ. However, this may not be the case, as is illustrated in Figures 5a and 5b, which show different identified rule sets for different σ, and where the logical XOR gate is represented by rule set 7; see Table 1.

Figure 5a shows the correct rule set (7) identified for $\sigma^2 = 0.1$. Now, fix $\sigma^2 = 0.5$ in all 16 rule sets for which the fit of the data is to be performed (Note: the I/O data was generated for the nominal σ^2). Figure 5b shows that rule 15 (with $\sigma^2 = .5$) now has the smallest output error, and hence the incorrect rule set has been identified.

The reason behind the above mis-identification plagues all identification methods that rely on output error minimization. Although not always well appreciated even in the conventional system ID community, this problem also surfaces in FLC ID. Indeed, in the presence of measurement noise, system identification algorithms require "good" excitation in order to work. In other words, the performance of system identification algorithms is dependent upon how well the data represents or spans the input and output spaces of the system. The input data for Figures 5a and 5b were randomly generated from the universe of discourse and then ran through the fuzzy XOR. This is equivalent to taking passive, or uneducated, noisy input and output measurements to use in identification. Unfortunately, the data points are not spread about and do not cover the entire input space. Thus they do not properly "excite" the entire dynamic range of the system and actually misrepresent the plant. Indeed, notice from Table 1 that spanning, or covering the input space requires sufficient points from the four input antecedents (s,s), (s,b), (b,s), (b,b). When input pairs are purposely chosen in an attempt to span the input space over [-1,2], error surfaces similar to Figures 6a and 6b are obtained. That is, for any σ, rule set seven is identified, and then minimizing the output error with respect to the variance parameter gives $\sigma^2 = 0.1$.

Hence, as is well recognized in classical system identification, good excitation is a prerequisite for successful identification. The same applies to system ID in FLC. If the data points available do not appear to span the spaces very well, heuristically one should start with a very small σ. Then, for the fixed σ find the best rule set and its goodness of fit. Increase σ and repeat until the optimal rule set changes or until σ increases beyond a reasonable value for the system. For each rule set identified as optimal for some σ, minimize with respect to a varying σ. Choose the best global fit as the "model".

It is believed that the identified rule set corresponding to the smallest σ is the global optimum with respect to the hypothesized membership functions, but no proof is provided.

This has been observed in all simulations performed. The results are found to be invariant with respect to varying the variance as well as the rule set used in generating the truth data. In this carefully controlled experiment the performance of the FLC ID algorithm is gauged by directly comparing the identified rule set to the known underlying rule set (# 7).

6   Concluding Remarks

Rule identification as presented in this paper will clearly play a central role in the application of fuzzy logic control, where a human expert's plant and/or control knowledge needs to be captured. Although classical system ID methods also operate on input/output data, FLC modeling and ID efficiently incorporate prior and/or side information on the plant, which improves the identification performance. Thus, one strives to encode all available knowledge of the system into the fuzzy model at the onset. FLC efficiently performs the critical (in ID) function of prior information incorporation into the model. Hence, FLC modeling has the potential of playing an important role in system ID.

The FLC modelling process has been illustrated in the example problems. This information adds structure to the problem, allowing a more efficient and therefore correct ID, by ameliorating the effects of poor excitation in system identification. Indeed, the inclusion of prior and/or side information has the beneficial effect of excitation enhancement required to conduct good ID. If properly posed, the optimization problem associated with FLC rule ID may lend itself to an exhaustive search, guaranteeing an optimal solution. If an exhaustive search for optimization is too expensive, the use of Genetic Algorithms is suggested, as they very naturally fit the proposed paradigm [2], [5]. The use of such a search in some illustrative examples has been presented.

In this paper, measurement noise issues are addressed and identification experiments have been performed. In our work, the quality of the ID algorithm can be validated, since the underlying truth model (which generated the data used by the fuzzy ID algorithms) is available in the experiment. That is, demonstrating a small output error is <u>not</u> relied upon as *proof of a "good"* identification. Instead, ID performance is accurately and reliably gauged by the degree of truth model recovery. The use of "real" test data and the absence of a known underlying model to "validate" ID algorithms is of little value.

Indeed, the requirement for good excitation is also shown in the context of system identification in FLC. Without it, the ID results can be misleading and have dramatic effects when the fuzzy model is used away from the operating point, where the ID experiment was performed. Hence, a word of caution that applies to all identification work is in order: Obtaining a small output error is no guarantee for a correct ID, except in the case where the identification experiment is performed under conditions of "good" excitation. Our experimental investigation of FLC ID confirms what is known from classical system identification, where the following dictum should be adhered to: If the excitation is poor, shut down the ID process. A challenging problem in its own right is the independent, viz input/output data driven, determination of the excitation level in FLC ID.

References

[1] Berenji, H. R. and P. Khedkar, "Learning and Tuning Fuzzy Logic Controllers Through Reinforcements," *IEEE Transactions on Neural Networks, 3:5* pp.724-40 (September 1992).

[2] Goldberg, David E., *Genetic Algorithms in Search, Optimization, and Machine Learning*, Reading, MA; Addison-Wesley Pub. Co.; 1989

[3] Hung, C-C and B. Fernandez., "Comparative Analysis of Control Design Techniques for a Cart-Inverted-Pendulum in Real-Time Implementation," *Proceedings of the American Control Conference,* pp.1870-73 (June 1993).

[4] Jang, J. R., "Self-Learning Fuzzy Controllers Based on Temporal Back Propagation," *IEEE Transactions on Neural Networks, 3:5* pp.714-23 (September 1992).

[5] Karr, C. L. and E. J. Gentry., "Fuzzy Control of pH Using Genetic Algorithms," *IEEE Transactions on Fuzzy Systems, 1:1* pp.46-53 (February 1993).

[6] Kosko, Bart, *Neural Networks and Fuzzy Systems: A Dynamical Systems Approach to Machine Intelligence*; Englewood, NJ; Prentice Hall, 1992.

[7] Lai, J and Y. Lin., "Fuzzy Model-Based Control of a Pneumatic Chamber," *Proceedings of the American Control Conference,* pp.1162-66 (June 1993).

[8] *MATLAB® Reference Guide*, The MathWorks Inc., 1992.

[9] Wang, L. and M. Mendel., "Generating Fuzzy Rules by Learning from Examples," *Proceedings of the IEEE Conference on Intelligent Control,* pp.263-68 (August 1991).

WSSIAA 5 (1995) pp. 221–238
© World Scientific Publishing Company

221

Control Strategies for an Endemic Disease in the Presence of Uncertainty

Lee, C.S.
*Division of Mathematics, School of Science,
National Institute of Education, Nanyang Technological University,
469, Bukit Timah Road, Singapore 1025,
Republic of Singapore*

and

Leitmann, G
*College of Engineering,
University of California, Berkeley, CA 94720,
U.S.A.*

1. Introduction

Until very recently, little research effort has been devoted to the design of bounded feedback controllers that stabilize a dynamical system at some specified level in the presence of uncertain disturbances. Recently, Corless and Leitmann [1-2] made some contributions in this area. Employing their results on bounded controllers for robust exponential convergence, mathematical modelers can now model many real life problems in which the control constraints cannot be neglected. In particular, Lee and Leitmann [3-4] have utilized these results to investigate harvesting problems of ecological systems with bounded parameter and input uncertainties. In this paper, we endeavour to demonstrate that the above mentioned results of Corless and Leitmann have applications in the modeling and control of endemic diseases as well.

As pointed out by Cromer [5], gonorrhea is the most reported communicable disease in the United States; see for example [6-12]. Those researchers have been reasonably successful in modeling the feature of seasonal fluctuation in the infected population of the disease. As for control strategies for the disease, Cromer [5] examined the effect of using a seasonal screening program and noted some decrease in the infected population.

In this paper, we consider a single population model similar to that of Cromer [5] except that all the system and input parameters contain uncertainty elements. It is desired to seek a control strategy that will stabilize the infected population at certain prescribed tolerable level.

2. Problem Formulation

Let $I(t)$ and $S(t)$ denote the fraction of infected individuals and the fraction of susceptible individuals at time t respectively. Since gonorrhea confers no immunity, immediately upon recovery, an infected individual becomes a susceptible again. We assume that the population is uniformly mixed and consists only of infected individuals and susceptibles as the incubation period of the disease is negligible. Hence $I(t) + S(t) = 1$. If it is assumed that the fraction of infected individuals increases at a rate proportional to the product of the sizes of the infected and susceptible fractions, and decreases at a rate proportional to the size of the infected

222

fraction due to medical treatment sought by infected individuals, and not by the intervention of the public health authority, then the following model results.

$$\frac{dI(t)}{dt} = a(t)I(t)\left[\,1-I(t)\,\right]-b(t)I(t) \tag{1}$$

where $a(t)$ and $b(t)$ are, respectively, the contact rate and the rate at which the fraction of infected individuals is being cured. They are time-varying parameters which are not known to us except that they are of the form

$$a(t) = a^* + \Delta a(t), \qquad b(t) = b^* + \Delta b(t)$$

with the uncertain parts $\Delta a(t)$ and $\Delta b(t)$ satisfying

$$|\Delta a(t)| \le \overline{\Delta a} < a^*, \qquad |\Delta b(t)| \le \overline{\Delta b} < b^*$$

where $a^*, b^*, \overline{\Delta a}$, and $\overline{\Delta b}$ are known positive constants.

Currently no vaccine is available for gonorrhea, so a control program of the public health authority would be aimed toward identifying infected individuals and curing them. Let $C(t)$ denote the fraction of the population screened at time t with infected individuals cured immediately. Thus our simple model with public health intervention becomes

$$\frac{dI(t)}{dt} = a(t)I(t)[1-I(t)]-b(t)I(t)-C(t)I(t) \tag{2}$$

Ideally, we would like to choose $C(t)$ to wipe out the disease. However, the cost of the control program is assumed to be proportional to the screening rate. This assumption is reasonable because, as pointed out by Cromer[5], testing is more expensive than treatment, and furthermore, most individuals screened will not need treatment ! Thus a more practical control program would be one that keeps the fraction of infected individuals always below certain prescribed low and tolerable level, say I^*. Suppose that for reasons of economy , shortage of public health workers or limited laboratory facilities, the fraction of the population screened at time t is constrained by $0 \le C(t) \le \overline{C}$, where $\overline{C}$ is a prescribed bound.

Consider the transformation

$$x(t) = \ln\frac{I(t)}{I^*}, \qquad u(t) = C(t) - \tfrac{1}{2}\overline{C} \ . \tag{3}$$

Then (2) becomes[1]

$$\frac{dx}{dt} = (a^* + \Delta a)(1 - I^* e^x) - (b^* + \Delta b) - (u + \tfrac{1}{2}\overline{C}) \tag{4}$$

which may be put in the form

$$\frac{dx}{dt} = Ax + B(u + \hat{e})$$

where

---

[1]Henceforth, for the sake of brevity, we omit the argument t.

$A = const. < 0$, $B = -1$, and $\hat{e} = a^* I^* e^x + \Delta a I^* e^x + b^* + \Delta b + \frac{1}{2}\overline{C} + Ax - a^* - \Delta a$.

Thus

$$|\hat{e}| \le k_0 + k_1(|x|) \cdot |x|$$

where

$$k_0 = \left|b^* - a^* + \tfrac{1}{2}\overline{C} + \overline{\Delta a} + \overline{\Delta b}\right|, \qquad k_1(|x|) = -A + \frac{\left(a^* + \overline{\Delta a}\right)I^* e^{|x|}}{|x|}.$$

The constraint $0 \le C(t) \le \overline{C}$ results in $|u| \le \frac{1}{2}\overline{C}$.

3. Stabilizing Control

In this section, we employ the results of [1,2] (see Appendix) to obtain a bounded, non-negative control strategy that keeps the fraction of infected individuals always at or below a certain prescribed and tolerably low level I^* regardless of the realization of the uncertain elements Δa and Δb.

Let

$$u = p(x) = -\rho\, s\left(\varepsilon^{-1} BPx\right) - \tilde{\rho}\, sat\left(\tilde{\rho}^{-1}\gamma(|x|)BPx\right) \tag{5}$$

where

$$\rho = k_0, \quad \tilde{\rho} = \overline{\rho} - \rho, \quad \overline{\rho} = \tfrac{1}{2}\overline{C}, \quad and \quad \gamma(|x|) = \tfrac{1}{2}\left[\sigma + \mu^{-1}k_1(|x|)^2\right]$$

and, σ and μ are two arbitrarily chosen positive numbers. The functions $s(\cdot)$ and $sat(\cdot)$ respectively, are given by

$$s(z) = \frac{z}{1+|z|}, \qquad sat(z) = \begin{cases} z, & if \ |z| \le 1 \\ \dfrac{z}{|z|}, & if \ |z| > 1 \end{cases}$$

and the solution P of the Riccati equation in the Appendix is

$$P = \frac{(\alpha + A) + \sqrt{(\alpha + A)^2 + \sigma\mu}}{\sigma}$$

The positive scalar ε is chosen sufficiently small to satisfy $\varepsilon < \dfrac{\alpha c^2}{k_0}$ where c is a positive number that satisfies

$$\frac{\sqrt{P} \cdot \left[\sigma\mu + k_1 \left(\frac{c}{\sqrt{P}} \right)^2 \right] c}{2\mu} \leq \tilde{\rho} \tag{6}$$

It suffices to constrain $x(0)$ to $A = \left\{ x \mid Px^2 \leq c^2 \right\}$ to ensure uniform exponential convergence of x to the neighborhood $B(r_\varepsilon) = \left\{ x \mid |x| \leq r_\varepsilon \right\}$ of the origin with rate α, where

$$r_\varepsilon = \left[\frac{\varepsilon k_0}{\alpha P} \right]^{\frac{1}{2}}$$

4. Simulation Results

For simulation purposes, we employ three sets of parameter values which are given in Table 1.

Table 1: Three sets of parameter values

	Set 1	Set 2	Set 3
a^*	0.0625	0.1250	0.250
b^*	0.0182	0.0182	0.0182
I^*	0.10	0.10	0.10
$\overline{C}$	0.35	0.50	0.75
$\overline{\Delta a}$	0.0063	0.0125	0.0250
$\overline{\Delta b}$	0.0018	0.0018	0.0018
α	0.15	0.15	0.15
σ	1	1	2
A	-0.20	-0.20	-0.20
μ	2	2	2
ε	0.0025	0.005	0.005

Simulations were carried out for three realizations of the parameter uncertainties:

$$\begin{cases} \Delta a(t) = \overline{\Delta a} \\ \Delta b(t) = -\overline{\Delta b} \end{cases} \tag{7}$$

$$\begin{cases} \Delta a(t): \text{random variable } \in \left[-\overline{\Delta a}, \overline{\Delta a} \right] \\ \Delta b(t): \text{random variable } \in \left[-\overline{\Delta b}, \overline{\Delta b} \right] \end{cases} \tag{8}$$

$$\begin{cases} \Delta a(t) = \overline{\Delta a}\cos 2\pi t \\ \Delta b(t) = -\overline{\Delta b}\cos 2\pi\left(t - \frac{1}{12}\right) \end{cases} \tag{9}$$

and for four initial values of x:

$$x(0) = \pm 0.0856, \quad x(0) = 1.6094379, \quad and \quad x(0) = 2.1972246.$$

The first two initial values of x, namely, $x(0) = 0.0856$ $(I(0) = 0.108937)$ and

$x(0) = -0.0856$ $(I(0) = 0.0917961)$ are both in $A = \left\{x \mid Px^2 \leq c^2\right\}$ with $c = 0.1$. The

latter two initial values of x, namely, $x(0) = 1.6094379$ ($I(0) = 0.5$) and $x(0) = 2.1972246$

$(I(0) = 0.9)$ do not belong to A and are included to illustrate the conservativeness of A.

Figures corresponding to the various parameter sets and the various realizations of parameter uncertainties are given in Table 2 . Figures 1(a) and 1(b) show the time histories of the fraction of infected individuals with initial values $I(0) = 0.109$ and $I(0) = 0.5$, respectively, and subject to no control program of the public health authority. Uncertain disturbances are absent in Figure 1(a), but present in Figure 1(b). In the latter, the time histories of the fraction of infected individuals correspond to the parameter Set 3 and the realization of parameter uncertainty (7).

Table 2: Figures corresponding to the various parameter sets and the various realizations of parameter uncertainties

	Set 1	Set 2	Set 3
realization of parameter uncertainty (7)	Fig.3	Figs.2 & 4	Figs.1(b) & 5
realization of parameter uncertainty (8)		Fig.8	
realization of parameter uncertainty (9)		Fig.6	Fig.7

Figures 2(a)-(b), 3(a)-(b), 4(a)-(b), and 5(a)-(b) also correspond to the realizations of the parameter uncertainties (7), but with the parameter values chosen from different parameter sets (see Table 2). Figure 2(a) shows the time histories of the fraction of infected individuals with initial values $I(0) = 0.109$ and $I(0) = 0.092$, respectively. The time histories of the corresponding constrained stabilizing screening rates are depicted in Figure 2(b). Figures 3(a), 4(a), and 5(a) correspond to $I(0) = 0.5$, each of Figures 3(a), 4(a), and 5(a) displays the time histories of both the fraction of infected individuals and the constrained stabilizing screening rate used. Similarly, Figures 3(b), 4(b), and 5(b) correspond to $I(0) = 0.9$, and each of Figures 3(b), 4(b), and 5(b) shows the time

226

histories of both the fraction of infected individuals and the constrained stabilizing screening rate employed.

Figures 6(a)-(b) and 7(a)-(b) correspond to the realizations of the parameter uncertainties (9). Again, Figures 6(a) and 7(a) correspond to $I(0) = 0.5$, and each of Figures 6(a) and 7(a) depicts the time histories of both the fraction of infected individuals and the constrained stabilizing screening rate employed. Figures 6(b) and 7(b) correspond to $I(0) = 0.9$, and each of Figures 6(b) and 7(b) shows the time histories of both the fraction of infected individuals and the constrained stabilizing screening rate used.

Finally, Figures 8(a)-(b) correspond to the realizations of the parameter uncertainties (8). Figure 8(a) shows the time histories of both the fraction of infected individuals and the corresponding constrained stabilizing screening rate with $I(0) = 0.5$, while Figure 8(b) depicts the time histories of both the fraction of infected individuals and the corresponding constrained stabilizing screening rate with $I(0) = 0.9$.

From Figures 1(a)-(b), we note that whether uncertain disturbances are present or not, the fraction of infected individuals in our model tends to increase with time if there is no control program of the public health authority. Figures 2(a)-(b) illustrate that when $x(0) \in A = \left\{ x \mid |x| \le 0.0856 \right\}$, uniform exponential convergence of x to the neighborhood $B(r_\varepsilon) = \left\{ x \mid |x| \le 0.062 \right\}$ of the origin with rate α is assured. In Figures 3(a)-(b), it is noted that a relatively low screening rate takes a much longer time to bring large initial fractions of infected population to the desired prescribed low level I^*. We also observe in Figures 7(a)-(b) that corresponding to sinusoidal realizations of uncertain disturbances in the contact rates and recovery rates, the stabilizing screening rate also displays very mild sinusoidal behavior. Furthermore, as depicted in both Figures 5(a)-(b) and 7(a)-(b), for a high contact rate, the fraction of population to be screened has to be more than the desired fraction of the infected population in order to achieve the goal of keeping the infected population at or below the desired level I^*.

5. Conclusion

In this paper, we employed recent results in robust control theory to devise screening strategies which assure stabilization of the infected population at an acceptable level in spite of uncertain knowledge of the model parameters. Simulations were carried out for three sets of parameters and three realizations of parameter uncertainties. As expected, a bigger screening rate is required to control the infected population at or below a prescribed desired level for a bigger contact rate. Equally expected is the result that corresponding to the realizations of sinusoidal parameter uncertainties (9), the constrained screening rate also displays sinusoidal behavior.

Finally, gonorrhea is a communicable disease, and we assume that men infect only women, and women infect only men. As noted by Cromer [5], the present model does not take into consideration the different responses to the disease by males and females. Under certain assumptions, Hethcote and Yorke [8] obtained the result that screening females for gonorrhea is much more effective than screening males! Thus, an extension of the single population model discussed above would be to consider a model incorporating the sex differences. Another extension of interest is one that allows for same sex contacts.

6. APPENDIX

Appendix A

Consider the system

$$\frac{dx(t)}{dt} = f(\,t, x(t)\,), \tag{A.1}$$

where $t \in R$, $x(t) \in R^n$. For any scalar $r \geq 0$, the ball of radius r is defined by

$$B(r) = \{x \in R^n : \|x\| \leq r\}.$$

Consider a scalar $\alpha > 0$ and a set $A \subset R^n$ containing a neighborhood of 0.

Definition A.1. System (A.1) is uniformly exponentially convergent to $B(r)$ with rate α and region of attraction A if there exists a scalar $\beta \geq 0$ such that the following hold.

(i) <u>Existence of solutions</u>. For each $t_0 \in R$ and $x_0 \in A$ there exists a solution $x(\cdot) : [t_0, t_1) \to R^n$, $t_0 < t_1$, of (A.1) with $x(t_0) = x_0$.

(ii) <u>Indefinite extension of solutions</u>. Each solution $x(\cdot) : [t_0, t_1) \to R^n$ of (A.1), with $x_0 \in A$, has an extension $\bar{x}(\cdot) : [t_0, \infty) \to R^n$, i.e., $\bar{x}(t) = x(t) \; \forall \, t \in [t_0, t_1)$ and $\bar{x}(\cdot)$ is a solution of (A.1).

(iii) <u>Uniform exponential convergence of solutions</u>. If $x(\cdot) : [t_0, \infty) \to R^n$ is any solution of (A.1) with $x(t_0) \in A$, then

$$\|x(t)\| \leq r + \beta\|x(t_0)\|e^{-\alpha(t-t_0)} \; \forall \; t \geq t_0.$$

Definition A.2. System (A.1) is <u>globally uniformly exponentially convergent</u> to $B(r)$ if it is exponentially convergent with R^n as a region of attraction.

Appendix B

Consider uncertain systems described by

$$\frac{dx(t)}{dt} = Ax(t) + \sum_{i=1}^{l}\left[B_i u_i(t) + \Delta F_i(t, x(t), u_i(t))\right] \tag{B.1}$$

where $t \in R$ is the time variable, $x(t) \in R^n$ is the state and $u_i(t) \in R^{m_i}$, $i = 1, \cdots, l$ are control inputs. The terms ΔF_i, $i = 1, \cdots, l$ are assumed to be continuous; they represent all the uncertainty, nonlinearity and time-dependence in the system. The constant matrices A and B_i, $i = 1, \cdots, l$, are known; they define the nominal system

$$\frac{dx(t)}{dt} = Ax(t) + Bu(t) \tag{B.2}$$

where $u = (u_1, u_2, \cdots, u_l)^T$, $B = (B_1, B_2, \cdots, B_l)$. Each control input u_i is subject to the constraint $\|u_i\| \le \overline{\rho_i}$ where the bound $\overline{\rho_i} > 0$ is prescribed.

Before we state a theorem of [2] on exponential convergence of all closed loop state trajectories originating from a bounded region to a neighborhood of the origin with a desired rate of convergence and with every control input bounded, we make the following assumptions.

Assumption B.1. The nominal system (B.2) is controllable.

Assumption B.2. For each $i = 1, 2, \cdots, l$, there is a function e_i such that $\Delta F_i = B_i e_i$.

Assumption B.3. For each $i = 1, 2, \cdots, l$, there exist scalars k_{0i}, k_{2i} with $k_{2i} < 1$, and a continuous non-decreasing function k_{1i} such that

$$\left\|e_i(t, x, u_i)\right\| \le k_{0i} + k_{1i}(\|x\|)\|x\| + k_{2i}\|u_i\| \quad \forall t \in R, \ x \in R^n, \ u_i \in R^{m_i}.$$

Assumption B.4. For each $i = 1, 2, \cdots, l$, $\overline{\rho_i} > \dfrac{k_{0i}}{1 - k_{2i}}$.

Theorem B.1 Consider an uncertain system described by (B.1) satisfying Assumptions B.1 - B.4. The bounded control given by

$$u_i^c = p_i^c(x) = -\rho_i s(\varepsilon^{-1} B_i^T Px) - \tilde{\rho}_i sat(\tilde{\rho}_i^{-1}\gamma_i(\|x\|)B_i^T Px) \tag{B.3}$$

ensures uniform exponential convergence of all state trajectories of the closed loop system

$$\dot{x}(t) = Ax(t) + \sum_{i=1}^{l}\left[B_i p_i(x) + \Delta F_i(t, x, p_i(x)) \right]$$

to $B(r_\varepsilon)$ with rate α and region of attraction $A = \left\{ x \in R^n : x^T Px \leq c^{*2} \right\}$ where

$$\gamma_i\left(\|x\|\right) = \frac{\sigma + \frac{l}{\mu}k_{1i}\left(\|x\|\right)^2}{2\left(1 - k_{2i}\right)}, \quad \rho_i = \frac{k_{0i}}{1 - k_{2i}}, \quad \tilde{\rho}_i = \overline{\rho}_i - \rho_i$$

and P is a positive definite symmetric matrix satisfying the Riccati equation

$$P(A + \alpha I) + (A + \alpha I)^T P - \sigma PBB^T P + \mu I = 0$$

for σ, α, $\mu > 0$. The saturation function $sat(\cdot)$ is given by

$$sat(z) = \begin{cases} z & if \ \ \|z\| \leq 1 \\ \dfrac{z}{\|z\|} & if \ \ \|z\| > 1 \end{cases}$$

The positive real scalar ε is chosen sufficiently small to satisfy

$$\varepsilon < \varepsilon^* = \frac{\alpha c^{*2}}{k_0} \quad \text{with} \ \ c^* = \min\left\{ c_i, \ i = 1, 2, \cdots l \right\}$$

where $c_i > 0$ satisfies

$$\phi_i(c_i) \leq \tilde{\rho}_i,$$

$$\phi_i(c_i) = \frac{\lambda_i\left[\sigma\mu + lk_{1i}\left(\lambda c_i\right)^2 \right]c_i}{2\mu\left(1 - k_{2i}\right)},$$

$$\lambda_i = \lambda_{max}\left(B_i^T PB_i\right)^{\frac{1}{2}}$$

$$\lambda = \lambda_{min}\left(P\right)^{-\frac{1}{2}}$$

7. References

[1] Corless, M., and Leitmann, G., Bounded controllers for robust exponential convergence, Journal of Optimization Theory and Applications, Vol.76, pp.1-12, 1993.

230

[2] Corless, M., and Leitmann, G., Componentwise bounded controllers for robust exponential convergence, Proceedings of the Variable Structure and Lyapunov Theory Workshop, Sept. 7-9, 1994, Benevento, Italy, pp.64-69.

[3] Lee, C.S., and Leitmann, G., A bounded harvest strategy for an ecological system in the presence of uncertain disturbances, Proceedings of the International Workshop on Intelligent Systems and Innovative Computations -- The 6th Bellman Continuum, August 1-2, 1994, Hachioji, Tokyo, Japan.

[4] Leitmann, G., and Lee, C.S., Bounded harvest strategies for a two species system based on an uncertain model, Seventh Workshop on Dynamics and Control, July 17-20, 1994, Ulm, Germany.

[5] Cromer, T.L., Seasonal control for an endemic disease with seasonal fluctuations, Theoretical Population Biology, Vol.33, pp.115-125, 1988.

[6] Hethcote, H.W., Asymptotic behavior in a deterministic epidemic model, Bulletin of Mathematical Biology, Vol.35, pp.607-614, 1973.

[7] Lajmanovich, A., and Yorke, J.A. A deterministic model for gonorrhea in a nonhomogeneous population, Mathematical Biosciences, Vol.28, pp.221-236, 1976.

[8] Hethcote, H.W., and Yorke, J.A., Gonorrhea Transmission Dynamics and Control, Springer-Verlag, New York, 1985.

[9] Cooke, K.L., and Kaplan, J.L., A periodicity threshold theorem for epidemics and population growth, Mathematical Biosciences, Vol.31, pp.87-104, 1976.

[10] Smith, H.L., On periodic solutions of a delay integral equation modeling epidemics, Journal of Mathematical Biology, Vol.4, pp.69-80, 1977.

[11] Nussbaum, R.D., A periodicity threshold theorem for some nonlinear integral equations SIAM Journal of Mathematical Analysis, Vol.9, pp.356-376, 1978.

[12] Cromer, T.L., Asymptotically periodic solutions to Volterra integral equations in epidemic models, Journal of Mathematical Analysis and Applications, Vol.110, No.2, pp.483-494,1985.

[13] Yorke, J.A., Hethcote, H.W., and Nold, A. Dynamics and control of the transmission of gonorrhea, Sex. Transm. Dis. Vol.5, No.2, pp.51-56, 1978.

1.0
0.9
0.8
0.7
0.6
0.5
0.4
0.3
0.2
0.1
0
I(t)
I(t)
I*
0
1
2
3
4
5
6
TIME IN YEARS

FIG.1(a)

1.0
0.9
0.8
0.7
0.6
0.5
0.4
0.3
0.2
0.1
0
I(t)
I(t)
I*
0
1
2
3
4
5
6
TIME IN YEARS

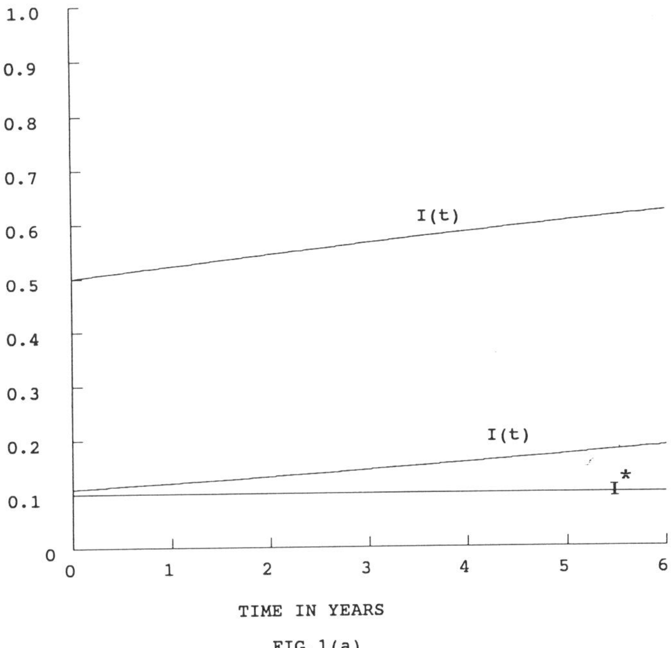

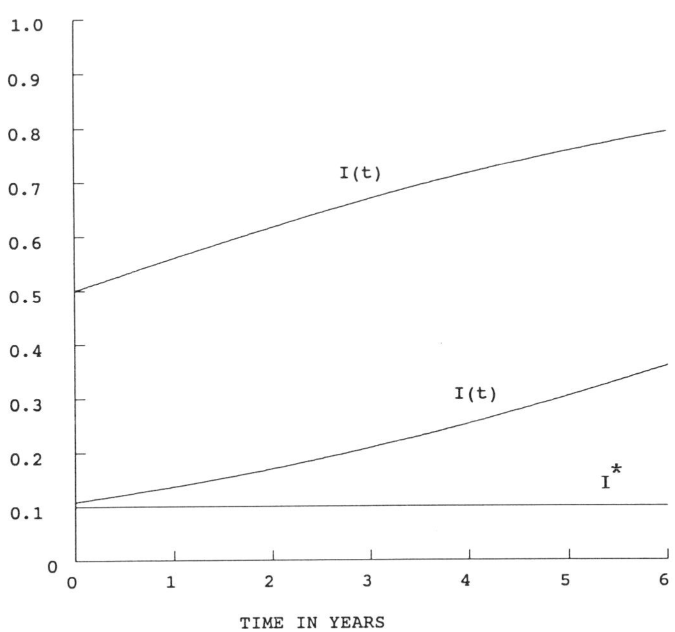

FIG.1(b)

232

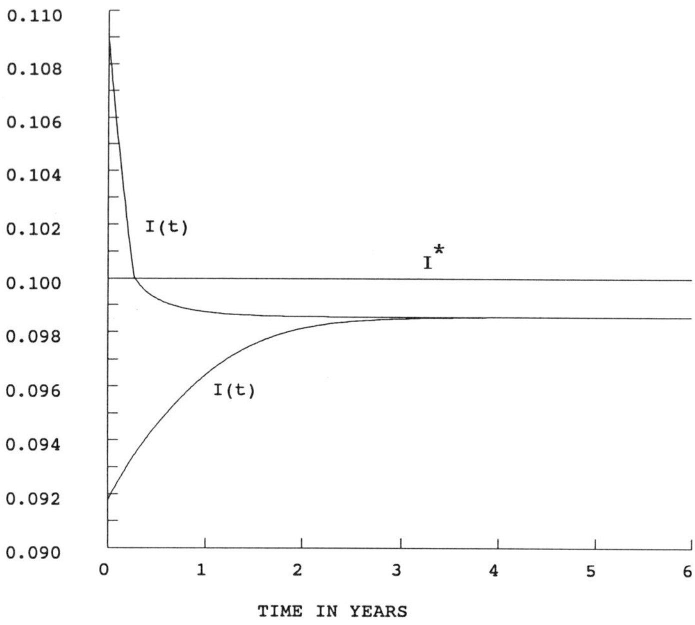

FIG.2(a)

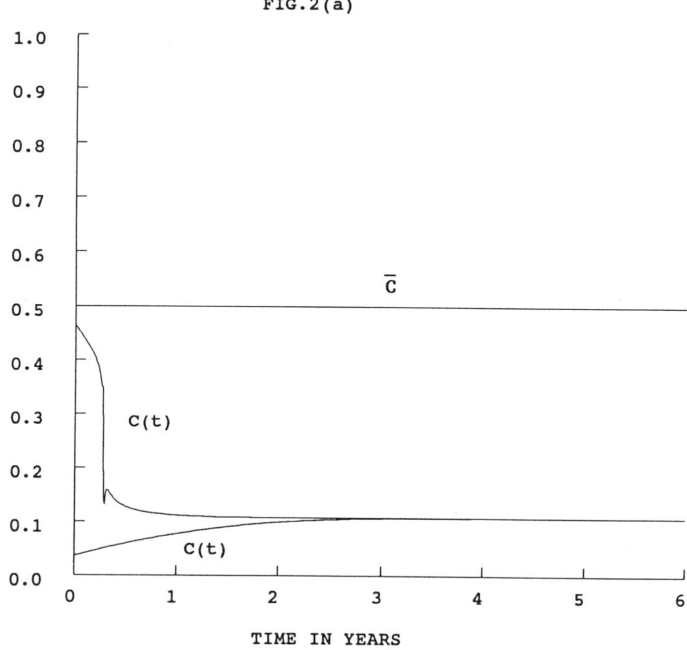

FIG.2(b)

1.0
0.9
0.8
0.7
0.6
0.5
0.4
0.3
0.2
0.1
0
$\overline{C}$
$I(t)$
$C(t)$
I^*
0
1
2
3
4
5
6
TIME IN YEARS

FIG.3(a)

1.0
0.9
0.8
0.7
0.6
0.5
0.4
0.3
0.2
0.1
0
$I(t)$
$\overline{C}$
$C(t)$
I^*
0
1
2
3
4
5
6
TIME IN YEARS

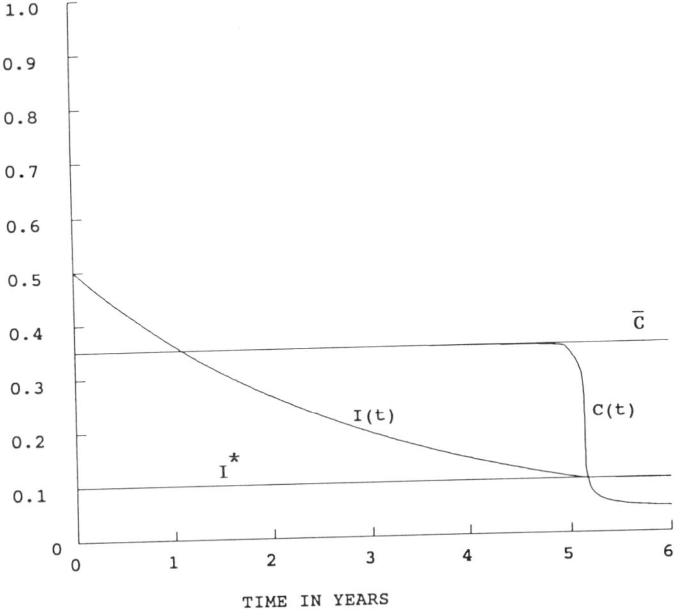

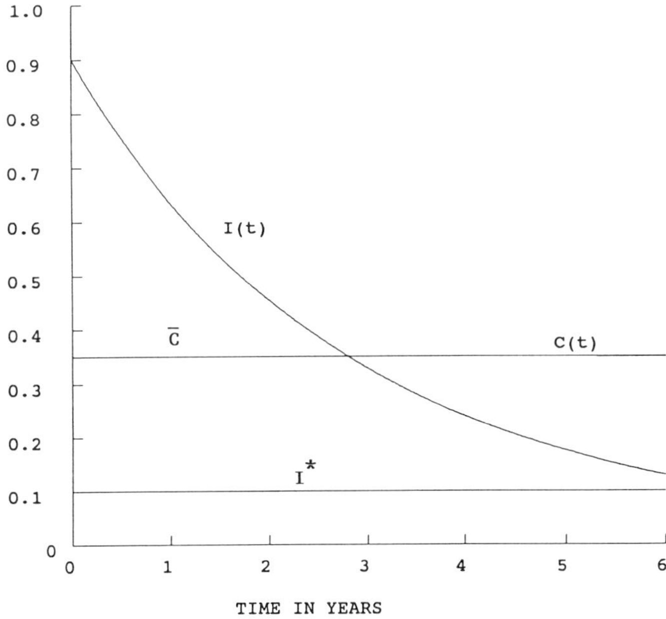

FIG.3(b)

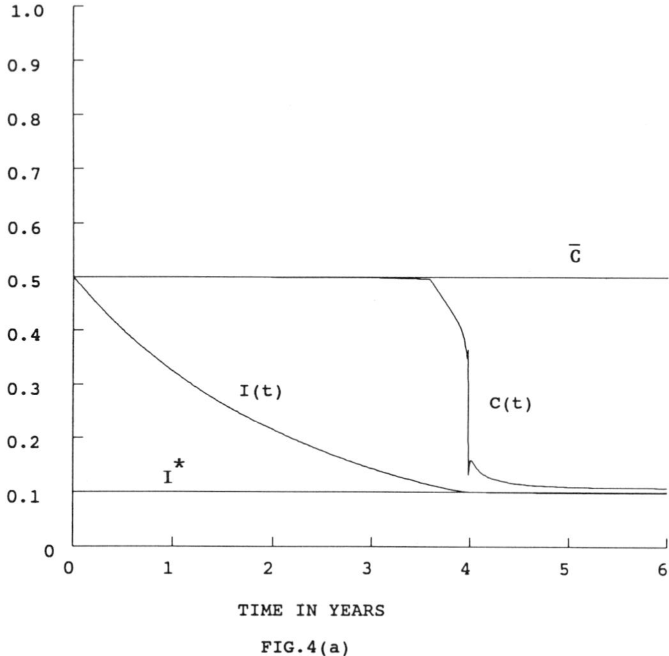

FIG.4(a)

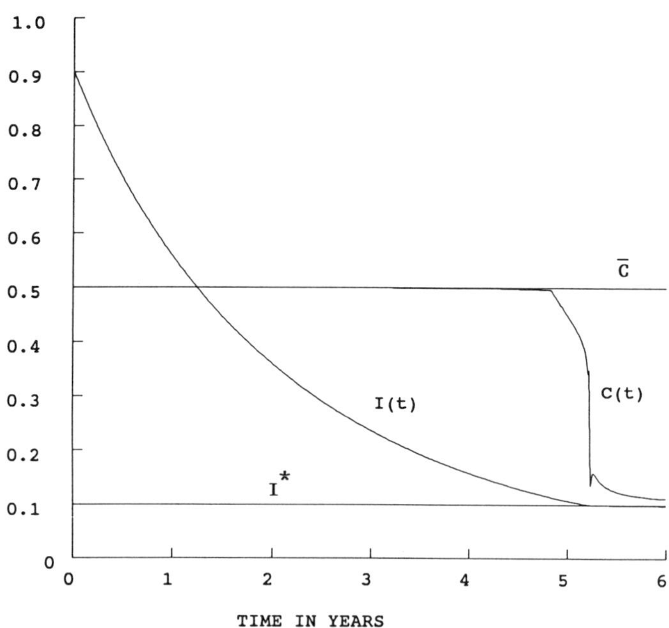

1.0
0.9
0.8
0.7
0.6
C̄
0.5
0.4
0.3
I(t)
0.2
C(t)
I*
0.1
0
0   1   2   3   4   5   6
TIME IN YEARS

1.0
0.9
0.8
0.7
0.6
C̄
0.5
0.4
0.3
I(t)
C(t)
0.2
I*
0.1
0
0   1   2   3   4   5   6
TIME IN YEARS

FIG.4(b)

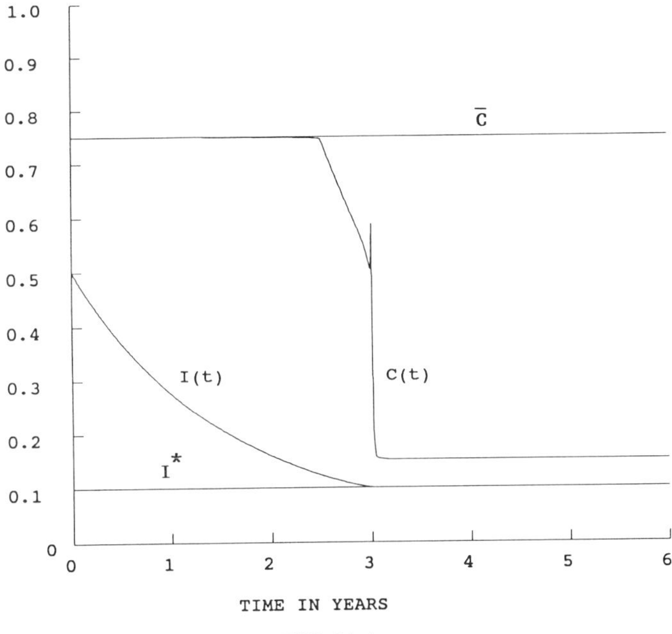

FIG.5(a)

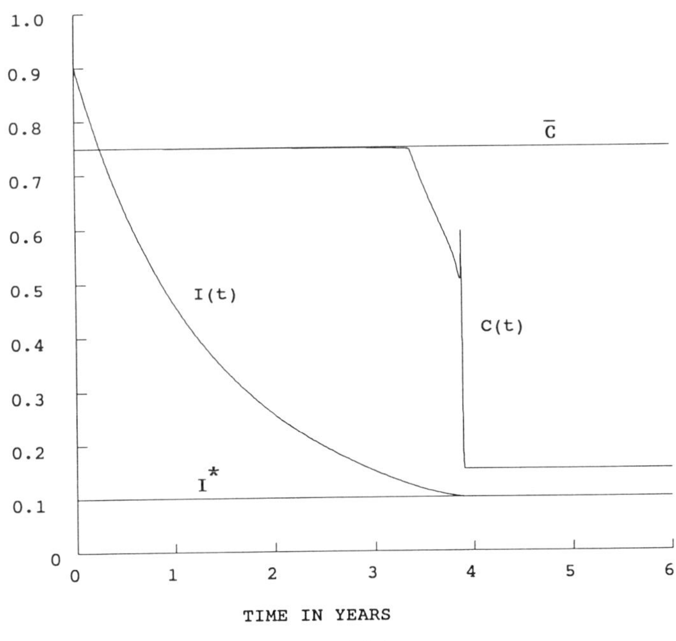

FIG.5(b)

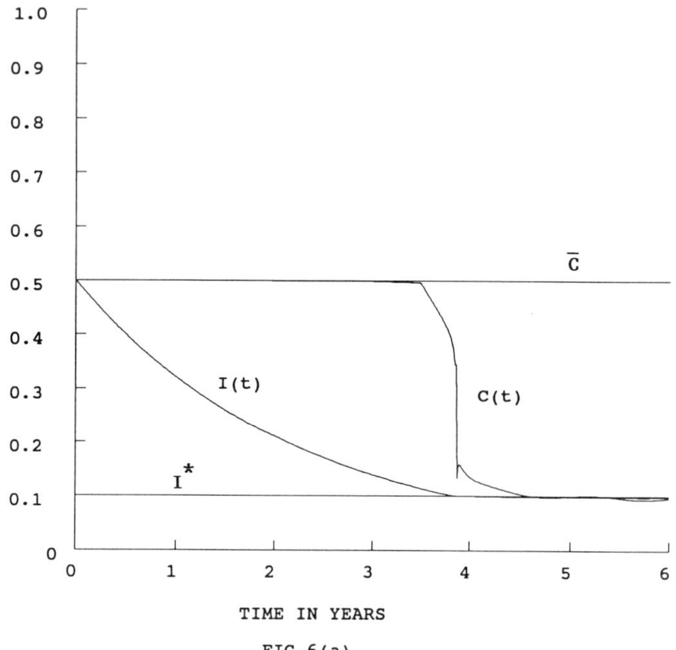

FIG.6(a)

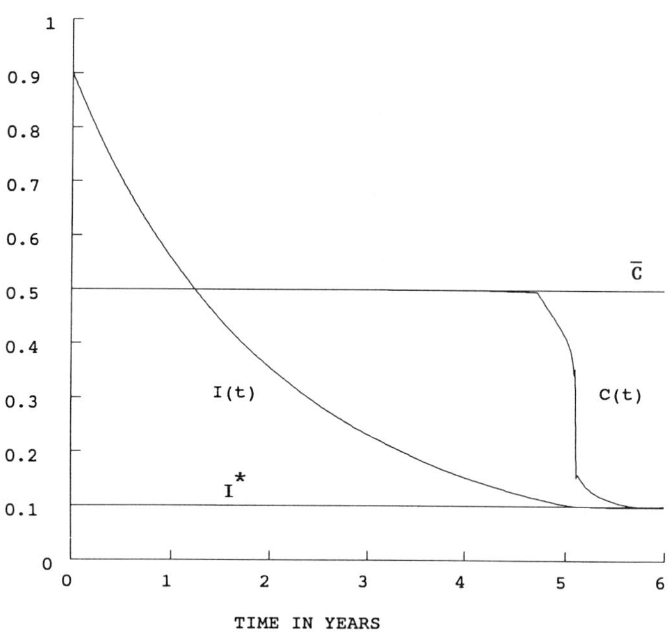

FIG.6(b)

1.0
0.9
0.8
0.7
0.6
0.5
0.4
0.3
0.2
0.1
0
C̄
C(t)
I(t)
I*
0 1 2 3 4 5 6
TIME IN YEARS

FIG.7(a)

1
0.9
0.8
0.7
0.6
0.5
0.4
0.3
0.2
0.1
0
C̄
C(t)
I(t)
I*
0 1 2 3 4 5 6
TIME IN YEARS

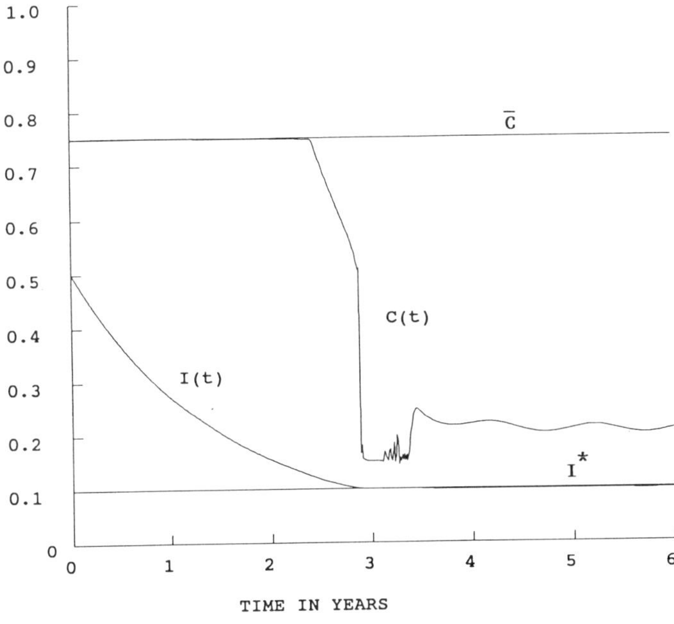

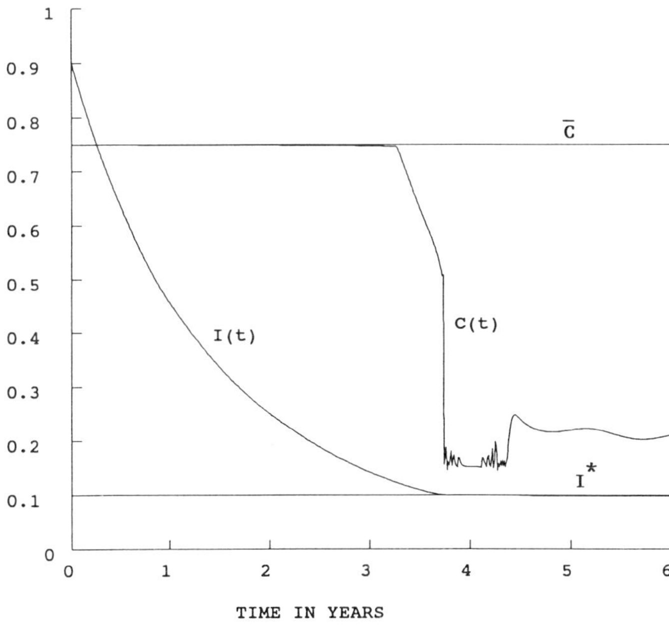

FIG.7(b)

238

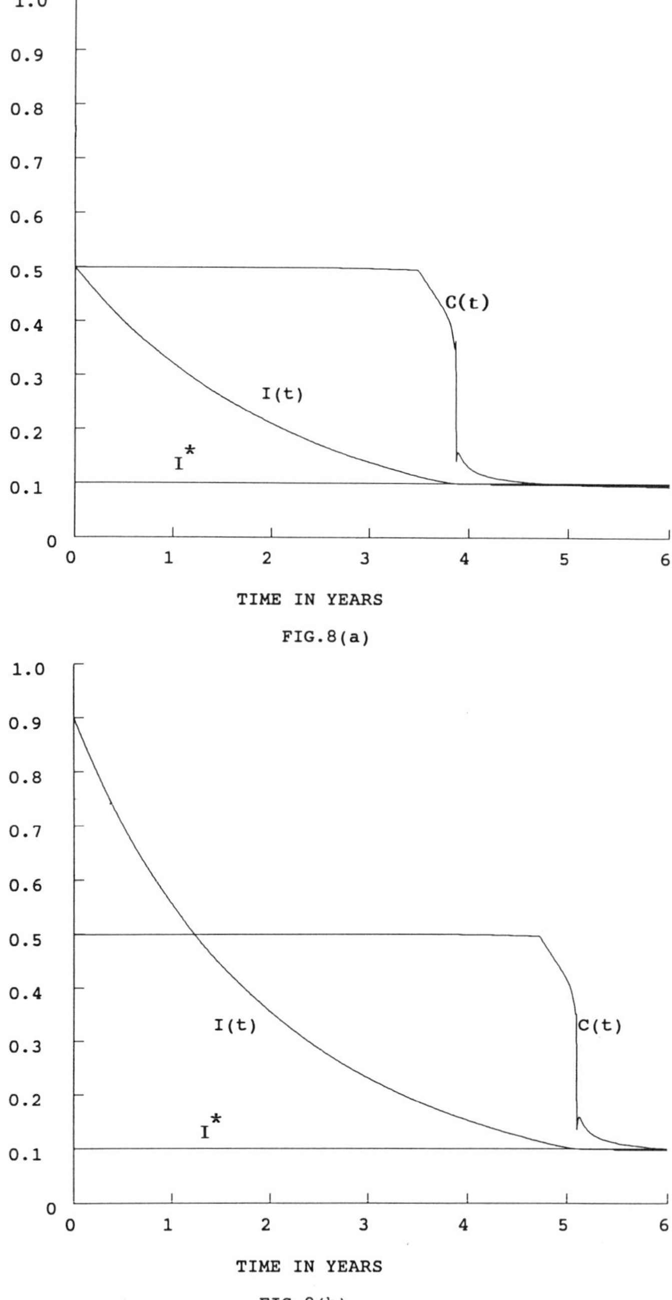

FIG.8(a)

FIG.8(b)

WSSIAA 5 (1995) pp. 239–254
© World Scientific Publishing Company

ON SECOND-ORDER SUFFICIENT CONDITIONS
IN SMOOTH NONLINEAR PROGRAMS

Jiming Liu
Department of Operations Research
George Washington University
Washington, DC 20052, USA

Bao-Guang Liu
Department of Applied Mathematics
Beijing Institute of Technology
Beijing, P.R. China

June 4, 1992
(Revised October 31, 1994)

Abstract

There have been a lot of efforts made in deriving various second-order sufficient conditions for local minima, strict local minima and isolated local minima in nonlinear programming. In the paper we aim at trying to synthesize ideas scattering in the literature concerning second-order sufficient conditions in smooth nonlinear programs, and in a sense to exhaust several approaches from a systematic investigation. Sharp second-order sufficient conditions for local minima, strict local minima, and isolated local minima are derived. The results obtained here recover many known second-order sufficient conditions in the literature.

Key Words. Smooth nonlinear programming, second-order sufficient condition, local minimum, strict local minimum, isolated local minimum.

1. Introduction

This paper is concerned with second-order sufficient conditions for local minima, strict local minima, and isolated local minima of a general nonlinear programming problem of the form:

$$\begin{aligned}
&\text{minimize } f(x), \\
&\text{subject to } g(x) \in Q^\circ, \\
&\qquad\qquad x \in C,
\end{aligned} \tag{1.1}$$

where f and g are functions from R^n to R and R^m respectively, $C \subset R^n$ is a closed convex set, $Q \subset R^n$ is a closed convex cone, and Q° is the polar cone of Q:

$$Q^\circ := \{y \in R^m : \langle q, y \rangle \leq 0 \text{ for each } q \in Q\}.$$

This formulation was first proposed and investigated by Robinson [28,30] and includes the so-called standard nonlinear programming problem as a special case in which $C = R^n$ and $Q = R_+^k \times R^\ell$, i.e.,

$$\begin{aligned}
&\text{minimize } f(x), \\
&\text{subject to } g_i(x) \leq 0 \quad (i = 1, \ldots, k), \\
&\qquad\qquad g_i(x) = 0 \quad (i = k+1, \ldots, k+\ell).
\end{aligned} \tag{1.2}$$

For convenience of discussion, we make the blanket assumption that f and g are continuous F-differentiable in some neighborhood of the point of interest. For a set $S \subset R^t$ and a point $y \in R^t$, we denote the normal cone and the tangent cone of S at y by $N_s(y)$ and $T_s(y)$, respectively.

The second-order sufficient conditions in nonlinear programs are old issues and have been extensively studied by many authors. Roughly speaking, these conditions involve the Hessians of the Lagrangian functions, utilize the quadratic information of the problem functions, and provide not only as sufficient conditions for local minima, strict local minima and isolated local minima but as basic assumptions for the convergence analysis of algorithms. The second-order sufficient conditions in their early forms go back at least to McShane [26], Hestenes [13] and Penissi [27]. Mainly due to the pioneer work of Fiacco and McCormick [9], second-order sufficient conditions have become standard regularity conditions and numerical variants have been derived. Existing literature concerning this topic is vast. For references on the topic, we recommend [3] for convex programs, [4,9,12,30] for smooth nonlinear programs, [15,24] for infinite-dimensional programming problems, and [5,7,31] for nonsmooth programs.

We are truly impressed by not only so much efforts have been made in deriving various second-order sufficient conditions in smooth nonlinear programs, often for the standard problem (1.2), but also by their roles as in the studies of diverse local convergence analyses and stability and sensitivity analysis. In the paper we aim at trying to synthesize ideas scattering in the literature concerning second-order sufficient conditions in smooth nonlinear programs, and in a sense to exhaust several approaches from a systematic investigation. Nevertheless, some ideas were not included, for example, asymptotic conditions of Fiacco and McCormick [10], primal sufficient conditions of Ben-Tal [4], and others. Recently, there are increasing interests in the exploration of second-order sufficient conditions for nonsmooth programs and the results being obtained appear to follow essentially early approaches for smooth nonlinear programs. We hope this work would also be a guide for this kind of study.

We should mention that historically there was some confusion surrounding the distinction between strict and isolated local minima under assumptions of second-order sufficient conditions. For instance many authors called a strict local minimum an isolated local minimum. It seems that Robinson [30] is the first who pointed out that a strict local minimum under standard second-order sufficient conditions need not be an isolated local minimum. He also obtained somewhat strong conditions for isolated local minima. We shall emphasize this issue and establish rather weak conditions to assure the isolatedness of local minima. It is hoped that our work would provide a deep insight into the optimality conditions for the constrained case.

The organization of the paper is as follows: In Section 2, we shall briefly review the first-order optimality conditions and some important subsets of the feasible set as well as some important sets of directions on which behaviors of the problem functions are crucial in the local study of minima. Our investigation is divided into three main sections, from Section 3 to Section 5, concentrating on sufficient conditions for local minima, strict local minima, and isolated local minima, respectively. In Section 3, we shall provide second-order sufficient conditions for local minima which are not necessarily strict local minima. The feature of the section is the use of neighborhood information to derive second-order sufficient conditions, which apply to degenerate cases. Then, in the fourth section, second-order sufficient conditions to ensure strict local minima and the second-order growth condition are considered. The derivation of weakest condition is our main concern in the first part of the section, attempting to include most standard second-order sufficient conditions for smooth programs in the literature. In the second part of the section second-order sufficient conditions utilizing neighborhood information will be provided, which complement the results in the first part in the same section. Finally, in Section 5 we obtain very general sufficient conditions for isolated local minima. In particular, we obtain a shape result that for the polyhedral case where Q is a convex polyhedral cone (as is usually the case in applications), under the assumptions of the constraint regularity and the second-order growth condition (which is weaker than standard second-order sufficient conditions), a strict local minimum is indeed an isolated local minimum. To make the reader easier to grasp the nature of the problems, we prefer to presenting the proofs of results in the appendix instead of giving the proofs just after stating them.

2. Preliminaries

In this section we shall present the basic definitions and various regularity conditions that will be used in the subsequent development.

We denote the feasible set of (1.1) by K, i.e., $K = \{x \in R^n : g(x) \in Q^\circ, x \in C\}$. A point x_0 is a local minimum of (1.1) if $f(x) \geq f(x_0)$ for all $x \in K \cap N(x_0)$, where $N(x_0)$ is some neighborhood of x_0. It is a strict local minimum of (1.1) if $f(x) > f(x_0)$ for all $x \in K \cap N(x_0)$, $x \neq x_0$. Finally, $x_0 \in K$ is said to be an isolated local minimum of (1.1) if it is the unique local minimum of (1.1) in some neighborhood $N(x_0)$. A local minimum may not be a strict local minimum. An isolated local minimum is also a strict local minimum, but not vice versa.

Before getting involved in detail on second-order conditions, we shall briefly review first-order conditions in nonlinear programs. A basic first-order optimality condition for (1.1) is the so-called Fritz-John type conditions [28]: If x_0 is a local minimum of (1.1), then there exist multipliers $u_0 \in R_+$,

$u = (u_1, \ldots, u_m) \in Q$ such that $(u_0, u) \neq 0$ and

$$0 \in \begin{bmatrix} \nabla_x L(x_0, u_0, u) \\ -g(x_0) \end{bmatrix} + N_{C \times Q}(x_0, u), \tag{2.1}$$

where $L(x, u_0, u) = u_0 f(x) + \langle u, g(x) \rangle$. We shall call such x_0 a Fritz-John stationary point, and call the vector (x_0, u_0, u) a Fritz-John point of (1.1). The set of multipliers satisfying (2.1) and its normalized set will be denoted by $FJ(x_0)$ and $NFJ(x_0)$, respectively. Furthermore, if certain regularity conditions, called constraint qualifications, are satisfied at x_0, then (2.1) holds with some multipliers (u_0, u) such that $u_0 = 1$. This is the so-called Karush-Kuhn-Tucker conditions for the general nonlinear programming problem (1.1). A point x_0 that satisfies (2.1) with some (u_0, u) such that $u_0 = 1$ is called a Karush-Kuhn-Tucker (KKT) stationary point, or for simplicity, a stationary point of (1.1) and the pair (x_0, u) a Karush-Kuhn-Tucker (KKT) point. In this case, we shall use $KKT(x_0)$ to denote $\{u : (1, u) \in FJ(x_0)\}$ and simply write $L(x, u)$ to denote $L(x, 1, u)$. A popular qualification for (1.1) is the requirement that the constraints

$$\begin{aligned} g(x) &\in Q^{\circ}, \\ x &\in C \end{aligned}$$

of (1.1) be regular at x_0 in the sense of [29]; that is, that

$$0 \in \text{int}\{g(x_0) + \nabla g(x_0)(C - x_0) - Q^{\circ}\}. \tag{2.2}$$

For the standard problem (1.2), the condition (2.1) is equivalent to the well-known Mangasarian-Fromovitz constraint qualification.

Suppose x_0 is a Fritz-John stationary point. We shall relate it with several subsets of the feasible set and some subsets of the set of normalized directions in R^n. These sets play an important role in the derivation of second-order sufficient conditions and are defined as follows:

$$R(x_0) := \{x \in K : f(x) \leq f(x_0)\},$$

$$R(x_0; u) := \{x \in K : f(x) - f(x_0) \geq 0, \ g(x) \in N_Q(u)\},$$

$$R(x_0; u_0, u) := \{x \in K : u_0(f(x) - f(x_0)) = 0, \ g(x) \in N_Q(u)\},$$

where $(u_0, u) \in FJ(x_0)$ for some u_0. We shall call $R(x_0)$ the critical domain at x_0.

The normalized tangent cone of $R(x_0)$ at x_0 is denoted as $S(x_0)$. The normalized linearized cone of $R(x_0)$ is defined to be

$$\begin{aligned} D(x_0) := \{z \in R^n : \ &z \in T_C(x_0), \ \nabla_x f(x_0) z \leq 0, \\ &\nabla_x g(x_0) z \in T_{Q^{\circ}}(g(x_0)), \ \|z\| = 1\}. \end{aligned}$$

An element of the above set is usually called a critical direction at x_0 in the literature.

Similarly, we use $S(x_0; u)$ and $S(x_0; u_0, u)$ to denote the normalized tangent cones of $R(x_0; u)$ and $R(x_0; u_0, u)$ for some $(u_0, u) \in FJ(x_0)$, respectively, and define the linearized cones of $R(x_0; u)$ and $R(x_0, u_0, u)$ as

$$\begin{aligned} D(x_0; u) := \{z \in R^n : \ &z \in T_C(x_0), \ \nabla_x f(x_0) z \geq 0, \ \langle u, \nabla_x g(x_0) z \rangle = 0, \\ &\nabla_x g(x_0) z \in T_{Q^{\circ}}(g(x_0)), \ \|z\| = 1\}, \end{aligned}$$

and

$$\begin{aligned} D(x_0; u_0, u) := \{z \in R^n : \ &z \in T_C(x_0), \ u_0 \nabla_x f(x_0) z = 0, \ \langle u, \nabla_x g(x_0) z \rangle = 0, \\ &\nabla_x g(x_0) z \in T_{Q^{\circ}}(g(x_0)), \ \|z\| = 1\}, \end{aligned}$$

respectively. These cones have been used to derive second-order sufficient conditions for the standard problem (1.2), see [4,8,9,18,25,27,30].

Relations between these sets just defined are summarized in the following.

Proposition 2.1. *Let x_0 be a Fritz-John stationary point of* (1.1). *Then,*
(1) $S(x_0) \subset D(x_0)$, $S(x_0;u) \subset D(x_0;u)$ *and* $S(x_0;u_0,u)$ *for all* $(u_0,u) \in FJ(x_0)$.
(2) $D(x_0) \subset D(x_0;u_0,u)$ *for all* $(u_0,u) \in FJ(x_0)$. *Furthermore,* $D(x_0) = D(x_0;u_0,u)$ *if* $u_0 > 0$.

For the readers who are not familiar with the general nonlinear form (1.1), they might simply take these sets as those corresponding sets for the standard problem (1.2), as follows:

$$K = \{x \in R^n: g_i(x) \le 0, \ i=1,\ldots,k; \ g_i(x)=0, \ i=k+1,\ldots,k+\ell.\},$$
$$R(x_0) = \{x \in K: f(x) \le f(x_0)\},$$
$$R(x_0;u) = \{x \in K: f(x)-f(x_0) \ge 0, \ g_i(x) \le 0, \ i \in I(x_0) \backslash J(x_0);$$
$$g_j(x)=0, \ j \in J(x_0); \ g_i(x)=0, \ i=k+1,\ldots,k+\ell.\},$$
$$R(x_0;u_0,u) = \{x \in K: u_0(f(x)-f(x_0))=0, \ g_i(x) \le 0, \ i \in I(x_0) \backslash J(x_0);$$
$$g_j(x)=0, \ j \in J(x_0); \ g_i(x)=0, \ i=k+1,\ldots,k+\ell.\},$$

where $I(x_0) = \{i=1,\ldots,k: g_i(x_0)=0\}$, $J(x_0) = \{i=1,\ldots,k: u_i > 0\}$.

$$D(x_0) = \{z \in R^n: \nabla_x f(x_0)z \le 0, \ \nabla_x g_i(x_0)z \le 0, \ i \in I(x_0);$$
$$\nabla_x g_i(x_0)z=0, \ i=k+1,\ldots,k+\ell; \ \|z\|=1\},$$
$$D(x_0,u) = \{z \in R^n: \nabla_x f(x_0)z \ge 0, \ \nabla_x g_i(x_0)z \le 0, \ i \in I(x_0) \backslash J(x_0);$$
$$\nabla_x g_i(x_0)z=0, \ i \in J(x_0); \ \nabla_x g_i(x_0)z=0, \ i=k+1,\ldots,k+\ell; \ \|z\|=1\}.$$
$$D(x_0,u_0,u) = \{z \in R^n: u_0 \nabla_x f(x_0)z=0, \ \nabla_x g_i(x_0)z \le 0, \ i \in I(x_0) \backslash J(x_0);$$
$$\nabla_x g_i(x_0)z=0, \ i \in J(x_0); \ \nabla_x g_i(x_0)z=0, \ i=k+1,\ldots,k+\ell; \ \|z\|=1\}.$$

3. Sufficient conditions for local minima

Second-order sufficient condition for local minima that are not necessarily strict local minima was first obtained by Fiacco [8] using neighborhood information rather than the local information about the point in question, hence, is particularly useful in some degenerate situations. Here we extend the result of Fiacco [8] to more general problems and also weaken the assumptions.

The following result essentially resembles the case in unconstrained minimization in which local convexity (not necessarily strict convexity) suffices to ensure a critical point (having zero gradient) to be a local minimum.

Lemma 3.1. *Consider a stationary point x_0 of* (1.1). *Assume that f and g are twice differentiable in a neighborhood of x_0. It is a local minimum if:*
there exist $\gamma, \delta > 0$ such that for any $z \in S(x_0;\gamma,\delta)$, one has

$$\max\{\langle z, \nabla_x^2 L(x_0+\lambda\delta_z z, u)z \rangle: u \in KKT(x_0)\} \ge 0 \text{ for all } \lambda \in (0,1), \tag{3.1}$$

where $S(x_0;\gamma,\delta) := \{z \in R^n: \|z-z'\| \le \gamma \text{ for some } z' \in S(x_0), \ x_0+\delta_z z \in K \text{ for some } \delta_z \in (0,\delta), \ \|z\|=1\}$.

Several useful second-order sufficient conditions can be derived from this lemma.

Proposition 3.2. *Suppose x_0 is a stationary point of* (1.1). *Assume that f and g are twice differentiable in a neighborhood of x_0. Then x_0 is a local minimum if one of the following second-order conditions holds:*
(1) *There exist $\gamma, \delta > 0$ such that for any $z \in D(x_0;\gamma,\delta)$ or any $z \in D(x_0;\gamma)$,*

$$\max\{\langle z, \nabla_x^2 L(x_0+\lambda\delta_z z, u)z \rangle: u \in KKT(x_0)\} \ge 0 \quad \text{for all } \lambda \in (0,1), \tag{3.2}$$

where $D(x_0;\gamma,\delta) := \{z \in R^n: \|z-z'\| \le \gamma \text{ for some } z' \in D(x_0), \ x_0+\delta_z z \in K \text{ for some } \delta_z \in (0,\delta), \ \|z\|=1\}$ *and* $D(x_0;\gamma) := \{z \in R^n: \|z-z'\| \le \gamma \text{ for some } z' \in S(x_0), \ \|z\|=1\}$, $\delta_z \in (0,\delta)$.
(2) *There exist $\gamma, \delta > 0$ and a $u \in KKT(x_0)$ such that for all $z \in D(x_0;\gamma)$,*

$$\langle z, \nabla_x^2 L(x_0+\lambda\delta_z z, u)z \rangle \ge 0 \quad \text{for all } \lambda \in (0,1), \tag{3.3}$$

where $D(x_0;\gamma) := \{z \in R^n: \|z-z'\| \le \gamma \text{ for some } z' \in S(x_0), \ \|z\|=1\}$, $\delta_z \in (0,\delta)$.

Remark 3.3. (1) The logical relations between these conditions are $D(x_0;\gamma) \supset D(x_0;\gamma,\delta)$ and $D(x_0;\gamma,\delta) \supset S(x_0;\gamma,\delta)$ for any $\gamma,\delta > 0$. Thus, $(3.3) \Rightarrow (3.2)$. However, the degree of difficulty to verify these conditions possesses a converse order. (2) For the standard problem (1.2), if we restrict the multipliers to be fixed, then the condition (3.2) reduce to that of Fiacco [8].

A simple example to demonstrate the usefulness of neighborhood information is in the following.

Example 3.4. Consider the problem

$$\text{minimize } f(x) = a \text{ (constant)},$$
$$\text{subject to } x \in R^n.$$

Let $x_0 = 0$. It follows from anyone of the above second-order sufficient conditions of neighborhood type that x_0 is a local minimum. However, all standard second-order sufficient conditions fail to detect the optimality.

4. Sufficient conditions for strict local minima

There are many available second-order sufficient conditions for strict local minima. The conditions require the Hessian of the Lagrange function to possess some kind of restricted positive definite with respect to different cones. When confined to unconstrained case, they amount to the condition that the Hessian of the objective function be positive definite, i.e., the objective function be strict convex locally (if the function is twice continuously differentiable near the point in question). These conditions are not only sufficient for strict optimality, but indeed enough to ensure the so-called second-order growth condition. We shall start with an investigation to this class of second-order sufficient conditions, then deal with the case where some degeneracy exists by utilizing neighborhood information.

The first lemma below is the base for the first half part of the section.

Lemma 4.1. *Let x_0 be a Fritz-John stationary point. Assume that f and g are twice differentiable at x_0. If for any $z \in S(x_0)$, one has*

$$\max\{\langle z, \nabla_x^2 L(x_0,u_0,u)z\rangle : (u_0,u) \in NFJ(x_0)\} > 0, \tag{4.1}$$

then x_0 is a strict local minimum of (1.1).

Most known second-order sufficient conditions for strict local minima can be derived directly from Lemma 4.1, present diverse features, as we shall see in the following.

Proposition 4.2. *Let x_0 be a Fritz-John stationary point. Assume that f and g are twice differentiable at x_0. Consider the following conditions:*
(1) There is a $\mu > 0$ such that, for any $z \in D(x_0)$, one has

$$\max\{\langle z, \nabla_x^2 L(x_0,u)z\rangle : u \in KKT(x_0)\} \geq \mu \|z\|^2. \tag{4.2}$$

(2) There are $(u_0,u) \in NFJ(x_0)$ such that

$$\langle z, \nabla_x^2 L(x_0,u_0,u)z\rangle > 0 \text{ for all } z \in D(x_0 ;u_0,u). \tag{4.3}$$

(3) There are $(u_0,u) \in NFJ(x_0)$ such that

$$\langle z, \nabla_x^2 L(x_0,u_0,u)z\rangle > 0 \text{ for all } z \in D(x_0). \tag{4.4}$$

(4) For any $z \in D(x_0)$, one has

$$\max\{\langle z, \nabla_x^2 L(x_0,u_0,u)z\rangle : (u_0,u) \in NFJ(x_0)\} > 0. \tag{4.5}$$

Then,
(1) $(4.3) \Rightarrow (4.4) \Rightarrow (4.5) \Rightarrow (4.1)$, and
(2) The condition (4.2) implies the so-called the second-order growth condition, i.e., there exist a neighborhood $N(x_0)$ of x_0 and an $\alpha > 0$ such that

$$f(x) \geq f(x_0) + \alpha \| x - x_0 \|^2 \tag{4.6}$$

for all $x \in K \cap N(x_0)$.

Remark 4.3. (1) The conclusion (2) is new even for the standard problem (1.2). The second-order growth condition is known to an important role in the analysis of convergence rate for Newton's method, see Sachs [32]. (2) The condition (4.2) is weaker than the classical second-order sufficient condition [30]: There exist multipliers $u \in KKT(x_0)$ and a $\mu > 0$ such that

$$\langle z, \nabla_x^2 L(x_0, u)z \rangle \geq \mu \| z \|^2 \quad \text{for all } z \in D(x_0).$$

Remark 4.4. (1) If one is confined to the standard problem, then (4.1) was obtained in Kyparisis and Fiacco [18]; (4.3) was established by Pennisi [27]; (4.4) was used in Han and Mangasarian [12]; (4.5) was obtained by Ben-Tal [4] and Hettich and Jongen [14]. Recent extensions of some results presented in Proposition 4.2 to nonsmooth optimization may be found in Chaney [7], Rockafellar [31], Burke [6], and Burke and Poliquim [5]. (2) The facts that the implications $(4.5) \Rightarrow (4.4)$ and $(4.4) \Rightarrow (4.3)$ even for the standard problem (1.2) do not hold were shown in Ben-Tal [4] and Kyparisis and Fiacco [18], respectively.

Remark 4.5. The distinguished feature of conditions (4.1), (4.2) and (4.5) is the use of the entire set of multipliers rather than a single vector of multipliers as in the classical theory of second-order sufficient conditions.

We should mention that the boundedness of the set of Fritz-John multipliers is equivalent to the regularity condition (2.2). The following result taken from Liu [20] may highlight the relation between the condition (2.2) and the Fritz-John multipliers.

Proposition 4.6. *For a feasible point x_0 of (1.1), the regularity condition (2.2) holds if and only if the system*

$$0 \in \begin{bmatrix} \nabla_x g(x_0)^* u \\ -g(x_0) \end{bmatrix} + N_{C \times Q}(x_0, u) \tag{4.7}$$

has $u = 0$ as the unique solution.

Proposition 4.6 simply says that if the regularity condition (2.2) holds, then all multipliers $(u_0, u) \in FJ(x_0)$ (assuming x_0 be a Fritz-John stationary point) satisfy $u_0 > 0$. Hence in this case we have the simple relation $FJ(x_0) = \{\lambda(1, u): u \in KKT(x_0), \lambda > 0\}$.

Notice that for the standard problem (1.2), the regularity condition (2.2) reduces to Mangasarian-Fromovitz constraint qualification and the condition that system (4.7) has 0 as the unique solution is equivalent to the "positive" linear independence constraint qualification that, precisely stated, the system

$$\sum_{i=1}^{k} u_i \nabla_x g_i(x_0) + \sum_{i=k+1}^{k+l} u_i \nabla_x g_i(x_0) = 0,$$
$$u_i g(x_0) = 0, \quad u_i \geq 0 \quad (i = 1, \ldots, k).$$

has $u = 0$ as the unique solution. In the case of $\ell = 0$, that is, the problem (1.2) does not include equality constraints, Proposition 4.6 reduces to the well-known Gordan theorem (see [2]).

As pointed out in Fiacco and McCormick [9], for the standard problem (1.2) the condition (4.3) is only slightly more general than (4.2).

Proposition 4.7. *Suppose x_0 is a Fritz-John stationary point. Assume that f and g are twice differentiable at x_0. If the following condition holds: There exist $(u_0, u) \in FJ(x_0) \backslash KKT(x_0)$ such that*

$$\langle z, \nabla_x^2 L(x_0, u_0, u)z \rangle > 0 \text{ for all } z \in D(x_0 ; u_0, u), \tag{4.8}$$

then x_0 is an isolated feasible point, i.e., $K \cap N(x_0) = \{x_0\}$ for some neighborhood $N(x_0)$ of x_0.

We now proceed to establish neighborhood type second-order sufficient conditions for strict local minima. These conditions are useful since in the case where standard second-order sufficient conditions do not hold, it is usually more efficient to verify second-order neighborhood conditions rather than high order conditions. This class of conditions usually assume a little more differentiability.

Lemma 4.8. *Let x_0 be a Fritz-John stationary point of (1.1). Assume that f and g are twice differentiable in a neighborhood of x_0. It is a strict local minimum if:*
There exist $\gamma, \delta > 0$ such that for any $z \in S(x_0 ; \gamma, \delta)$, one has

$$\max\{\langle z, \nabla_x^2 L(x_0 + \lambda \delta_z z, u_0, u)z\rangle : (u_0, u) \in NFJ(x_0)\} > 0 \text{ for all } \lambda \in (0,1), \tag{4.9}$$

where $S(x_0 ; \gamma, \delta) := \{z \in R^n : \|z - z'\| \leq \gamma \text{ for some } z' \in S(x_0), x_0 + \delta_z z \in K \text{ for some } \delta_z \in (0, \delta), \|z\| = 1\}$.

The above lemma is strong in logical sense and acts as a basic deriving result. More verifiable results are as follows.

Proposition 4.9. *Let x_0 be a Fritz-John stationary point of (1.1). Assume that f and g are twice differentiable in a neighborhood of x_0. Then it is a strict local minimum if there exist $\gamma, \delta > 0$ such that for any $z \in D(x_0 ; \gamma, \delta)$ or any $z \in D(x_0 ; \gamma)$, one has*

$$\max\{\langle z, \nabla_x^2 L(x_0 + \lambda \delta_z z, u_0, u)z\rangle : (u_0, u) \in NFJ(x_0)\} > 0 \text{ for all } \lambda \in (0,1), \tag{4.10}$$

where $D(x_0 ; \gamma, \delta) := \{z \in R^n : \|z - z'\| \leq \gamma \text{ for some } z' \in S(x_0), x_0 + \delta_z z \in K \text{ for some } \delta_z \in (0, \delta), \|z\| = 1\}$ and $D(x_0 ; \gamma) := \{z \in R^n : \|z - z'\| \leq \gamma \text{ for some } z' \in S(x_0), \|z\| = 1\}, \delta_z \in (0, \delta)$.

It is easy to present quite simple examples to show that in some case the above conditions are applicable, while standard conditions in Proposition 4.2 lose efficacy.

5. Sufficient conditions for isolated local minima

It is known that most available nonlinear optimization softwares only find a local minimum. In fact, some development on the theory of global optimization is based on the assumption that every local minimum is isolated. From both theoretical and computational points of view, it is important to know if a local minimum is indeed an isolated one. However, until recently many authors still called a strict local minimum an isolated local minimum. Part of the reason may be that in the unconstrained case a strict local minimum of a sufficiently smooth function is indeed an isolated one. However, this is no longer true for the constrained case. Robinson [30] gave the following simple example which demonstrates such a situation:

$$\text{minimize} \quad \frac{1}{2} x^2, \tag{5.1}$$
$$\text{subject to} \quad x^6 \sin(1/x) = 0 \quad [\sin(1/0) := 0].$$

Note that the second-order sufficient condition (4.2) holds at $x_0 = 0$, hence x_0 is a strict local minimum of the problem (5.1); however, it is a cluster point of the local minimums rather than an isolated local minimum.

In order to exclude pathological problems, like (5.1), Robinson [30] established sufficient conditions for an isolated local minimum by imposing the constraint qualification (2.2) at the point in question and strengthening the second-order sufficient condition (4.3) to the general second-order sufficient condition, i.e., the classical second-order sufficient condition holds for all associated multipliers. Kyparisis and Fiacco [18] slightly weakened Robinson's condition for isolated local minima to the standard problem (1.2) by using the tangent cone of the critical domain with respect to some multipliers instead of using the set of critical directions. These assumptions seem to be very strong as compared with those for strict local minima. This motivates the following question: What are the weakest conditions for isolated local minima? We shall concentrate on this issue in the section.

Motivated by the fact that a local minimum of a strict convex function is also an isolated local minimum, we shall first derive second-order sufficient conditions for isolated local minima along this line. This type of conditions are applicable to some degenerate situations, however, difficult to verify. We then impose the constraint regularity on the local minimum in question, and show that even the weakest KKT type second-order sufficient condition (4.2) is sufficient for the isolatedness. This result considerably weakens that of Robinson [30], and provides a deep insight into the optimality conditions for the constrained case.

The following lemma gives a basic condition to guarantee an isolated local minimum.

Lemma 5.1. *Let x_0 be a local minimum of* (1.1). *Assume that f and g are twice differentiable in a neighborhood of x_0. It is an isolated local minimum if there exist $\gamma, \delta > 0$ such that for any $(u_0, u) \in NFJ(x_0)$, one has*

$$\min\{\langle z, \nabla_x^2 L(x_0 + \lambda \delta_z z, u_0, u)z\rangle : z \in S(x_0; u; \gamma, \delta)\} > 0 \text{ for all } \lambda \in (0,1), \tag{5.1}$$

where $S(x_0; u; \gamma, \delta) := \{z \in R^n: \|z - z'\| \leq \gamma \text{ for some } z' \in S(x_0; u), x_0 + \delta_z z \in K \text{ for some } \delta_z \in (0, \delta), \|z\| = 1\}$.

Example 5.2. Consider the following minimization problem,

$$\text{minimize } x^4,$$
$$x \in R^n.$$

The unique local minimum is $x_0 = 0$. Evidently, second-order condition (5.1) holds at x_0. Thus x_0 is an isolated local minimum from Lemma 5.1.

Remark 5.3. (1) It is easy to see that the isolatedness of a local minimum in the constrained case depends not only on the behavior of the objective function but heavily on the structure of the feasible set. A reasonable set of conditions sufficiently to ensure isolatedness of local minima has to impose conditions either implicitly or explicitly on the functions that define the feasible set. When restricted in unconstrained minimization problems, (5.1) essentially assumes the functions being minimized to have strict local convexity. Hence, one might not expect a substantial improvement over (5.1) if conditions are imposed explicitly on the defining functions. (2) Example 4.1 in Kyparisis and Fiacco [18] suggests that the assumption that x_0 is a local minimum of (1.1) cannot be omitted in Lemma 5.1.

A direct consequence of Lemma 5.1 is the following result.

Proposition 5.4. *Let x_0 be a local minimum of* (1.1). *Assume that f and g are twice differentiable in a neighborhood of x_0. It is an isolated local minimum if there exist $\gamma, \delta > 0$ such that for any $(u_0, u) \in NFJ(x_0)$, one has*

$$\min\{\langle z, \nabla_x^2 L(x_0 + \lambda \delta_z z, u_0, u)z\rangle : z \in D(x_0; u; \gamma, \delta)\} > 0 \text{ for all } \lambda \in (0,1), \tag{5.2}$$

where $D(x_0; u; \gamma, \delta) := \{z \in R^n: \|z - z'\| \leq \gamma \text{ for some } z' \in D(x_0; u), x_0 + \delta_z z \in K \text{ for some } \delta_z \in (0, \delta), \|z\| = 1\}$.

The following theorem provides a more convenient condition using the neighborhood information.

Proposition 5.5. *Let x_0 be a Fritz-John stationary point of* (1.1). *Assume that f and g are twice differentiable in a neighborhood of x_0. It is an isolated local minimum if there exist $\gamma, \delta > 0$ such that for any $(u_0, u) \in NFJ(x_0)$, one has*

$$\min\{\langle z, \nabla_x^2 L(x_0 + \lambda \delta_z z, u_0, u)z\rangle : z \in D(x_0; u_0, u; \gamma, \delta)\} > 0 \text{ for all } \lambda \in (0,1), \tag{5.3}$$

where $D(x_0; u_0, u; \gamma, \delta) := \{z \in R^n: \|z - z'\| \leq \gamma \text{ for some } z' \in D(x_0; u_0, u), x_0 + \delta_z z \in K \text{ for some } \delta_z \in (0, \delta), \|z\| = 1\}$.

If the problem functions possess more differentiability some useful conditions along essentially Robinson's approach can be derived as follows.

Lemma 5.6. *Let x_0 be a local minimum of* (1.1). *Assume that f and g are twice continuously* *differentiable in a neighborhood of x_0. It is an isolated local minimum if for any $(u_0,u) \in NFJ(x_0)$, one has*

$$\min\{\langle z, \nabla_x^2 L(x_0,u_0,u)z\rangle: z \in S(x_0;u)\} > 0. \tag{5.4}$$

The more verifiable conditions are the following.

Proposition 5.7. *Let x_0 be a Fritz-John stationary point of (1.1). Assume that f and g are twice continuously differentiable in a neighborhood of x_0. It is an isolated local minimum if for any $(u_0,u) \in NFJ(x_0)$, one has*

$$\min\{\langle z, \nabla_x^2 L(x_0,u_0,u)z\rangle: z \in D(x_0;u_0,u)\} > 0. \tag{5.5}$$

Remark 5.8. (1) Proposition 4.7 indicates that (5.5) appears to be only slightly more general than (5.6) and can be rephrased as follows: A Fritz-John point is an isolated local minimum if either it is an isolated feasible point or the condition (5.6) holds at the point. (2) The following sufficient conditions for isolated local minima were obtained by Robinson [30]: In the above differentiability setting, assume that x_0 is a stationary point and the constraint regularity holds at x_0. Suppose that for any $u \in KKT(x_0)$, one has

$$\min\{\langle z, \nabla_x^2 L(x_0,u_0,u)z\rangle: z \in D(x_0)\} > 0. \tag{5.6}$$

Then x_0 is an isolated local minimum.

So far we have obtained several second-order sufficient conditions for isolated local minima. They share two common features: One, they are Fritz-John type and do not require any constraint qualifications. Two, they all impose conditions on all multipliers, and as a consequence, difficult to verify.

Finally, we present an elegant result. It is the weakest known standard conditions for isolatedness of local minima in the polyhedral case.

Lemma 5.9. *Let x_0 be a KKT stationary point of (1.1). Assume that Q is a polyhedral convex cone, f and g are twice continuously differentiable in a neighborhood of x_0. Suppose that the constraint regularity (2.2) and the second-order growth condition (4.6) hold at x_0. Then it is an isolated local minimum.*

Note that under the assumptions of Lemma 5.9, the constraint regularity holds in a neighborhood of x_0. Therefore, a direct consequence of the above lemma and Proposition 4.2 is the following result.

Proposition 5.10. *Let x_0 be a KKT stationary point of (1.1). Assume that Q is a polyhedral convex cone, f and g are twice continuously differentiable in a neighborhood of x_0. Suppose that the constraint regularity (2.2) and the second-order sufficient condition (4.2) hold at x_0. Then it is an isolated local minimum.*

The above theorem indicates that pathologies like Example 5.2 can take place in the constrained case mainly due to the structure of the feasible set. We conclude by remarking that variations that further weaken the constraint regularity assumption in Proposition 5.10 can be proved. But in view of the position of the constraint regularity (2.2) in the class of constraint qualifications, they are not intrinsically interesting and certainly less useful.

Acknowledgement
We thank O.L. Mangasarian for helpful references.

Appendix. Proofs of results

We shall give proofs of the results presented in the paper in the sequel.

Proof of Position 2.1. (1) We first show $S(x_0) \subset D(x_0)$. Suppose $z \in S(x_0)$ and $\|z\| = 1$. Then, there exist sequences of strictly positive numbers $\{h_j\}$ converging to 0 and $\{z_j\}$ converging to z such that $x_0 + h_j z_j \in R(x_0)$, i.e., $g(x_0 + h_j z_j) \in Q°$, $x_0 + h_j z_j \in C$ and $f(x_0 + h_j z_j) \le f(x_0)$. Obviously, $z \in T_C(x_0)$. We have for each j

$$g(x_0 + h_j z_j) = g(x_0) + \nabla_x g(x_0)(h_j z_j) + o(\|h_j z_j\|) \in Q°,$$

thus

$$\nabla_x g(x_0)z = \lim_{j \to \infty} [g(x_0 + h_j z_j) - g(x_0)]/h_j \in T_{Q°}(g(x_0)).$$

We have for each j

$$f(x_0) \ge f(x_0 + h_j z_j) = f(x_0) + \nabla_x f(x_0)(h_j z_j) + o(\|h_j z_j\|),$$

so

$$0 \ge \nabla_x f(x_0)(h_j z_j) + o(\|h_j z_j\|).$$

Dividing the above inequality by h_j and passing to the limit, we obtain $\nabla_x f(x_0)z \le 0$ and therefore, z is a critical direction at x_0, i.e., $z \in D(x_0)$.

We now prove that $S(x_0;u) \subset D(x_0;u)$. Choose any $z \in S(x_0;u)$. By definition, there exist sequences of strictly positive numbers $\{h_j\}$ converging to 0 and $\{z_j\}$ converging to z such that $x_0 + h_j z_j \in R(x_0;u)$, i.e., $g(x_0 + h_j z_j) \in Q°$, $x_0 + h_j z_j \in C$, $g(x_0 + h_j z_j) \in N_Q(u)$, and $f(x_0 + h_j z_j) \ge f(x_0)$. Similar to the above proof, we can deduce that $z \in T_C(x_0)$ and $\nabla_x g(x_0)z \in T_{Q°}(g(x_0))$. Since $g(x_0)$, $g(x_0 + h_j z_j) \in N_Q(u)$ and Q is a cone, we have

$$\langle u, g(x_0) \rangle = 0 \text{ and } \langle u, g(x_0 + h_j z_j) \rangle = 0.$$

So

$$\langle u, \nabla_x g(x_0)z \rangle = \lim_{j \to \infty} \langle u, g(x_0 + h_j z_j) - g(x_0) \rangle / \|h_j\| = 0.$$

From $f(x_0 + h_j z_j) \ge f(x_0)$ we have $\nabla_x f(x_0)z \ge 0$. Hence $z \in D(x_0;u)$.

Finally, the inclusion $S(x_0;u_0,u) \subset D(x_0;u_0,u)$ can be established by using the above technique again. The details are omitted.

(2) For the proof of the first part of the conclusions, we need to show that $\langle u, \nabla_x g(x_0)z \rangle = 0$ and $u_0 \nabla_x f(x_0)z = 0$ for any $z \in D(x_0)$. Note that the Fritz-John conditions imply $g(x_0) \in N_Q(u)$, so that $u \in N_{Q°}(g(x_0))$ since Q is a closed convex cone. From the inclusion $\nabla_x g(x_0)z \in T_{Q°}(g(x_0))$ we deduce that $\langle u, \nabla_x g(x_0)z \rangle \le 0$. Since $z \in T_C(x_0)$ and $-[u_0 \nabla_x f(x_0) + \nabla_x g(x_0)^* u] \in N_C(x_0)$, we have

$$0 \le [u_0 \nabla_x f(x_0) + \nabla_x g(x_0)^* u]z \tag{A1}$$
$$= u_0 \nabla_x f(x_0)z + \langle u, \nabla_x g(x_0)z \rangle.$$

Relation (A1) together with $\nabla_x f(x_0)z \le 0$ implies $\langle u, \nabla_x g(x_0)z \rangle \ge 0$. Thus $\langle u, \nabla_x g(x_0)z \rangle = 0$. Using (A1) again we can deduce $u_0 \nabla_x f(x_0)z = 0$.

Now suppose $u_0 > 0$. Then $u_0 \nabla_x f(x_0)z = 0$ implies $\nabla_x f(x_0)z = 0$. So that the second part of the conclusions holds trivially. $\blacksquare$

Proof of Lemma 3.1. By contraposition. Suppose that x_0 is not a local minimum. Then, there exists a sequence $\{x_j\} \subset C$ converging to x_0 with $x_j \ne x_0$, $g(x_j) \in Q°$ and $f(x_j) < f(x_0)$ for each j. Without loss of generality we may assume that the sequence $\{z_j = (x_j - x_0)/\alpha_j\}$ with $\alpha_j = \|x_j - x_0\|$ converges to some $z \in R^n$. Obviously, $z \in S(x_0)$ and $z_j \in S(x_0;\gamma,\delta)$ for any fixed $\gamma > 0$, $\delta > 0$ provided that j is sufficiently large. For any $u \in KKT(x_0)$ we have that since $\langle u, g(x_j) \rangle \le 0$

$$f(x_0) > f(x_j) \ge L(x_j, u)$$
$$= L(x_0, u) + \nabla_x L(x_0, u)(x_j - x_0) + \tfrac{1}{2} \langle x_j - x_0, \nabla_x^2 L(x_0 + \lambda_j \alpha_j z_j, u)(x_j - x_0) \rangle$$
$$\ge f(x_0) + \tfrac{1}{2} \alpha_j^2 \langle z_j, \nabla_x^2 L(x_0 + \lambda_j \alpha_j z_j, u) z_j \rangle,$$

where $\lambda_j \in (0,1)$. This contradicts (3.1). $\blacksquare$

Proof of Proposition 3.2. Since $S(x_0) \subset D(x_0)$ (Proposition 2.1), it follows that $S(x_0;\gamma,\delta) \subset D(x_0;\gamma,\delta)$ for any γ, $\delta > 0$. This proves (3.2). Similarly, we can show (3.3) and (3.4) without difficulty. ∎

Proof of Lemma 4.1. By contraposition. Suppose that x_0 is not a strict local minimum. Then, there exists a sequence $\{x_j\} \subset C$ converging to x_0 with $x_j \neq x_0$, $g(x_j) \in Q^\circ$ and $f(x_j) \leq f(x_0)$ for each j. Without loss of generality we may assume that the sequence $\{(x_j\text{-}x_0)/\|x_j\text{-}x_0\|\}$ converges to some $z \in R^n$; obviously, $\|z\| = 1$ and $z \in S(x_0)$. Choose any $(u_0,u) \in NFJ(x_0)$. Since $\langle u,g(x_j)\rangle \leq 0$, we deduce that

$$u_0 f(x_0) \geq u_0 f(x_j) \geq L(x_j,u_0,u)$$
$$= L(x_0,u_0,u) + \nabla_x L(x_0,u_0,u)(x_j - x_0) + \tfrac{1}{2}\langle x_j - x_0, \nabla_x^2 L(x_0,u_0,u)(x_j - x_0)\rangle + o(\|x_j - x_0\|^2)$$
$$\geq u_0 f(x_0) + \tfrac{1}{2}\langle x_j - x_0, \nabla_x^2 L(x_0,u_0,u)(x_j - x_0)\rangle + o(\|x_j - x_0\|^2).$$

Subtracting $u_0 f(x_0)$, dividing by $(1/2)\|x_j\text{-}x_0\|^2$ and passing to the limit, we find that

$$0 \geq \langle z, \nabla_x^2 L(x_0,u_0,u)z\rangle.$$

This contradicts the condition (4.1). ∎

Proof of Proposition 4.2. For the conclusion (1), the implication $(4.3) \Rightarrow (4.4)$ follows from $D(x_0) \subset D(X_0;u_0,u)$. Obviously, $(4.4) \Rightarrow (4.5)$. Finally, $(4.5) \Rightarrow (4.6)$ because of $S(x_0) \subset D(x_0)$.

We use the same technique of Theorem 2.2 of Robinson [30] to prove the conclusion (2). Suppose that there exist an $\alpha > 0$ and a sequence $\{x_i\} \subset C$ converging to x_0 with $x_i \neq x_0$, $g(x_i) \in Q^\circ$ and $f(x_i) \leq f(x_0) + (\alpha/2)\|x_i\text{-}x_0\|^2$ for each i. Then it suffices to prove our conclusion by showing that $\alpha \geq \mu$. We assume without loss of generality that $\{(x_i\text{-}x_0)/\|x_i\text{-}x_0\|\}$ converges to some $z \in R^n$. Then we have $\|z\| = 1$ and $z \in T_C(x_0)$. Furthermore, for each i one has

$$g(x_i) = g(x_0) + \nabla_x g(x_0)(x_i\text{-}x_0) + o(\|x_i\text{-}x_0\|) \in Q^\circ.$$

It follows that

$$\nabla_x g(x_0)z = \lim_{i\to\infty} \|x_i\text{-}x_0\|^{-1}(g(x_i)\text{-}g(x_0)) \in T_{Q^\circ}(g(x_0)).$$

On the other hand, we have

$$f(x_0) + (\alpha/2)\|x_i\text{-}x_0\|^2 \geq f(x_i) = f(x_0) + \nabla_x f(x_0)(x_i\text{-}x_0) + o(\|x_i\text{-}x_0\|).$$

Dividing the above inequality by $\|x_i\text{-}x_0\|$ and passing to the limit, we obtain that $\nabla_x f(x_0)z \leq 0$. Thus $z \in D(x_0)$. Note that the relation $f(x_i) \geq L(x_i,u)$ is always true provided that x_i is a feasible point and $u \in Q$. It follows that

$$f(x_0) + \tfrac{1}{2}\alpha\|x_i - x_0\|^2 \geq f(x_i) \geq \max\{L(x_i,u) : u \in KKT(x_0)\}$$
$$= \max\{L(x_0,u) + \nabla_x L(x_0,u)(x_i - x_0) + \tfrac{1}{2}\langle x_i - x_0, \nabla_x^2 L(x_0,u)(x_i - x_0)\rangle + o(\|x_i - x_0\|^2) : u \in KKT(x_0)\}$$
$$\geq f(x_0) + \max\{\tfrac{1}{2}\langle x_i - x_0, \nabla_x^2 L(x_0,u)(x_i - x_0)\rangle + o(\|x_i - x_0\|^2) : u \in KKT(x_0)\}.$$

Subtracting $f(x_0)$, dividing by $(1/2)\|x_i\text{-}x_0\|^2$ and passing to the limit, we have that from the assumption

$$\alpha \geq \max\{\langle z, L_x^2(x_0,u)z\rangle : u \in KKT(x_0)\} \geq \mu\|z\|^2 = \mu.$$

This completes the proof. ∎

Proof of Proposition 4.7. Suppose x_0 is not an isolated feasible point. Then there exists a sequence $\{x_j\} \in K$ converging to x_0 with $x_j \neq x_0$. Therefore, there is a subsequence of $\{x_j\}$ such that $\{(x_j\text{-}x_0)/\|x_j\text{-}x_0\|\}$ converges to some z. Using the technique in the proof of Proposition 2.1, we can deduce that $z \in T_C(x_0)$, $\nabla_x g(x_0)z \in T_{Q^\circ}(g(x_0))$, and $\|z\| = 1$. Suppose $(u_0,u) \in FJ(x_0)\backslash KKT(x_0)$. Then $u_0 = 0$. We now proceed to show $\langle u, \nabla_x g(x_0)z\rangle = 0$. The Fritz-John conditions imply

$$0 \in \nabla_x g(x_0)^* u + N_C(x_0), \tag{A.2}$$

$$g(x_0) \in N_Q(u). \tag{A3}$$

It follows from (A2) and $z \in T_C(x_0)$ that

$$\langle u, \nabla_x g(x_0)z \rangle \geq 0. \tag{A4}$$

On the other hand, (A3) can be expressed as $u \in N_{Q^\circ}(g(x_0))$. Since $\nabla_x g(x_0)z \in T_{Q^\circ}(g(x_0))$ we have

$$\langle u, \nabla_x g(x_0)z \rangle \leq 0. \tag{A5}$$

So that $\langle u, \nabla_x g(x_0)z \rangle = 0$ from (A4) and (A5). Consequently, we find that $z \in D(x_0; u_0, u)$.

Using Taylor's expansion and the fact that $\langle u, g(x_j) \rangle \leq 0$, we deduce

$$\begin{aligned}
0 &\geq L(x_j, u_0, u) \\
&= L(x_0, u_0, u) + \nabla_x L(x_0, u_0, u)(x_j - x_0) + \tfrac{1}{2}\langle x_j - x_0, \nabla_x^2 L(x_0, u_0, u)(x_j - x_0) \rangle + o(\| x_j - x_0 \|^2) \\
&= \tfrac{1}{2}\langle x_j - x_0, \nabla_x^2 L(x_0, u_0, u)(x_j - x_0) \rangle + o(\| x_j - x_0 \|^2).
\end{aligned}$$

Dividing by $(1/2)\| x_j - x_0 \|^2$ and passing to the limit, we find that

$$0 \geq \langle z, \nabla_x^2 L(x_0, u_0, u)z \rangle.$$

This contradicts the condition (4.8). ∎

Proof of Lemma 4.8. The proof is similar to that of Lemma 3.1. Suppose that x_0 is not a strict local minimum. Then, there exists a sequence $\{x_j\} \subset C$ converging to x_0 with $x_j \neq x_0$, $g(x_j) \in Q^\circ$ and $f(x_j) \leq f(x_0)$ for each j. Without loss of generality we may assume that the sequence $\{z_j = (x_j - x_0)/\alpha_j\}$ with $\alpha_j = \| x_j - x_0 \|$ converges to some $z \in R^n$. Obviously, $z \in S(x_0)$ and $z_j \in S(x_0; \gamma, \delta)$ for any fixed $\gamma > 0$, $\delta > 0$ provided that j is sufficiently large. For any $u \in KKT(x_0)$ we have that since $\langle u, g(x_j) \rangle \leq 0$

$$\begin{aligned}
u_0 f(x_0) &\geq u_0 f(x_j) \geq L(x_j, u) \\
&= L(x_0, u) + \nabla_x L(x_0, u)(x_j - x_0) + \tfrac{1}{2}\langle x_j - x_0, \nabla_x^2 L(x_0 + \lambda_j \alpha_j z_j, u)(x_j - x_0) \rangle \\
&\geq u_0 f(x_0) + \tfrac{1}{2}\alpha_j^2 \langle z_j, \nabla_x^2 L(x_0 + \lambda_j \alpha_j z_j, u)z_j \rangle,
\end{aligned}$$

where $\lambda_j \in (0, 1)$. This contradicts (4.9). ∎

Proof of Proposition 4.9. $(4.10) \Rightarrow (4.9)$ follows from the fact that $S(x_0; \gamma, \delta) \subset D(x_0; \gamma, \delta)$. $(4.11) \Rightarrow (4.9)$ is true since $D(x_0; \gamma, \delta) \subset D(x_0; \gamma)$ for any $\gamma > 0$, $\delta > 0$. ∎

Proof of Lemma 5.1. By the Fritz-John conditions, it suffices to show that x_0 is an isolated Fritz-John stationary point. Suppose it is not true. Then there is a sequence of Fritz-John stationary points $\{x_j\} \subset K$ converging to x_0 with $x_j \neq x_0$ and $f(x_j) \geq f(x_0)$. Hence there exist nonzero $(u_0^j, u_j) \in R_+ \times Q$ such that the point (x_j, u_0^j, u_j) satisfies the system

$$\begin{aligned}
0 &\in u_0^j \nabla_x f(x_j) + \nabla_x g(x_j)^* u_j + N_C(x_j), \\
g(x_j) &\in N_Q(u_j)
\end{aligned} \tag{A6}$$

for each j. Without loss of generality we may assume that

$$(u_0^j, u_j)/\| (u_0^j, u_j) \| \to (u_0, u).$$

It follows that $(u_0, u) \in R^+ \times Q$ and $\| (u_0, u) \| = 1$. Furthermore, using the fact that N_C is a closed map, we deduce that $(u_0, u) \in FJ(x_0)$. Similarly, we assume without loss of generality that

$$z_j \to z \text{ with } z_j = (x_j - x_0)/\alpha_j, \ \alpha_j = \| x_j - x_0 \|.$$

Evidently, we have $z \in S(x_0; u)$ and $z_j \in S(x_0; u; \gamma, \delta)$ for any fixed $\gamma, \delta > 0$ if j is sufficiently large.

We now proceed to deduce a contradiction to the second-order condition (5.1) in two cases.

Case I: $u_0 > 0$.

We may assume without loss of generality that $u_0^j = 1$; otherwise we can normalize the sequence $\{(u_0^j, u_j)\}$

by dividing (u_0^j, u_j) by u_0^j for sufficiently large j, and obtain a sequence of normalized multipliers. For each fixed j we define a scalar function s over [0,1] by

$$s(t) := [\nabla_x f(x_t) + \nabla_x g(x_t)^* u_t](x_j - x_0) - \langle u_j - u, g(x_t)\rangle,$$

where $x_t := (1-t)x_0 + tx_j$, $u_t := (1-t)u + tu_j$. By Fritz-John conditions, we can find that $s(0) \geq 0 \geq s(1)$. Utilizing the mean value theorem, we claim that for each j, there is a $t_j \in (0,1)$ such that

$$0 \geq s'(t_j) = \langle x_j - x_0, \nabla_x^2 L(x_{t_j}, u_{t_j})(x_j - x_0)\rangle = \alpha_j^2 \langle z_j, \nabla_x^2 L(x_0 + \lambda_j \alpha_j z_j, u_{t_j})z_j\rangle.$$

This contradicts condition (5.1) since $u_j \to u$.

Case II: $u_0 = 0$.

Similar to Case I, for each fixed j we define a scalar function s over [0,1] by

$$s(t) := [u_0^t \nabla_x f(x_t) + \nabla_x g(x_t)^* u_t](x_j - x_0) - (u_0^j - u_0)f(x_t) - \langle u_j - u, g(x_t)\rangle,$$

where $x_t := (1-t)x_0 + tx_j$, $u_0^t := (1-t)u_0 + tu_0^j$, and $u_t := (1-t)u + tu_j$. The Fritz-John conditions imply that $s(0) \geq -u_0^j f(x_0)$ and $s(1) \leq -u_0^j f(x_j)$. By the same reasoning as did in Case I we can show that there is a $t_j \in (0,1)$ such that

$$\begin{aligned}
s'(t_j) &= \langle x_j - x_0, \nabla_x^2 L(x_{t_j}, u_0^{t_j}, u_{t_j})(x_j - x_0)\rangle \\
&= \alpha_j^2 \langle z_j, \nabla_x^2 L(x_0 + \lambda_j \alpha_j z_j, u_0^{t_j}, u_{t_j})z_j\rangle \\
&= s(1) - s(0) \\
&\leq u_0^j (f(x_0) - f(x_j)) \\
&\leq 0,
\end{aligned}$$

for sufficiently large j. Then, similar to Case I, we can find a contradiction. ∎

Proof of Proposition 5.4. It is easy to see that $S(x_0; u; \gamma, \delta) \subset D(x_0; u; \gamma, \delta)$ for any $\gamma, \delta > 0$. So that the conclusion is proved by applying Lemma 5.1. ∎

Proof of Proposition 5.5. We first claim that x_0 is a strict local minimum of (1.1). This assertion is a direct consequence of (1) of Proposition 4.9 since $D(x_0; \gamma, \delta) \subset D(x_0; u_0, u; \gamma, \delta)$ for any $\gamma, \delta > 0$.

We shall prove the conclusion by contraposition. Suppose x_0 is not an isolated local minimum of (1.1). Then there is a sequence of Fritz-John stationary points $\{x_j\} \subset K$ converging to x_0 with $x_j \neq x_0$ and $f(x_j) > f(x_0)$. Hence there exist nonzero $(u_0^j, u_j) \in R_+ \times Q$ such that the point (x_j, u_0^j, u_j) satisfies the system

$$\begin{aligned}
0 &\in u_0^j \nabla_x f(x_j) + \nabla_x g(x_j)^* u_j + N_C(x_j), \\
g(x_j) &\in N_Q(u_j),
\end{aligned} \tag{A7}$$

for each j. Without loss of generality we may assume that

$$(u_0^j, u_j)/\|(u_0^j, u_j)\| \to (u_0, u).$$

It follows that $(u_0, u) \in R^+ \times Q$ and $\|(u_0, u)\| = 1$. Furthermore, using the fact that N_C is a closed map, we deduce that $(u_0, u) \in FJ(x_0)$. Similarly, we assume without loss of generality that

$$z_j \to z \text{ with } z_j = (x_j - x_0)/\alpha_j, \quad \alpha_j = \|x_j - x_0\|.$$

We shall show that $z \in D(x_0; u_0, u)$, so that $z_j \in D(x_0; u; \gamma, \delta)$ for any fixed $\gamma, \delta > 0$ if j is sufficiently large. From the proof of Proposition 2.1 we know that $z \in T_C(x_0)$, $\nabla_x g(x_0)z \in T_{Q^\circ}(g(x_0))$, and $\|z\| = 1$. What we have left is to show that $u_0 \nabla_x f(x_0)z = 0$ and $\langle u, \nabla_x g(x_0)z\rangle = 0$.

We show $\langle u, \nabla_x g(x_0)z\rangle = 0$ first. By the differentiability of g at x_0, we have

$$g(x_j) = g(x_0) + \nabla_x g(x_0)(x_j - x_0) + o(\|x_j - x_0\|). \tag{A8}$$

Noting that $\langle u_j, g(x_0) \rangle \leq 0$ we deduce from relations (A7) and (A8) that

$$\langle u_j, \nabla_x g(x_0)(x_j - x_0) + o(\| x_j - x_0 \|) \rangle \geq 0.$$

Dividing the above inequality by $\| (u_0^j, u_j) \| \, \| x_j - x_0 \|$ and passing to the limit, we obtain

$$\langle u, \nabla_x g(x_0) z \rangle \geq 0. \tag{A9}$$

On the other hand, the same reasoning yields

$$\langle u, \nabla_x g(x_0) z \rangle \leq 0 \tag{A10}$$

since $\langle u, g(x_j) \rangle \leq 0$ and $\langle u, g(x_0) \rangle = 0$. Inequalities (A9) and (A10) imply

$$\langle u, \nabla_x g(x_0) z \rangle = 0. \tag{A11}$$

Now we shall show $u_0 \nabla_x f(x_0) z = 0$. If $u_0 = 0$, then nothing needs to be proved; if $u_0 > 0$, it follows from (A6) and $z \in T_C(x_0)$ that

$$u_0 \nabla_x f(x_0) z \leq 0. \tag{A12}$$

In order to prove the converse inequality, noting that $u_0^j > 0$ for sufficiently large j, so that we infer from (A6) that

$$u_0^j \nabla_x f(x_j)(x_0 - x_j) + \langle u_j, \nabla_x g(x_j)(x_0 - x_j) \rangle \geq 0.$$

Dividing the above inequality by $\| (u_0^j, u_j) \| \, \| x_j - x_0 \|$ and passing to the limit, we find

$$u_0 \nabla_x f(x_0) z + \langle u, \nabla_x g(x_0) z \rangle \leq 0.$$

The above inequality together with (A11) and (A12) implies

$$u_0 \nabla_x f(x_0) z = 0.$$

Therefore we have proved $z \in D(x_0; u_0, u)$.

Similar to the second part of the proof of Lemma 5.1, we can deduce a contradiction to the condition (5.3). ∎

Proof of Lemma 5.6. Since we assume that f and g are twice continuously differentiable near x_0, therefore, the implication $(5.4) \Rightarrow (5.1)$ can be proved by a simple standard continuity argument. ∎

Proof of Proposition 5.7. The implication $(5.5) \Rightarrow (5.3)$ follows from the fact that f and g are twice continuously differentiable in a neighborhood of x_0. ∎

Proof of Lemma 5.9. By contraposition. Suppose the conclusion is not true. Then there are sequences $\{x_i\} \subset C$ with $x_i \neq x_0$ such that $\{x_i\}$ converges to x_0 and for each i, x_i is a KKT stationary point, i.e., for some u_i,

$$0 \in \nabla_x f(x_i) + \nabla_x g(x_i)^* u_i + N_C(x_i), \tag{A13}$$

$$g(x_i) \in N_Q(u_i). \tag{A14}$$

Note that the constraint regularity implies that the set $KKT(x_i)$ is upper semicontinuous at x_0. So that we can assume without loss of generality that
(i) $(x_i - x_0) / \| x_i - x_0 \| \to$ some z;
(ii) $u_i \to$ some $u^0 \in KKT(x_0)$.
Similar to the proof of Proposition 5.5 we can establish that $z \in D(x_0)$.

We first show that for sufficiently large i one has $\langle u^0, g(x_i) \rangle = 0$. Since u_i converges to u^0, and Q is a polyhedral convex cone, it is easy to see that for sufficiently large i, there is a $\bar{u}_i \in Q$ such that $u_i \in [u^0, \bar{u}_i]$. For each such i, it follows from (A14) that

$$0 = \langle u_i, g(x_i) \rangle = \langle \lambda_i u^0 + (1 - \lambda_i) \bar{u}_i, g(x_i) \rangle \quad \text{(for some } \lambda_i \in [0,1])$$

$$= \lambda_i \langle u^0, g(x_i) \rangle + (1-\lambda_i) \langle \overline{u}_i, g(x_i) \rangle. \tag{A15}$$

However, we have $\langle u^0, g(x_i) \rangle \leq 0$ and $\langle \overline{u}_i, g(x_i) \rangle \leq 0$ from the fact that $u^0, \overline{u}_i \in Q$. Hence we deduce from (A15) that $\langle u^0, g(x_i) \rangle = 0$.

We now prove that

$$\langle z, \nabla_x^2 L(x_0, u^0) z \rangle > 0. \tag{A16}$$

Using the second-order growth condition, we can infer that

$$
\begin{aligned}
f(x_i) &= L(x_i, u^0) \\
&= L(x_0, u^0) + \nabla_x L(x_0, u^0)(x_i - x_0) + \tfrac{1}{2}\langle x_i - x_0, \nabla_x^2 L(x_0, u^0)(x_i - x_0) \rangle + o(\|x_i - x_0\|^2) \\
&= f(x_0) + \tfrac{1}{2}\langle x_i - x_0, \nabla_x^2 L(x_0, u^0)(x_i - x_0) \rangle + o(\|x_i - x_0\|^2) \\
&\geq f(x_0) + \alpha \|x_i - x_0\|^2.
\end{aligned}
$$

Subtracting $f(x_0)$, dividing by $\|x_i - x_0\|^2$ and passing to the limit, we obtain the desired result.

Finally, we can deduce a contradiction. For each fixed i we define a scalar function s over $[0,1]$ by

$$s(t) := [\nabla_x f(x_t) + \nabla_x g(x_t)^* u_t](x_i - x_0) - \langle u_i - u^0, g(x_t) \rangle,$$

where $x_t := (1-t)x_0 + tx_i$, $u_t := (1-t)u^0 + tu_i$. Since x_0 is a KKT stationary point and (x_i, u_i) satisfies (A13) and (A14), we find that $s(0) \geq 0 \geq s(1)$. Utilizing the mean value theorem, we claim that for each i, there is a $t_i \in (0,1)$ such that

$$0 \geq s'(t_i) = \langle x_i - x_0, \nabla_x^2 L(x_{t_i}, u_{t_i})(x_i - x_0) \rangle.$$

Dividing the above inequality by $\|x_i - x_0\|^2$ and passing to the limit, we establish a contradiction to (A16).
∎

References

[1]   S.N. Afriat, "Theory of maxima and the method of Lagrange", SIAM J. Appl. Math., 20 (1971) 343-357.

[2]   M.S. Bazaraa and C.M. Shetty, Nonlinear programming: Theory and algorithms (John Wiley & Sons, New York, 1979).

[3]   A. Ben-Israel, A. Ben-Tal, and S. Zlobec, Optimality in nonlinear programming: A feasible direction approach, (John Wiley, New York 1981).

[4]   A. Ben-Tal, "Second-order and related extremality conditions in nonlinear programming", Journal of Optimization Theory and Applications, 31 (1980) 143-165.

[5]   J.V. Burke and R.A. Poliquin, "Optimality conditions for nonfinite valued convex composite functions", Mathematical Programming, forthcoming.

[6]   J.V. Burke, "An exact penalization viewpoint of constrained optimization", SIAM J. Control and Optim., 29 (1991) 968-998.

[7]   R.W. Chaney, "Second-order sufficient conditions in nonsmooth optimization", Mathematics of Operations Research, 13 (1988) 660-673.

[8]   A.V. Fiacco, "Second order sufficient conditions for weak and strict constrained minima", SIAM Journal of Applied Mathematics, 16 (1968) 105-108.

[9]   A.V. Fiacco and G.P. McCormick, Nonlinear programming: Sequential unconstrained minimization techniques (Wiley, New York, 1968).

[10]   A.V. Fiacco and G.P. McCormick, "Asymptotic conditions for constrained minimization", Technical Paper RAC-TP-340, Research Analysis Corporation, 1968.

[11]   B. Gollan, "Higher order necessary conditions for an abstract optimization problem", Math. Programming study, 14 (1981) 69-76.

[12] S.P. Han and O.L. Mangasarian, "Exact penalty functions in nonlinear programming", Mathematical Programming, 17 (1979) 251-269.

[13] M.R. Hestenes, "An indirect sufficiency proof for the problem of Bolza in nonparametric form", Ibid, 62 (1947) 509-535.

[14] R. Hettich and H.Th. Jongen, "On first and second order conditions for local optim for optimization problems in finite dimensions", Methods of Operations Research, 23 (1977) 82-97.

[15] A.D. Ioffe, "Necessary and sufficient conditions for a local minimum, 1: A reduction theorem and order conditions", SIAM J. Control Optim. 17 (1979) 245-250.

[16] A.D. Ioffe, "Necessary and sufficient conditions for a local minimum, 2: Conditions of Levitin-Miljutin-Osmolovskii type", SIAM J. Control Optim. 17 (1979) 251-265.

[17] A.D. Ioffe, "Necessary and sufficient conditions for a local minimum, 3: Second order conditions and augmented duality", SIAM J. Control Optim. 17 (1979) 266-288.

[18] J. Kyparisis and A.V. Fiacco, "Second order sufficient conditions for strict and isolated local minima in nonlinear programming", Technical Paper T-504, Department of Operations Research, George Washington University, Washington DC (1985).

[19] E.S. Levitin, A.A. Miljutin and N.P. Osmolovskii, "On conditions for a local minimum in a problem with constraints", In: Mathematical Economics and Functional Analysis, B.S. Mitjagin, ed., Nauka, Moscow, 1974.

[20] B. Liu, "Second order sufficient conditions for isolated local minima", Proceedings of The Second Conference of The Association of Asian-Pacific Operations Research Societies, forthcoming.

[21] A. Majthay, "Optimality conditions for quadratic programming", Math. Programming, 1 (1971) 359-365.

[22] O.L. Mangasarian, "Local unique solutions of quadratic programs, linear and nonlinear complementarity problems", mathematical Programming, 19 (1980) 200-212.

[23] O.L. Mangasarian and S. Fromovitz, "The Fritz John necessary optimality conditions in the presence of equality and inequality constraints", Journal of Mathematical Analysis and Applications, 17 (1967) 73-74.

[24] H. Maurer, "First and second order sufficient optimality conditions in mathematical programming and optimal control", Math. Programming Study, 14 (1981) 163-177.

[25] G.P. McCormick, "Second-order conditions for constrained minima", SIAM Journal on Applied Mathematics, 15 (1967) 641-652.

[26] E.J.McShane, "Sufficient conditions for a weak relative minimum in the problem of Bolza", Trans. Amer. Math. Soc., 52 (1942) 344-379.

[27] E.J. Penissi, "An indirect sufficiency proof for the problem of Lagrange with differential inequalities as added side conditions", Transactions of American mathematical Society, 74 (1953) 177-198.

[28] S.M. Robinson, "First order conditions for general nonlinear optimization", SIAM Journal on Applied Mathematics 30 (1976) 597-607.

[29] S.M. Robinson, "Stability theory for systems of inequalities, Part II: Differentiable nonlinear systems", SIAM Journal on Numerical Analysis 13 (1976) 497-513.

[30] S.M. Robinson, "Generalized equations and their solutions, Part II: Applications to nonlinear programming", Mathematical Programming Study 19 (1982) 200-221.

[31] R.T. Rockafellar, "Second-order optimality conditions in nonlinear programming obtained by way of epi-derivatives", Math. Oper. Res., to appear.

[32] E.W. Sachs, "Convergence of algorithms for perturbed optimization problems", Annals of Operations Research 27 (1990) 311-342.

[33] C. Singh, "Sufficient optimality criteria in nonlinear programming for generalized equality-inequality constraints", Journal of Optim. Appl., 22 (1977) 631-635.

WSSIAA 5 (1995) pp. 255–270
© World Scientific Publishing Company

CONVERGENCE ANALYSIS OF PRIMAL-DUAL INTERIOR POINT ALGORITHMS FOR CONVEX QUADRATIC PROGRAMS

Zhi-Quan Luo
Department of Electrical and Computer Engineering
McMaster University, Hamilton, Ontario, Canada L8S 4K1

ABSTRACT

In this paper, we study the iterate convergence of a class of primal-dual interior point algorithms for solving convex quadratic programs. We establish the linear convergence of the iterate sequence without any nondegeneracy assumptions (such as the strict complementarity or the Jacobian being nonsingular). Our proof is based on an error bound for polynomial systems. Our work extends the iterate convergence results for linear programming by Tapia, Zhang and Ye.

1. Introduction

Consider the following convex quadratic programming problem

$$\begin{aligned}
\text{Minimize} \quad & c^T x + \frac{1}{2} x^T Q x \\
\text{Subject to} \quad & Ax = b, \\
& x \geq 0,
\end{aligned} \tag{1.1}$$

where $c, x \in \Re^n$, $b \in \Re^m$, $A \in \Re^{m \times n}$ $(m < n)$ and has full row rank, and $Q \in \Re^{n \times n}$ is symmetric positive semi-definite. Let $X = diag(x)$, $Y = diag(y)$ and $e \in \Re^n$ be the vector of 1's. Then, the first order optimality condition for (1.1) can be written as

$$\begin{bmatrix} Ax - b \\ A^T \lambda - Qx + y - c \\ XYe \end{bmatrix} = 0, \qquad (x, y) \geq 0,$$

where λ and y are the dual variables. It is well known that, for the problem (1.1), the above condition is both necessary and sufficient for optimality. Let $B \in \Re^{(n-m) \times n}$ be any matrix such that the columns of B^T form a basis for the null space of A. Using $BA^T = 0$ to eliminate λ from the above system yields the following equivalent optimality condition for (1.1)

$$\begin{bmatrix} Ax - b \\ -BQx + By - Bc \\ XYe \end{bmatrix} = 0, \qquad (x, y) \geq 0. \tag{1.2}$$

We assume throughout that the feasible set of (1.1)

$$\Omega = \{(x, y) \; : \; x, y \in \Re^n, \; Ax = b, \; -BQx + By - Bc = 0, \; (x, y) \geq 0\}$$

has a nonempty relative interior. That is, there exists some $(x, y) \in \Omega$ such that $(x, y) > 0$.

The objective of this paper is to study the asymptotic behavior of a general class of primal-dual interior point algorithms for solving (1.1). Previously, there have been several studies of local convergence behavior of primal-dual interior algorithms. In particular, assuming that the iterates converge to some optimal solution (x^*, y^*), Kojima, Megiddo and Noma[5] analyzed, for a class of complementarity problems, Q-linear, superlinear and quadratic convergence of the duality gap. However, their analysis is based on the restrictive assumption that a certain Jacobian matrix is nonsingular at (x^*, y^*). Recently, Zhang, Tapia and Potra[25] established the Q-superlinear convergence of the duality gap under the assumption that the iterates converge to some optimal solution satisfying the strict complementarity condition holds at (x^*, y^*). Unfortunately, most of the real world problems are degenerate and do not satisfy the strict complementarity condition. In fact, there exist convex programs which do not have any primal-dual solution solution satisfying strict complementarity. Also, assuming the convergence of iterates in the analysis is much less satisfying from a theoretical standpoint. To the best of our knowledge, no iterate convergence or rate of convergence results are available for (1.1) when nondegeneracy assumptions are removed.

For linear programming (i.e., $Q = 0$), the iterate convergence of primal-dual interior point algorithms has been recently obtained by Tapia, Zhang and Ye[18] without any nondegeneracy assumption. The only assumptions needed in their work are (1) the iterates stay centered; (2) the duality decays sufficiently fast; (3) the centering step is phased out sufficiently fast. The analysis of Tapia, Zhang and Ye[18] uses Hoffman's error bound and an interesting fact that the strict complementarity holds for any cluster point of the iterate sequence (see Güler and Ye[3]). (For a rather general account of Q-superlinear convergence of duality gap for linear programming, the readers are referred Zhang, Tapia and Dennis[24].)

In this paper, we establish the convergence of the iterates for a large class of primal-dual interior point algorithms for solving (1.1) under the same three assumptions listed above (See the statement of Theorem 3.1 for details). In particular, we do not require any nondegeneracy assumptions, and the convergence of the iterates is established analytically rather than assumed in our analysis. Our proof is based on a generalized Hoffman's error bound for polynomial systems. In particular, this error bound allows us to bound the distance from the iterate to the optimal solution set by the duality gap. The latter typically converges to zero (super)linearly.

We remark that the error bound approach used in this paper has been particularly effective in the analysis of iterative descent algorithms for solving degenerate convex optimization problems with linear constraints. In fact, using this approach, Luo and Tseng[8,9] (and see the references therein) recently proved the convergence of several well known algorithms for solving linear complementarity problems, convex quadratic programs and a general class of linearly constrained convex minimization problems whereby the cost function is the composition of a strongly convex function with an affine mapping. The algorithms studied include the matrix splitting algorithms, the gradient projection algorithms and the coordinate descent algorithms. All previous convergence results for these algorithms only assert that each cluster point of the iterate sequence, if exists, is an optimal solution.

We briefly describe our notations below. For any vector $x \in \Re^n$ and any $I \subseteq \{1, ..., n\}$, we denote by x_i the ith component of x and x_I the vector with components $x_i \in I$ (with the x_i's arranged in the same order as in x). For any two vectors x and y of the same dimension, we denote by $\|x\|$ the Euclidean norm of x, that is, $\|x\| = \sqrt{x^T x}$. For any $m \times n$ matrix M and any $I \subseteq \{1, ..., n\}$, we denote by M_i the ith column of M and by M_I the submatrix of M obtained by removing all columns $i \notin I$. Finally, for any $I \subseteq \{1, 2, ..., n\}$, we denote by $|I|$ the number of elements of I and by $\tilde{I}$ the complement of I in $\{1, 2, ..., n\}$.

2. Algorithm description

For notational convenience, we let

$$F(x, y) = \begin{bmatrix} Ax - b \\ -BQx + By - Bc \\ XYe \end{bmatrix} \tag{2.1}$$

and let the Jacobian of $F(x, y)$ be denoted by $F'(x, y)$. It is clear that

$$F'(x, y) = \begin{bmatrix} A & 0 \\ -BQ & B \\ Y & X \end{bmatrix}. \tag{2.2}$$

Algorithm 1.

Given a pair $(x^0, y^0) \in \Omega$ such that $(x^0, y^0) > 0$. For $k = 0, 1, 2, ...,$ do

(1) Choose $\sigma_k \in [0, 1)$ and $\tau_k \in (0, 1)$. Set $\mu_k = \sigma_k (x^k)^T y^k / n$.

(2) Solve the following system for $(\triangle x^k, \triangle y^k)$:

$$F'(x^k, y^k) \begin{bmatrix} \triangle x \\ \triangle y \end{bmatrix} = -F(x^k, y^k) + \begin{bmatrix} 0 \\ \mu_k e \end{bmatrix}. \tag{2.3}$$

(3) Compute the step-size:

$$\alpha_k = \min \left\{ 1, \frac{-\tau_k}{\min \left((X^k)^{-1} \triangle x^k, -1/2 \right)}, \frac{-\tau_k}{\min \left((Y^k)^{-1} \triangle y^k, -1/2 \right)} \right\}. \tag{2.4}$$

where we have used the convention $\min(v) = \min_i v_i$ for any vector $v = (v_1, ..., v_n)^T$.

(4) Update the iterates:

$$x_{k+1} = x_k + \alpha_k \triangle x_k, \qquad y_{k+1} = y_k + \alpha_k \triangle y_k. \tag{2.5}$$

The step-size formula (2.4) is so designed to ensure that (x^{k+1}, y^{k+1}) remains positive for all $k \geq 0$. It is known from Zhang, Tapia and Potra[25] (see Proposition 2.1 therein) that $F'(x, y)$ is nonsingular for all $(x, y) > 0$. Thus, the linear system (2.3) has a unique solution for each $k \geq 0$ and consequently Algorithm 1 is well defined. Also, using (2.3) and a simple induction, we see that $(x^k, y^k) \in \Omega$ for all k. In other words, (x^k, y^k) satisfies $Ax^k - b = 0$ and $-BQx^k + By^k - Bc = 0$, $\forall k \geq 0$.

The above algorithmic framework covers a large number of existing interior point algorithms for linear programming and convex quadratic programming. For example, the algorithms by Gonzaga and Todd[2], Kojima, Mizuno and Yoshise[6], Lustig[10], Mizuno, Todd and Ye[13,14], Monteiro and Adler[15,16], Todd and Ye[19] can all be cast into the above framework by choosing specific values of σ_k and τ_k at each iteration k. For a detailed discussion, the readers are referred to Zhang, Tapia and Dennis[24] or Zhang, Tapia and Potra[25].

3. Convergence Analysis

In this section, we establish the iterate convergence of primal-dual interior point algorithms (as described by Algorithm 1). The proof is quite intricate and is divided into several technical lemmas. The first lemma is well known (see Proposition 2.1 in Tapia, Zhang and Ye[18] for a proof in the case of linear programming). Its proof is based on the fact that the inner product of primal and dual incremental vectors $\triangle x^k$, $\triangle y^k$ is nonnegative.

Lemma 3.1. Let $\{(x^k, y^k)\}$ be the iterate sequence generated by Algorithm 1. Suppose that the duality sequence $\{(x^k)^T y^k\}$ is bounded. Then, the sequence $\{(x^k, y^k)\}$ is bounded.

Proof. In light of Algorithm 1, both (x^0, y^0) and (x^k, y^k) belong to the feasible set Ω, so that they satisfy the linear system of equations

$$Ax = b, \quad -BQx + By = Bc.$$

Therefore, we have

$$A(x^k - x^0) = 0, \quad -BQ(x^k - x^0) + B(y^k - y^0) = 0.$$

Since the columns of B^T form a basis for the null space of A, there exists some vector u^k such that $(x^k - x^0) = B^T u^k$. Combining this with above two relations yields

$$\begin{aligned}(x^k - x^0)^T(y^k - y^0) &= (B^T u^k)^T(y^k - y^0) = (u^k)^T B(y^k - y^0) \\ &= (u^k)^T \left(BQ(x^k - x^0)\right) = (u^k)^T(BQB^T)u^k \geq 0,\end{aligned}$$

where the last step follows from the positive semi-definiteness of Q. By expanding $(x^k - x^0)^T(y^k - y^0)$ and rearranging terms, we obtain

$$(y^0)^T x^k + (x^0)^T y^k \leq (x^k)^T y^k + (x^0)^T y^0.$$

Since $(x^0, y^0) > 0$, $(x^k, y^k) > 0$ and $(x^k)^T y^k$ is bounded, the above relation implies the boundedness of (x^k, y^k). **Q.E.D.**

Next we use the duality gap $(x^k)^T y^k$ to bound the distance from (x^k, y^k) to the optimal solution set $X^* \times Y^*$. To do this, we need the following error bound for polynomial systems (see Luo and Luo[7]).

Proposition 3.1 (Error bound for polynomial systems). Let P denote the set of $z \in \Re^n$ satisfying the following polynomial system

$$f_1(z) \leq 0, ..., f_r(z) \leq 0, \quad g_1(z) = 0, ..., g_s(z) = 0,$$

where each f_i and g_j is a polynomial. Let ρ be a positive number such that P contains an element z with $\|z\| \leq \rho$. Then, there exist some constants $\tau > 0$ and $\gamma > 0$ such that

$$\mathrm{dist}(z, P) \leq \tau \left(\|[f(z)]_+\| + \|g(z)\| \right)^\gamma, \qquad \forall z \text{ with } \|z\| \leq \rho,$$

where $[\cdot]_+$ denotes the orthogonal projection to the nonnegative orthant $[0, \infty)^r$, and $f(z)$ (respectively, $g(z)$) denotes the vector function whose component functions are $f_i(z)$, $i = 1, ..., r$ ($g_j(z)$, $j = 1, ..., s$, respectively).

The above error bound basically says that the solution set of a polynomial system is Hölder continuous as the right-hand side is perturbed. This is a generalization of the well-known error bound by Hoffman[4] which says that the solution set of a *linear* system is *Lipschitz* continuous (i.e., $\gamma = 1$) as the right-hand side varies. We now apply the above generalized error bound to the polynomial system $F(x, y) = 0$, $(x, y) \geq 0$ to obtain an error estimate for the distance from (x^k, y^k) to the optimal solution set $X^* \times Y^*$.

Lemma 3.2. Suppose that $(x^k)^T y^k \to 0$. Then, there exists some $(x^*, y^*) \in X^* \times Y^*$ such that

$$\|(x^k, y^k) - (x^*, y^*)\| \leq \tau \left((x^k)^T y^k \right)^\gamma, \qquad \forall k \geq k_0, \tag{3.1}$$

where $\tau > 0$, $\gamma > 0$ are some positive constants independent of k, and k_0 is an nonnegative integer. Furthermore, the optimal primal-dual pair (x^*, y^*) in (3.1) can be chosen to satisfy $x^*_{J^k} = 0$, $y^*_{\bar{J}^k} = 0$, where J^k denotes the set of indices j for which $x^k_j \leq y^k_j$.

Proof. Let J^k be defined as above. We claim that there exists some k' such that for each $k \geq k'$ $X^* \times Y^*$ contains a pair (x^*, y^*) satisfying $x^*_{J^k} = 0$, $y^*_{\bar{J}^k} = 0$. Suppose the contrary, so there exists a subsequence K such that there is no optimal solution pair (x^*, y^*) with the property $x^*_{J^k} = 0$, $y^*_{\bar{J}^k} = 0$, for all $k \in K$. By further passing onto a subsequence if necessary, we assume $J^k = J$ for some fixed J, $\forall k \in K$. By Lemma 3.1, the sequence $\{(x^k, y^k)\}_K$ is bounded. Let (x^∞, y^∞) be a cluster point of $\{(x^k, y^k)\}_K$. Then, $(x^\infty)^T y^\infty = \lim_{k \to \infty, k \in K} (x^k)^T y^k = 0$. Moreover, since

$(x^k, y^k) \in \Omega$ for all k and Ω is closed, we have $(x^\infty, y^\infty) \in \Omega$. This implies that $(x^\infty, y^\infty) \in X^* \times Y^*$. Also, by $(x^k)^T y^k \to 0$ and $(x^k, y^k) > 0$, we have $\min\{x_j^k, y_j^k\} \to 0$ for all j. Hence, $x_j^\infty = \lim_{k \to \infty, k \in K} x_j^k = \lim_{k \to \infty, k \in K} \min\{x_j^k, y_j^k\} = 0$. Similarly, we have $y_{\bar{j}}^\infty = 0$. This contradicts the assumption that there does not exist an optimal solution pair (x^*, y^*) with the property $x_j^* = 0$, $y_{\bar{j}}^* = 0$.

Since $\|X^k Y^k e\| \to 0$, there exists some $k'' \geq 0$ such that

$$\|X^k Y^k e\| \leq 1, \qquad \forall k \geq k''. \tag{3.2}$$

Define $k_0 = \max\{k', k''\}$. Fix any $k \geq k_0$ and let $J^k = J$. Consider the polynomial system

$$Ax - b = 0, \quad -BQx + By - Bc = 0, \quad XY = 0, \quad x_J = 0, \quad y_{\bar{j}} = 0, \quad (x, y) \geq 0. \tag{3.3}$$

In light of the optimality condition (1.2), each (x, y) satisfying the above polynomial system must be an optimal primal-dual solution pair. Moreover, by the argument in the preceding paragraph, the polynomial system (3.3) has a solution, say (x^J, y^J). Let

$$\rho \overset{\triangle}{=} \max_{J \subseteq \{1, \ldots, n\}} \left\{ \limsup_k \|x^k\|, \limsup_k \|y^k\|, \|(x^J, y^J)\| \right\}.$$

Since the sequence $\{(x^k, y^k)\}$ is bounded (cf. Lemma 3.1) and there are only a finite number of different subsets in $\{1, \ldots, n\}$, it follows that $\rho < \infty$. Now, notice that (x^k, y^k) satisfies

$$Ax^k - b = 0, \quad -BQx^k + By^k - Bc = 0, \quad x^k \geq 0, \quad y^k \geq 0.$$

By applying Proposition 3.1 to (3.3) with ρ defined as above, we see that there exist some $\tau_J > 0$, $\gamma_J > 0$ such that

$$\|(x^k, y^k) - (x^*, y^*)\| \leq \tau_J \left(\|x_J^k\| + \|y_{\bar{j}}^k\| + \|X^k Y^k e\| \right)^{\gamma_J} \tag{3.4}$$

for some (x^*, y^*) satisfying (3.3), where $X^k = diag(x^k)$ and $Y^k = diag(y^k)$. Clearly, $(x^*, y^*) \in X^* \times Y^*$. Since $J^k = J$, we have

$$\|x_J^k\| = \|\min\{x_J^k, y_J^k\}\| \leq n^{1/4} \|(X^k Y^k e)_J\|^{1/2}$$

and

$$\|y_{\bar{j}}^k\| = \|\min\{x_{\bar{j}}^k, y_{\bar{j}}^k\}\| \leq n^{1/4} \|(X^k Y^k e)_{\bar{j}}\|^{1/2}.$$

Substituting the above two relations into (3.4) and using (3.2) yields

$$\begin{aligned}
\|(x^k, y^k) - (x^*, y^*)\| &\leq \tau_J (2n^{1/4} + 1)^{\gamma_J} \|X^k Y^k e\|^{\gamma_J/2} \\
&\leq \tau_J (2n^{1/4} + 1)^{\gamma_J} \rho^{\gamma_J/2} \left((x^k)^T y^k \right)^{\gamma_J/4},
\end{aligned}$$

where the last step follows from $\|x^k\| \le \rho$ and $\|y^k\| \le \rho$. Now let

$$\tau = \max_{J \subseteq \{1,\dots,n\}} \tau_J (2n^{1/4}+1)^{\gamma_J} \rho^{\gamma_J/2}, \qquad \gamma = \min_{J \subseteq \{1,\dots,n\}} (\gamma_J/4).$$

The proof is complete. **Q.E.D.**

Our next lemma shows an interesting property of the Jacobian matrix $F'(x,y)$. This property will be used later to estimate certain entries of $(F'(x,y))^{-1}$ in the proof of Theorem 3.1. To facilitate the presentation, we need to fix some notations. For any index set $I \subseteq \{1,\dots,n\}$ and any vector $z \in \Re^n$, we use $z(I)$ to denote $\prod_{i \in I} z_i$. We shall use the convention that $z(I) = 1$ if I is the empty set. Thus, if $I = \{1,2,\dots,n\}$, then $z(I) = \prod_{i=1}^{n} z_i$ and $z(\tilde{I}) = 1$.

Lemma 3.3. Let $x > 0$ and $y > 0$. There holds

$$|\det[F'(x,y)]| = \sum_{I \subseteq \{1,\dots,n\}} a_I x(I) y(\tilde{I}), \tag{3.5}$$

where each a_I is a nonnegative number. Moreover, the (i,j)th cofactor of $F'(x,y)$, denoted by $\det[F'(x,y)]_{ij}$, can be written in the form

$$\det[F'(x,y)]_{ij} = \sum_{\substack{I \subset \{1,\dots,n\} \\ i-n \notin I}} d_{I,j} x(I) y(\tilde{I}\backslash\{i-n\}), \quad \forall\, n+1 \le i \le 2n,\ \forall\, 1 \le j \le 2n,$$

$$\tag{3.6}$$

where each $d_{I,j}$ is some constant (depending on the problem data only) satisfying:

$$\text{if } d_{I,j} \ne 0, \text{ then } a_I a_{I \cup \{i-n\}} > 0. \tag{3.7}$$

Proof. We first show (3.5). By the definition of $F'(x,y)$ (cf. (2.2)), we have

$$\det[F'(x,y)] = \begin{vmatrix} A & 0 \\ -BQ & B \\ Y & X \end{vmatrix}.$$

Notice that X and Y are both diagonal matrices. Using Cauchy–Binet Theorem to expand the above determinant according to the last n rows, we see $\det[F'(x,y)]$ is a homogeneous polynomial (in terms of x and y) of degree n and it can be put in the form

$$\det[F'(x,y)] = \sum_{I \subseteq \{1,\dots,n\}} a'_I x(I) y(\tilde{I}), \tag{3.8}$$

for some constants a'_I. Thus, to prove (3.5), we only need to argue that all of a'_I's have the same sign. Without loss of generality, we suppose that $a'_I > 0$ for some I. Then, for each $\epsilon > 0$, we let (x^ϵ, y^ϵ) be given by

$$x_i^\epsilon = \begin{cases} 1, & \text{if } i \in I, \\ \epsilon, & \text{if } i \in \tilde{I}; \end{cases} \qquad y_i^\epsilon = \begin{cases} 1, & \text{if } i \in \tilde{I}, \\ \epsilon, & \text{if } i \in I. \end{cases}$$

262

Thus, $x^\epsilon(I)y^\epsilon(\tilde{I}) = 1$ and $x^\epsilon(J)y^\epsilon(\tilde{J}) = O(\epsilon)$ for all $J \neq I$. By (3.8), we have $\det[F'(x^\epsilon, y^\epsilon)] \to a'_I$ as $\epsilon \to 0$. Thus, choosing ϵ sufficiently small, we have $\det[F'(x^\epsilon, y^\epsilon)] > 0$. Now, if the coefficients $\{a'_I\}$ do not have the same sign, so $a'_I < 0$ for some index set I, then by an argument similar to the above, we can find a point $(x, y) > 0$ such that $\det[F'(x, y)] < 0$. However, this is impossible, since $\det[F'(x, y)]$ is a continuous function of (x, y), so if $\det[F'(x, y)]$ changes sign over $(0, \infty)^{2n}$ then $\det[F'(x, y)]$ must vanish at some point in $(0, \infty)^{2n}$ (Mean Value Theorem). This would contradict the fact that $F'(x, y)$ is nonsingular for all $(x, y) > 0$ (see Proposition 2.1 in Zhang, Tapia and Potra[25]).

Equation (3.6) can be argued analogously. In particular, let us fix some $n + 1 \leq i \leq 2n$ and $1 \leq j \leq 2n$. By definition, the (i, j)th cofactor of $\det[F'(x, y)]$ is the determinant of the submatrix of $F'(x, y)$ after removing its ith row and jth column. Applying Cauchy–Binet Theorem to expand the determinant of this submatrix yields

$$\det[F'(x, y)]_{ij} = \sum_{\substack{I \subset \{1, \ldots, n\} \\ i - n \notin I}} d_{I,j} x(I) y(\tilde{I} \setminus \{i - n\}),$$

where each $d_{I,j}$ is some constant.

It remains to show (3.7). Let us make some observations. First, if $i - n = j$ (or $i = j$), then (3.7) holds trivially since we have in this case $a_I = |d_{I,j}|$ (or $a_{I \cup \{i-n\}} = |d_{I,j}|$ respectively). So we can assume, in the remainder of the proof, that $i - n \neq j$ and $i \neq j$. Let I be some index set such that $d_{I,j} \neq 0$. We now proceed to prove that $a_I > 0$ when $j \leq n$. The proof of $a_{I \cup \{i-n\}} > 0$ when $j \geq n + 1$ is symmetrical and thus omitted. Since $j \leq n$, the variables y_{i-n}, y_j do not appear in the above expression of $\det[F'(x, y)]_{ij}$ because they are removed before the expansion. Moreover, since the determinant $\det[F'(x, y)]_{ij}$ contains a row with x_j as the only nonzero entry, x_j must be in each term of the expansion of $\det[F'(x, y)]_{ij}$. In other words, $j \in I$.

Let

$$M = \begin{bmatrix} A \\ -BQ \end{bmatrix}, \qquad N = \begin{bmatrix} 0 \\ B \end{bmatrix}.$$

Then, by a careful examination of the expansion of the determinant $\det[F'(x, y)]_{ij}$, we can see that $d_{I,j}$ is equal to (up to a possible sign change) the determinant of the matrix formed by those columns of M indexed by $I_1 = (I \cup \{i - n\}) \setminus \{j\}$ and those columns of N indexed by $\tilde{I}$. In other words, we have

$$d_{I,j} = \det[M_{I_1}, N_{\tilde{I}}] \neq 0.$$

Similarly, we have (up to a possible sign change) $a_I = \det[M_I, N_{\tilde{I}}]$. Since the matrix

$$[M_{I_1}, N_{\tilde{I}}] = \begin{bmatrix} A_{I_1} & 0 \\ -(BQ)_{I_1} & B_{\tilde{I}} \end{bmatrix}$$

is nonsingular, it follows that both A_{I_1} and $B_{\tilde{I}}$ must be square matrices and nonsingular. Thus, $|I_1| = n - m$ and $|\tilde{I}| = m$. Since $I_1 = (I \cup \{i - n\}) \setminus \{j\}$, I is identical to I_1 except the element j of I is replaced by the element of $i - n$. Note that $d_{I,j} = \det[M_{I_1}, N_{\tilde{I}}] = \det[A_{I_1}] \det[B_{\tilde{I}}] \neq 0$, and $a_I = \det[M_I, N_{\tilde{I}}] = \det[A_I] \det[B_{\tilde{I}}]$. So, to show $a_I \neq 0$ we only need to prove $\det[A_I] \neq 0$, or equivalently A_I is nonsingular. In other words, we have to show that the matrix A_{I_1} remains full rank after we replace its column A_{i-n} by the column A_j.

We argue by contradiction. Suppose A_I is singular, so A_j can be written as a linear combination of the columns of $A_{I_1 \setminus \{i-n\}} = A_{I \setminus \{j\}}$. Thus, there exists some $C \in \Re^{(n-m-1) \times (n-m)}$ such that $A_I = A_{I \setminus \{j\}} C$. By reindexing if necessary, we assume that A and B are partitioned as $A = [A_I, A_{\tilde{I}}]$ and $B = [B_I, B_{\tilde{I}}]$. Since $B^T A = 0$, we have

$$
\begin{aligned}
0 &= [B_I, B_{\tilde{I}}] [A_I, A_{\tilde{I}}]^T \\
&= [B_I, B_{\tilde{I}}] \left[A_{I \setminus \{j\}} C, A_{\tilde{I}}\right]^T \\
&= [B_I, B_{\tilde{I}}] \begin{bmatrix} C^T & 0 \\ 0 & I \end{bmatrix} \left[A_{I \setminus \{j\}}, A_{\tilde{I}}\right]^T \\
&= \left[B_I C^T, B_{\tilde{I}}\right] \left[A_{I \setminus \{j\}}, A_{\tilde{I}}\right]^T .
\end{aligned}
$$

Therefore, the row vectors of $\left[A_{I \setminus \{j\}}, A_{\tilde{I}}\right]$ lie in the null space of the matrix $\left[B_I C^T, B_{\tilde{I}}\right]$. The latter matrix has full row rank (cf. $B_{\tilde{I}}$ has full rank) and has dimension $m \times (n-1)$. Therefore the null space of $\left[B_I C^T, B_{\tilde{I}}\right]$ has dimension at most $n - m - 1$. However, since

$$
\text{rank} \left[A_{I \setminus \{j\}}, A_{\tilde{I}}\right] = \text{rank} [A_I, A_{\tilde{I}}] = n - m,
$$

the matrix $\left[A_{I \setminus \{j\}}, A_{\tilde{I}}\right]$ has $n - m$ linearly independent rows. They cannot all lie in the null space of $\left[B_I C^T, B_{\tilde{I}}\right]$, a contradiction! **Q.E.D.**

The next lemma is known for the case of linear programming, i.e., when $Q = 0$, (see [Lemma 3.2, 17]). It remains true when $Q \neq 0$.

Lemma 3.4. For any $(x, y) \in \Re^n \times \Re^n$ and for any $(x^*, y^*) \in X^* \times Y^*$, there holds

$$
F'(x, y) \begin{bmatrix} x - x^* \\ y - y^* \end{bmatrix} - F(x, y) = \begin{bmatrix} 0 \\ (X - X^*)(Y - Y^*)e \end{bmatrix}. \tag{3.9}
$$

Proof. Use the definitions of $F(x, y)$ and $F'(x, y)$ (cf. (2.1)–(2.2)) and the fact that $F(x^*, y^*) = 0$. **Q.E.D.**

Using Lemmas 3.1–3.4, we are now ready to show the following main convergence result. The key proof idea is to use the error estimate (3.1) to bound the distance from (x^k, y^k) to the optimal solution set by the duality gap $(x^k)^T y^k$; then, using Lemmas 3.3 and 3.4 to show that the difference $\|(x^{k+1}, y^{k+1}) - (x^k, y^k)\|$ can be bounded by the distance from (x^k, y^k) to the optimal solution set and the size of σ_k. Therefore, if the duality gap and σ_k decreases sufficiently fast, then the sequence $\{(x^k, y^k)\}$ is convergent.

Theorem 3.1. Let $\{(x^k, y^k)\}$ be a sequence of iterates generated by Algorithm 1. Assume

A1. Centering condition: $\min_{1 \leq i \leq n} x_i^k y_i^k / (x^k)^T y^k \geq \eta$, for some constant $\eta > 0$.

Then, there exist some constants $\beta > 0$, $\theta > 0$ (independent of k) such that

$$\|(x^{k+1}, y^{k+1}) - (x^k, y^k)\| \leq \beta \left((x^k)^T y^k\right)^\gamma + \theta \sigma_k, \qquad \forall k \geq k_0, \tag{3.10}$$

where the constant $\gamma > 0$ and the integer k_0 are given by Lemma 3.2. Therefore, if we further assume

A2. Sufficient decay of the duality gap: $(x^k)^T y^k \to 0$ at least R-linearly,

A3. Phasing out of the centering step: $\sigma^k \to 0$ at least R-linearly,

then the iterate sequence $\{(x^k, y^k)\}$ converges at least R-linearly to an optimal primal-dual pair $(x^*, y^*) \in X^* \times Y^*$. [See Ortega and Rheinboldt[17] for a definition of R-linear convergence.]

Proof. Fix some $k \geq k_0$, where k_0 is the integer given in Lemma 3.2. For notational convenience, we replace the superscript $k+1$ by $+$ and drop the superscript k. By Lemma 3.2, there exist some positive constants $\tau > 0$, $\gamma > 0$ such that

$$\|(x, y) - (x^*, y^*)\| \leq \tau (x^T y)^\gamma, \qquad \text{for some } (x^*, y^*) \in X^* \times Y^*. \tag{3.11}$$

Moreover, (x^*, y^*) satisfies $x_J^* = 0$, $y_{\bar{J}}^* = 0$, where J denotes the set of indices j such that $x_j \leq y_j$. We claim

$$\frac{|x_j - x_j^*||y_j - y_j^*|}{y_j} \leq \|(x, y) - (x^*, y^*)\|, \qquad \frac{|x_j - x_j^*||y_j - y_j^*|}{x_j} \leq \|(x, y) - (x^*, y^*)\|, \quad \forall j. \tag{3.12}$$

By symmetry, we only show the first relation. We consider two cases. If $x_j \leq y_j$, then $x_j^* = 0$. Hence, we have

$$\frac{|x_j - x_j^*||y_j - y_j^*|}{y_j} = \frac{x_j |y_j - y_j^*|}{y_j} \leq |y_j - y_j^*|.$$

If, on the other hand, $x_j > y_j$, then $y_j^* = 0$. In this case, we have

$$\frac{|x_j - x_j^*||y_j - y_j^*|}{y_j} = \frac{|x_j - x_j^*| y_j}{y_j} = |x_j - x_j^*|.$$

Combining the above two relations yields the desired result.

Let $(\hat{x}, \hat{y}) = (x, y) + (\triangle x, \triangle y)$. Since $(x^+ - x^*, y^+ - y^*) = (x - x^*, y - y^*) + \alpha(\triangle x, \triangle y)$ (cf. (2.5)) and $\alpha \in (0, 1]$, we have

$$\|(x^+ - x^*, y^+ - y^*)\| \leq \|(x - x^*, y - y^*)\| + \|(\hat{x} - x^*, \hat{y} - y^*)\|.$$

It then follows

$$
\begin{aligned}
\|(x^+ - x, y^+ - y)\| &\leq \|(x - x^*, y - y^*)\| + \|(x^+ - x^*, y^+ - y^*)\| \\
&\leq 2\|(x - x^*, y - y^*)\| + \|(\hat{x} - x^*, \hat{y} - y^*)\| \\
&\leq 2\tau(x^T y)^\gamma + \|(\hat{x} - x^*, \hat{y} - y^*)\|,
\end{aligned}
\tag{3.13}
$$

where the last step follows from (3.11). Thus, to show (3.10), we only need to estimate $\|(\hat{x} - x^*, \hat{y} - y^*)\|$.

By Lemma 3.4 and relation (2.3), we easily obtain

$$(\hat{x}, \hat{y}) - (x^*, y^*) = F'(x, y)^{-1} \begin{bmatrix} 0 \\ w \end{bmatrix}, \tag{3.14}$$

where

$$w = (X - X^*)(Y - Y^*)e + \frac{\sigma x^T y}{n} e. \tag{3.15}$$

Fix some $1 \leq j \leq 2n$. From (3.14) and Cramer's rule, we have

$$(\hat{x} - x^*, \hat{y} - y^*)_j = \sum_{i=n+1}^{2n} (-1)^{i+j} \frac{\det[F'(x, y)]_{ij}}{\det[F'(x, y)]} w_{i-n}. \tag{3.16}$$

We now estimate the size of $\det[F'(x, y)]_{ij} / \det[F'(x, y)]$. By Lemma 3.3, we have

$$
\begin{aligned}
\frac{|\det[F'(x, y)]_{ij}|}{|\det[F'(x, y)]|} &= \frac{\left| \sum_{\substack{I \subseteq \{1,\dots,n\} \\ i-n \notin I}} d_{I,j} x(I) y(\tilde{I} \setminus \{i - n\}) \right|}{\sum_{I \subseteq \{1,\dots,n\}} a_I x(I) y(\tilde{I})} \\
&\leq \frac{\sum_{d_{I,j} \neq 0} |d_{I,j}| x(I) y(\tilde{I} \setminus \{i - n\})}{\sum_{I \subseteq \{1,\dots,n\}} a_I x(I) y(\tilde{I})}.
\end{aligned}
$$

For each I for which $d_{I,j} \neq 0$, we have from Lemma 3.3 that either $a_I > 0$ or $a_{I \cup \{i-n\}} > 0$, depending on whether $j \leq n$ or $j \geq n + 1$. Thus, we have

$$
\begin{aligned}
\frac{|d_{I,j}| x(I) y(\tilde{I} \setminus \{i - n\})}{\sum_{I \subseteq \{1,\dots,n\}} a_I x(I) y(\tilde{I})} &\leq
\begin{cases}
\dfrac{|d_{I,j}| x(I) y(\tilde{I} \setminus \{i-n\})}{a_I x(I) y(\tilde{I})} & \text{if } j \leq n \\[2ex]
\dfrac{|d_{I,j}| x(I) y(\tilde{I} \setminus \{i-n\})}{a_{I \cup \{i-n\}} x(I \cup \{i-n\}) y(\tilde{I} \setminus \{i-n\})} & \text{if } j \geq n + 1
\end{cases} \\[2ex]
&= \begin{cases}
|d_{I,j}| / (a_I y_{i-n}) & \text{if } j \leq n \\
|d_{I,j}| / (a_{I \cup \{i-n\}} x_{i-n}) & \text{if } j \geq n + 1
\end{cases}
\end{aligned}
$$

where the inequality follows from $a_I \geq 0$ for all I (cf. Lemma 3.3).

We shall prove that

$$|(\hat{x} - x^*, \hat{y} - y^*)_j| \leq n\beta_j (x^T y)^\gamma + \theta_j \sigma, \qquad (3.17)$$

for some constants $\beta_j > 0$, $\theta_j > 0$. Clearly, (3.17) and (3.13) imply the desired error estimate (3.10). We shall prove (3.17) for $j \leq n$. The case where $j \geq n + 1$ can be handled symmetrically. It follows from the preceding paragraph (note that $j \leq n$)

$$\frac{|\det[F'(x,y)]_{ij}|}{|\det[F'(x,y)]|} \leq \sum_{d_{I,j} \neq 0} \frac{|d_{I,j}|}{a_I y_{i-n}}. \qquad (3.18)$$

We now use (3.18) to bound $(\hat{x} - x^*, \hat{y} - y^*)_j$. Using (3.15) and (3.16), we have

$$(\hat{x} - x^*, \hat{y} - y^*)_j = \sum_{i=n+1}^{2n} \left(S_i' + \frac{\sigma}{n} S_i'' \right), \qquad (3.19)$$

where

$$S_i' = (-1)^{i+j} \frac{\det[F'(x,y)]_{ij}}{\det[F'(x,y)]} (x_{i-n} - x_{i-n}^*)(y_{i-n} - y_{i-n}^*),$$

$$S_i'' = (-1)^{i+j} \frac{\det[F'(x,y)]_{ij}}{\det[F'(x,y)]} (x^T y).$$

We shall bound S_i' and S_i'' separately. In particular, using (3.18), we obtain

$$
\begin{aligned}
|S_i'| &= \frac{|\det[F'(x,y)]_{ij}|}{|\det[F'(x,y)]|} |x_{i-n} - x_{i-n}^*||y_{i-n} - y_{i-n}^*| \\
&\leq \sum_{d_{I,j} \neq 0} \frac{|d_{I,j}|}{a_I} \frac{|x_{i-n} - x_{i-n}^*||y_{i-n} - y_{i-n}^*|}{y_{i-n}} \leq \sum_{d_{I,j} \neq 0} \frac{|d_{I,j}|}{a_I} \|(x,y) - (x^*,y^*)\| \\
&\leq \beta_j \|(x,y) - (x^*,y^*)\|, \qquad \text{for some constant } \beta_j > 0
\end{aligned}
$$

where the second inequality follows from (3.12). Next, we estimate S_i''. From (3.18), we have

$$
\begin{aligned}
|S_i''| &= \frac{|\det[F'(x,y)]_{ij}|}{|\det[F'(x,y)]|} (x^T y) \\
&\leq \sum_{d_{I,j} \neq 0} \frac{|d_{I,j}|(x^T y)}{a_I y_{i-n}} \\
&\leq \sum_{d_{I,j} \neq 0} \frac{|d_{I,j}| x_{i-n} y_{i-n}}{\eta a_I y_{i-n}} \\
&= \sum_{d_{I,j} \neq 0} \frac{|d_{I,j}| x_{i-n}}{\eta a_I} \\
&\leq \theta_j, \qquad \text{for some constant } \theta_j > 0,
\end{aligned}
$$

where the second inequality is due to Assumption 1, and the last step follows from the boundedness of (x, y). Combining above two estimates for S_i' and S_i'' with (3.19) yields

$$
\begin{aligned}
|(\hat{x} - x^*, \hat{y} - y^*)_j| &\leq \sum_{i=n+1}^{2n} \left(|S_i'| + \frac{\sigma}{n}|S_i''|\right) \\
&\leq n\beta_j \|(x, y) - (x^*, y^*)\| + \theta_j \sigma \\
&\leq n\beta_j \tau (x^T y)^\gamma + \theta_j \sigma,
\end{aligned}
$$

where the last step follows from (3.11). This completes the proof of (3.17). Combining this with (3.13) yields immediately the desired error estimate (3.10).

Now the convergence of the iterate sequence $\{(x^k, y^k)\}$ follows immediately from (3.10). In particular, suppose that the duality gap $(x^k)^T y^k$ and the centering step σ_k tend to zero at least R-linearly in the sense that $(x^k)^T y^k \leq \kappa(\nu)^k$, $\sigma_k \leq \kappa(\nu)^k$, $\forall k \geq 0$, where $\kappa > 0$ and $\nu \in (0, 1)$ are some constants. Then, it follows from (3.10) that

$$
\|(x^{k+1}, y^{k+1}) - (x^k, y^k)\| \leq \beta \left(\kappa(\nu)^k\right)^\gamma + \theta\kappa(\nu)^k \leq (\beta\kappa^\gamma + \theta\kappa)\omega^k, \qquad \forall k \geq k_0,
$$

where the last step follows from letting $\omega = \max\{\nu, \nu^\gamma\}$. Since $\nu \in (0, 1)$, $\gamma > 0$, we have $\omega \in (0, 1)$. Thus, the above relation shows that $\|(x^{k+1}, y^{k+1}) - (x^k, y^k)\|$ converges to zero at least R-linearly. Thus, the sequence $\{(x^k, y^k)\}$ is Cauchy, so it must converge. Moreover, the rate of convergence is at least linear. By Lemma 3.2, the vector to which $\{(x^k, y^k)\}$ converges is in $X^* \times Y^*$. **Q.E.D.**

Notice that if the duality gap $(x^k)^T y^k$ and the centering stepsize σ^k converge to zero R-quadratically in the sense that there exists some $\nu \in (0, 1)$ and some constant κ such that

$$
(x^k)^T y^k \leq \kappa\nu^{2^k}, \qquad \sigma^k \leq \kappa\nu^{2^k}, \qquad \forall k \geq 0,
$$

then, the error estimate (3.10) implies that

$$
\|(x^{k+1}, y^{k+1}) - (x^k, y^k)\| \leq \beta\kappa^\gamma \nu^{\gamma 2^k} + \theta\kappa\nu^{2^k}, \qquad \forall k \geq k_0.
$$

Therefore, the iterate sequence $\{(x^k, y^k)\}$ is convergent and it converges quadratically.

As a direct consequence of Theorem 3.1, we obtain the iterate convergence of the primal-dual interior point algorithms for solving linear programming (with $Q = 0$), thus recovering the recent result of Tapia, Zhang and Ye[18]. We remark that the techniques used in Tapia, Zhang and Ye[18] to handle degeneracy is very different from ours. While our analysis is based on the generalized Hoffman's error bound and some careful estimation of the iteration mapping, the analysis of Tapia *et. al.* [18] relies on an interesting property established by Güler-Ye[3]: every cluster point of the iterate sequence satisfies the strict complementarity.

Most of the interior point algorithms use a fixed centering stepsize σ, so Assumption 3 is not usually satisfied. For example, a number of practical primal-dual interior point algorithms (see, for example, Choi *et. al.* [1], McShane *et. al.* [12], Lustig *et. al.* [11]) use small fixed σ, while some of the theoretical primal-dual interior point algorithms (see, for example, the algorithms by Mizuno-Todd-Ye[13,14], Monteiro and Adler[15,16], Todd and Ye[19]) use large centering stepsizes. Assumption 2 is pretty mild, since they are satisfied by most of the primal-dual interior point algorithms for linear programming or convex quadratic programming.

The compatibility of Assumption 1 with Assumptions 2–3 may be of some concern. Can we choose the parameters σ_k, τ_k in such a way that ensures Assumptions 1–3 are satisfied simultaneously? In the recent work[20,21,22], Ye and his co-workers modified the Mizuno-Todd-Ye's predictor-corrector algorithm by setting $\sigma^k = 0$ for all k sufficiently large. They showed the quadratic convergence of the duality gap and the convergence of the iterates. It can be easily verified that this modified predictor-corrector algorithm satisfies Assumptions 1–3 simultaneously. Thus, by using Theorem 3.1 (and the discussion that follows), we can recover their result that the iterate sequence is (quadratically) convergent. For Algorithm 1 in its general form, it is not clear at this point how to systematically choose σ_k and τ_k so to satisfy Assumptions 1–3. Some numerical experiments reported in Zhang, Tapia and Potra[25] indicate that these assumptions are typically satisfied in practice. In the case of linear programming (i.e., $Q = 0$), Zhang and Tapia[23] has recently demonstrated this compatibility, but under another unrealistic assumption that the iterate sequence $\{(x^k, y^k)\}$ converges. [There is a logic cycle here.]

Acknowledgement

Research supported by the Natural Sciences and Engineering Research Council of Canada, Grant No. OPG009031.

References

1. I.C. Choi, C.L. Monma and D.F. Shanno, Further development of a primal-dual interior point method for linear programming, *Technical Report RRR#60–88* (1988), RUTCOR, Rutgers University.

2. C.C. Gonzaga and M.J. Todd, An $O(\sqrt{n}L)$-iteration large-step primal-dual affine algorithm for linear programming, *Technical Report 862* (1989), School of Operations Research and Industrial Engineering, Cornell University.

3. O. Güler and Y. Ye, Convergence behavior of some interior point algorithms, *Working paper series* No. 91-4 (1991), the College of Business Administration, The University of Iowa.

4. A.J. Hoffman, On approximate solutions of systems of linear inequalities, *J. Res. Natl. Bur. Standards*, 49 (1952), 263–265.

5. M. Kojima, N. Megiddo and T. Noma, *Homotopy continuation methods for complementarity problems, 1988*, manuscript, IBM Almaden Research Center, San Jose, California.

6. M. Kojima, S. Mizuno and A. Yoshise, A primal-dual interior point method for linear programming, In Nimrod Megiddo, editor, *Progress in Mathematical programming, interior-point and related methods,* pp. 29–47. Springer-Verlag, New York, 1989.

7. X.D. Luo and Z.-Q. Luo, Extension of Hoffman's error bound to polynomial systems, *SIAM Journal of Optimization*, 4 (1994), 383-392.

8. Z.-Q. Luo and P. Tseng, Error bound and convergence analysis of matrix splitting algorithms for the affine variational inequality problem, *SIAM Journal of Optimization*, 2 (1992), 43–54.

9. Z.-Q. Luo and P. Tseng, On the linear convergence of descent methods for convex Essentially smooth minimization, *SIAM J. Contr. & Optim.*, 30 (1992), 408–425.

10. I.J. Lustig, A generic primal-dual interior point algorithm, *Technical Report SOR 88-3* (1988), Dept. Civil Eng. and O.R., Princeton University.

11. I.J. Lustig, R.E. Marsten and D.F. Shanno, Computational experience with a primal-dual interior point method for linear programming, *Technical Report SOR 89-17* (1989), Dept. Civil Eng. and O.R., Princeton University.

12. K.A. McShane, C.L. Monma and D.F. Shanno, An implementation of a primal-dual interior point method for linear programming, *ORSA J. Computing,* 1 (1989), 70–83.

13. S. Mizuno, M.J. Todd and Y. Ye, Anticipated behavior of long-step algorithms for linear programming, *Technical Report*, Working paper series No. 89-33 (1989), College of Business Administration, The University of Iowa.

14. S. Mizuno, M.J. Todd and Y. Ye, Anticipated behavior of long-step algorithms for linear programming, *Technical Report 878* (1989), School of Operations Research and Industrial Engineering, Cornell University.

15. R.C. Monteiro and I. Adler, Interior path-following primal-dual algorithms, Part I: linear programming, *Math. Prog.*, 44 (1989), 27-41.

16. R.C. Monteiro and I. Adler, Interior path-following primal-dual algorithms, Part II: convex quadratic programming, *Math. Prog.*, 44 (1989), 43-66.

17. J.M. Ortega and W.C. Rheinboldt, *Iterative Solution of Nonlinear Equations in Several Variables*, Academic Press, New York, 1970.

18. R.A. Tapia, Y. Zhang and Y. Ye, On the convergence of the iteration sequence in primal-dual interior point methods, *Technical Report TR91-24* (1991), Dept. Mathematical Sciences, Rice University.

19. M.J. Todd and Y. Ye, A centered projective algorithm for linear programming, *Math. of O.R.*, 15 (1990), 508–529.

20. Y. Ye, On the Q-order of convergence of interior-point algorithms for linear programming, Dept. of Management Sciences, University of Iowa, Iowa City, IA 52242.

21. Y. Ye and K. Anstreicher, On quadratic and $O(\sqrt{n}L)$ convergence of a predictor-corrector algorithm for LCP, Dept. Management Sciences, University of Iowa, Iowa City, IA 52242.

22. Y. Ye, O. Güler, R.A. Tapia and Y. Zhang, A quadratically convergent $O(\sqrt{n}L)$-iteration algorithm for liner programming, Dept. Management Sciences, University of Iowa, Iowa City, IA 52242.

23. Y. Zhang and R.A. Tapia, A quadratically convergent polynomial primal-dual interior point algorithm for linear programming, *Technical Report TR90-40* (1990), Dept. Mathematical Sciences, Rice University.

24. Y. Zhang, R.A. Tapia and J.E. Dennis, On the superlinear and quadratic convergence of primal-dual interior point linear programming algorithms, *Technical Report TR90-6* (1990), Dept. Mathematical Sciences, Rice University.

25. Y. Zhang, R.A. Tapia and F. Potra, On the superlinear Convergence of Interior point algorithms for a general class of problems, *Technical Report TR90-9* (1990), Dept. Mathematical Sciences, Rice University.

WSSIAA 5 (1995) pp. 271–296
© World Scientific Publishing Company

On Direct Relaxation of Optimal Material Design Problems for Plates[*]

Konstantin A. Lurie
Department of Mathematical Sciences
Worcester Polytechnic Institute
100 Institute Road
Worcester, MA 01609, USA

Abstract

The technique of direct relaxation developed in[1,2] is applied to problems of optimal layout for plates. Optimal microstructures are explicitly indicated for the case when the original and conjugate strain tensors are coaxial. The paper develops and extends results obtained in[2].

1. Statement of the Problem

In this paper, we consider non-self-adjoint optimization problems for thin anisotropic plates subjected to transverse load. The state of equilibrium of a plate Σ is described by the equation

$$\nabla\nabla \cdot \cdot \mathcal{D} \cdot \cdot \nabla\nabla w = q, \quad (x, y) \in \Sigma, \tag{1}$$

where w denotes the normal displacement, $\mathcal{D}$ the tensor of stiffness, q the transverse load desnity, and the symbol $\cdot\cdot$ denotes a double convolution. The symbol ∇ is traditionally defined as $i\partial/\partial x + j\partial/\partial y$. The boundary $\partial\Sigma$ of a plate will be assumed clamped, this property being expressed by the boundary conditions

$$w|_{\partial\Sigma} = \partial w/\partial n|_{\partial\Sigma} = 0. \tag{2}$$

The 4th rank tensor $\mathcal{D} = \mathcal{D}(x, y)$ will play the role of control; it may take one of two admissible values $\mathcal{D}_1$ or $\mathcal{D}_2$ at each point of the plate. The materials 1 and 2, with stiffness tensors $\mathcal{D}_1$ and $\mathcal{D}_2$, will both be assumed isotropic, i.e.,

$$\mathcal{D}_i = k_i a_1 a_1 + \mu_i(a_2 a_2 + a_3 a_3), \quad i = 1, 2. \tag{3}$$

---

[*]This paper is dedicated to Professor George Leitmann on the occasion of his 70th birthday.

Here and below, a_1, a_2, a_3 represent an orthonormal basis in the space of 2nd rank symmetric tensors in the plane, i.e.,

$$a_1 = (1/\sqrt{2})(ii + jj), \quad a_2 = (1/\sqrt{2})(ii - jj), \quad a_3 = (1/\sqrt{2})(ij + ji). \tag{4}$$

Introduce the characteristic function $\chi_1(x, y)$ of the domain occupied by material 1, with stiffness tensor $\mathcal{D}_1$, and a similar function $\chi_2(x, y)$ for material 2; obviously, $\chi_1 + \chi_2 = 1$. It is required to find the distribution

$$\mathcal{D}(x, y) = \chi_1(x, y)\mathcal{D}_1 + \chi_2(x, y)\mathcal{D}_2 \tag{5}$$

of the stiffness tensor throughout Σ which maximizes some weakly continuous functional $I(w)$ of solution to the boundary-value problem (1), (2). Weak continuity is assumed to be with respect to $\overset{\circ}{W}{}_2^2(\Sigma)$, this space naturally associated with (1), (2). Specifically, as a typical example, we will consider the functional

$$I(w) = -\int_\Sigma [w(x, y) - w_0(x, y)]^2 dx\, dy.$$

where $w_0(x, y) \in L_2(\Sigma)$.

This and similar optimization problems are known to be ill-posed and therefore requiring relaxation, i.e., the construction of an appropriate minimal extension of the initial set $U = \{\mathcal{D}_1, \mathcal{D}_2\}$ of admissible controls. Such an extension is currently proposed on the basis of a precise knowledge of the G-closure of U, i.e., the set GU of invariants of the effective stiffness tensors $\mathcal{D}_0$ of all composites assembled from the elements of U^3. However, the G-closures are known only for a few particular examples[4], and the plate problem is not among them. Yet for these selected examples, the construction of GU presents difficulties, and for the plate problem these difficulties are still not overcome.

At the same time, for many applications we do not need to know the GU-set in full. Instead, it is often enough to specify some linear combination of components of $\mathcal{D}_0$; for our problem, this is the combination $\mathcal{D}_0 \cdot\cdot \nabla\nabla w$ which only matters in view of Hooke's law. To determine this combination, we apply a direct approach, free from any reference to the G-closure.

Similar problems for the 2nd order equation $\nabla \cdot \mathcal{D} \cdot \nabla w = f$ have been discussed in[1,5-7].

2. An Equivalent sup inf Problem

In[2], the problem

$$\max_{\mathcal{D}, w \in (1)-(5)} I \tag{6}$$

was reduced to an equivalent sup inf problem:

$$\sup_{\mathcal{D},w} \inf_{\lambda} J, \tag{7}$$

$$J = I + \int_{\Sigma} (\nabla\nabla\lambda \cdot \cdot \mathcal{D} \cdot \cdot \nabla\nabla w - \lambda q)dx\, dy, \tag{8}$$

where $\mathcal{D} \in (5), w \in (2)$ and λ - the conjugate variable (Lagrange multiplier) - satisfies the boundary conditions

$$\lambda|_{\partial\Sigma} = \partial\lambda/\partial n|_{\partial\Sigma} = 0.$$

The problem (7), (8) is ill-posed, and its relaxation is constructed with the aid of two-sided bounds. An *upper* bound will be built through an appropriate mathematical construction whereas the *lower* bound will be generated by a suitable composite assembled from the original constituents.

3. Upper Bound for $\sup\limits_{\mathcal{D},w} \inf\limits_{\lambda} J$

This functional possesses the following upper bound:

$$\sup_{\mathcal{D},w} \inf_{\lambda} J = \sup_{w} \sup_{\mathcal{D}} \inf_{\lambda} J \leq \sup_{w} \inf_{\lambda} \sup_{\mathcal{D}} J$$

$$= \sup_{w} \inf_{\lambda} \left[-\int_{\Sigma} (w - w_0)^2 dx\, dy - \int_{\Sigma} \lambda q\, dx\, dy \right.$$

$$\left. + \int_{\Sigma} G(\nabla\nabla w, \nabla\nabla\lambda)dx\, dy \right], \tag{9}$$

where

$$G(\xi, \eta) = \begin{cases} \xi \cdot \cdot \mathcal{D}_1 \cdot \cdot \eta, & \xi \cdot \cdot \mathcal{D}_1 \cdot \cdot \eta \geq \xi \cdot \cdot \mathcal{D}_2 \cdot \cdot \eta, \\ \xi \cdot \cdot \mathcal{D}_2 \cdot \cdot \eta, & \xi \cdot \cdot \mathcal{D}_1 \cdot \cdot \eta \leq \xi \cdot \cdot \mathcal{D}_2 \cdot \cdot \eta. \end{cases} \tag{10}$$

The notation

$$\xi = \nabla\nabla w, \quad \eta = \nabla\nabla\lambda$$

will be used below. The function $G(\xi, \eta)$ is convex with respect to any of its arguments, but nonconvex with respect to their union.

The problem

$$\sup_{w} \inf_{\lambda} \left[-\int_{\Sigma} (w - w_0)^2 dx\, dy - \int_{\Sigma} \lambda q dx\, dy + \int_{\Sigma} G(\nabla\nabla w, \nabla\nabla\lambda)dx\, dy \right] \tag{11}$$

is still ill-posed. It would be well-posed if the integrand $G(\xi, \eta)$ were a saddle function, i.e., concave in ξ for fixed η and convex in η for fixed ξ. Then the solution would exist and the operations sup and inf would commute. For our problem, this is obviously not the case. However, the requirement that the function $G(\xi, \eta)$ be a saddle is too restrictive now that ξ and η are gradients; to ensure the existence of the sup inf for this case, it is enough to require that this function be only a quasisaddle[1]. The quasisaddle envelope $G^{**}(\xi,\eta)$ of $G(\xi,\eta)$ will be constructed applying the so-called polysaddlification transform introduced in[1]. This transform plays the same role for sup inf problems as the polyconvexification transform[8-10] plays for the minimum problems. For the problem considered, the polysaddlification transform is given by the formula[1]

$$G^{**}(\xi,\eta) = \inf_d \sup_\omega \sup_b \inf_a \{a \cdot \cdot \xi + b \cdot \cdot \eta + \omega \cdot \cdot (\xi \times \eta) + d\xi \cdot \cdot T \cdot \cdot \eta$$
$$- \inf_\xi \sup_\eta [a \cdot \cdot \xi + b \cdot \cdot \eta + \omega \cdot \cdot (\xi \times \eta) + d\xi \cdot \cdot T \cdot \cdot \eta - G(\xi,\eta)]\} \quad (12)$$

Here, we introduced the notation T for a tensor

$$T = a_1 a_1 - a_2 a_2 - a_3 a_3; \quad (13)$$

the terms $\omega \cdot \cdot \xi \times \eta$ and $d\xi \cdot \cdot T \cdot \cdot \eta$ represent the null Lagrangians

$$\xi \times \eta = (\xi_2 \eta_3 - \xi_3 \eta_2)a_1 + (\xi_3 \eta_1 - \xi_1 \eta_3)a_2 + (\xi_1 \eta_2 - \xi_2 \eta_1)a_3$$

and $\xi \cdot \cdot T \cdot \cdot \eta^{4,8-10}$ taken into account with the aid of Lagrange multipliers ω and d. The symbols $\xi_1, \ldots, \eta_3$ denote the components of ξ and η in the basis a_1, a_2, a_3.

The transform $G^{**}(\xi,\eta)$ defined by Eq. 12 satisfies the inequality

$$G^{**}(\xi,\eta) \geq G(\xi,\eta), \quad (14)$$

for any $G(\xi,\eta)$ convex in η and arbitrary in ξ[1]. Applying $G^{**}(\xi,\eta)$ instead of $G(\xi,\eta)$, we arrive at the upper bound

$$\sup_\omega \inf_\lambda \left[-\int_\Sigma (\omega - \omega_0)^2 dx\, dy - \int_\Sigma \lambda q\, dx\, dy + \int_\Sigma G^{**}(\xi,\eta) dx\, dy \right] \quad (15)$$

for Eq. 11, and consequently for the original functional Eq. 7.

4. The Formula for $G^{**}(\xi,\eta)$

This function is computed as

$$G^{**}(\xi,\eta) = \inf_d \sup_\omega \sup_{m_1} \phi(\xi,\eta) \quad (16)$$

---

[1]In[2], the function $G^{**}(\xi,\eta)$ was defined by the formula Eq. 15 differing from Eq. 12 in one aspect: the operation $\sup_d \sup_\omega$ was applied instead of $\inf_d \sup_\omega$. The version of[2] also provides with an upper estimate for $G(\xi,\eta)$, this estimate being, however, weaker than the one given by Eq. 12. This inaccuracy, nevertheless, does not undermine the results of[2] since these results are related to the *stationary* values of $G^{**}(\xi,\eta)$ with regard to ω and d.

with (see[2])

$$\phi(\xi, \eta) = \eta \cdot \cdot \langle S^{-1} \rangle^{-1} \cdot \cdot \xi + \omega \cdot \cdot (\xi \times \eta) + d\xi \cdot \cdot T \cdot \cdot \eta \tag{17}$$

Here,

$$\begin{aligned}
\langle S^{-1} \rangle &= m_1 S_1^{-1} + m_2 S_2^{-1}, \\
S_i &= \mathcal{D}_i - dT + \omega \cdot \cdot \in, \\
m_1, m_2 &\geq 0, \quad m_1 + m_2 = 1, \\
\in &= -E \times E, \quad E = a_1 a_1 + a_2 a_2 + a_3 a_3, \\
\omega \cdot \cdot \in &= -\omega \cdot \cdot E \times E = -\omega \times E = -E \times \omega = \in \cdot \cdot \omega
\end{aligned} \tag{18}$$

The following formulas apply[2]

$$\phi_\omega = -m_1 m_2 (\Delta S^{-1} \cdot \cdot \langle S^{-1} \rangle^{-1} \cdot \cdot \xi) \times (\eta \cdot \cdot \langle S^{-1} \rangle^{-1} \cdot \cdot \Delta S^{-1}), \tag{19}$$

$$\phi_d = -m_1 m_2 (\Delta S^{-1} \cdot \cdot \langle S^{-1} \rangle^{-1} \cdot \cdot \xi) \cdot \cdot T \cdot \cdot (\eta \cdot \cdot \langle S^{-1} \rangle^{-1} \cdot \cdot \Delta S^{-1}), \tag{20}$$

$$\Delta S^{-1} = S_2^{-1} - S_1^{-1}.$$

Equations 16-20 contain all the information necessary to compute $G^{**}(\xi, \eta)$. The operation $\sup_{m_1} \phi$ will be particularly responsible for the appearance of certain ranges for the admissible value of d in the course of a subsequent $\inf_{d}$-operation.

For the general analysis of Eq. 16, stationary values of ϕ with respect to ω and d are important. Some of those values were investigated in[2] where they were shown to be saturated by appropriate microstructures. Additional stationary points will be examined below; also a classification of optimal ranges (i.e., microstructures saturating the upper bounds specified by $G^{**}(\xi, \eta)$) will be given for the case when tensors ξ and η are coaxial.

5. Stationary Values of ϕ

We first will list the stationary values of ϕ with respect to ω and d found in[2] for general values of m_1. The microstructures saturating the stationary values of ϕ will also be indicated.

The analysis of[2] was related to two possible situations: (a) case of free variations $\delta\omega, \delta d$; (b) case when either $det S_1 = 0$ or $det S_2 = 0$; the significance of the latter case will be motivated below.

Assuming that the tensors ξ, η are general, we apply the formulas

$$\begin{aligned}
\xi &= \xi_1 a_1 + \xi_2 a_2 + \xi_3 a_3, \\
\eta &= \eta_1 a_1 + \eta_2 a_2 + \eta_3 a_3,
\end{aligned} \tag{21}$$

where the basis a_1, a_2, a_3 (see Eq. 4) will be appropriately specified below.

276

(a) Case of free variations $\delta\omega, \delta d$; tensors ξ and η are arbitrary.

The basis a_1, a_2, a_3 will be defined by the equation

$$\xi_3(\Delta k\eta_1 \pm \Delta\mu\eta_2) + \eta_3(\Delta k\xi_1 \pm \Delta\mu\xi_2) = 0, \tag{22}$$

where $\Delta k = k_2 - k_1$, $\Delta\mu = \mu_2 - \mu_1$.

With this basis we associate the unit vectors i, j determined by Eq. 4. The stationary bilinear form turns out to be equal to

$$\phi(\xi, \eta) = \xi \cdot \cdot \langle \mathcal{D} \rangle \cdot \cdot \eta - m_1 m_2[\xi \cdot \cdot \Delta\mathcal{D} \cdot \cdot nn][\eta \cdot \cdot \Delta\mathcal{D} \cdot \cdot nn]/(ii \cdot \cdot \tilde{\mathcal{D}} \cdot \cdot ii) \tag{23}$$

where we applied the notation

$$\langle \mathcal{D} \rangle = m_1\mathcal{D}_1 + m_2\mathcal{D}_2, \ \ \tilde{\mathcal{D}} = m_1\mathcal{D}_2 + m_2\mathcal{D}_1, \ \ \Delta\mathcal{D} = \mathcal{D}_2 - \mathcal{D}_1.$$

The values Eq. 23_+, Eq. 23_- related to plus (minus) signs in Eq. 22, are defined by equation Eq. 23 in which we set $n = i$ and $n = j$, respectively. In the first of these cases, layers are oriented along j, in the second-along i.

Remark 1 Eq. 22 defines several stationary bases a_1, a_2, a_3.

Remark 2 The Lagrange multipliers ω, d are eliminated from Eqs. 22, 23. We will need, however, formulas expressing for this case the stationary values $\xi_2, \eta_2, \xi_3, \eta_3$ in terms of ω, d. With tensor ω represented as

$$\omega = \omega_1 a_1 + \omega_2 a_2 + \omega_3 a_3, \tag{24}$$

the following stationarity conditions hold:

$$\begin{aligned}
\xi_2 &= \pm(\Delta k/\Delta\mu)\xi_1(\tilde{M} - \omega_3)/(\tilde{K} + \omega_3), \\
\eta_2 &= \pm(\Delta k/\Delta\mu)\eta_1(\tilde{M} + \omega_3)/(\tilde{K} - \omega_3), \\
\xi_3 &= \pm(\Delta k/\Delta\mu)\xi_1(\omega_2 - \omega_1)/(\tilde{K} + \omega_3), \\
\eta_3 &= \mp(\Delta k/\Delta\mu)\eta_1(\omega_2 - \omega_1)/(\tilde{K} - \omega_3), \tag{25}
\end{aligned}$$

with

$$\tilde{M} = m_1M_2 + m_2M_1, \ \ \tilde{K} = m_1K_2 + m_2K_1, \ \ K_i = k_i - d, \ \ M_i = \mu_i + d, \ \ i = 1, 2.$$

The expressions Eq. 25 identically satisfy Eq. 22. The formula 23 for $\phi(\xi, \eta)$ can now be rewritten as

$$\phi(\xi, \eta) = \xi \cdot \cdot \langle \mathcal{D} \rangle \cdot \cdot \eta - m_1 m_2(\Delta k)^2 \xi_1\eta_1(\tilde{k} + \tilde{\mu})/(\tilde{K}^2 - \omega_3^2) \tag{26}$$

where

$$\tilde{k} = m_1k_2 + m_2k_1, \ \ \tilde{\mu} = m_1\mu_2 + m_2\mu_1.$$

b) Case of linked variations $\delta w, \delta d$; tensors ξ and η are coaxial.

The Lagrange multipliers w, d are assumed linked either through the relationship

$$det S_1 = K_1 M_1^2 + K_1 w_1^2 + M_1(w_2^2 + w_3^2) = 0 \text{ (case 1)} \tag{27}$$

or through the relationship

$$det S_2 = K_2 M_2^2 + K_2 w_1^2 + M_2(w_2^2 + w_3^2) = 0 \text{ (case 2)} . \tag{28}$$

When ξ and η are coaxial, it may be assumed that $w_1 = w_2 = 0$, so that

$$w = w_3 a_3$$

and

$$\begin{aligned} K_1 M_1^2 + M_1 w_3^2 &= 0 \text{ in case 1,} \\ K_2 M_2^2 + M_2 w_3^2 &= 0 \text{ in case 2.} \end{aligned} \tag{29}$$

In^2, the stationary ranges were obtained assuming that

$$\begin{aligned} s_1 &= K_1 M_1 + w_3^2 = 0 \text{ in case 1,} \\ s_2 &= K_2 M_2 + w_3^2 = 0 \text{ in case 2.} \end{aligned} \tag{30}$$

Another possibility, i.e.

$$\begin{aligned} M_1 &= 0 \text{ in case 1,} \\ M_2 &= 0 \text{ in case 2} \end{aligned} \tag{31}$$

will be considered below.

The results related to Eqs. 30 may be summarized as follows. Let i and j be the unit vectors of (common) main axes of ξ and η; then the stationary form $\phi(\xi, \eta)$ is defined as

$$\begin{aligned} \phi(\xi, \eta) = \xi_1 \eta_1 \{ \mu_1 \zeta \sigma + k_1 + [m_2 \Delta k \Delta \mu (k_1 + \mu_1)/2] [\zeta \sigma / \bar{u} + (1/\bar{v}) \\ + (1/\bar{u})\sqrt{(\zeta^2 - \bar{u}/\bar{v})(\sigma^2 - \bar{u}/\bar{v})}] \} \text{ in case 1 } (K_1 M_1 + w_3^2 = 0), \end{aligned} \tag{32_1}$$

$$\begin{aligned} \phi(\xi, \eta) = \xi_1 \eta_1 \{ \mu_2 \zeta \sigma + k_2 - [m_1 \Delta k \Delta \mu (k_2 + \mu_2)/2] [\zeta \sigma / u + (1/v) \\ + (1/u)\sqrt{(\zeta^2 - u/v)(\sigma^2 - u/v)}] \} \text{ in case 2 } (K_2 M_2 + w_3^2 = 0). \end{aligned} \tag{32_2}$$

In these formulas, the following notation is used:

$$\begin{aligned} \zeta &= \xi_2/\xi_1, \ \sigma = \eta_2/\eta_1, \\ u &= (k_2 + \tilde{\mu})\Delta k, \ v = (\tilde{k} + \mu_2)\Delta \mu, \\ \bar{u} &= (k_1 + \tilde{\mu})\Delta k, \ \bar{v} = (\tilde{k} + \mu_1)\Delta \mu. \end{aligned} \tag{33}$$

The square roots are assumed here and below to take their arithmetic values. The values Eq. 32 of ϕ are saturated by rank 2 matrix laminates with layers parallel to the (common) main axes of ξ and η. In case 1, material $\mathcal{D}_1$ forms the environment of the laminate whereas $\mathcal{D}_2$ is concentrated in the core; in case 2, the materials are switched.

The stationary ranges associated with Eq. 31, were never considered in[2]. The corresponding results are obtained in Appendix A and summarized below.

As before let i and j be the unit vectors of (common) main axes of ξ and η; then the stationary form $\phi(\xi, \eta)$ is defined as

$$\phi(\xi, \eta) = \xi_1 \eta_1 \left\{ \langle \mu \rangle \zeta \sigma + \langle k \rangle - (m_2 \Delta\mu/2) \left[\zeta\sigma + m_1(\Delta k)^2/\bar{v} \right. \right.$$
$$\left. \left. + \sqrt{(\zeta^2 + m_1(\Delta k)^2/\bar{v})(\sigma^2 + m_1(\Delta k)^2/\bar{v})} \right] \right\} \quad \text{in case 1} \quad (M_1 = 0), \qquad (34_1)$$

$$\phi(\xi, \eta) = \xi_1 \eta_1 \left\{ \langle \mu \rangle + \langle k \rangle \zeta\sigma + (m_1 \Delta\mu/2) \left[\zeta\sigma - m_2(\Delta k)^2/v \right. \right.$$
$$\left. \left. + \sqrt{(\zeta^2 - m_2(\Delta k)^2/v)(\sigma^2 - m_2(\Delta k)^2/v)} \right] \right\} \quad \text{in case 2} \quad (M_2 = 0), \qquad (34_2)$$

where

$$\langle k \rangle = m_1 k_1 + m_2 k_2, \quad \langle \mu \rangle = m_1 \mu_1 + m_2 \mu_2.$$

The values Eqs. 34 of ϕ are saturated by rank 2 matrix laminates with layers of different scales making the appropriately specified angle θ; this angle is bisected by the (common) main axis of ξ, η. As before, in case 1 material $\mathcal{D}_1$ forms the environment whereas $\mathcal{D}_2$ is concentrated in the core; in case 2, the materials are switched.

In the following sections, we will classify the stationary ranges on the basis of operation Eq. 16 which is the only criterion for such a classification. We begin with the self-adjoint case $\eta = c\xi$, $c = const$ considered before in[11,12]; the corresponding results (along with some new ones) will be obtained below as consequences of Eq. 16.

6. The Self-Adjoint Case $\eta = c\xi$

The operation Eq. 16 is now reformulated as

$$G^{**}(\xi, c\xi) = \inf_d \sup_\omega \sup_{m_1} \phi(\xi, c\xi) = \inf_d \sup_\omega \sup_{m_1} c[\xi \cdot \cdot \langle S^{-1} \rangle^{-1} \xi + d\xi \cdot \cdot T \cdot \cdot \xi]$$
$$= \begin{cases} c \inf_d \sup_\omega \sup_{m_1} \phi(\xi, \xi), & c \geq 0, \\ c \sup_d \inf_\omega \inf_{m_1} \phi(\xi, \xi), & c < 0. \end{cases} \qquad (35)$$

Without any loss of generality we may assume that the basis Eq. 4 is so chosen as to make $\omega_2 = 0$:

$$\omega = \omega_1 a_1 + \omega_3 a_3.$$

With this assumption, the following formulas will hold for components of $\langle S^{-1}\rangle^{-1}$:

$$\begin{aligned}
\langle S^{-1}\rangle_{11}^{-1} &= (\tilde{q})^{-1}[K_1 K_2(\tilde{M}^2 + \omega_1^2) + \omega_3^2 \tilde{M}\langle K\rangle], \\
\langle S^{-1}\rangle_{22}^{-1} &= (\tilde{q})^{-1}[\tilde{K}\tilde{M} M_1 M_2 + (\tilde{K}\omega_1^2 + \tilde{M}\omega_3^2)\langle M\rangle], \\
\langle S^{-1}\rangle_{33}^{-1} &= (\tilde{q})^{-1}[\tilde{K}\tilde{M} M_1 M_2 + \tilde{K}\langle M\rangle\omega_1^2 + M_1 M_2 \omega_3^2], \\
\langle S^{-1}\rangle_{12}^{-1} &= (\tilde{q})^{-1}\omega_3[\tilde{M}(KM) + \tilde{K}\omega_1^2 + \tilde{M}\omega_3^2] = -\langle S^{-1}\rangle_{21}^{-1}, \\
\langle S^{-1}\rangle_{23}^{-1} &= (\tilde{q})^{-1}\omega_1[\tilde{K}(M^2) + \tilde{K}\omega_1^2 + \tilde{M}\omega_3^2] = -\langle S^{-1}\rangle_{32}^{-1}, \\
\langle S^{-1}\rangle_{31}^{-1} &= -m_1 m_2(\tilde{q})^{-1}\omega_1\omega_3\Delta k\Delta\mu = \langle S^{-1}\rangle_{13}^{-1}
\end{aligned} \tag{36}$$

Here we used the notation

$$\begin{aligned}
\tilde{q} &= \tilde{K}\tilde{M}^2 + \tilde{K}\omega_1^2 + \tilde{M}\omega_3^2, \\
(KM) &= m_1 K_2 M_2 + m_2 K_1 M_1, \quad (M^2) = m_1 M_2^2 + m_2 M_1^2.
\end{aligned} \tag{37}$$

With Eq. 36 it is clear that $\phi(\xi,\xi)$ will be stationary with respect to ω if $\omega_1 = \omega_3 = 0$; with this assumption, $\phi(\xi,\xi)$ reduces to

$$\phi(\xi,\xi) = \langle k\rangle\xi_1^2 + \langle\mu\rangle\xi_2^2 + \left(\frac{M_1 M_2}{\tilde{M}} - d\right)\xi_3^2 - \frac{m_1 m_2}{\tilde{K}\tilde{M}}(\tilde{M}\Delta k^2\xi_1^2 + \tilde{K}\Delta\mu^2\xi_2^2) \tag{38}$$

It is easy to see that neither $\tilde{M}$ nor $\tilde{K}\tilde{M}$ may ever change sign for any $m_1 \in (0,1)$. Assuming the contrary, we allow for $\sup\limits_{\omega,m_1}\phi(\xi,\xi)$ to become $+\infty$ ($\inf\limits_{\omega,m_1}\phi(\xi,\xi)$ to become $-\infty$) which never yields $G^{**}(\xi,c\xi)$. This observation means particularly that K_1 and K_2 should never be of opposite signs, and M_1 and M_2 should not possess opposite signs, too. Therefore, when $\omega_3 = 0$, only those values of d should be admitted that satisfy inequalities

$$d \le \min(k_1, k_2) \text{ or } d \ge \max(k_1, k_2)$$

as well as inequalities

$$d \ge \max(-\mu_1, -\mu_2) \text{ or } d \le \min(-\mu_1, -\mu_2).$$

Particularly, if materials $\mathcal{D}_1, \mathcal{D}_2$ are regularly ordered, i.e.

$$k_1 \le k_2, \quad \mu_1 \le \mu_2, \tag{39}$$

then either

$$-\mu_1 \le d \le k_1 \tag{40}$$

or

$$d \le -\mu_2, \tag{41}$$

or

$$d \ge k_2. \tag{42}$$

280

If materials $\mathcal{D}_1$ and $\mathcal{D}_2$ are ordered irregularly, i.e.

$$k_1 \geq k_2, \quad \mu_1 \leq \mu_2, \tag{43}$$

then either

$$-\mu_1 \leq d \leq k_2 \tag{44}$$

or

$$d \leq -\mu_2 \tag{45}$$

or

$$d \geq k_1. \tag{46}$$

1. Assume that $c < 0$ and consider the regular case Eq. 39. With $\omega_1 = \omega_3 = 0$, the values of d making $\phi(\xi, \xi)$ stationary are given by $d = \pm\infty$ and by

$$d_{1,2} = -\frac{\tilde{\mu}\Delta k \mp t\tilde{k}\Delta\mu}{\Delta k \pm t\Delta\mu}, \quad t = |\sqrt{\xi_2^2 + \xi_3^2}/\xi_1|; \tag{47}$$

the value d_1 (upper sign) is a local *maximum* of $\phi(\xi, \xi)$ as the function of d whereas the value d_2 (lower sign) is a local *minimum* of it no matter which of admissible ranges Eq. 40-42 is chosen. This choice will be made if we require that the stationary values $\omega_1 = \omega_3 = 0$ of $\phi(\xi, \xi)$ should *minimize* this function with regard to ω. Applying Eq. 36 we show that

$$\phi_{\omega_1\omega_1}|_{\omega_1=\omega_3=0} = 2\tilde{M}^{-3}m_1 m_2 (\Delta\mu)^2 (\xi_2^2 + \xi_3^2), \tag{48}$$

and the inequality $\phi_{\omega_1\omega_1}|_{\omega_1=\omega_3=0} \geq 0$ yields $\tilde{M} > 0$ which prohibits Eq. 41 and implies either Eq. 40 or Eq. 42. The second possibility is ruled out since $k_2 - d \geq 0$ for $d = d_1$(see Eq. 47). We conclude that Eq. 40 represents the only admissible range for d; with this range, the values $\omega_1 = \omega_3 = 0$ can be shown to minimize $\phi(\xi, \xi)$ (we apply the second derivative test using Eq. 36).

A similar analysis holds for the values $d = k_1$ and $d = -\mu_1$, so ultimately we arrive at three upper bounds effective for all of these ranges. The bounds will be given by:

(a1) Eq. 23_+ with i and j defined as the unit vectors of the main axes of ξ and η - for the range $d = d_1 \in$ Eq. 40;

(a2) Eq. 32_1 - for the range $d = k_1$;

either

(a3) Eq. 32_1

or

(a4) Eq. 34_1 for the range $d = -\mu_1$.

Remark 3 Of course, in the Equations 23_+, 32_1, 34_1 for bounds one should apply $\zeta = \sigma$ and $\eta_1 = c\xi_1$ in the self-adjoint case.

Remark 4 The ranges a1-a3 were listed as optimal in[11]; the range a4 will be shown in Appendix A to be equivalent to a3.

2. Assume now that $c > 0$ and consider the regular case Eq. 39. With $\omega_1 = \omega_3 = 0$, the root $d = d_2$ (see Eq. 47) will be chosen to minimize $\phi(\xi, \xi)$ regardless of the range in d. These ranges will be specified, as before, by the requirement that the stationary values $\omega_1 = \omega_3 = 0$ should ensure the required extremum of $\phi(\xi, \xi)$. This extremum should now be *maximum* (see Eq. 35); we see from Eq. 48 that $\phi_{\omega_1\omega_1}|_{\omega_1=\omega_3=0}$ will be non-positive if $\tilde{M} \leq 0$ which prohibits both Eq. 40 and Eq. 42 and allows for Eq. 41. We arrive at two stationary values of $\phi(\xi, \xi)$ suspicious for optimality, i.e. those given by:

(b1) Eq. 23 with i and j defined as the unit vectors of the main axes of ξ and η - for the range $d = d_2 \in$ Eq. 41;

(b2) Eq. 32_2 - for the range $d = -\mu_2$.

These values will generate the required bounds if we show that they are in fact the ω-maximums of $\phi(\xi, \xi)$, each for its own d-range.

To this end we apply the second derivative test. Before we do so, we observe that for d_2 varying from $-\infty$ to $-\mu_2$, the parameter t (see Eq. 47) will vary, respectively, from $\Delta k / \Delta \mu$ to $m_2 \Delta k / (\tilde{k} + \mu_2)$:

$$\frac{\Delta k}{\Delta \mu} \geq t \geq \frac{m_2 \Delta k}{\tilde{k} + \mu_2}.$$

For this range in t, the case b1 applies; as to the case b2, the appropriate values of t should be less than $m_2 \Delta k / (\tilde{k} + \mu_2)$:

$$0 \leq t \leq \frac{m_2 \Delta k}{\tilde{k} + \mu_2}. \tag{49}$$

With this information, we implement the second derivative test. Along with Eq. 48, we also have the formulas

$$\phi_{\omega_3\omega_3}(\xi, \xi)|_{\omega_1=\omega_3=0} = \frac{2m_1 m_2}{\tilde{K}^2 \tilde{M}^2}[\tilde{M}\Delta k^2 \xi_1^2 + \tilde{K}\Delta\mu^2 \xi_2^2]$$

$$\phi_{\omega_1\omega_3}(\xi, \xi)|_{\omega_1=\omega_3=0} = -\frac{2m_1 m_2}{\tilde{K}\tilde{M}^2}\xi_1\xi_3$$

The second derivative test requires that $\phi_{\omega_1\omega_1}|_{\omega_1=\omega_3=0} \leq 0$ and $\phi_{\omega_1\omega_1}\phi_{\omega_3\omega_3} - \phi_{\omega_1\omega_3}^2 \geq 0$. The first condition was satisfied above by requiring that $\tilde{M} \leq 0$; the second is reduced to

$$4m_1^2 m_2^2 \frac{\Delta\mu^2 \xi_2^2}{\tilde{K}^2 \tilde{M}^5}[\tilde{M}\Delta k^2 \xi_1^2 + \tilde{K}\Delta\mu^2(\xi_2^2 + \xi_3^2)] \geq 0 \tag{50}$$

For $d = d_2$ (case b1), the expression in the square brackets is reduced to $-(\tilde{k} + \tilde{\mu})\xi_1^2 t^2 \Delta k \Delta \mu$, and Eq. 50 holds since $\tilde{M} \leq 0$. For $d = -\mu_2$, (case b2) the left hand side of Eq. 50 reduces to

$$4m_1^2 \Delta k \frac{\xi_1^2}{(\tilde{k} + \mu_2)m_2^3 \Delta \mu^2} \left[\frac{m_2 \Delta k}{\tilde{k} + \mu_2} - \frac{\Delta \mu}{\Delta k} t^2 \right]. \tag{51}$$

But in case b2, Ineq. 49 applies, and Eq. 51 allows for a lower bound

$$4m_1^2 \Delta k^2 \frac{\xi_1^2}{(\tilde{k} + \mu_2)^2 m_2^2 \Delta \mu^2} \left[1 - \frac{\Delta \mu}{\Delta k} \frac{m_2 \Delta k}{\tilde{k} + \mu_2} \right].$$

The expression in the square bracket is ≥ 0, therefore, the one of Eq. 51 is non-negative, and Ineq. 50 holds. We also have to mention the stationary value $d = -\infty$ which will minimize $\phi(\xi, \xi)$ given by Eq. 38; the minimum will be equal to

$$\phi(\xi, \xi) = \langle k \rangle \xi_1^2 + \langle \mu \rangle (\xi_2^2 + \xi_3^2). \tag{52}$$

(Notice that the value $d = \infty$, also stationary, will maximize $\phi(\xi, \xi)$). This minimum generates an additional upper bound given by:

(b3) Eq. 23 with $\eta = c\xi$ and the basis a_1, a_2, a_3 chosen in accordance with the rule Eq. 22_. When $\eta = c\xi$ we choose the basis generated by the formula $\Delta k \xi_1 - \Delta \mu \xi_2 = 0$.

7. Case of Coaxial Tensors: Classification of Ranges in Terms of Lagrange Multipliers d and ω_3

In the previous section, we considered the case $\eta = c\xi$ and listed the ranges stationary in d and ω for a fixed value of m_1. The operations $\inf_d \sup_\omega$ and $\sup_{m_1}$ appearing in Eq. 35 are in this case interchangeable:

$$\inf_d \sup_\omega \sup_{m_1} \phi(\xi, c\xi) = \sup_{m_1} \inf_d \sup_\omega \phi(\xi, c\xi),$$

and the operation

$$\inf_d \sup_\omega \phi(\xi, c\xi)$$

can be interpreted as the upper bound that applies to $\phi(\xi, c\xi)$ for any fixed value of m_1.

The ranges listed above are optimal for $\eta = c\xi$; they also apply as stationary ranges for a more general setting when tensors ξ and η are merely coaxial.

We claim, however, that *these ranges are also optimal* in this more general situation, i.e. the operation

$$\inf_d \sup_\omega \phi(\xi, \eta)$$

represents then the upper bound for $\phi(\xi, \eta)$ for any fixed value of m_1.

The reason for this statement is that the interchangeability property holds when $\phi(\xi, \eta)$ is applied instead of $\phi(\xi, c\xi)$, and ξ and η are assumed coaxial:

$$\inf_d \sup_\omega \sup_{m_1} \phi(\xi, \eta) = \sup_{m_1} \inf_d \sup_\omega \phi(\xi, \eta). \tag{53}$$

This property follows from the observation similar to that made in section 6 right after Eq. 38. The stationary ranges Eq. 30 and Eq. 31 stay in harmony with this observation.

The relevant analysis will be based on the formula

$$\phi(\xi, \eta) = \langle k \rangle \xi_1 \eta_1 + \langle \mu \rangle \xi_2 \eta_2 + \left(\frac{M_1 M_2}{\tilde{M}} - d \right) \xi_3 \eta_3 - \frac{m_1 m_2}{\tilde{K} \tilde{M} + \omega_3^2} (\tilde{M} \Delta k^2 \xi_1 \eta_1 + \tilde{K} \Delta \mu^2 \xi_2 \eta_2)$$

$$+ \frac{\omega_3 [\tilde{K} \tilde{M} - (KM)]}{\tilde{K} \tilde{M} + \omega_3^2} (\xi_1 \eta_2 - \xi_2 \eta_1) \tag{54}$$

generalizing that of Eq. 38 to the case $\omega_1 = 0, \omega_3 \neq 0$: as in a similar case in section 6, we conclude that neither $\tilde{M}$ nor $\tilde{K} \tilde{M} + \omega_3^2$ may ever change sign for any $m_1 \in (0, 1)$. To this end it is necessary that both M_1 and M_2, as well as s_1 and s_2 (see Eq. 30) should be of same sign. This requirement is also sufficient but the proof of it is elaborate and therefore reproduced in Appendix B. We now apply this result to classify the possible ranges.

Consider the regular case Eq. 39 and assume that $\Delta k > \Delta \mu$; on Fig. 1 there are shown the open regions I and III (hatched horizontally) where s_1 and s_2 are both negative, and the open region II (hatched vertically) where these quantities are both positive. As to M_1 and M_2, these are both positive to the right of the line $d = -\mu_1$, and both negative to the left of the line $d = -\mu_2$. The figure gives a general idea of possible contacts between various stationary ranges. Before we list these possibilities, let us introduce the following notation for the ranges related to the case when ξ and η are coaxial.

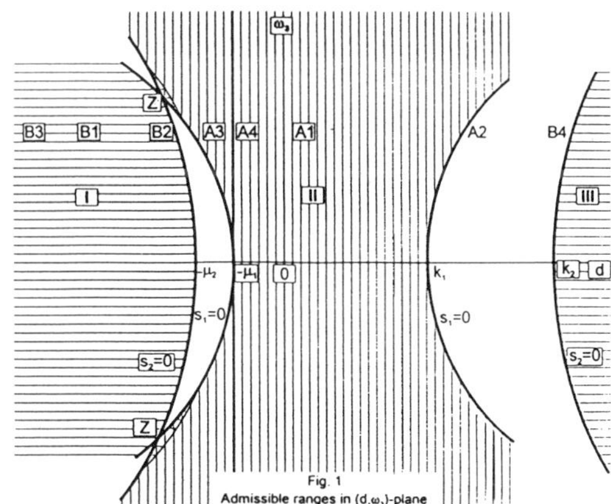

A1 will be defined by Eq. 23_+, this time with d and ω_3 belonging to that part of domain II on Fig. 1, for which $d > -\mu_1$; when $\omega_3 = 0$ and $d \in$ Eq. 40, then A1 becomes identical with a1;

A2 will be defined as Eq. 32_1, this time with (d, ω_3) belonging to hyperbola $s_1 = 0$ (right branch on Fig. 1) passing through the point $d = k_1$, $\omega_3 = 0$. When $\omega_3 = 0$, this point only matters, and A2 becomes the same as a2;

A3 will be defined by Eq. 32_1, this time with (d, ω_3) belonging to hyperbola $s_1 = 0$ (left branch on Fig. 1) passing through the point $d = -\mu_1$, $\omega_3 = 0$. When $\omega_3 = 0$, this point only matters, and A3 then reduces to a3;

A4 will be defined by Eq. 34_1, with $d = -\mu_1$, and ω_3 arbitrary;

B1 will be defined by Eq. 23, this time with (d, ω_3) belonging to domain I on Fig. 1; when $\omega_3 = 0$ and $d \in$ Eq. 41, then B1 reduces to b1;

B2 will be defined by Eq. 32_2, this time with (d, ω_3) belonging to hyperbola $s_2 = 0$ (left branch on Fig. 1) passing through the point $d = -\mu_2, \omega_3 = 0$; when $\omega_3 = 0$, this point only matters, and B2 reduces to b2;

B3 will be defined by Eq. 23 with the basis a_1, a_2, a_3 generated by an appropriate solution of Eq. 22_- (see Remark 1). When $\eta = c\xi, c > 0$, then Eq. 22 reduces to $\Delta k\xi_1 - \Delta\mu\xi_2 = 0$, and B3 becomes b3.

We should also consider the stationary range B4, defined by Eq 32_2 (case 2), with (d, ω_3) belonging to hyperbola $s_2 = 0$ (right branch on Fig.. 1) passing through the point $d = u_2, \omega_3 = 0$.

With this notation, it is easy to characterize the possible contacts. We see from Fig. 1 that A1 may contact either of A2, A3 and A4 with a direct contact possible between A3 and A4 at the point $d = \mu_1, \omega_3 = 0$, but not between A2 and A3 or A2 and A4. Also, B1 may contact both B2 and B3 with no direct contact possible between them. There is also a possibility of contact between B1 and A3, and also between B1, B2 and A3 at the point Z. As to B4, this range never appears in this case though Fig. 1 shows that the right branch of hyperbola $s_2 = 0$ contacts the region III that may correspond to rank 1 laminates. We shall see that this region is in some sense unattainable.

The regular case Eq. 39 with $\Delta k < \Delta\mu$ as well as the irregular case Eq. 43 allow for a similar qualitative analysis.

8. Case of Coaxial Tensors: Classification of Ranges in Terms of Fields ξ and η

The classification given in the previous section with the aid of Lagrange multipliers should be translated into the language of fields. To this end we apply Eqs 23, $32_{1,2}$, $34_{1,2}$ defining the form $\phi(\xi, \eta)$ for various stationary ranges. We will describe the optimal pattern for the regular case Eq. 39 assuming that $\Delta k > \Delta \mu$.

The bilinear form $\phi(\xi, \eta)$ will be normalized against the product $\xi_1 \eta_1$. We will consider two possible situations: $\xi_1 \eta_1 < 0$ and $\xi_1 \eta_1 > 0$; they will particularly include the relevant self-adjoint cases, namely, those of a) and b), section 7.

Referring to Eqs. 21, we introduce the symbols

$$\zeta_2 = \xi_2/\xi_1, \quad \zeta_3 = \xi_3/\xi_1, \quad \sigma_2 = \eta_2/\eta_1, \quad \sigma_3 = \eta_3/\eta_1. \tag{55}$$

It will be assumed that the basis a_1, a_2, a_3 is so chosen as to make $\sigma_2 + \zeta_2 \geq 0$; evidently, this assumption does not affect the generality. For most of the optimal ranges, the unit vectors i, j generating the a-basis, will be those of the main axes of ξ and η, so that $\zeta_3 = \sigma_3 = 0$. Exceptions from this rule will be specified explicitly.

Case $\xi_1 \eta_1 < 0$.

Consider the interfaces separating A1 from A2 and A3. To find them, we equalize $\phi(\xi, \eta)$ given by Eq. 23_+ and those given by Eq. 32_1. After some algebra we arrive at the equation

$$\left(x_1 - \bar{\alpha}_1\right)^2 - x_2^2 = \bar{\alpha}_2^2 \tag{56}$$

where

$$x_1 = \sigma_2 + \zeta_2, \quad x_2 = \sigma_2 - \zeta_2, \tag{57}$$

$$\bar{\alpha}_1 = \bar{u}/(m_1 \Delta k \Delta \mu) + (m_1 \Delta k \Delta \mu)/\bar{v}, \quad \bar{\alpha}_2 = \bar{u}/(m_1 \Delta k \Delta \mu) - (m_1 \Delta k \Delta \mu)/\bar{v}. \tag{58}$$

Both branches of hyperbola Eq. 56 are shown on Fig. 2. The right branch separates A1 from A2, the left separates A1 from A3.

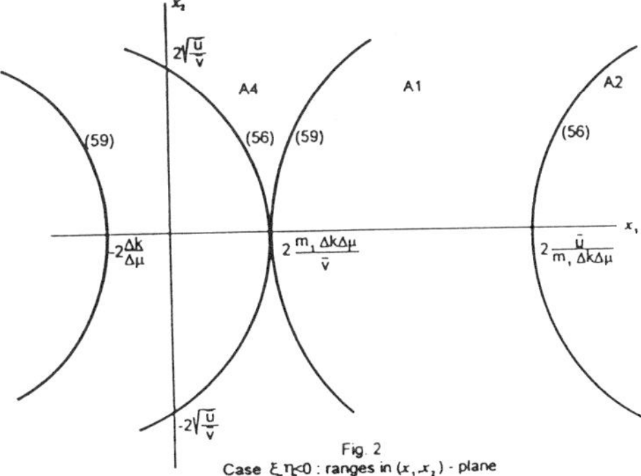

Fig. 2
Case $\xi_1 \eta_1 < 0$: ranges in (x_1, x_2) - plane

Remark 5 The x_1-intercepts of Eq. 56 can also be defined as points at which the value d_1 (see Eq. 47, upper sign) equals either $-\mu_1$ or k_1.

Consider now the interface separating A1 from A4. Equalizing the expressions Eq. 23_+ and Eq. 34_1 or, more easily, the expressions Eq. 26 and Eqs. A.19, A.20,

Appendix A, we arrive at the equation

$$\left(x_1 - \frac{\Delta k}{\Delta \mu}\frac{\tilde{M} - \tilde{K}}{\tilde{K}}\right)^2 - x_2^2 = \left(\frac{\Delta k}{\Delta \mu}\frac{\tilde{M} + \tilde{K}}{\tilde{K}}\right)^2 \tag{59}$$

in which we have to use $d = -\mu_1 (M_1 = 0$, see Eq. 34_1). The hyperbola Eq. 59 has two branches, both shown on Fig. 2. Of these, only the right branch belongs to the halfplane $x_2 \geq 0$. In the domain bounded by this branch and the x_2-axis, the range A4 will be preferred to A1. The left branch of hyperbola Eq. 56 now looses its significance as an interface participating in the optimal layout because as we move from within the domain occupied by A1, we meet the right branch of Eq. 59 *before* we cut the left branch of Eq. 56. The hyperbolas Eq. 59 and Eq. 56 touch each other, and this contact corresponds to that between the vertical line $d = -\mu_1$ and hyperbola $s_1 = 0$ (left branch) on Fig. 1. Observe that as we move on Fig. 1 from within the domain II towards the left branch of $s_1 = 0$, we meet the line $d = -\mu_1$ *before* we reach this branch. We then stay on this line and never leave it eventually.

The range A4 will be preferable to A3 in the region to the left of the right branch Eq. 59 up to the x_2-axis; these ranges will be equivalent only for points of the x_1-axis belonging to the interval $[0, 2m_1 \Delta k \Delta \mu / \bar{v}]$, that is for a self-adjoint case. This equivalence is confirmed in Appendix A. We thus arrive at the following final classification of optimal ranges (Fig. 2).

A4 for the region bounded by the x_2-axis and the right branch of hyperbola Eq. 59;

A1 for the region bounded by this branch and the right branch of hyperbola Eq. 56;

A2 for the region located to the right of the right branch of Eq. 56.

The following example illustrates this classification. Assume that we wish to minimize the functional

$$I = \int_\Sigma ii \cdot \cdot \mathcal{D} \cdot \cdot jj\, dxdy.$$

In this case,

$$\xi = \xi_1 a_1 + \xi_2 a_2 = 2^{-1/2}[\xi_1 + \xi_2)ii + (\xi_1 - \xi_2)jj] = ii,$$
$$\eta = \eta_1 a_1 + \eta_2 a_2 = 2^{-1/2}[(\eta_1 + \eta_2)ii + (\eta_1 - \eta_2)jj] = -jj,$$

and

$$\xi_1 = \xi_2 = 2^{-1/2}, \quad \eta_1 = -\eta_2 = -2^{-1/2},$$

that is

$$\xi_1\eta_1 = -1/2 < 0, \quad \zeta_2 = -\sigma_2 = 1,$$

i.e. $x_1 = 0$. Referring to our classification, we conclude that the range A4 applies: the corresponding microstructure is the rank 2 matrix composite with mutually perpendicular layers bisecting the right angle between the main axes i, j of ξ and η.

A similar example was handled by Cherkaev in[13].

Case $\xi_1 \eta_1 > 0$.

Consider the interfaces separating B1 from B2 and B4. To find them, we equalize $\phi(\xi, \eta)$ defined by Eq. 23_ and that given by Eq. 32_2. After calculations we obtain the equation

$$(x_1 + \alpha_1)^2 - x_2^2 = \alpha_2^2 \tag{60}$$

where

$$\alpha_1 = -u/(m_2 \Delta k \Delta \mu) - (m_2 \Delta k \Delta \mu)/v, \quad \alpha_2 = -u/(m_2 \Delta k \Delta \mu) + (m_2 \Delta k \Delta \mu)/v. \tag{61}$$

Both branches of hyperbola Eq. 60 are shown on Fig 3. The right branch separates B1 from B2.

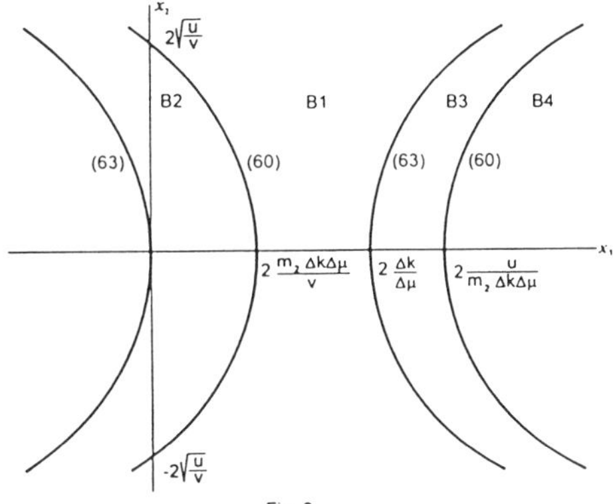

Fig. 3
Case $\xi_1 \eta_1 > 0$: ranges in (x_1, x_2) - plane

Remark 6 The x_1-intercepts of Eq. 60 can be found as points at which the value d_2 (see Eq. 47, lower sign) equals either $-\mu_2$ or k_2.

Consider now the interface separating B1 from B3. Either of these ranges is associated with rank 1 laminate. If the unit vectors i, j are directed, respectively, along and across the layers, then Eq. 22 (lower sign) holds for both ranges. When ξ and η are coaxial, then Eq. 22 is satisifed by $\xi_3 = \eta_3 = 0$ for B1, and by another solution (Remark 1) for B3. On the interface, both solutions come into a smooth contact.

Considering the range B3, we apply Eq. 22_ (the layers in rank 1 laminate oriented along the unit vector i, the basis a_1, a_2, a_3 defined by Eq. 4). Introducing the standard representations for ξ and η as

$$\xi = b_1 e_1 e_1 + b_2 e_2 e_2, \quad \eta = \beta_1 e_1 e_1 + \beta_2 e_2 e_2$$

with eigenvalues $b_1, b_2, \beta_1, \beta_2$ and unit vectors e_1, e_2 of the main axes, we obtain the formulas (Fig. 4)

$$\xi_1 = \xi \cdot \cdot a_1 = 2^{1/2}(b_1 + b_2), \qquad \eta_1 = 2^{1/2}(\beta_1 + \beta_2),$$
$$\xi_2 = \xi \cdot \cdot a_2 = 2^{1/2}(b_1 - b_2)\cos 2\phi, \qquad \eta_2 = 2^{1/2}(\beta_1 - \beta_2)\cos 2\phi,$$
$$\xi_3 = \xi \cdot \cdot a_3 = -2^{1/2}(b_1 - b_2)\sin 2\phi, \quad \eta_3 = -2^{1/2}(\beta_1 - \beta_2)\sin 2\phi.$$

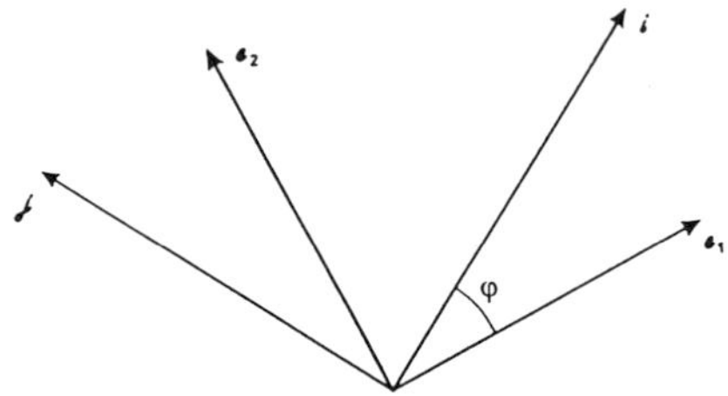

Fig. 4
The main axes $\mathbf{a}_1, \mathbf{a}_2$, and the axes i, j.

As we apply these formulas to transform Eq. 22, we get

$$\frac{b_1 - b_2}{b_1 + b_2}\sin 2\phi\left(\frac{\Delta k}{\Delta \mu} - \frac{\beta_1 - \beta_2}{\beta_1 + \beta_2}\cos 2\phi\right) + \frac{\beta_1 - \beta_2}{\beta_1 + \beta_2}\sin 2\phi\left(\frac{\Delta k}{\Delta \mu} - \frac{b_1 - b_2}{b_1 + b_2}\cos 2\phi\right) = 0.$$

Apparently, $\sin 2\phi = 0$ is a solution generating $\xi_3 = \eta_3 = 0$. We want another solution, though, i.e. the one given by

$$\cos 2\phi = (\Delta k/\Delta \mu)(c + \gamma)/2c\gamma \tag{62}$$

where

$$c = (b_1 - b_2)/(b_1 + b_2), \quad \gamma = (\beta_1 - \beta_2)/(\beta_1 + \beta_2).$$

We now apply Eq. 62 to points located on the interface separating B3 from B1; for any such point, $\phi = 0$, and Eq. 62 yields

$$\frac{c + \gamma}{c\gamma} = 2\frac{\Delta \mu}{\Delta k},$$

or, given the fact that for $\phi = 0$ $\quad c = \xi_2/\xi_1 = \zeta_2$, $\gamma = \eta_2/\eta_1 = \sigma_2$, we get (see Eq. 57)

$$\left(x_1 - \frac{\Delta k}{\Delta \mu}\right)^2 - x_2^2 = \left(\frac{\Delta k}{\Delta \mu}\right)^2. \tag{63}$$

Both branches of this hyperbola are shown on Fig. 3. Of them, only the right branch belongs to the half-plane $x_2 \geq 0$. In the domain bounded by this branch and the left branch of Eq. 60, the range B1 applies. On the right from the right branch of Eq. 63, the range B3 applies, and the right branch of hyperbola Eq. 60 now looses its significance as a possible interface participating in the optimal layout. Like in a similar situation above, once we start moving from within the domain occupied by B1 in the positive x_1-direction, we meet the right branch of Eq. 63 *before* we cut the right branch of Eq. 60. The range B4 never applies in this domain; instead, we have the range B3 there.

Remark 7 We do not consider in this paper the contacts between the ranges B1 and A3, as well as those between B1, B2 and A3 at the point Z on Fig. 1. Evidently, these contacts are related to the case $\xi_1\eta_1 > 0$. Some of them assume that Eqs. (30), i.e. $s_1 = 0$ and $s_2 = 0$, should hold simultaneously. The description of such contacts in the language of fields ξ, η should introduce some new microstructures produced as mixtures of rank 2 matrix laminates appearing in cases 1 and 2, section 5. Since we do not consider these contacts, the classification of ranges offered by Fig. 3 is incomplete.

Appendix A: The range $M_2 = 0$ The necessary conditions of stationarity reduce to the requirement $\phi_\omega = 0$. Given Eq. 19 and Eqs. 51-54 of[2] and assuming that $\omega_1 = \omega_2 = 0$, we represent the stationarity conditions in the form (see Eqs. 55)

$$\zeta_2 = \frac{\Delta k}{\Delta \mu} \frac{\tilde{M}p - \omega_3}{\tilde{K} + \omega_3 p}, \quad \zeta_3 = \frac{\Delta k}{\Delta \mu} \frac{\tau}{\tilde{K} + \omega_3 p}, \tag{A.1}$$

$$\sigma_2 = \frac{\Delta k}{\Delta \mu} \frac{\tilde{M}p + \omega_3}{\tilde{K} - \omega_3 p}, \quad \sigma_3 = \frac{\Delta k}{\Delta \mu} \frac{\tau}{\tilde{K} - \omega_3 p}. \tag{A.2}$$

Here p and τ denote parameters, the value d should be taken equal to $-\mu_2$ in accordance with the requirement $M_2 = 0$.

Because ξ and η are coaxial, the following relation holds:

$$\frac{\tau}{\tilde{M}p - \omega_3} = \frac{\tau}{\tilde{M}p + \omega_3}.$$

This means that either $\tau = 0$ or $p = \infty$, or $\omega_3 = 0$. The latter possibility strongly specializes both ξ and η making them proportional to one another. The assumption $p = \infty$ is less general than $\tau = 0$, so we consider the latter one.

We now apply Eqs. 36, 37 to compute $\phi(\xi, \eta)$ defined by Eq. 17. After calculations, we obtain

$$\phi(\xi, \eta) = \xi \cdot \cdot \langle \mathcal{D} \rangle \cdot \cdot \eta - m_1 m_2 (\Delta k)^2 \xi_1 \eta_1 \frac{\tilde{K} + \tilde{M}p^2}{\tilde{K}^2 - \omega_3^2 p^2}. \tag{A.3}$$

This formula is similar to Eq. 26; here, as before, we define (recall that $\mu_2 + d = 0$)

$$\tilde{K} = \tilde{k} - d = \tilde{k} + \mu_2, \quad \tilde{M} = \tilde{\mu} + d = \tilde{\mu} - \mu_2 = -m_2 \Delta \mu. \tag{A.4}$$

290

The expression

$$\sup_{m_1} \phi(\xi, \eta)$$

with $\phi(\xi, \eta)$ given by Eq. A.3 represents the upper bound for $G^{**}(\xi, \eta)$.

To obtain the *lower* bound, we consider a special microstructure. This will be a matrix rank 2 lamination with layers of different rank making an angle θ specified below (see Fig. 5).

The effective tensor $\mathcal{D}_0$ of stiffness of this structure is given by the formula[14]

$$\mathcal{D}_0 = \mathcal{D}_2 + m_1 A^{-1}$$

with tensor A defined as

$$A = (\mathcal{D}_2 - \mathcal{D}_1)^{-1} + 2\delta(\alpha_1 nnnn + \alpha_2 \bar{n}\bar{n}\bar{n}\bar{n}).$$

The parameters $\alpha_1, \alpha_2 \geq 0, \alpha_1 + \alpha_2 = 1$ are linked with geometric parameters f, ρ of the microstructure by the formulas (see Fig. 5)

$$\alpha_1 = f(1 - \rho)/m_2, \quad \alpha_2 = \rho/m_2,$$

the symbol δ is defined as

$$\delta = \frac{m_2}{k_2 + \mu_2}, \tag{A.5}$$

and the unit vectors $n, \bar{n}$ are normal to the layers of different ranks shown on the figure.

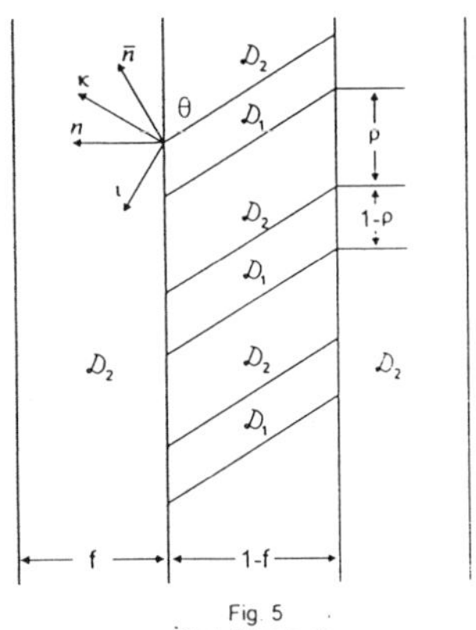

Fig. 5
Rank 2 laminate

Introduce now the unit vectors

$$\kappa = \frac{n + \bar{n}}{2\cos\frac{\theta}{2}}, \quad \iota = \frac{n - \bar{n}}{2\sin\frac{\theta}{2}} \tag{A.6}$$

shown on the figure, and define a tensor basis

$$h_1 = (1/\sqrt{2})(\kappa\kappa + \iota\iota), \quad h_2 = (1/\sqrt{2})(\kappa\kappa - \iota\iota), \quad h_3 = (1/\sqrt{2})(\kappa\iota + \iota\kappa) \tag{A.7}$$

associated with these vectors.

In this basis, the tensor A is given by the formula

$$A = A_{11}h_1h_1 + A_{22}h_2h_2 + A_{33}h_3h_3 + A_{12}(h_1h_2 + h_2h_1) + A_{23}(h_2h_3 + h_3h_2) + A_{13}(h_3h_1 + h_1h_3) \tag{A.8}$$

with the table of components

$$\begin{aligned}
A_{11} &= \delta - (\Delta k)^{-1}, & A_{12} &= \delta\cos\theta, & A_{13} &= \delta(\alpha_1 - \alpha_2)\sin\theta, \\
& & A_{22} &= \delta\cos^2\theta - (\Delta\mu)^{-1}, & A_{23} &= \delta(\alpha_1 - \alpha_2)\sin\theta\cos\theta, \\
& & & & A_{33} &= \delta\sin^2\theta - (\Delta\mu)^{-1}.
\end{aligned} \tag{A.9}$$

Introduce the bilinear form $\xi \cdot\cdot\mathcal{D}_0 \cdot\cdot\eta$ associated with $\mathcal{D}_0$ and consider its stationarity conditions with respect to α_1, θ as well as to the basis h_1, h_2, h_3 (i.e. the unit vectors $n, \bar{n}$). The first of these conditions is given by

$$\xi \cdot\cdot(A^{-1})_{\alpha_1} \cdot\cdot\eta = -\xi \cdot\cdot A^{-1} \cdot\cdot A_{\alpha_1} \cdot\cdot A^{-1} \cdot\cdot\eta = -X \cdot\cdot A_{\alpha_1} \cdot\cdot Y = 0 \tag{A.10}$$

where

$$X = A^{-1} \cdot\cdot\xi, \quad Y = A^{-1} \cdot\cdot\eta; \tag{A.11}$$

another two conditions are given by similar formulas. We finally arrive at the relationships:

$$\begin{aligned}
X \cdot\cdot(h_1h_3 + h_3h_1) \cdot\cdot Y &= 0, \\
X \cdot\cdot(h_2h_3 + h_3h_2) \cdot\cdot Y &= 0, \\
X \cdot\cdot(h_1h_2 + h_2h_1) \cdot\cdot Y &= 2\cos\theta[(X \cdot\cdot h_3)(Y \cdot\cdot h_3) - (X \cdot\cdot h_2)(Y \cdot\cdot h_2)].
\end{aligned} \tag{A.12}$$

We will try to satisfy these conditions assuming that $\alpha_1 = \alpha_2 = 1/2$. With this assumption, the matrix A is easy to reverse; we obtain the formulas

$$(A^{-1})_{11} = \frac{A_{22}A_{33}}{\det A}, \quad (A^{-1})_{12} = -\frac{A_{12}A_{33}}{\det A}, \quad (A^{-1})_{13} = 0,$$

$$(A^{-1})_{22} = \frac{A_{11}A_{33}}{\det A}, \quad (A^{-1})_{23} = 0, \tag{A.13}$$

$$(A^{-1})_{33} = \frac{1}{A_{33}}.$$

For tensors X and Y, the following representations hold

$$X = A^{-1} \cdot \cdot \xi = \frac{A_{33}}{det A}[(A_{22}\xi_{(1)} - A_{12}\xi_{(2)})h_1$$
$$+ (A_{11}\xi_{(2)} - A_{12}\xi_{(1)})h_2] + \frac{1}{A_{33}}\eta_{(3)}h_3,$$
$$Y = A^{-1} \cdot \cdot \eta = \frac{A_{33}}{det A}[(A_{22}\eta_{(1)} - A_{12}\eta_{(2)})h_1$$
$$+ (A_{11}\eta_{(2)} - A_{12}\eta_{(1)})h_2] + \frac{1}{A_{33}}\eta_{(3)}h_3. \qquad (A.14)$$

Here, $\xi_{(1)}, \ldots, \eta_{(3)}$ denote the components of ξ and η in the h-basis (A.7).

Desiring to satisfy Eq. A.12, assume that the common main axes of ξ and η are directed along the unit vectors κ and ι. Then $\xi_{(3)} = \eta_{(3)} = 0$, and the first two Eq. A.12

$$(A_{22}\xi_{(1)} - A_{12}\xi_{(2)})\eta_{(3)} + (A_{22}\eta_{(1)} - A_{12}\eta_{(2)})\xi_{(3)} = 0,$$
$$(A_{11}\xi_{(2)} - A_{12}\xi_{(1)})\eta_{(3)} + (A_{11}\eta_{(2)} - A_{12}\eta_{(1)})\xi_{(3)} = 0,$$

are satisfied. As to the third Eq. A.12, this one with $\xi_{(3)} = \eta_{(3)} = 0$ is reduced to

$$[(A_{22} - A_{12}\cos\theta)\xi_{(1)} + (A_{11}\cos\theta - A_{12})\xi_{(2)}](A_{11}\eta_{(2)} - A_{12}\eta_{(1)})$$
$$+ [(A_{22} - A_{12}\cos\theta)\eta_{(1)} + (A_{11}\cos\theta - A_{12})\eta_{(2)}](A_{11}\xi_{(2)} - A_{12}\xi_{(1)}) = 0 \quad (A.15)$$

Referring to Eq. A.9, we obtain the formulas

$$A_{22} - A_{12}\cos\theta = \delta\cos^2\theta - (\Delta\mu)^{-1} - \delta\cos^2\theta = -(\Delta\mu)^{-1},$$

$$A_{11}\cos\theta - A_{12} = \delta\cos\theta - (\Delta k)^{-1}\cos\theta - \delta\cos\theta = -(\Delta k)^{-1}\cos\theta,$$

and Eq. A.15 is now reduced to

$$[(\Delta\mu)^{-1}\xi_{(1)} + (\Delta k)^{-1}\cos\theta\xi_{(2)}]\{[\delta - (\Delta k)^{-1}]\eta_{(2)} - \delta\cos\theta\eta_{(1)}\}$$
$$+ [(\Delta\mu)^{-1}\eta_{(1)} + (\Delta k)^{-1}\cos\theta\eta_{(2)}]\{[\delta - (\Delta k)^{-1}]\xi_{(2)} - \delta\cos\theta\xi_{(1)}\} = 0.$$

This is a quadratic equation that will determine θ for given values of $\xi_{(1)}, \ldots, \eta_{(2)}$:

$$(\Delta k)^{-1}\delta(\xi_{(1)}\eta_{(2)} + \xi_{(2)}\eta_{(1)})\cos^2\theta + 2\{(\Delta\mu)^{-1}\delta\xi_{(1)}\eta_{(1)} - (\Delta k)^{-1}[\delta - (\Delta k)^{-1}]\xi_{(2)}\eta_{(2)}\} \cdot$$
$$\cdot \cos\theta - (\Delta\mu)^{-1}[\delta - (\Delta k)^{-1}](\xi_{(1)}\eta_{(2)} + \xi_{(2)}\eta_{(1)}) = 0. \qquad (A.16)$$

We now identify the h-basis Eq. A.7 with the a-basis Eq. 4 introduced originally; this means that $\xi_{(i)} = \xi_i$, $\eta_{(i)} = \eta_i$, $i = 1, 2, 3$.

Eqs. A.1, A.2 allow us to express the coefficients in Eq. A.16 as functions of ω_3 and p. We obtain the formulas:

$$\xi_2\eta_2 = \xi_1\eta_1 \left(\frac{\Delta k}{\Delta\mu}\right)^2 \frac{\tilde{M}^2 p^2 - \omega_3^2}{\tilde{K}^2 - \omega_3^2 p^2},$$

$$\xi_1\eta_2 + \xi_2\eta_1 = 2\xi_1\eta_1\frac{\Delta k}{\Delta\mu}\frac{\tilde{K}\tilde{M} + \omega_3^2}{\tilde{K}^2 - \omega_3^2 p^2}p.$$

By virtue of these formulas, Eq. A.16 is reduced to

$$(\tilde{M}p\cos\theta + \tilde{K})(\cos\theta - p) = 0,$$

and its roots are found to be

$$\cos\theta_1 = p \tag{A.17}$$

and

$$\cos\theta_2 = -\frac{\tilde{K}}{\tilde{M}p}. \tag{A.18}$$

We now apply these roots to compute the bilinear form $\xi\cdot\cdot\mathcal{D}_0\cdot\cdot\eta$. More specifically, we will compute a related expression

$$\xi\cdot\cdot\mathcal{D}_0\cdot\cdot\eta - \xi\cdot\cdot\langle\mathcal{D}\rangle\cdot\cdot\eta = m_1\xi\cdot\cdot(\Delta\mathcal{D} + A^{-1})\cdot\cdot\eta. \tag{A.19}$$

This calculation can be carried out with the aid of Eqs A.1, A.2, A.9 and A.13. The root $\cos\theta_1 = p$ then generates the formula

$$m_1\xi\cdot\cdot(\Delta\mathcal{D} + A^{-1})\cdot\cdot\eta = -m_1 m_2(\Delta k)^2\xi_1\eta_1\frac{\tilde{K} + \tilde{M}p^2}{\tilde{K}^2 - \omega_3^2 p^2}, \tag{A.20}$$

whereas the root $\cos\theta_2 = -\dfrac{\tilde{K}}{\tilde{M}p}$ results in the expression

$$m_1\xi\cdot\cdot(\Delta\mathcal{D} + A^{-1})\cdot\cdot\eta = m_1 m_2(\Delta k)^2\xi_1\eta_1\frac{\omega_3^2}{\tilde{K}\tilde{M}}\frac{\tilde{K} + \tilde{M}p^2}{\tilde{K}^2 - \omega_3^2 p^2}. \tag{A.21}$$

We see that Eq. A.20 matches that of A.3 and therefore generates a lower bound saturating the upper bound A.3. The result given by Eq. A.21 does not possess this property unless $\omega_3^2 + \tilde{K}\tilde{M} = 0$. This latter case is special because Eqs. A.1, A.2 show that ζ_2 and σ_2 will then become independent of p.

Eq. 34_2 will now appear as we eliminate $\tilde{K}, \tilde{M}, \omega_3$ and p from Eqs. A.19, A.20 with the aid of Eqs. A.1, A.2, and A.4.

A similar argument applies if we consider the range $M_1 = 0$. The relevant microstructure is the same as the one above with materials $\mathcal{D}_1$ and $\mathcal{D}_2$ interchanged. For $\phi(\xi,\eta)$, we arrive at the expression given by Eq. 34_1.

If we in this latter case apply Eq. A.1 and assume that $\omega_3 = 0$ (self-adjoint case), then we will see that

$$p = \frac{\tilde{k} + \mu_1}{m_1\Delta k}\zeta_2.$$

Ineq. $0 \leq p \leq 1$ will hold provided that

$$0 \leq \zeta_2 \leq \frac{m_1 \Delta k}{\tilde{k} + \mu_1}$$

which corresponds to the segment $[0, 2m_1 \Delta k \Delta \mu / \bar{v}]$ of the x_1-axis on Fig. 2. The relevant expression Eq. 34_1, for ϕ will show that

$$\phi = -\xi_1^2 \left\{ \mu_1 \zeta_2^2 + \left[\left(\frac{m_1}{k_1 + \mu_1} + \frac{m_2}{k_2 + \mu_1} \right)^{-1} - \mu_1 \right] \right\}$$

which is the same as that given by Eq. 32_1. We see that the ranges (a3) and (a4), section 6, are indistinguishable in this case.

Appendix B

Assume that $s_1 = K_1 M_1 + \omega_3^2$ and $s_2 = K_2 M_2 + \omega_3^2$ are of the same sign; then the expression $\tilde{s} = \tilde{K}\tilde{M} + \omega_3^2$ treated as the function of m_1 will preserve the same sign as that of s_1 and s_2 in the actual material pattern.

The function

$$\tilde{s}(m_1) = (\tilde{k} - d)(\tilde{\mu} + d) + \omega_3^2$$

has the stationary value m_1 defined by the relation

$$\tilde{s}'(m_1) = \Delta k(\tilde{\mu} + d) + \Delta \mu(\tilde{k} - d) = 0; \tag{B.1}$$

this stationary value being a minimum for $\Delta k \Delta \mu > 0$. For this reason the above statement holds should the values of s_1 and s_2 be both negative.

When these values are positive, additional analysis must be applied. The minimal value of $\tilde{s}(m_1)$ is equal to

$$\tilde{s}_{min} = -\frac{1}{4\Delta k \Delta \mu}(K_2 M_1 - K_1 M_2)^2 + \omega_3^2,$$

this value will be achieved when (see Eq. B.1)

$$d = d_* \equiv -\frac{\tilde{\mu}\Delta k + \tilde{k}\Delta \mu}{\Delta k - \Delta \mu},$$

this value being negative for $\Delta x - \Delta \mu > 0$. The value d_* should be generated by $m_1 \in [0, 1]$, that is

$$d_- \leq d_* \leq d_+ \tag{B.2}$$

where

$$d_- = -\frac{\mu_2 \Delta k + k_2 \Delta \mu}{\Delta k - \Delta \mu}, \quad d_+ = -\frac{\mu_1 \Delta k + k_1 \Delta \mu}{\Delta k - \Delta \mu}.$$

The condition $\tilde{s}_{min} \geq 0$ will be violated provided that

$$| \omega_3 | \leq \frac{| K_2 M_1 - K_1 M_2 |}{2\sqrt{\Delta k \Delta \mu}} = \frac{d_*(\Delta k + \Delta \mu) + \mu_1 \Delta k - k_1 \Delta \mu}{2\sqrt{\Delta k \Delta \mu}} \qquad (B.3)$$

with $d_* \in$ (B.2). The region on the (d, ω_3)-plane specified by Eqs. B.2, B.3 and inequalities $s_1 > 0, s_2 > 0$ is cross-hatched on Fig. 1. In sections 7 and 8 it is shown that this region can never be achieved.

Acknowledgements

The research has been supported by NSF Grant DMS-9305840. The author acknowledges discussions with Ray V. Adams, Andrei V. Cherkaev and Leonid V. Gibiansky.

References

1. Lurie, K.A., The Extension of Optimization Problems Containing Controls in Coefficients, Proceedings of the Royal Society of Edinburgh, vol. 114A, pp. 87-97, 1990.

2. Lurie, K.A., Direct Relaxation of Optimal Layout Problems for Plates, Journal of Optimization Theory and Applications, vol. 80, no. 1, pp. 93-116, January 1994.

3. Lurie, K.A., Fedorov, A.V., and Cherkaev, A.V., Regularization of Optimal Design Problems for Bars and Plates, parts 1 and 2, Journal of Optimization Theory and Applications, vol. 37, pp. 499-521, 1982, and vol. 37, pp. 523-543, 1982.

4. Lurie, K.A., and Cherkaev, A.V., Effective Characteristics of Composite Materials and Optimal Design of Structural Elements (in Russian), Advances in Mechanics (Poland), vol. 9, no. 2, pp. 3-81, 1986 (an English translation to appear in "Topics in Material Design," Birkhauser Verlag, 1994).

5. Lurie, K.A., and Lipton, R., Direct Solution of an Optimal Layout Problem for Isotropic Heat Conductors in Three Dimensions, Theoretical Aspects of Industrial Design, ed. by David A. Field and Vadim Komkov, SIAM, Philadelphia, pp. 1-11, 1992.

6. Lurie, K.A., Direct Solution of an Optimal Layout Problem for Isotropic Heat Conductors on a Plane, Journal of Optimization Theory and Applications, vol. 72, pp. 553-575, 1992.

7. Adams, R.V., Direct Solution of an Optimal Layout Problem for Isotropic Heat Conductors with a Volume Fraction Constraint, Journal of Optimization Theory and Applications, to be published.

8. Ball, J.V., Convexity Conditions and Existence Theorems in Nonlinear Elasticity, Archive for Rational Mechanics and Analysis, vol. 63, pp. 337-403, 1977.

9. Strang, G., The Polyconvexification of $F(\Delta u)$, Research Report CMA-R09-83, Australian National University, 1983.

10. Kohn, R.V., and Strang, G., Optimal Design and Relaxation of Variational Problems, parts 1,2,3, Communications on Pure and Applied Mathematics, vol. 39, pp. 113-137, 1986; vol. 39, pp. 139-182, 1986; vol. 39, pp. 353-377, 1986.

11. Gibiansky, L.V. and Cherkaev, A.V., Design of Composite Plates of Extremal Stiffness (in Russian), A. F. Ioffe Institute Report 914, Leningrad, 1984.

12. Allaire, G., and Kohn, R.V., Optimal Bounds on the Effective Behavior of a Mixture of two Well-Ordered Elastic Materials, Quarterly of Applied Mathematics, vol. LI, No. 4, pp. 675-699, 1993.

13. Cherkaev, A.V., Relaxation of Problems of Optimal Structural Design, International Journal of Solids and Structures, vol. 31, pp. 2251-2280, 1994.

14. Francfort, G., and Murat, F., Homogenization and Optimal Bounds in Linear Elasticity, Archive for Rational Mechanics and Analysis, vol. 94, pp. 307-334, 1986.

WSSIAA 5 (1995) pp. 297–310

UPPER AND LOWER BOUNDS IN QUADRATIC MAXIMIZATION WITH INTEGER CONSTRAINTS

PIERLUIGI MAPONI

Dipartimento di Matematica e Fisica,
Università di Camerino, 62032 Camerino, Italy

GRAZIELLA PACELLI, MARIA CRISTINA RECCHIONI

Istituto di Matematica e Statistica,
Università di Ancona, 60100 Ancona, Italy

FRANCESCO ZIRILLI

Dipartimento di Matematica "G.Castelnuovo",
Università di Roma "La Sapienza", 00185 Roma, Italy

ABSTRACT

In this paper the problem of maximizing a quadratic function defined in $\{-1,1\}^n$ is considered. We propose a technique to obtain an upper bound and a lower bound to the maximum of a quadratic function on the set $\{-1,1\}^n$ and a feasible point where the lower bound is attained. The problem of the *approximability* of the quadratic maximization problem with integer constraints by the method proposed here is studied and solved negatively. Moreover a special class of matrices such that the feasible point obtained with our method is the solution of the maximization problem considered is given. Numerical implementation of the method proposed and related numerical experience are shown.

Keywords: Quadratic maximization problems, Nonlinear Programming, NP-hard problems

1. Introduction

Let $\mathbf{R}^n$ be the n-dimensional real Euclidean space, and $\mathbf{y} = (y_1, y_2, \ldots, y_n)^T \in \mathbf{R}^n$, $\mathbf{z} = (z_1, z_2, \ldots, z_n)^T \in \mathbf{R}^n$ be generic vectors where the superscript T means transposed. We denote with $\mathbf{y}^T \mathbf{z}$ the usual Euclidean inner product.

Let $\mathcal{S}_n$ be the discrete set defined by

$$\mathcal{S}_n = \{-1, 1\}^n = \{\, \mathbf{y} = (y_1, y_2, \ldots, y_n)^T \in \mathbf{R}^n \mid y_i \in \{-1, 1\}, \, i = 1, 2, \ldots, n\}. \quad (1)$$

The set $\mathcal{S}_n$ contains 2^n points. Let $Q \in \mathbf{R}^{n \times n}$ be an $n \times n$ real symmetric matrix and $f(\mathbf{y}) = \mathbf{y}^T Q \mathbf{y}$, $\mathbf{y} \in \mathbf{R}^n$ be the associated quadratic form.

In this paper we consider the following quadratic optimization problem:

$$\max_{\mathbf{y}} f(\mathbf{y}) \quad (2)$$

subject to:

$$\mathbf{y} \in \mathcal{S}_n \quad (3)$$

and we denote the optimum value solution of problem (2), (3) with $f_{max}(Q)$.

We note that the quadratic form $f(\mathbf{y})$ can be transformed in a negative definite form by changing Q into $Q - cI$ for some suitable $c > 0$. These two quadratic forms restricted to $\mathcal{S}_n$ differ only by the constant cn because we have $\mathbf{y}^T \mathbf{y} = n \quad \forall \mathbf{y} \in \mathcal{S}_n$. So that the maximization problem (2), (3) corresponding to the two quadratic forms can be easily reduced one to the other.

In this work we make no assumptions about the sign of the quadratic form $f(\mathbf{y})$.

The solution of the quadratic optimization problem (2), (3) is related to some graph partitioning problems [1,2,3] and it is well known that some of these problems are NP hard. In this paper we propose a technique to obtain a closed interval $[l(Q), u(Q)] \subset \mathbf{R}$, which contains $f_{max}(Q)$ and a feasible point $\mathbf{y}^* \in \mathcal{S}_n$ such that $f(\mathbf{y}^*) = l(Q)$.

To obtain the upper bound $u(Q)$ we use a technique introduced in [4,5] based on an auxiliary continuous optimization problem. The upper bound $u(Q)$ is obtained integrating numerically an initial value problem for a suitable system of ordinary differential equations. The lower bound $l(Q)$ and the feasible point $\mathbf{y}^*$ are obtained using an algorithm that computes the components of $\mathbf{y}^*$ one at a time. This algorithm is independent of the auxiliary continuous problem that gives $u(Q)$.

Let $\mathbf{0} = (0, 0, \ldots, 0)^T \in \mathbf{R}^n$, $\mathbf{e} = (1, 1, \ldots, 1)^T \in \mathbf{R}^n$, let $\mathbf{w} = (w_1, w_2, \ldots, w_n)^T \in \mathbf{R}^n$ and let $W = Diag(\mathbf{w}) = Diag(w_1, w_2, \ldots, w_n) \in \mathbf{R}^{n \times n}$ be the diagonal matrix whose diagonal entries are w_1, w_2, $\ldots$, w_n respectively.

Let $\mathcal{U}$ be the following set:

$$\mathcal{U} = \{\, \mathbf{w} = (w_1, w_2, \ldots, w_n)^T \in \mathbf{R}^n \mid \mathbf{e}^T \mathbf{w} = 1, \, w_i > 0, \, i = 1, \ldots, n\} \quad (4)$$

We embed $\mathcal{S}_n$ in the ellipsoid $\mathcal{E}(\mathbf{w})$ defined by

$$\mathcal{E}(\mathbf{w}) = \{\, \mathbf{y} = (y_1, y_2, \ldots, y_n)^T \in \mathbf{R}^n \mid \mathbf{y}^T W \mathbf{y} \le 1, \, \mathbf{w} \in \mathcal{U}\}. \quad (5)$$

Let $\mathbf{w} \in \mathcal{U}$, $W^{1/2} = Diag(w_1^{1/2}, w_2^{1/2}, \ldots, w_n^{1/2})$, and $W^{-1/2}$ be the inverse of $W^{1/2}$, we can see that if λ is the greatest eigenvalue of the matrix $W^{-1/2}QW^{-1/2}$ we have

$$\lambda = \max_{\mathbf{y} \neq \mathbf{0}} \frac{\mathbf{y}^T W^{-1/2}QW^{-1/2}\mathbf{y}}{\mathbf{y}^T\mathbf{y}} = \max_{\mathbf{z} \neq \mathbf{0}} \frac{\mathbf{z}^T Q\mathbf{z}}{\mathbf{z}^T W\mathbf{z}} \geq \max_{\substack{\mathbf{z} \neq \mathbf{0} \\ \mathbf{z} \in \mathcal{E}(\mathbf{w})}} \frac{\mathbf{z}^T Q\mathbf{z}}{\mathbf{z}^T W\mathbf{z}}. \tag{6}$$

Let $\mu^* = \max\{0, \lambda\}$, equation (6) implies

$$\mathbf{z}^T Q\mathbf{z} \leq \mu^*, \qquad \forall \mathbf{z} \in \mathcal{E}(\mathbf{w}), \tag{7}$$

i.e. since $\mathcal{E}(\mathbf{w}) \supset \mathcal{S}_n$ μ^* is an upper bound of $f_{max}(Q)$. We define the set:

$$\mathcal{L} = \{(\mu, \mathbf{w}^T)^T \in \mathbf{R}^{n+1} \mid \mu \geq 0, \mathbf{w} \in \mathcal{U}, \frac{\mathbf{y}^T Q\mathbf{y}}{\mathbf{y}^T W\mathbf{y}} \leq \mu, \forall \mathbf{y} \in \mathbf{R}^n \setminus \{\mathbf{0}\}\}. \tag{8}$$

The auxiliary continuous optimization problem that we introduce is the following one:

$$\inf_{(\mu, \mathbf{w}^T)^T} \mu \tag{9}$$

subject to:

$$(\mu, \mathbf{w}^T)^T \in \mathcal{L}. \tag{10}$$

To problem (9), (10) we associate a system of ordinary differential equations [5,6,7,8], that are the steepest descent equations in a suitable metric of the objective function μ. The upper bound $u(Q)$ is obtained integrating (numerically) an initial value problem for this system of differential equations. Moreover we obtain a suboptimal point $\mathbf{y}^* \in \mathcal{S}_n$ using a polynomial time algorithm and we choose

$$l(Q) = f(\mathbf{y}^*). \tag{11}$$

The algorithm that constructs $\mathbf{y}^*$ is independent from the procedure that gives the upper bound $u(Q)$, and applies in our context a rather general set of ideas [9].

The suboptimal point $\mathbf{y}^*$ is optimal when some special classes of matrices Q are considered, one of these classes is defined in section 4 and the optimality of $\mathbf{y}^*$ is proved.

Moreover we show that the quantities:

$$\alpha = \frac{f_{max}(Q) - l(Q)}{|f_{max}(Q)|}, \tag{12}$$

and

$$\beta = \frac{u(Q) - f_{max}(Q)}{|f_{max}(Q)|}, \tag{13}$$

can be arbitrarily large for a suitable choice of the matrix Q such that $f_{max}(Q) \neq 0$, that is we show that the algorithms described here cannot be used to prove the *approximability* of the optimization problem (2), (3).

In section 2 we recall the results obtained in [5], we introduce the steepest descent system of differential equations and we obtain an upper bound $u(Q)$ to $f_{max}(Q)$. In section 3 we construct a lower bound $l(Q)$ to $f_{max}(Q)$ and a feasible point $\mathbf{y}^* \in \mathcal{S}_n$ that approximates the optimizer of problem (2), (3) such that (11) holds. In section 4 we show that α and β can be arbitrarily large and we introduce a special class of matrices such that for these matrices the suboptimal point $\mathbf{y}^*$ constructed in section 3 is optimal, that is $\mathbf{y}^*$ is a solution of (2), (3). In section 5 we present the numerical implementation of our method and some numerical experience obtained on a wide class of test problems.

2. An upper bound to $f_{max}(Q)$

Let us rewrite problem (9),(10). We note that since the weights w_1, w_2, ...,w_n are positive the condition

$$\frac{\mathbf{y}^T Q \mathbf{y}}{\mathbf{y}^T W \mathbf{y}} \leq \mu, \quad \forall \mathbf{y} \in \mathbf{R}^n \setminus \{\mathbf{0}\}, \tag{14}$$

can be expressed as

$$\mathbf{y}^T (\mu I - W^{-1/2} Q W^{-1/2}) \mathbf{y} \geq 0, \quad \forall \mathbf{y} \in \mathbf{R}^n, \tag{15}$$

i.e. the matrix $\mu I - W^{-1/2} Q W^{-1/2}$ is positive semidefinite.

Let $\mathbf{x} = (x_0, x_1, \ldots, x_{n+1})^T = (\mu, w_1, w_2, \ldots, w_n, \xi)^T \in \mathbf{R}^{n+2}$ we define the vector function $\mathbf{g}(\mathbf{x}) = (g_0(\mathbf{x}), g_1(\mathbf{x}))^T \in \mathbf{R}^2$ and the vector $\mathbf{c} = (c_0, c_1, \ldots, c_{n+1})^T \in \mathbf{R}^{n+2}$ as follows:

$$g_0(\mathbf{x}) = \lambda_{min}(\mu I - W^{-1/2} Q W^{-1/2}) - \xi \tag{16}$$
$$g_1(\mathbf{x}) = \mathbf{e}^T \mathbf{w} - 1 \tag{17}$$

and

$$c_0 = 1, \quad c_i = 0, \ i = 1, \ldots, n+1. \tag{18}$$

where $\lambda_{min}(\cdot)$ denotes the smallest eigenvalue of the matrix $(\cdot)$.

Problem (9), (10) can be rewritten as:

$$\inf \mathbf{c}^T \mathbf{x} \tag{19}$$

subject to:

$$\mathbf{g}(\mathbf{x}) = \mathbf{0} \tag{20}$$
$$\mu \geq 0, \ w_i > 0, \ i = 1, \ldots, n, \ \xi \geq 0. \tag{21}$$

Let $X = Diag(\mathbf{x}) = Diag(\mu, w_1, \ldots, w_n, \xi)$ and let $J_{\mathbf{g}}(\mathbf{x}) \in \mathbf{R}^{2\times(n+2)}$ be the Jacobian matrix associated to $\mathbf{g}$. We consider the following initial value problem:

$$\frac{d\mathbf{x}}{dt} = -X\mathbf{c} + XJ_{\mathbf{g}}^T(\mathbf{x})(J_{\mathbf{g}}(\mathbf{x})XJ_{\mathbf{g}}^T(\mathbf{x}))^{-1}J_{\mathbf{g}}(\mathbf{x})X\mathbf{c} \tag{22}$$

$$\mathbf{x}(0) = \mathbf{x}_0 \tag{23}$$

where $\mathbf{x}_0$ is a feasible point for problem (19)-(21). We note that since $g_0(\mathbf{x})$ is not always differentiable in the elementary sense, the jacobian matrix $J_{\mathbf{g}}(\mathbf{x})$ is not always defined in the elementary sense. Here to keep things simple we consider only cases where the right hand side of equation (22) can be interpreted in the elementary sense. That is we assume that $J_{\mathbf{g}}(\mathbf{x})$ exists in the elementary sense and that $J_{\mathbf{g}}(\mathbf{x})X^{1/2}$ has full rank along the trajectory $\mathbf{x}(t, \mathbf{x}_0)$, $0 \leq t \leq t^* < +\infty$ solution of (22), (23). Let us denote with $\mathbf{h}(\mathbf{x}) = (h_0(\mathbf{x}), h_1(\mathbf{x}), \ldots, h_{n+1}(\mathbf{x}))^T \in \mathbf{R}^{n+2}$ the vector field on the right hand side of (22).

Let Γ be the set

$$\Gamma = \{\, \mathbf{x} = (x_0, x_1, \ldots, x_{n+1})^T \in \mathbf{R}^{n+2} \,|\, x_0 \geq 0,\ x_{n+1} \geq 0,\ x_i > 0,\ i = 1, \ldots, n \,\} \tag{24}$$

and $\tilde{\Gamma}$ be the positive orthant

$$\tilde{\Gamma} = \{\, \mathbf{x} \in \mathbf{R}^{n+2} \,|\, \mathbf{x} > \mathbf{0} \,\}, \tag{25}$$

where the inequality in (25) is understood componentwise. Let $\mathcal{F}$ be the set

$$\mathcal{F} = \{\, \mathbf{x} \in \mathbf{R}^{n+2} \,|\, \mathbf{g}(\mathbf{x}) = \mathbf{0} \,\}. \tag{26}$$

We have:

Lemma 2.1 The vector field $\mathbf{h}(\mathbf{x})$ is the steepest descent vector field associated to the objective function $b(\mathbf{x}) = \mathbf{c}^T\mathbf{x}$ of the optimization problem (19)-(21) restricted to $\mathcal{F} \cap \tilde{\Gamma}$ with respect to the Riemannian metric $G(\mathbf{x}) = X^{-1} = Diag(\mathbf{x}^{-1}) = Diag(\mu^{-1}, w_1^{-1}, \ldots, w_n^{-1}, \xi^{-1})$, defined on the positive orthant $\tilde{\Gamma}$.

Proof: The proof is analogous to the one in [7], p. 270, Lemma 2.9.

Lemma 2.2 Let $\mathbf{h}(\mathbf{x})$ be the right hand side of (22) and $\mathbf{x}_0 \in \mathcal{F} \cap \tilde{\Gamma}$, and let $\mathbf{x}(t; \mathbf{x}_0)$ be the solution of the initial value problem:

$$\frac{d\mathbf{x}}{dt} = \mathbf{h}(\mathbf{x}) \tag{27}$$

$$\mathbf{x}(0) = \mathbf{x}_0. \tag{28}$$

Then: $\mathbf{x}(t; \mathbf{x}_0) \in \mathcal{F} \cap \tilde{\Gamma}$ for the values of t where $\mathbf{x}(t; \mathbf{x}_0)$ is defined, that is $0 \leq t \leq t^* < +\infty$ and the function $b(\mathbf{x}) = \mathbf{c}^T\mathbf{x}$ is monotonically non increasing along the trajectory $\mathbf{x}(t; \mathbf{x}_0)$.

Proof: From the expression of the vector field $\mathbf{h}(\mathbf{x})$ we have:

$$J_{\mathbf{g}}(\mathbf{x})\mathbf{h}(\mathbf{x}) = 0 , \qquad \mathbf{x} \in \mathcal{F} \cap \tilde{\Gamma} \tag{29}$$

We have:

$$J_{\mathbf{g}}(\mathbf{x})\frac{d\mathbf{x}}{dt} = 0 \tag{30}$$

so that $\mathbf{x}_0 \in \mathcal{F} \cap \tilde{\Gamma}$ implies $\mathbf{x}(t;\mathbf{x}_0) \in \mathcal{F} \cap \tilde{\Gamma}$ for $0 \le t \le t^* < +\infty$. Moreover we can see that:

$$\frac{d}{dt}[\mathbf{c}^T\mathbf{x}(t;\mathbf{x}_0)] = -\mathbf{c}^T X^{1/2}\left(I - X^{1/2}J_{\mathbf{g}}^T(\mathbf{x})(J_{\mathbf{g}}(\mathbf{x})XJ_{\mathbf{g}}^T(\mathbf{x}))^{-1}J_{\mathbf{g}}(\mathbf{x})X^{1/2}\right)X^{1/2}\mathbf{c} \le 0 \tag{31}$$

which concludes the proof.

We note that Lemma 2.2 implies that starting from a feasible point $\mathbf{x}_0$ and following the trajectory $\mathbf{x}(t;\mathbf{x}_0)$ we obtain an upper bound

$$u(Q) = \lim_{t \to t^*} \mathbf{c}^T\mathbf{x}(t;\mathbf{x}_0) \tag{32}$$

to the optimum of problem (2), (3) which improves the upper bound given by $\mathbf{c}^T\mathbf{x}_0$. We note that in general the upper bound (32) will depend on $\mathbf{x}_0$.

3. The construction of a suboptimal point $\mathbf{y}^*$ and of a lower bound $l(Q)$ to $f_{max}(Q)$

Let us define the algorithm that gives the suboptimal point $\mathbf{y}^*$. This algorithm determines the components of the vector $\mathbf{y}^*$ one at a time.

Let $\mathbf{y}_p \in \mathcal{S}_p, 1 \le p \le n$, and let $\boldsymbol{\eta} \in \mathcal{S}_{n-p}$ we have $\mathbf{y} = (\mathbf{y}_p^T, \boldsymbol{\eta}^T)^T \in \mathcal{S}_n$. Partitioning the matrix Q in four submatrices

$$Q = \begin{pmatrix} Q_p & R \\ R^T & Q_{n-p} \end{pmatrix} \tag{33}$$

where $Q_p \in \mathbf{R}^{p \times p}, Q_{n-p} \in \mathbf{R}^{(n-p) \times (n-p)}, R \in \mathbf{R}^{p \times (n-p)}$ we can rewrite the quadratic function $f(\mathbf{y})$ of problem (2), (3) as follows:

$$f(\mathbf{y}) = \boldsymbol{\eta}^T Q_{n-p}\boldsymbol{\eta} + 2\mathbf{y}_p^T R\boldsymbol{\eta} + \mathbf{y}_p^T Q_p\mathbf{y}_p \tag{34}$$

and we define

$$g_p(\mathbf{y}_p) = \mathbf{y}_p^T Q_p\mathbf{y}_p. \tag{35}$$

The algorithm that gives $\mathbf{y}^* = (y_1^*, y_2^*, \ldots, y_n^*)^T$ is the following:

Step 1 Let $\xi_1 = (y_1) \in \{-1, 1\}$. Solve by enumeration problem:

$$\max_{y_1 \in \mathcal{S}_1} g_1(\xi_1) \tag{36}$$

Let y_1^* be an optimizer of problem (36).
Let $i = 2$.
Step 2 Let $\xi_i = (y_1^*, \ldots, y_{i-1}^*, y_i)^T$. Solve by enumeration problem:

$$\max_{y_i \in \mathcal{S}_1} g_i(\xi_i) \tag{37}$$

Let y_i^* be an optimizer of problem (37).
Step 3 If $i = n$ stop otherwise set i equal to $i + 1$, go to Step 2.

When the maximization problems (36), (37) are degenerate we choose always $y_i^* = 1$.

4. Some mathematical properties of the previous algorithms

Let A be an algorithm that gives an approximation $\tilde{f}_{max}(Q)$ to $f_{max}(Q)$. We require the input of the algorithm A to be given by the matrix Q.

Definition 4.1 The quadratic optimization problem (2), (3) is said to be *approximable* if there exists an algorithm A, which works in polynomial time, and a positive constant K, independent of n, such that the following inequality holds:

$$|\tilde{f}_{max}(Q) - f_{max}(Q)| \leq K |f_{max}(Q)| \tag{38}$$

for every symmetric matrix $Q \in \mathbf{R}^{n \times n}$.

We note that Definition 4.1 is weaker than the usual definition of approximability (see [1,2,3,9]). In fact Definition 4.1 does not involve the optimizer of problem (2), (3) but only the optimum value $f_{max}(Q)$.

Let

$$\mathcal{T}_1 = \left\{ Q \in \mathbf{R}^{2 \times 2} | Q = ((q_{i,j})), i, j = 1, 2; \right.$$

$$\left. q_{12} = q_{21} = 0, q_{11} < 0, q_{22} > 0, q_{11} + q_{22} = \epsilon, \epsilon > 0 \right\} \tag{39}$$

and let $v(Q)$ denote the solution of the auxiliary continuous problem (19)-(21) associated to problem (2), (3). We have:

$$v(Q) \leq u(Q). \tag{40}$$

We have the following lemma:

Lemma 4.2 For any positive constant K there exists $Q \in \mathcal{T}_1$ such that the following inequality holds:

$$v(Q) - f_{max}(Q) > K|f_{max}(Q)|. \tag{41}$$

Proof: For $Q \in \mathcal{T}_1$ we have:

$$f_{max}(Q) = q_{11} + q_{22} + 2|q_{12}| = q_{11} + q_{22}. \tag{42}$$

If $Q \in \mathcal{T}_1$ from (6) we obtain:

$$\lambda(Q) = \frac{1}{2w_1 w_2}(q_{11}w_2 + q_{22}w_1 + |q_{11}w_2 + q_{22}w_1|), \tag{43}$$

and the optimization problem (9), (10) can be written as

$$\inf_{\substack{w_1+w_2=1 \\ w_1,w_2>0}} \lambda(Q). \tag{44}$$

An easy computation shows that the solution $v(Q)$ of the problem (44) is given by:

$$v(Q) = q_{22}. \tag{45}$$

From (42) and (45) we have:

$$\beta = \frac{v(Q) - f_{max}(Q)}{|f_{max}(Q)|} = \frac{-q_{11}}{\epsilon}, \quad q_{11} + q_{22} = \epsilon, q_{22} < 0, q_{22} > 0, \tag{46}$$

so that choosing $q_{11} = -1$ and $\epsilon < 1/K$ we have:

$$v(Q) - f_{max}(Q) > K|f_{max}(Q)|. \tag{47}$$

This concludes the proof.

Lemma 4.2 and relation (40) show that the method described in section 2 to obtain an upper bound of $f_{max}(Q)$ cannot be used to show the *approximability* of problem (2), (3).

Let

$$\mathcal{T}_2 = \Big\{ Q \in \mathbf{R}^{3\times 3} \,|\, Q = ((q_{ij})), i,j = 1,2,3, q_{11} + q_{22} = -5 + \epsilon, \epsilon > 0,$$

$$q_{12} = q_{21} = -1/2, q_{13} = q_{31} = -3/2, q_{23} = q_{32} = -1, q_{33} = 1 \Big\}. \tag{48}$$

For $Q \in \mathcal{T}_2$ let $l(Q)$ be the lower bound computed with the algorithm described in section 3, we have:

Lemma 4.3 For any positive constant K there exists $Q \in T_2$ such that the following inequality holds:

$$f_{max}(Q) - l(Q) > K|f_{max}(Q)|. \tag{49}$$

Proof: For $Q \in T_2$ we have:

$$f_{max}(Q) = q_{11} + q_{22} + 5 = \epsilon. \tag{50}$$

Let $\boldsymbol{\xi}_1 = (y_1) \in S_1$, the first step of the algorithm of section 3 solves the problem:

$$\max_{y_1 \in S_1} q_{11} y_1^2 . \tag{51}$$

Problem (51) is degenerate so that we choose $y_1 = y_1^* = 1$. Let $\boldsymbol{\xi}_2 = (1, y_2)^T \in S_2$, the second step of the algorithm of section 3 solves the problem:

$$\max_{y_2 \in S_1} q_{11} + q_{22} y_2^2 - y_2, \tag{52}$$

so that we have $y_2 = y_2^* = -1$. Let $\boldsymbol{\xi}_3 = (1, -1, y_3)^T$, the last step of the algorithm of section 3 solves the problem:

$$\max_{y_3 \in S_1} q_{11} + q_{22} - y_3 + y_3^2 + 1, \tag{53}$$

so that we have $y_3 = y_3^* = -1$. So that we have $y^* = (1, -1, -1)^T$ and $l(Q) = f(\mathbf{y}^*) = q_{11} + q_{22} + 3$. From (50) we have that when $\epsilon < 2/K$ the inequality (49) holds. This concludes the proof.

Lemma 4.3 shows that the algorithm described in section 3 to obtain a lower bound to $f_{max}(Q)$ cannot be used to show the *approximability* of problem (2), (3).

Let

$$T_n^* = \Big\{ Q \in \mathbf{R}^{n \times n}, Q = ((q_{ij})), i, j = 1, 2, \ldots, n, \text{ such that } Q \text{ is symmetric}$$

$$\text{and } f_{max}(Q) = \sum_{i=1}^{n} q_{ii} + 2 \sum_{i=1}^{n-1} \sum_{j=i+1}^{n} |q_{ij}| \Big\}. \tag{54}$$

We define:

$$\text{sign}(x) = \begin{cases} 1 & x > 0; \\ 0 & x = 0; \\ -1 & x < 0 . \end{cases}$$

It is easy to see that every 2×2 symmetric matrix belongs to T_2^*.

Moreover let $\tilde{\mathbf{x}} = (\tilde{x}_1, \tilde{x}_2, \ldots, \tilde{x}_n)^T$ be an optimizer of (2), (3) and $Q \in T_n^*$ then when $q_{ij} \neq 0$ and $i \neq j$ we have:

$$\text{sign}(q_{ij}) = \tilde{x}_i \tilde{x}_j \quad i, j = 1, 2, \ldots, n \tag{55}$$

Equation (55) implies:

$$\operatorname{sign}(q_{ik})\operatorname{sign}(q_{ij}) = \operatorname{sign}(q_{kj}), \quad q_{ik}, q_{ij}, q_{kj} \neq 0, \quad i < k < j.$$

Lemma 4.4 Let $Q \in T_n^*$ and let $l(Q)$ be the lower bound of $f_{max}(Q)$ obtained with the algorithm of section 3 and let $\mathbf{y}^* \in \mathcal{S}_n$ be the suboptimal point obtained with the algorithm of section 3 such that $f(\mathbf{y}^*) = l(Q)$. Then we have:

$$f_{max}(Q) = \max_{\mathbf{y} \in \mathcal{S}_n} \mathbf{y}^T Q \mathbf{y} = l(Q), \ Q \in T_n^*. \tag{56}$$

Proof: Let $Q_j \in \mathbf{R}^{j \times j}$, $j = 1, 2, \ldots, n-1$, be the submatrices of Q shown in (33). It is easy to see that if $Q \in T_n^*$ then $Q_j \in T_j^*$, $j = 1, 2, \ldots, n-1$.

We prove this lemma under the extra assumption that $q_{1j} \neq 0$, $j = 2, 3, \ldots, n$. Under this assumption we prove also than an optimizer of problem (2), (3) $\mathbf{y}^* = (y_1^*, y_2^*, \ldots, y_n^*)^T$ is given by

$$y_1^* = 1, \quad y_i^* = \operatorname{sign}(q_{1i}), \ i = 2, 3, \ldots, n. \tag{57}$$

The proof of (56) and (57) is by induction on n.

Let $n = 2$, $Q \in T_2^*$ let $\boldsymbol{\xi}_1 = (y_1) \in \mathcal{S}_1$, the algorithm of section 3 solves the problem:

$$\max_{y_1 \in \mathcal{S}_1} q_{11} y_1^2. \tag{58}$$

Problem (58) is degenerate so we choose $y_1 = y_1^* = 1$. Let $\boldsymbol{\xi}_2 = (1, y_2)^T$, step 2 of the algorithm of section 3 consists in solving the problem:

$$\max_{y_2 \in \mathcal{S}_1} q_{11} + q_{22} y_2^2 + 2 q_{12} y_2, \tag{59}$$

so that we have $y_2 = y_2^* = \operatorname{sign}(q_{12})$ and (56) holds. This concludes the proof when $n = 2$.

Now we assume that when $Q \in T_{n-1}^*$ we have:

$$l(Q) = f(\mathbf{y}_{n-1}^*) = f_{max}(Q) = \sum_{i=1}^{n-1} q_{ii} + 2 \sum_{i=1}^{n-2} \sum_{j=i+1}^{n-1} |q_{ij}|, \tag{60}$$

where $\mathbf{y}_{n-1}^* = (y_1^*, \ldots, y_{n-1}^*)^T \in \mathcal{S}_{n-1}$ is given by

$$y_1^* = 1, \ y_i^* = \operatorname{sign}(q_{1i}), \ i = 2, \ldots, n-1. \tag{61}$$

Let $Q \in T_n^*$, have:

$$Q = \begin{pmatrix} Q_{n-1} & R \\ R^T & q_{nn} \end{pmatrix}. \tag{62}$$

Let $\boldsymbol{\xi}_n = (y_1^*, \ldots, y_{n-1}^*, y_n)^T = (\mathbf{y}_{n-1}^{*T}, y_n)^T \in \mathcal{S}_n$ the last step of the algorithm of section 3 consists in solving the problem

$$\max_{y_n \in \mathcal{S}_1} \mathbf{y}_{n-1}^{*T} Q_{n-1} \mathbf{y}_{n-1}^* + 2\mathbf{y}_{n-1}^{*T} R y_n + q_{nn} y_n^2 . \tag{63}$$

From (60) and (61) we obtain

$$\boldsymbol{\xi}_n^T Q \boldsymbol{\xi}_n = \sum_{i=1}^{n} q_{ii} + 2\sum_{i=1}^{n-2}\sum_{j=i+1}^{n-1} |q_{ij}| + 2y_n\Big(q_{1n} + q_{2n}\text{sign}(q_{12}) + \cdots + q_{n-1\,n}\text{sign}(q_{1\,n-1})\Big). \tag{64}$$

Choosing $y_n = \text{sign}(q_{1n})$ and using

$$\text{sign}(q_{1i})\,\text{sign}(q_{1j}) = \text{sign}(q_{ij}), \tag{65}$$

from (64) we have the thesis since we have assumed $q_{1j} \neq 0$, $j = 2, 3, \ldots, n$.

This concludes the proof.

When the extra assumption made in the proof of Lemma 4.4 does not hold but the matrix Q is such that

$$\sum_{i=1}^{j-1} |q_{ij}| \neq 0 \quad j = 2, 3, \ldots, n \tag{66}$$

a proof analogous to the previous one but more tedious can be given. Finally when (66) does not hold we must preprocess the matrix Q and apply the algorithm of section 3 after this preprocessing. The preprocessing of the matrix Q consists in building a new symmetric matrix $\tilde{Q}$ such that:

   (i) the quadratic forms of Q and $\tilde{Q}$ coincide up to a permutation of the variables;

  (ii) the columns of Q such that (66) does not hold are moved and become the last columns of $\tilde{Q}$.

5. Numerical experiments

We present here the implementation of our algorithm and some numerical results obtained on some test problems with the algorithms of the previous sections. The eigenvalue $\lambda_{min}(\mu I - W^{-1/2}QW^{-1/2})$ has been evaluated using an IMSL routine that computes the extreme eigenvalue of a matrix by a modified QR method. An initial approximation of the gradient of $\lambda_{min}(\mu I - W^{-1/2}QW^{-1/2})$ respect to μ and to the diagonal elements of W is computed by finite differences. To enforce the constraints (20), (21) we restore the value of the variable ξ at each iteration, that is after the computation of $\lambda_{min}(\mu I - W^{-1/2}QW^{-1/2})$ we set:

$$\xi = \lambda_{min}(\mu I - W^{-1/2}QW^{-1/2}). \tag{67}$$

308

The initial value problem (22), (23) is solved numerically using Euler method with variable stepsize

More details about the implementation of our algorithm can be found in [5].

In the numerical experiments the matrices are $n \times n$ real symmetric matrices generated sampling their entries from a random variable uniformly distributed in $[-1, 1]$.

We have reported two experiments.

The first one considers only matrices Q of low order, i.e. $n = 5, 10, 15, 20$, so that problem (2), (3) can be solved easily by enumeration.

Let $f_{max}(Q)$ be the solution of problem (2), (3) and let $l(Q)$ be the approximation of $f_{max}(Q)$ found with the algorithm of section 3 and let α be the performance index defined in equation (12). Let us denote with $p\{\cdot\}$ the probability of $\cdot$. Given n we have generated a sample of 2000 random matrices Q and we have approximated the probability distribution function of the random variable α, that is:

$$f(t) = p\{\alpha < t\} \tag{68}$$

with the corresponding frequencies observed in the sample (i.e. number of cases such that $\alpha < t$ divided by 2000). In Fig. 1 we have reported the results obtained. Moreover the mean value $E(\alpha)$ of the random variable α is approximated using the probability distribution functions of Fig. 1. The results obtained are shown in Fig. 2.

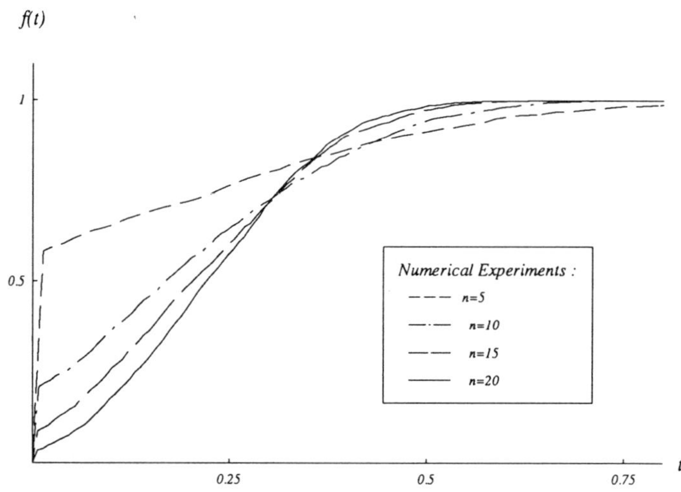

Fig 1. Probability distribution functions of α

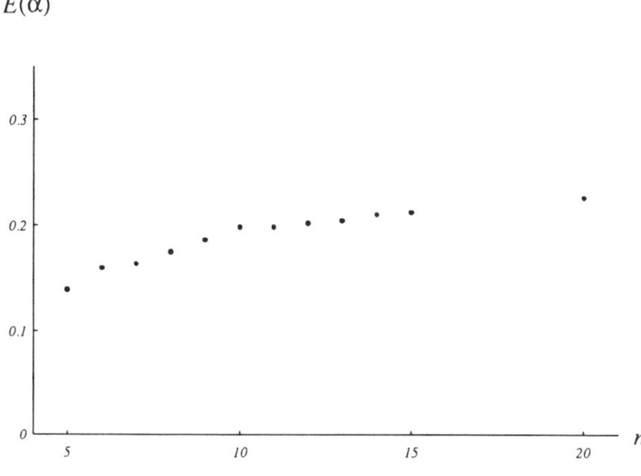

Fig 2. Mean value of α

Finally in Table 1 we report the results obained in the second experiment that is we show some results obtained with matrices Q of higher order, that is $n = 100, 200$. In Table 1 we have reported the order n of the matrix Q, the number n_s of Euler steps performed in the solution of the initial value problem (22), (23), the initial value μ_0 of the variable μ, the improved value $u(Q)$ obtained integrating numerically (22), (23) and the lower bound $l(Q)$ obtained with the algorithm of section 3. We note that for problems of the size considered in Table 1 the solution by enumeration of the optimization problem (2), (3) is computationally very expansive.

Table 1. Numerical experiments

n	n_s	μ_0	$u(Q)$	$l(Q)$
100	4503	1094.86	1054.19	490.18
100	7967	1109.18	1038.34	583.91
100	7586	1128.71	1063.90	602.37
100	7641	1090.88	1024.25	614.58
100	5431	1087.49	1038.90	573.34
200	11835	3141.14	3033.26	1686.02
200	11002	3259.27	3151.89	1793.50
200	10568	3175.89	3078.27	1624.06
200	12707	3187.57	3069.25	1716.70
200	8678	3217.07	3133.54	1792.80

310

The numerical results shown in Fig. 1, Fig. 2, Table 1 are very encouraging. We can conclude that the use of the ideas presented in this paper to study discrete optimization problems appears to be promising and deserves further investigation.

References

1. Garey, M.R., Johnson, D.S., *Computers and Intractability: A Guide to the Theory of NP-Completeness*, W.H. Freeman and Company, San Francisco, 1979.
2. Garey, M.R., Johnson, D.S., Stockmejer, L., *Some Simplified NP complete Graph Problems*, Theoretical Computer Science **1** (1976), 237–267.
3. Parker, G.R., Rardin R.L., *Discrete Optimization*, Academic Press, New York, 1991.
4. Kamath, A., Karmarkar, N., *A Continuous Approach to Compute Upper Bound in Quadratic Maximization Problems with Integer Constraints*, Recent Advances in Global Optimization, C.A. Floudas and P.M. Pardalos, Eds., Princeton University Press, Princeton N.J.(USA), (1991), 125–140.
5. Maponi, P., Recchioni, M.C., Zirilli, F., *The Use of Ordinary Differential Equations in Quadratic Maximization with Integer Constraints*, to appear.
6. Karmarkar, N., *A New Polynomial Time Algorithm for Linear Programming*, Combinatorica **4** (1984), 373-395.
7. Herzel, S., Recchioni, M.C., Zirilli, F., *A Quadratically Convergent Method for Linear Programming*, Linear Algebra and Its Appl. **152** (1991), 255–289.
8. Zirilli, F., *The Use of Ordinary Differential Equations in the Solution of Non Linear Systems of Equations*, Nonlinear Optimization, 1981, M.T.D. Powell, Ed., Academic Press, New York, (1982), 39–47.
9. Crescenzi, P., Panconesi, A., *Completeness in Approximation Classes*, Information and Computation **93** (1991), 241–262.

WSSIAA 5 (1995) pp. 311–326

311

RICCATI DIFFERENCE EQUATIONS FOR DISCRETE TIME SPECTRAL FACTORIZATION WITH UNIT CIRCLE ZEROS. [1]

JEREMY B. MATSON   BRIAN D.O. ANDERSON
Department of Systems Engineering,
Research School of Information
Sciences and Engineering,
Australian National University,
Canberra ACT 0200, AUSTRALIA

ALAN J. LAUB
Department of Electrical
and Computer Engineering,
University of California,
Santa Barbara, CA 93106-9560
U.S.A.

DAVID J. CLEMENTS
School of Electrical Engineering,
University of New South Wales,
P.O. Box 1,
Kensington 2033, AUSTRALIA

ABSTRACT

Spectral matrices that have unit circle transmission zeros arise in the consideration of H_∞ control, the bounded-real lemma and discrete spectral factorization problems. Spectral matrices with unit circle invariant (but not transmission) zeros arise when considering Kalman filtering for systems with unit circle modes which are not corrupted by process noise. It is well known that if a spectral matrix is generically nonsingular, minimum phase spectral factors can be constructed from a strong solution of an algebraic Riccati equation associated with a state-space realization of the spectral matrix. Closely associated with the algebraic equation is a Riccati difference equation whose iterates are shown to converge to the strong solution under fairly mild conditions on the realization of the spectral matrix. The key observations made in this paper concern the fine structure of the Riccati difference equation iterates from which a convergence rate of $\mathcal{O}(\frac{1}{k})$ is deduced.

1 Introduction.

1.1 Spectral Factorization Review.

In discrete time, a *spectral matrix* $\Psi(z)$ is a square real rational matrix-valued function of a complex variable z with the properties that $\Psi^T(z^{-1}) = \Psi(z)$ and $\Psi(e^{j\theta}) \geq 0$, for all $\theta \in [0, 2\pi)$. We consider only spectral matrices which are generically *nonsingular*, in the sense that $\det(\Psi(z)) \not\equiv 0$. Spectral matrices arise naturally in the description of stochastic processes, in the formulation of linear control and filtering problems and in the discrete bounded-real lemma.

[1] The authors wish to acknowledge the funding of the activities of the Cooperative Research Centre for Robust and Adaptive Systems by the Australian Commonwealth Government under the Cooperative Research Centres Program.

It is well known that the construction of spectral factors is central to the solution of the abovementioned problems. A *spectral factor* $\Omega(z)$ of $\Psi(z)$ is a real rational matrix-valued function of the complex variable z which satisfies $\Omega^T(z^{-1})\Omega(z) = \Psi(z)$. If, in addition, $\Omega^{-1}(z)$ exists and is analytic when $|z| > 1$, $\Omega(z)$ is called a *minimum phase* spectral factor.

It is also well known[3,11] that if $\Psi(z)$ is nonsingular, there exists a spectral decomposition of the form

$$\Psi(z) = \Theta^T(z^{-1})N\,\Theta(z) \tag{1}$$

where N is a positive definite symmetric matrix and $\Theta(z)$ is a square real rational transfer function matrix which is invertible, satisfies $\lim_{z\to\infty}\Theta(z) = I$ and, along with its inverse, is analytic when $|z| > 1$. Hence a minimum-phase spectral factor of $\Psi(z)$ can be constructed as $\Omega(z) = N^{\frac{1}{2}}\Theta(z)$.

In this paper, we consider a class of nonsingular spectral matrices, having a state-space realization of the form

$$\Psi(z) = U + G^T(z^{-1}I - F^T)^{-1}V(zI - F)^{-1}G \tag{2}$$

where each constant matrix is real and where the following assumptions hold:

A.1 (F, G) is stabilizable.

A.2 $V = V^T$, $U = U^T$ and U is nonsingular.

In fact, by applying appropriate preliminary transformations, most nonsingular spectral factorization problems can be treated via a spectral matrix of the above form.

It should be emphasized that no assumptions regarding the sign-definiteness of either V or U have been made. In H_2 linear-quadratic optimal control and Kalman filtering, spectral matrices arise which are special cases of the above class in which generally $U > 0$ and $V \geq 0$. Clearly these conditions preclude the possibility that the spectral matrix realization has unit circle transmission zeros. Note, however, that unit circle *invariant* zeros can appear if the realization of the spectral matrix is non-minimal: for example if $(F, V^{\frac{1}{2}})$ has unobservable unit-circle modes,[4,7,8] these become a subset of the invariant zeros of the spectral matrix realization.

Spectral matrices for which unit circle *transmission* zeros can occur arise in discrete time spectral factorization[1] and in the discrete time version of the bounded-real lemma,[10] which is relevant in the H_∞ control problem. Recall that a discrete-time transfer function matrix $L(z)$ is called *bounded real* if all poles of $L(z)$ are inside the unit circle and $\|L\|_\infty \leq 1$. Consider the spectral matrix $\Psi(z)$ defined by $\Psi(z) = I - L^T(z^{-1})L(z)$. Observe that with the state-space realization $L(z) = H_L(zI - F_L)^{-1}G_L$, $\Psi(z)$ is of the standard form given in Eq. 2 with $U = I$, $F = F_L$, $G = G_L$ and $V = -H_L^T H_L$. Should $\sigma_{\max}(L(e^{j\theta^*})) = 1$ for some θ^*, then $\Psi(e^{j\theta^*})$ loses rank at that point, corresponding to a transmission zero.

Real symmetric solutions of the discrete time algebraic Riccati equation (ARE)

$$\Phi = F^T\left(\Phi - \Phi G(U + G^T\Phi G)^{-1}G^T\Phi\right)F + V \tag{3}$$

enable the state-space construction of spectral factors of $\Psi(z)$. Such equations have been studied in many contexts including spectral factorization,[1] and infinite-horizon control and filtering problems.[2,3] It can be demonstrated using Eq. 3 that Eq. 1 is satisfied with the definitions $N = U + G^T \Phi G$ and $\Theta(z) = I + N^{-1} G^T \Phi F(zI - F)^{-1} G$. The resulting spectral factor $\Omega(z) = N^{\frac{1}{2}} \Theta(z)$ has an inverse

$$\Omega^{-1}(z) = \left(I - N^{-1} G^T \Phi F(zI - \hat{F})^{-1} G\right) N^{-\frac{1}{2}} \tag{4}$$

where $\hat{F}$ is the *closed-loop matrix* given by $\hat{F} = \left(I - GN^{-1}G^T\Phi\right) F$. A solution Φ of Eq. 3 is said to be *strong* if $\hat{F}$ has all eigenvalues either inside or on the unit circle. Note that the eigenvalues of $\hat{F}$ are also the invariant zeros of $\Omega(z)$ and thus spectral factors constructed from strong solutions of Eq. 3 have the minimum phase property.

Remark: It is not the purpose of the present paper to address the question of when a unique strong solution Φ of Eq. 3 exists. Henceforth we assume that such a solution exists for the realization of the spectral matrix at hand. $\qquad\square$

We now consider the Riccati difference equation (RDE) associated with Eq. 3

$$\Phi_{k+1} = F^T \left(\Phi_k - \Phi_k G(U + G^T \Phi_k G)^{-1} G^T \Phi_k\right) F + V \tag{5}$$

where this equation has some real symmetric initial condition Φ_0. The main result of this paper (which follows immediately) presents conditions under which iterates of the RDE Φ_k ($k \in \{0, 1, 2, \ldots\}$) converge to the strong solution of Eq. 3 and describes the convergence rate when the associated spectral matrix has unit circle invariant zeros.

1.2 Main Result.

Firstly, it is demonstrated that RDE convergence results previously established for linear-quadratic control and Kalman filtering problems[8] and for spectral factorization[1] hold for any spectral matrix of the form in Eq. 2 under assumptions **A.1** and **A.2** . The spectral matrix may have unit circle invariant zeros which arise due to non-minimal modes in its realization, transmission zeros, or any combination of these. Secondly, and most importantly, the fine structure of the Riccati difference equation iterates is investigated. This leads to new results concerning the *rates* at which the iterates of Eq. 5 converge to the strong solution of Eq. 3.

Theorem 1.1 *Given a realization as in Eq. 2 of a discrete time spectral matrix $\Psi(z)$ satisfying assumptions **A.1** and **A.2** , along with the strong solution Φ of the associated algebraic Riccati equation Eq. 3, then provided $\Phi_0 \geq \Phi$, iterates of Eq. 5 have the following properties:*

1. $\Phi_k \geq \Phi$.

2. $\lim_{k \to \infty} \Phi_k = \Phi$.

3. *If $\Psi(z)$ has an invariant zero on the unit circle, then there exist constants κ_1, κ_2 (depending on the realization of $\Psi(\cdot)$ and on Φ_0) with $\kappa_1 \geq \kappa_2 > 0$ such that:*

   a) For all $\epsilon > 0$, there exists a k_ϵ such that when $k \geq k_\epsilon$

$$\lambda_{\max}(\Phi_k - \Phi) \leq \frac{\kappa_1 + \epsilon}{k}. \tag{6}$$

   b) When $\Phi_0 > \Phi$, there exists a k_0 such that when $k \geq k_0$

$$\lambda_{\max}(\Phi_k - \Phi) \geq \frac{\kappa_2}{k}. \tag{7}$$

A proof of this result is delayed until the final section of the paper. Item *3 a)* of this theorem reports a worst-case $\frac{1}{k}$ convergence rate in the case of unit-circle invariant zeros. Item *3 b)* says that, with the exclusion of (non-generic) cases where $\Phi_0 - \Phi$ is singular, the convergence rate can be no better than $\frac{1}{k}$.

Remark: In cases where the spectral matrix has no invariant zeros on the unit circle, an exponential convergence rate has been reported:[1,6] there exist constants A, κ_3 such that $1 > \kappa_3 > 0$ and $\lambda_{\max}(\Phi_k - \Phi) \leq A\kappa_3^k$. □

2 Preliminary Results.

2.1 Notation.

Let O_m denote the $m \times m$ zero matrix and I_m the $m \times m$ identity matrix. Given a matrix M, let $\{\sigma_i(M)\}$ denote the singular values of M and $\sigma_{\max}$ be the largest of these; if M is square, denote its eigenvalues as $\{\lambda_i(M)\}$. Suppose M has an even number of rows and columns, consisting of a matrix of 2×2 matrix sub-blocks; for convenience we let $[M]_{ij}$ denote the $(i,j)^{\text{th}}$ 2×2 sub-block of M.

Given $f(l)$ and $g(l)$, both scalar functions of an integer variable l, we say $g(l) = \mathcal{O}(f(l))$ if there exists a constant $\kappa < \infty$ such that $\lim_{l \to \infty} \frac{|g(l)|}{|f(l)|} = \kappa$. Given $U(l)$, a square matrix-valued function of l, we say that $U(l) = \mathcal{O}(f(l))$ if $\sigma_{\max}(U(l)) = \mathcal{O}(f(l))$. Note that this definition has the following property: If $U(l)$ is such that each $(U(l))_{mn} = \mathcal{O}(f(l))$ or each $[U(l)]_{ij} = \mathcal{O}(f(l))$, then $U(l) = \mathcal{O}(f(l))$.

Real Jordan Form.

The following summarizes standard results concerning the real Jordan decomposition.[12] Any real square matrix B can be expressed as $B = TAT^{-1}$ where

$$A = \text{diag}\{A_1, \ \cdots \ , A_p\} \tag{8}$$

and p is the number of real Jordan blocks. For each $q \in \{1, \ldots, p\}$, A_q has one of the two forms described below.

In the first form, $A_q \in \mathbb{R}^{2n_q \times 2n_q}$ where $n_q \geq 1$ and

$$A_q = \begin{pmatrix} \Lambda_q & & & \\ I_2 & \Lambda_q & & \\ & \ddots & \ddots & \\ & & \Lambda_q & \\ & & I_2 & \Lambda_q \end{pmatrix} \tag{9}$$

where

$$\Lambda_q = \begin{pmatrix} \sigma_q & \omega_q \\ -\omega_q & \sigma_q \end{pmatrix}. \tag{10}$$

In this case, $\sigma_q \pm j\omega_q$ is a pair of complex conjugate eigenvalues of B. If λ_q is a real eigenvalue of B, then in the second form, $A_q \in \mathbb{R}^{n_q \times n_q}$ where $n_q \geq 1$ and

$$A_q = \begin{pmatrix} \lambda_q & & & \\ 1 & \lambda_q & & \\ & \ddots & \ddots & \\ & & \lambda_q & \\ & & 1 & \lambda_q \end{pmatrix}. \tag{11}$$

2.2 Asymptotic Behaviour of a Linear Matrix Difference Equation with Jordan Structure.

Due to its significance in describing the convergence behaviour of the Riccati difference equation to its strong solution, we examine the behaviour of the following linear matrix difference equation with initial condition $X_q(0) = 0$:

$$X_q(k+1) = A_q X_q(k) A_q^T + I. \tag{12}$$

In the first instance, we assume that A_q has the first real Jordan form as described in Eq. 9 which corresponds to a complex conjugate pair of eigenvalues. Similar arguments to those which follow for this case can be used in the second case and are thus not treated here. It follows by direct iteration of Eq. 12 that for $k \geq 1$, $X_q(k) = S_q(k)$ where

$$S_q(k) = \sum_{l=0}^{k-1} A_q^l (A_q^T)^l. \tag{13}$$

We focus here on the case where the Jordan block A_q corresponds to a complex conjugate pair of *unit circle* eigenvalues. It will be shown in the next section that it is the behaviour of iterates of this type that are the most important in establishing the convergence rate of RDEs associated with spectral matrices which have unit circle invariant zeros.

316

Lemma 2.1 *Let A_q be a Jordan block of size $2n_q \times 2n_q$ which has the form given in Eq. 9, corresponding to a complex conjugate pair of unit circle eigenvalues.*
For sufficiently large l, one has the following identity:

$$A_q^l (A_q^T)^l = C(l) + P(l) = C(l)(I + \mathcal{O}(l^{-1})) \tag{14}$$

where

$$C(l) = \begin{pmatrix} I_2 & l\Lambda_q & \frac{l^2}{2}\Lambda_q^2 & \cdots & \frac{l^{n_q-1}}{(n_q-1)!}\Lambda_q^{n_q-1} \\ l(\Lambda_q^T) & l^2 I_2 & \frac{l^3}{2}\Lambda_q & & \vdots \\ \frac{l^2}{2}(\Lambda_q^T)^2 & \frac{l^3}{2}(\Lambda_q^T) & \frac{l^4}{4}I_2 & & \\ \vdots & & & \ddots & \\ \frac{l^{n_q-1}}{(n_q-1)!}(\Lambda_q^T)^{n_q-1} & & & & \frac{l^{2n_q-2}}{(n_q-1)!(n_q-1)!}I_2 \end{pmatrix} \tag{15}$$

or equivalently

$$[C(l)]_{ij} = \begin{cases} \frac{l^{i+j-2}}{(i-1)!(j-1)!}\Lambda_q^{j-i} & \text{if } i \leq j \\ [C(l)]_{ji}^T & \text{if } i > j \end{cases}, \tag{16}$$

and

$$[P(l)]_{ij} = \begin{cases} \mathcal{O}(l^{i+j-3}) & \text{if } i+j \geq 3 \\ O_2 & \text{if } i = j = 1 \end{cases}. \tag{17}$$

Proof: A straightforward calculation based on approximation of each 2×2 sub-block of $A_q^l(A_q^T)^l$ leads to Eq. 14. $\qquad \square$

Lemma 2.2 *Let A_q be a Jordan block of size $2n_q \times 2n_q$ which has the form given in Eq. 9, corresponding to a complex conjugate pair of unit circle eigenvalues.*
With $S_q(k)$ defined in Eq. 13, the following identity holds:

$$S_q(k) = D(k) + G(k) = D(k)(I + \mathcal{O}(k^{-1})) \tag{18}$$

where

$$[D(k)]_{ij} = \begin{cases} \frac{k^{i+j-1}}{(i-1)!(j-1)!(i+j-1)}(\Lambda_q^T)^{i-j} & \text{if } i \geq j \\ [D(k)]_{ji}^T & \text{if } i < j \end{cases} \tag{19}$$

and

$$[G(k)]_{ij} = \begin{cases} \mathcal{O}(k^{i+j-2}) & \text{if } i+j \geq 3 \\ O_2 & \text{if } i = j = 1 \end{cases}. \tag{20}$$

Moreover, there exists a symmetric matrix $\Theta \in \mathbb{R}^{2n_q \times 2n_q}$ such that $\Theta > 0$ and

$$D(k) = k H^T(k) \Theta H(k) \tag{21}$$

where

$$H(k) = \text{diag}\{I_2, \ k\Lambda_q, \ \cdots, \ k^{(n_q-1)}\Lambda_q^{(n_q-1)}\}. \tag{22}$$

Proof: Follows by application of the results in Lemma 2.1 plus further 2×2 block approximations. $\qquad\square$

Lemma 2.3 *Let A_q be a Jordan block of the form given in Eq. 9, having size $2n_q \times 2n_q$, which corresponds to a complex conjugate pair of eigenvalues on the unit circle. Let $S_q(k)$ be defined as in Eq. 13. Then there exists a constant $\zeta > 0$ such that*

1.

$$\lambda_{\min}(S_q(k)) \leq k\frac{2n_q}{\zeta}, \tag{23}$$

2. for all $\epsilon > 0$, there exists a constant k_ϵ such that $k > k_\epsilon$ implies that

$$\lambda_{\min}(S_q(k)) \geq \frac{k}{\zeta + \epsilon}. \tag{24}$$

Proof: Recall from Lemma 2.2 that $S_q(k) = D(k) + G(k)$ with $D(k)$ given in (21). Since $H(k)$ is invertible, it follows that

$$S_q(k) = kH^T(k)\{\Theta + W(k)\}H(k) \tag{25}$$

where $W(k) = H^{-T}(k)\frac{G(k)}{k}H^{-1}(k)$ from which it can be verified fairly simply that $W(k) = \mathcal{O}(k^{-1})$.

That $S_q(k)$ is always a positive definite matrix can be seen from its definition in Eq. 13. In order to describe its eigenvalues as a function of k, we investigate those of $S_q^{-1}(k)$. Recall that for any positive definite matrix M, if $\lambda_{\max}(M)$ is the maximum eigenvalue of M, then the minimum eigenvalue of M^{-1} is $\lambda_{\min}(M^{-1}) = \lambda_{\max}^{-1}(M)$.

It follows from Eq. 25 that $\Theta + W(k)$ is positive definite. Observe that $(\Theta + W(k))^{-1} = \Theta^{-\frac{1}{2}}(I + \mathcal{O}(k^{-1}))^{-1}\Theta^{-\frac{1}{2}}$. Now $(I + \mathcal{O}(k^{-1}))^{-1} = I + \mathcal{O}(k^{-1})$ and hence

$$(\Theta + W(k))^{-1} = \Theta^{-\frac{1}{2}}(I + \mathcal{O}(k^{-1}))\Theta^{-\frac{1}{2}} = \Theta^{-1} + \mathcal{O}(k^{-1}). \tag{26}$$

Inverting Eq. 25 and employing Eq. 26 reveals that

$$kS_q^{-1}(k) = H^{-1}(k)\{\Theta^{-1} + \mathcal{O}(k^{-1})\}H^{-T}(k) \tag{27}$$

and since $H^{-1}(k) = \mathcal{O}(1)$,

$$kS_q^{-1}(k) = H^{-1}(k)\Theta^{-1}H^{-T}(k) + \mathcal{O}(k^{-1}). \tag{28}$$

With M a nonnegative definite matrix of dimension n_M, recall the standard identity $\lambda_{\max}(M) \leq \text{trace}\{M\} \leq n_M\lambda_{\max}(M)$. Applying this result to Eq. 28 reveals that

$$\lambda_{\max}(kS_q^{-1}(k)) \leq \text{trace}\left\{H^{-1}(k)\Theta^{-1}H^{-T}(k)\right\} + \mathcal{O}(k^{-1}) \leq 2n_q\lambda_{\max}(kS_q^{-1}(k)). \tag{29}$$

318

Observe that

$$\left[H^{-1}(k)\Theta^{-1}H^{-T}(k)\right]_{ij} = \frac{1}{k^{2-i-j}}\Lambda_q^{1-i}\left[\Theta^{-1}\right]_{ij}(\Lambda_q^T)^{1-j} \qquad (30)$$

and as a result that

$$\text{trace}\left\{H^{-1}(k)\Theta^{-1}H^{-T}(k)\right\} = \text{trace}\left\{\left[\Theta^{-1}\right]_{11}\right\} + \mathcal{O}(k^{-2}). \qquad (31)$$

With the definition $\zeta = \text{trace}\left\{[\Theta^{-1}]_{11}\right\}$, note firstly from Eq. 29 and Eq. 31 that

$$\lambda_{\max}(kS_q^{-1}(k)) \leq \zeta + \mathcal{O}(k^{-1}) \qquad (32)$$

and secondly that

$$\lambda_{\max}(kS_q^{-1}(k)) \geq \frac{\zeta}{2n_q}. \qquad (33)$$

The stated results follow immediately from Eq. 32 and Eq. 33. $\qquad\square$

3   Riccati Difference Equation Convergence.

3.1   Comparison Theorem for RDE Iterates.

We now state a minor extension of a well-known result which describes the way in which Riccati difference equation iterates behave under perturbations to the initial condition Φ_0. Having established this result, we will find it has several applications in the proof of RDE convergence.

Lemma 3.1 *Let the sequences $\{\Phi_k^1\}$ and $\{\Phi_k^2\}$ be defined by application of the Riccati Difference Equation, Eq. 5, with initial conditions Φ_0^1 and Φ_0^2 respectively. Define $\tilde{\Phi}_k = \Phi_k^2 - \Phi_k^1$. Then*

1. *The following recursions hold for all $k \geq 0$:*

$$\tilde{\Phi}_{k+1} = (\hat{F}_k^1)^T \tilde{\Phi}_k \hat{F}_k^1 - (\hat{F}_k^1)^T \tilde{\Phi}_k G(G^T\tilde{\Phi}_k G + G^T\Phi_k^1 G + U)^{-1}G^T\tilde{\Phi}_k\hat{F}_k^1 \quad (34)$$
$$\tilde{\Phi}_{k+1} = (\hat{F}_k^2)^T \tilde{\Phi}_k \hat{F}_k^2 + (\hat{F}_k^2)^T \tilde{\Phi}_k G(G^T\Phi_k^1 G + U)^{-1}G^T\tilde{\Phi}_k\hat{F}_k^2 \qquad (35)$$

   where
$$\hat{F}_k^1 = (I - G(G^T\Phi_k^1 G + U)^{-1}G^T\Phi_k^1)F$$
$$\hat{F}_k^2 = (I - G(G^T\Phi_k^2 G + U)^{-1}G^T\Phi_k^2)F.$$

2. *Suppose the RDE, Eq. 5, is associated with a nonsingular spectral matrix $\Psi(z)$ and the ARE, Eq. 3, has a strong solution Φ. Suppose also that both $\Phi_0^1 \geq \Phi$ and $\Phi_0^2 \geq \Phi$. Then if $\tilde{\Phi}_0 \geq 0$ it follows that $\tilde{\Phi}_k \geq 0$ for all $k \geq 0$.*

Proof: A more general version of the first difference equation in item *1* which also accounts for perturbations in V is well known.[9] The second difference equation can be obtained from the first simply by first reversing the superscripts and then multiplying the equation by -1.[5]

Item *2* has been established in the nonnegative definite cost case for LQ control and Kalman filtering.[5] A generalization of this result to the broader class of spectral matrices we consider here follows from the discussion below.

Observe first that with $\Phi_0^1 = \Phi$, then $\Phi_k^1 = \Phi$ for all subsequent k. By hypothesis, $\Phi_0^2 \geq \Phi$ and hence $\tilde{\Phi}_0 \geq 0$. Next observe from Eq. 35 that since $G^T \Phi G + U > 0$ (which follows from the assumed spectral property), it follows that $\Phi_k^2 \geq \Phi$ for all subsequent k.

Suppose now that one is given any $\Phi_0^1 \geq \Phi$. It follows from reversing subscripts in the argument immediately above that $\Phi_k^1 \geq \Phi$ for all subsequent k. Since $G^T \Phi G + U > 0$, it follows that $G^T \Phi_k^1 G + U > 0$ which together with Eq. 35 implies that $\tilde{\Phi}_k \geq 0$ for all subsequent k. $\qquad \Box$

3.2 A Preliminary Convergence Result.

In this subsection, a weakened version of the main theorem is proven in Lemma 3.2. In the following subsection, we show how the additional assumptions introduced in Lemma 3.2 may be relaxed.

Properties *1* and *2* in the following lemma have been stated in the literature.[1,7] One of the first observations of the convergence rate stated in item *3 a)* was in the context of a Kalman filtering example[4] in which the plant model has an identity state mapping, with no process noise and observations corrupted by Gaussian white noise. The worst-case convergence rate given in item *3 a)* has been stated[1] for a spectral factorization problem. Full proofs which spell out the mechanism and rate of convergence do not seem to be available in the literature, however. We now review the first steps towards a proof of the convergence result[1,7] and then present a novel and nontrivial completion of the proof which addresses the question of convergence rate.

Lemma 3.2 *Consider a realization Eq. 2 of a nonsingular discrete time spectral matrix* $\Psi(z)$ *which, in addition to assumptions* **A.1** *and* **A.2** *, satisfies the following two assumptions:*

A.3 (F, G) *is controllable.*

A.4 F *is nonsingular.*

Let Φ *be the strong solution of the associated algebraic Riccati equation Eq. 3. Then provided* $\Phi_0 > \Phi$*, iterates of Eq. 5 have the following properties:*

1. $\Phi_k > \Phi$.

2. $\lim_{k \to \infty} \Phi_k = \Phi$.

3. *If $\Psi(z)$ has an invariant zero on the unit circle, then there exist constants δ_1, δ_2 (depending on the realization of $\Psi(\cdot)$ and on Φ_0) with $\delta_1 \geq \delta_2 > 0$ such that:*

a) For all $\epsilon > 0$, there exists a k_ϵ such that when $k \geq k_\epsilon$

$$\lambda_{\max}(\Phi_k - \Phi) \leq \frac{\delta_1 + \epsilon}{k}. \tag{36}$$

b) There exists a k_0 such that when $k \geq k_0$

$$\lambda_{\max}(\Phi_k - \Phi) \geq \frac{\delta_2}{k}. \tag{37}$$

Proof: With the definitions $\Delta_k = \Phi_k - \Phi$ and $\hat{F} = (I - G(G^T \Phi G + U)^{-1} G^T \Phi)F$, one can apply Lemma 3.1 to obtain

$$\Delta_{k+1} = \hat{F}^T \Delta_k \hat{F} - \hat{F}^T \Delta_k G(G^T \Delta_k G + G^T \Phi G + U)^{-1} G^T \Delta_k \hat{F} . \tag{38}$$

Suppose $\Delta_k > 0$. Then since $N = U + G^T \Phi G > 0$, the so-called matrix inversion lemma may be applied to Eq. 38, revealing that

$$\Delta_{k+1}^{-1} = \hat{F}^{-1} \Delta_k^{-1} \hat{F}^{-T} + \hat{F}^{-1} G N^{-1} G^T \hat{F}^{-T}. \tag{39}$$

Invertibility of $\hat{F}$ is a consequence of the invertibility of F and N; application of the matrix inversion lemma yields $\hat{F}^{-1} = F^{-1}(I + GU^{-1}G^T\Phi)$. Since $N^{-1} > 0$, Eq. 39 implies that $\Delta_{k+1} > 0$ whenever $\Delta_k > 0$. Thus our assumption that $\Delta_0 > 0$ ensures $\Delta_k > 0$ for all $k \geq 0$. This establishes item *1* in the Lemma statement.

The proof of the convergence of iterates of Eq. 38 to zero is based on the following observation:[1,7] $\lambda_{\min}(\Delta_k^{-1}) \to \infty$ implies $\lambda_{\max}(\Delta_k) \to 0$. An explicit account of the divergent behaviour of $\lambda_{\min}(\Delta_k^{-1}) \to \infty$ is given which draws upon the preliminary results obtained in Section 2.

Observe that $\hat{F}$ is of the form $\hat{F} = F - GL$ (where $L = (G^T \Phi G + U)^{-1} G^T \Phi F$). It is a well known result that controllability of the pair (F, G) guarantees controllability of $(\hat{F}, G)$, which in turn implies the controllability of $(\hat{F}^{-1}, G)$. Since $(\hat{F}^{-1}, G)$ is a controllable pair, so is the pair $(\hat{F}^{-1}, GN^{-\frac{1}{2}})$. Hence the controllability Gramian for the latter pair satisfies

$$\hat{W} = \sum_{j=0}^{n-1} \hat{F}^{-j} G N^{-1} G^T (\hat{F}^T)^{-j} > 0,$$

with n the dimension of the state space.

Iteration of the identity Eq. 39 reveals that

$$\Delta_{k+n}^{-1} = \hat{F}^{-n} \Delta_k^{-1} (\hat{F}^T)^{-n} + \hat{W}. \tag{40}$$

With $\hat{A} = \hat{F}^{-n}$ and $\hat{X}_j = \Delta_{jn}^{-1}$, Eq. 40 reads

$$\hat{X}_{j+1} = \hat{A}\hat{X}_j\hat{A}^T + \hat{W}. \tag{41}$$

Let $\hat{A}$ have the real Jordan decomposition

$$\hat{A} = TAT^{-1} \tag{42}$$

where A has the structure described in Eq. 8.

Since Φ is a strong solution of the algebraic Riccati equation, we know that $|\lambda_i(\hat{F})| \leq 1$. It can be easily checked that $|\lambda_i(\hat{A})| \geq 1$ is a consequence of this.

With T the transformation in Eq. 42 and $W = T^{-1}\hat{W}T^{-T} > 0$, observe that one can express $\hat{X}_j$ as $\hat{X}_j = TX_jT^T$, where X_j are iterates defined by the equation

$$X_{j+1} \;=\; AX_jA^T + W \; , \; X_0 = T^{-1}\hat{X}_0T^{-T}. \tag{43}$$

We next define a sequence of matrices $\{Y_j\}$ which *under-bounds* $\{X_j\}$.

$$Y_{j+1} \;=\; AY_jA^T + \lambda_{\min}(W)I \; , \; Y_0 = 0. \tag{44}$$

It is trivial to show by induction that $Y_j \leq X_j$ for all j. Thus if we can show that $\{Y_j\}$ diverges, divergence of $\{X_j\}$ and $\{\hat{X}_j\}$ follow.

A closed-form expression for Y_j can be found immediately:

$$Y_j \;=\; \lambda_{\min}(W)\left\{\sum_{l=0}^{j-1} A^l(A^T)^l\right\} \tag{45}$$

$$=\; \lambda_{\min}(W)\mathrm{diag}\{S_1(j), \; \cdots \; , S_p(j)\} \tag{46}$$

where $S_q(j)$ is defined in Eq. 13.

Thus the set of eigenvalues of Y_k is simply the union of all the eigenvalues of $S_q(k)$ for all q. A straightforward but lengthy argument employing item *2* of Lemma 2.3 then establishes item *3 a)* of the lemma.

Provided $\mu \geq \lambda_{\max}(X_1)$, the sequence of matrices $\{Z_j\}$ defined below *over-bounds* X_j:

$$Z_{j+1} \;=\; AZ_jA^T + \mu I \; , \; Z_0 = 0. \tag{47}$$

It is trivial to show by induction that $Z_j \geq X_j$ for all $j \geq 1$.

In an identical manner to that employed in investigating Y_j, one can deduce the following expression for Z_j: $Z_j = \mu\mathrm{diag}\{S_1(j), \; \cdots \; , S_p(j)\}$. A straightforward but lengthy argument employing item *1* of Lemma 2.3 yields part *3 b)* of the lemma. $\square$

3.3 Proof of the Main Theorem.

Having established convergence and the associated rate under the preliminary assumptions **A.3** , **A.4** and $\Phi_0 > 0$ of Lemma 3.2, we now successively relax each of these assumptions to give Theorem 1.1. It has been shown[8] (albeit by different means to those proposed here) that in the case of Kalman filtering problems, these assumptions can be relaxed to extend previously established convergence results.[7] Here we consider the more general class of spectral matrices given in Eq. 2 and present a proof of convergence which as well as relaxing these assumptions, also enables statements to be made concerning the *convergence rate* of the RDE.

Relaxing assumption **A.3** *(that (F, G) is controllable.)*
This assumption has previously been relaxed[8] via a sequence of perturbations on the original problem, each of which has (F, G) controllable. The emphasis in the present paper is to investigate the *structure* of RDE iterates associated with the stable and uncontrollable modes of (F, G). These observations give rise to statements concerning the convergence *rate*.

We assume now that (F, G) is stabilizable and that, without loss of generality, $F = \begin{pmatrix} F_{11} & F_{12} \\ 0 & F_{22} \end{pmatrix}$ and $G = \begin{pmatrix} G_1 \\ 0 \end{pmatrix}$, where $|\lambda_i(F_{22})| < 1$ and (F_{11}, G_1) is a controllable pair. We partition Φ_k and V conformally: $\Phi_k = \begin{pmatrix} \Phi_{11}^k & \Phi_{12}^k \\ \Phi_{12}^{k~T} & \Phi_{22}^k \end{pmatrix}$ and $V = \begin{pmatrix} V_{11} & V_{12} \\ V_{12}^T & V_{22} \end{pmatrix}$. Expression of Eq. 5 in terms of this partitioning reveals that Φ_{11}^k satisfies the Riccati difference equation

$$\Phi_{11}^{k+1} = F_{11}^T \left(\Phi_{11}^k - \Phi_{11}^k G_1 (U + G_1^T \Phi_{11}^k G_1)^{-1} G_1^T \Phi_{11}^k \right) F_{11} + V_{11}. \tag{48}$$

With conformal partitioning of Φ (the strong solution of Eq. 3), it can be readily shown that Φ_{11} is a strong solution of the algebraic equation

$$\Phi_{11} = F_{11}^T \left(\Phi_{11} - \Phi_{11} G_1 (U + G_1^T \Phi_{11} G_1)^{-1} G_1^T \Phi_{11} \right) F_{11} + V_{11} \tag{49}$$

in the sense that the following matrix only has eigenvalues with magnitude less than or equal to unity: $\hat{F}_{11} = \left(I - G_1 N_{11}^{-1} G_1^T \Phi_{11} \right) F_{11}$ where $N_{11} = N = U + G_1^T \Phi_{11} G_1$.

Observe that in fact $\Psi(z) = U + G_1^T (z^{-1} I - F_1^T)^{-1} V_{11} (z I - F_1)^{-1} G_1$. Recall also that for the moment, we maintain the assumption that F is invertible, from which it follows that F_1 is also invertible. Since we also assume that $\Phi_0 > \Phi$ and therefore that $\Delta_0 > 0$, it follows that $\Delta_{11}^0 > 0$. Since (F_{11}, G_1) is controllable, we can apply Lemma 3.2 to deduce that iterates of the reduced-order RDE Eq. 48 satisfy

$$\Delta_{11}^k = \mathcal{O}(\frac{1}{k}). \tag{50}$$

It also follows from Eq. 5 that the partitions Φ_{12}^k of the iterates Φ_k satisfy

$$\Phi_{12}^{k+1} = (\hat{F}_{11}^k)^T \Phi_{12}^k F_{22} + W_{12}^k \tag{51}$$

where

$$\hat{F}_{11}^k = \left(I - G_1 (N_{11}^k)^{-1} G_1^T \Phi_{11}^k \right) F_{11} \tag{52}$$

$$N_{11}^k = U + G_1^T \Phi_{11}^k G_1 \tag{53}$$

$$W_{12}^k = \hat{F}_{11}^k \Phi_{11}^k F_{12} + V_{12}. \tag{54}$$

Lemma 3.3 *Let $\{\Upsilon_k\}$ be a bounded sequence of matrices defined for $k \geq 0$. Consider the linear matrix difference equation with (possibly non-square) iterates Ξ_k having a finite initial condition Ξ_0:*

$$\Xi_{k+1} = A_k \Xi_k B_k + \Upsilon_k. \tag{55}$$

Suppose the (square) matrix sequences $\{A_k\}$ and $\{B_k\}$ are such that $A_k \to A$ and $B_k \to B$ where $\|\lambda_i(A)\lambda_j(B)\| < 1$ for all i and j. Then if $\Upsilon_k = \mathcal{O}(\frac{1}{k})$, it follows that

$$\Xi_k = \mathcal{O}(\frac{1}{k}). \tag{56}$$

Proof: A number of standard stability results for difference equations can be applied to show this result. $\qquad\square$

Recall from Eq. 50 that $\Phi_{11}^k = \Phi_{11} + \mathcal{O}(\frac{1}{k})$. It follows from Eq. 53 that $N_{11}^k = N_{11} + \mathcal{O}(\frac{1}{k})$, from Eq. 52 that $\hat{F}_{11}^k = \hat{F}_{11} + \mathcal{O}(\frac{1}{k})$ and hence that $W_{12}^k = W_{12} + \mathcal{O}(\frac{1}{k})$ where $W_{12} = \hat{F}_{11}\Phi_{11}F_{12} + V_{12}$.

Observe that by hypothesis there exists a solution Φ_{12} of the algebraic equation

$$\Phi_{12} = \hat{F}_{11}^T \Phi_{12} F_{22} + W_{12}. \tag{57}$$

Subtracting this equation from Eq. 51 and simultaneously adding and subtracting the term $(\hat{F}_{11}^k)^T \Phi_{12} F_{22}$ yields the equation

$$\Delta_{12}^{k+1} = (\hat{F}_{11}^k)^T \Delta_{12}^k F_{22} + (\hat{F}_{11}^k - \hat{F}_{11})^T \Phi_{12} F_{22} + W_{12}^k - W_{12}. \tag{58}$$

Recall that $\hat{F}_{11}$ has all eigenvalues in the closed unit circle. Observe that F_{22} is stable. It follows that $\|\lambda_i(\hat{F}_{11})\lambda_j(F_{22})\| < 1$. We now identify Ξ_k with Δ_{12}^k, A_k with $(\hat{F}_{11}^k)^T$, B_k with F_{22} and Υ_k with the remaining terms in Eq. 58, which can be easily shown to be $\mathcal{O}(\frac{1}{k})$. We now apply Lemma 3.3 to Eq. 58 to conclude that

$$\Delta_{12}^k = \mathcal{O}(\frac{1}{k}). \tag{59}$$

Note that examination of the $(2,2)$ partition of Eq. 5 reveals the following iteration:

$$\Phi_{22}^{k+1} = F_{22}^T \Phi_{22}^k F_{22} + S_{22}^k \tag{60}$$

where

$$
\begin{aligned}
S_{22}^k &= F_{12}^T \Phi_{11}^k \hat{F}_{12}^k + (\hat{F}_{12}^k)^T \Phi_{12}^k F_{22} \\
&\quad + F_{22}^T (\Phi_{12}^k)^T \left(\hat{F}_{12}^k - G_1 N_{11}^{-k} G_1^T \Phi_{12}^k F_{22} \right) + V_{22}
\end{aligned} \tag{61}
$$

$$\hat{F}_{12}^k = \left(I - G_1 (N_{11}^k)^{-1} G_1^T \Phi_{11}^k \right) F_{12}. \tag{62}$$

Recall that by hypothesis, there exists a solution Φ_{22} of the equation

$$\Phi_{22} = F_{22}^T \Phi_{22} F_{22} + S_{22} \tag{63}$$

where S_{22} is given by taking the limit of Eq. 61. Subtracting this equation from Eq. 60 yields the equation

$$\Delta_{22}^{k+1} = F_{22}^T \Delta_{22}^k F_{22} + S_{22}^k - S_{22}. \tag{64}$$

From Eq. 50 and Eq. 59 it follows that $S_{22}^k = S_{22} + \mathcal{O}(\frac{1}{k})$. Since F_{22} is stable, we can apply Lemma 3.3 with $A_k = B_k = F_{22}$ and $\Upsilon_k = S_{22}^k - S_{22}$ to conclude that $\Delta_{22}^k = \mathcal{O}(\frac{1}{k})$.

Since $\Delta_{ij}^k = \mathcal{O}(\frac{1}{k})$ for each partition of Φ_k, it follows that $\Delta_k = \mathcal{O}(\frac{1}{k})$. This establishes the worst-case convergence result in item $3\ a)$ of Theorem 1.1.

We now establish the best-case result in item $3\ b)$. Choose any Φ_0 such that $\Delta_0 > 0$ and note therefore that $\Delta_{11}^0 > 0$. Recall that the invariant zeros of the minimum phase spectral factor $\Omega(z)$ are the eigenvalues of $\hat{F}$. It is easy to check that in the new basis, $\hat{F}$ has diagonal blocks $\hat{F}_{11}$ and F_{22}. Since F_{22} is stable, all unit circle invariant zeros of $\Psi(z)$ are eigenvalues of $\hat{F}_{11}$. Note that we can apply item $3\ b)$ of Lemma 3.2 to deduce that there exists a k_0 such that when $k \geq k_0$, $\lambda_{\max}(\Phi_{11}^k - \Phi_{11}) \geq \frac{\delta_2}{k}$. Note now that the positive definite matrix Δ_{11}^k is a partition of the larger positive definite matrix Δ_k and hence that $\lambda_{\max}(\Phi_k - \Phi) \geq \lambda_{\max}(\Phi_{11}^k - \Phi_{11}) \geq \frac{\delta_2}{k}$, which establishes item $3\ b)$.

First strengthening of Lemma 3.2:
With the additional assumptions **A.4** and $\Phi_0 > \Phi$, each item of Theorem 1.1 holds.

Relaxing the assumption: $\Phi_0 > \Phi$.
Suppose now that $\Phi_0 \geq \Phi$ but not $\Phi_0 > \Phi$. It is well known that Eq. 38, the difference equation for Δ_k, holds also when Δ_k is singular. In particular, from item 2 of Lemma 3.1, it follows that $\Delta_k \geq 0$ for all $k \geq 0$ which establishes item 1 of the Lemma statement.

Suppose we have any $\bar{\bar{\Phi}}_0$ such that $\bar{\bar{\Phi}}_0 \geq \Phi_0 \geq \Phi$ and $\bar{\bar{\Phi}}_0 > \Phi$. From item 2 of Lemma 3.1 it follows that $\bar{\bar{\Phi}}_k \geq \Phi_k \geq \Phi$ for all $k \geq 0$ (where $\bar{\bar{\Phi}}_k$ are iterates of the RDE with initial condition $\bar{\bar{\Phi}}_0$). Since $\bar{\bar{\Phi}}_k \geq \Phi_k$ and the convergence of $\{\bar{\bar{\Phi}}_k\}$ is guaranteed by item 2 of the first strengthening of Lemma 3.2, item 2 in the theorem statement is established.
Item $3\ a)$ in the first strengthening of Lemma 3.2 establishes a worst-case bound for the convergence rate of $\{\bar{\bar{\Phi}}_k\}$ which, by virtue of the above observations, guarantees the same convergence rate for $\{\Phi_k\}$ which is stated in item $3\ a)$ of the theorem.
Since the restriction $\Phi_0 > \Phi$ is maintained in item $3\ b)$ of the theorem, clearly this worst-case convergence result of item $3\ b)$ in the first strengthening of Lemma 3.2 remains.

Second strengthening of Lemma 3.2:
With the additional assumption **A.4** , the statements in Theorem 1.1 hold.

Relaxing assumption **A.4** *(that F is nonsingular).*
If F is singular then $\hat{F} = (I - GN^{-1}G^T\Phi)F$ will be also. Suppose it has a Jordan

canonical form $\hat{F} = M^{-1}\tilde{F}M$ where

$$\tilde{F} = \mathrm{diag}\left\{\bar{F} \quad F_z\right\} \tag{65}$$

and $\bar{F}$ and F_z are block diagonal and contain Jordan blocks corresponding to the non-zero and zero eigenvalues of $\hat{F}$, respectively.

Note that the difference equation for Δ_k given in Eq. 38 still holds under the new assumptions (i.e., those stated in Theorem 1.1). Now express this equation in the coordinate basis introduced above and define $\tilde{\Delta}_k = M^T \Delta_k M$. Let $\tilde{F}_q$ be any Jordan block of $\tilde{F}$ of the form in Eq. 11, corresponding to a zero eigenvalue. We now investigate the RDE evolution in the subspace corresponding to this Jordan block. It can be shown fairly easily via Eq. 38 that

$$(\tilde{\Delta}_k)_q = \mathrm{diag}\{D_k \quad O_k\} \tag{66}$$

where $(\tilde{\Delta}_k)_q$ is the $n_q \times n_q$ diagonal sub-block of $\tilde{\Delta}_k$ and $D_k \in \mathbb{R}^{(n_q-k)\times(n_q-k)}$ is a nonzero matrix in general. Observe that $(\tilde{\Delta}_k)_q = O_{n_q}$ for all iterations $k \geq n_q$. This reasoning can be applied to each Jordan block which has a zero eigenvalue. It follows that there exists an integer n_s (the size of the largest zero-eigenvalue Jordan block) such that when $k \geq n_s$,

$$\tilde{\Delta}_k = \mathrm{diag}\left\{\bar{\Delta}_k \quad O_{n_z}\right\} \tag{67}$$

where $\bar{\Delta}_k \geq 0$ and n_z is the size of the whole invariant subspace corresponding to an eigenvalue of zero. Now define a lower dimensional problem by considering iterates of $\bar{\Delta}_k$ only. Let G in this basis be partitioned as follows: $G^T = \left(\bar{G}^T \quad G_z^T\right)$.

Note that for $n \geq n_s$, it follows from Eq. 38 that $\bar{\Delta}_k$ satisfies the following recursion:

$$\bar{\Delta}_{k+1} = \bar{F}^T \bar{\Delta}_k \bar{F} - \bar{F}^T \bar{\Delta}_k \bar{G} \left(\bar{G}^T \bar{\Delta}_k \bar{G} + G^T \Phi G^T + U\right)^{-1} \bar{G}^T \bar{\Delta}_k \bar{F}. \tag{68}$$

Now consider the above RDE as being associated with the factorization of a spectral matrix $\bar{\Psi}(z) = U + G^T \Phi G$ of the form in Eq. 2 with "F" replaced by $\bar{F}$, "G" replaced by $\bar{G}$, "U" replaced by $U + G^T \Phi G$ and "V" replaced by zero. Clearly the factorization of this matrix from the original state-space realization is trivial and the strong solution of the algebraic equation associated with Eq. 68 is $\bar{\Delta} = 0$.

It can be easily checked that if (F, G) is stabilizable then $(\bar{F}, \bar{G})$ is also. Since we assume that $\Delta_0 \geq 0$, it follows that $\bar{\Delta}_0 \geq 0$. By construction, $\bar{F}$ is invertible. Convergence of $\bar{\Delta}_k$ then follows by application of the second strengthening of Lemma 3.2 to $\bar{\Psi}(z)$.

Recall that the parts of the RDE iterates associated with invariant subspaces corresponding to zero eigenvalues converge in a finite number of iterations. The best and worst case convergence behaviour of $\bar{\Delta}_k$ are therefore inherited by Δ_k, as stated in items *3 a)* and *3 b)* of the theorem. $\qquad\square$

Acknowledgements.

The authors wish to gratefully acknowledge helpful correspondence and discussions with Professor F.M. Callier, Dr. Robert Bitmead, Dr. Michael Green, Dr. Michel Gevers and Dr. Leonid Gurvits.

References

[1] B.D.O. Anderson, K.L. Hitz, N.D. Diem, Recursive Algorithm for Spectral Factorization, *IEEE Trans. Circuits Syst.*, vol. **21**, no. 6, 1974, pp. 742–750.

[2] B.D.O. Anderson, J.B. Moore, *Optimal Control – Linear Quadratic Methods*, Prentice-Hall, Englewood Cliffs, NJ, 1990.

[3] B.D.O. Anderson, J.B. Moore, *Optimal Filtering*, Prentice-Hall, Englewood Cliffs, NJ, 1979.

[4] B.D.O. Anderson, Stability Properties of Kalman-Bucy Filters, *Journal of the Franklin Institute*, vol. **291**, No. 2, 1971, pp. 137–144.

[5] R.R. Bitmead, M. Gevers, Riccati Difference and Differential Equations: Convergence, Monotonicity and Stability, *The Riccati Equation*, S. Bittanti, A.J. Laub, J.C. Willems (eds.), Springer-Verlag, Berlin, 1991.

[6] P.E. Caines, *Linear Stochastic Systems*, Wiley, New York, 1988.

[7] S.W. Chan, G.C. Goodwin, K.S. Sin, Convergence Properties of the Riccati Difference Equation in Optimal Filtering of Nonstabilizable Systems, *IEEE Trans. Automat. Contr.*, vol. **AC-29**, February 1984, pp. 110–118.

[8] C.E. de Souza, M.R. Gevers, G.C. Goodwin, Riccati Equations in Optimal Filtering of Nonstabilizable Systems Having Singular State Transition Matrices, *IEEE Trans. Automat. Contr.*, vol. **AC-31**, September 1986, pp. 831–838.

[9] C.E. de Souza, On Stabilizing Properties of Solutions of the Riccati Difference Equation, *IEEE Trans. Automat. Contr.*, vol. **34**, September 1989, pp. 1313–1316.

[10] C.E. de Souza and L. Xie, On the Discrete-time Bounded Real Lemma with application in the characterization of static state feedback H_∞ controllers, *Systems & Control Letters*, vol. **18**, 1992, pp. 61–71.

[11] B.P. Molinari, The Stabilizing Solution of the Discrete Algebraic Riccati Equation, *IEEE Trans. Automat. Contr.*, vol. **AC-20**, June 1975, pp. 396–399.

[12] W.M. Wonham, *Linear Multivariable Control. A Geometric Approach*, Springer-Verlag, New York, 1974.

WSSIAA 5 (1995) pp. 327–335
© World Scientific Publishing Company

On Global Convergence of an Algorithm for Optimal Control of Discrete-Time Systems

YOSHIYUKI SAKAWA

Department of Intelligent Mechanics

Kinki University, Uchita, Naga-gun, Wakayama, 649-64, Japan

Abstract

This paper presents a simple algorithm for solving the optimal control problems of discrete-time systems with constraints on control, but without constraints on the trajectory or the terminal state. In this algorithm, monotonous reduction of cost values at each iteration is guaranteed. It is proved that the converged control satisfies the Pontryagin type necessary conditions for optimality under some conditions.

1. Introduction

Optimal control problems of continuous-time systems have been solved numerically by first discretizing the system, then reducing it to a large scale mathematical programming problem [1], [2]. Another method is to use Pontryagin's maximum principle. Sakawa and Shindo [3] presented the following algorithm for the system governed by ordinary differential equations : Given a control and a corresponding state, compute the associated costate, then compute simultaneously a new state and control such that the state equation holds and the new control minimizes, at each time instant, the sum of the Hamiltonian function and a weighted quadratic term penalizing the difference between the new and old controls. It is proved in [3] that, under some assumptions, the values of cost function decrease monotonously and that, if a subsequence of controls converges

328

to some control $\hat{u}$, then $\hat{u}$ satisfies the optimality conditions of Pontryagin's maximum principle.

In this paper, we consider the optimal control problem for the discrete-time system. We propose the similar algorithm as in [3], and we prove that almost the same results can be obtained for the discrete-time systems.

2. Optimal Control Problem and Computational Procedure

We consider the discrete -time system described by

$$x(t+1) = f(x(t), u(t), t), \qquad x(0) = x_0 \tag{1}$$

where x is an n-dimensional state vector, u is an r-dimensional control vector, and x_0 gives the initial condition. The vector-valued function $f(x, u, t)$ should satisfy some conditions which will be stated later. The control vectors are required to satisfy the constraint

$$u(t) \in U, \quad \forall t \in T \tag{2}$$

where U is a compact and convex subset of the r-dimensional Euclidean space, and $T = \{0, 1, ..., N-1\}$.

The problem is to find the optimal control sequence $\mathbf{u} = \{u(t), t \in T\}$ that minimizes the cost function

$$J(\mathbf{u}) = \sum_{t=0}^{N-1} L(x(t), u(t), t) + \theta(x(N)) . \tag{3}$$

In order that the discrete maximum (or minimum) principle holds, which will be stated below [4], [5], we need the following assumption:

<u>Assumption 1</u>: For any $x \in R^n$, $u, v \in U$, $t \in T$ and $0 \le \alpha \le 1$, there exists a $w \in U$ such that

$$L(x, w, t) \le \alpha L(x, u, t) + (1-\alpha)L(x, v, t)$$

$$f(x, w, t) = \alpha f(x, u, t) + (1-\alpha)f(x, v, t). \tag{4}$$

Let $u(t)$ be the optimal control and let $x(t)$ be the corresponding optimal trajectory satisfying (1). Then it is necessary that there exist a sequence of n-dimensional row-vectors $\lambda(t)$, $t=0, ..., N-1$, such that

$$\lambda(t) = \frac{\partial H(x(t),\ u(t),\ \lambda(t+1),\ t)}{\partial x}\ , \qquad \lambda(N) = \frac{\partial \theta(x(N))}{\partial x} \tag{5}$$

where

$$H(x,\ u,\ \lambda,\ t) = L(x,\ u,\ t) + \lambda f(x,\ u,\ t) \tag{6}$$

and that for all $t \in T$ the function $H(x(t),\ u,\ \lambda(t+1),\ t)$ of the variable $u \in U$ attains its minimum at the point $u = u(t)$, namely

$$H(x(t),\ u(t),\ \lambda(t+1),\ t) = \min_{u \in U} H(x(t),\ u,\ \lambda(t+1),\ t)\ . \tag{7}$$

For seeking the optimal pair $(x(t),\ u(t))$ satisfying the above conditions, we consider the following algorithm.

Step 0: Select a nominal control $u^0(t) \in U$, $t \in T$. Let $x^0(t)$, $t \in T$ be the corresponding nominal trajectory. Set $i = 1$.

Step 1: Compute $\lambda^{i-1}(t)$ by solving the difference equation

$$\left.\begin{array}{c} \lambda^{i-1}(t) = \dfrac{\partial H(x^{i-1}(t),\ u^{i-1}(t),\ \lambda^{i-1}(t+1),\ t)}{\partial x},\quad t \in T \\[3mm] \lambda^{i-1}(N) = \partial \theta(x^{i-1}(N)) / \partial x \end{array}\right\} \tag{8}$$

backward in time starting at $t=N$.

Step 2: Define the function

$$K(x,\ u,\ \lambda,\ t;\ v,\ c) = H(x,\ u,\ \lambda,\ t) + \frac{c}{2}|u - v|^2. \tag{9}$$

Select a nonnegative constant c^i properly. Given $x^i(t)$, determine $u^i(t)$ which satisfies

$$K(x^i(t),\ u^i(t),\ \lambda^{i-1}(t+1),\ t;\ u^{i-1}(t),\ c^i) = H(x^i(t),\ u^i(t),\ \lambda^{i-1}(t+1),\ t) + \frac{c^i}{2}\left\|u^i(t) - u^{i-1}(t)\right\|^2$$

$$= \min_{u \in U} K(x^i(t),\ u,\ \lambda^{i-1}(t+1),\ t;\ u^{i-1}(t),\ c^i).$$

$$(10)$$

Then calculate

$$x^i(t+1) = f(x^i(t),\ u^i(t),\ t), \tag{11}$$

for $t = 0, 1, ..., N-1$ forward in time. The initial condition is given by $x^i(0) = x_0$.

Step 3: Calculate

$$J(u^i) = \sum_{t=0}^{N-1} L(x^i(t),\ u^i(t),\ t) + \theta(x^i(N)). \tag{12}$$

If $J(u^i) - J(u^{i-1}) > 0$, select a larger c^i and go to Step 2. Otherwise, set $i := i + 1$ and go to Step 1.

Stop the computation if the sequence $\{c^i\}$ is bounded and the sequence $\{u^i(t),\ t \in T\}$ of the controls converges.

In this algorithm we minimize K instead of H. Since the function K contains a penalizing term $(c/2)\left\|u - u^{i-1}\right\|^2$ for the possible large change of control, instability of the algorithm at the first stage of computation can be avoided by taking large c^i [3].

3. Reduction of the Cost

We need the following assumption throughout this paper.

Assumption 2: Each component functions $f_j(x, u, t)$, $j = 1, 2, ..., n$, of f and $L(x, u, t)$ and their partial derivatives f_{jx}, f_{jxx}, f_{ju}, f_{juu}, f_{jxu}, L_x, L_{xx}, L_u, L_{uu}, L_{xu} are continuous in x, u and t, and the function $\theta(x)$ is also twice continuously differentiable.

Since the control sequence $\mathbf{u} = \{u(t),\ t \in T\}$ is bounded, the solutions $\{x^i(t)\}$ and $\{\lambda^i(t)\}$ of the difference equations (11) and (8) are also bounded, i.e., there is a constant M_1 such that

$$\left\| x^i(t) \right\| \le M_1, \qquad \left\| \lambda^i(t) \right\| \le M_1, \qquad \forall i, \ \forall t \in T. \tag{13}$$

<u>Theorem 1.</u> There is a constant $M > 0$ independent of i such that the inequality

$$J(\mathbf{u}^i) - J(\mathbf{u}^{i-1}) \le -(c^i - M) \sum_{t=0}^{N-1} \left\| u^i(t) - u^{i-1}(t) \right\|^2 \tag{14}$$

holds for any i. If we choose c^i such that $c^i \ge c_0 > M$, then the sequence $\{J(\mathbf{u}^i)\}$ of the cost functions decreases monotonically and converges.

<u>Proof.</u> It can be easily seen that

$$J(\mathbf{u}^i) - J(\mathbf{u}^{i-1}) = \sum_{t=0}^{N-1} [H(x^i(t),\ u^i(t),\ \lambda^{i-1}(t+1),\ t) - H(x^{i-1}(t),\ u^{i-1}(t),\ \lambda^{i-1}(t+1),\ t)$$

$$- \lambda^{i-1}(t+1)(f(x^i(t),\ u^i(t),\ t) - f(x^{i-1}(t),\ u^{i-1}(t),\ t))]$$

$$+ \theta(x^i(N)) - \theta(x^{i-1}(N)). \tag{15}$$

Define
$$\delta x^i(t) = x^i(t) - x^{i-1}(t), \qquad \delta u^i(t) = u^i(t) - u^{i-1}(t). \tag{16}$$

It is obvious that

$$H(x^i(t),\ u^i(t),\ \lambda^{i-1}(t+1),\ t) - H(x^{i-1}(t),\ u^{i-1}(t),\ \lambda^{i-1}(t+1),\ t)$$

$$= H(x^i(t),\ u^i(t),\ \lambda^{i-1}(t+1),\ t) - H(x^i(t),\ u^i(t) - \delta u^i(t),\ \lambda^{i-1}(t+1),\ t)$$

$$+ H(x^{i-1}(t) + \delta x^i(t),\ u^{i-1}(t),\ \lambda^{i-1}(t+1),\ t) - H(x^{i-1}(t),\ u^{i-1}(t),\ \lambda^{i-1}(t+1),\ t). \tag{17}$$

From Assumption 2 and (13), we see that there are constants M_2 and M_3 such that

$$H(x^i(t),\ u^i(t),\ \lambda^{i-1}(t+1),\ t) - H(x^i(t),\ u^i(t) - \delta u^i(t),\ \lambda^{i-1}(t+1),\ t)$$

$$\leq +\frac{\partial H(x^i(t),\ u^i(t),\ \lambda^{i-1}(t+1),\ t)}{\partial u}\delta u^i(t) + M_2\left\|\delta u^i(t)\right\|^2,$$

$$H(x^{i-1}(t) + \delta x^i(t),\ u^{i-1}(t),\ \lambda^{i-1}(t+1),\ t) - H(x^{i-1}(t),\ u^{i-1}(t),\ \lambda^{i-1}(t+1),\ t)$$

$$\leq \frac{\partial H(x^{i-1}(t),\ u^{i-1}(t),\ \lambda^{i-1}(t+1),\ t)}{\partial x}\delta x^i(t) + M_2\left\|\delta x^i(t)\right\|^2,$$

$$\theta(x^i(N)) - \theta(x^{i-1}(N)) \leq \frac{\partial \theta(x^{i-1}(N))}{\partial x}\delta x^i(N) + M_3\left\|\delta x^i(N)\right\|^2.$$

Using these inequalities and (8) for (15) gives

$$J(\mathbf{u}^i) - J(\mathbf{u}^{i-1})$$

$$\leq \sum_{t=0}^{N-1}\left[\frac{\partial H(x^i(t),\ u^i(t),\ \lambda^{i-1}(t+1),\ t)}{\partial u}\delta u^i(t) + M_2\left\|\delta u^i(t)\right\|^2\right.$$

$$\left. + \lambda^{i-1}(t)\delta x^i(t) + M_2\left\|\delta x^i(t)\right\|^2 - \lambda^{i-1}(t+1)\delta x^i(t+1)\right]$$

$$+ \frac{\partial \theta(x^{i-1}(N))}{\partial x}\delta x^i(N) + M_3\left\|\delta x^i(N)\right\|^2. \tag{18}$$

Since $u^i(t)$ minimizes K on the convex set U, it follows that [6]

$$\frac{\partial K(x^i(t),\ u^i(t),\ \lambda^{i-1}(t+1);\ u^{i-1}(t),\ c^i)}{\partial u}(u^i(t) - u)$$

$$= \frac{\partial H(x^i(t),\ u^i(t),\ \lambda^{i-1}(t+1),\ t)}{\partial u}(u^i(t) - u)$$

$$+ c^i(u^i(t) - u^{i-1}(t))^T(u^i(t) - u) \leq 0, \qquad \forall u \in U. \tag{19}$$

From (19) we see that

$$\frac{\partial H(x^i(t),\ u^i(t),\ \lambda^{i-1}(t+1),\ t)}{\partial u}\delta u^i(t) \leq -c^i\left\|\delta u^i(t)\right\|^2. \tag{20}$$

Using (20) and the boundary condition of (8) in (18) gives

$$J(\mathbf{u}^i) - J(\mathbf{u}^{i-1}) \le \sum_{t=0}^{N-1} \left[(-c^i + M_2)\|\delta u^i(t)\|^2 + M_2\|\delta x^i(t)\|^2 \right] + M_3|\delta x^i(N)|^2.$$

(21)

By Assumption 2, there is a constant C such that

$$|\delta x^i(t+1)| = \|f(x^i(t),\, u^i(t),\, t) - f(x^{i-1}(t),\, u^{i-1}(t),\, t)\|$$
$$\le C\|\delta x^i(t)\| + C\|\delta u^i(t)\|.$$

(22)

Assume that $C > 1$. it is easily calculated that

$$\|\delta x^i(t)\| \le C\sqrt{\frac{C^{2t}-1}{C^2-1}}\sqrt{\sum_{k=0}^{N-1}|\delta u^i(k)|^2}, \qquad t = 0,\, 1,\, ...,N\,.$$

(23)

Therefore, we see that there are constants M_4 and M_5 such that

$$\sum_{t=0}^{N-1}\|\delta x^i(t)\|^2 \le M_4 \sum_{t=0}^{N-1}|\delta u^i(t)|^2, \qquad \|\delta x^i(N)\|^2 \le M_5 \sum_{t=0}^{N-1}|\delta u^i(t)|^2.$$

(24)

Substituting (24) into (21) yields (14). □

4. Convergence of the Control

We need the following additional assumption:

<u>Assumption 3</u>: The function $H(x,\, u,\, \lambda,\, t)$ defined by (6) is convex with respect to u.

<u>Theorem 2</u>. Suppose that the sequence $\{c^i\}$ is bounded. If the control sequence $\{u^i(t),\ t \in T\}$ converges, i. e.,

$$\lim_{i \to \infty} u^i(t) = \hat{u}(t), \qquad \forall t \in T,$$

(25)

then, the control $\hat{u}(t)$ and the corresponding state $\hat{x}(t)$ satisfy the necessary conditions (5) and (7) for optimality.

<u>Proof.</u> We see that

$$\|\hat{x}(t+1) - x^i(t+1)\| \le C\|\hat{x}(t) - x^i(t)\| + C\|\hat{u}(t) - u(t)\|.$$

(26)

In the same way as in (23), we obtain

$$\left\| \hat{x}(t) - x^i(t) \right\| \le C \sqrt{\frac{C^{2t} - 1}{C^2 - 1}} \sqrt{\sum_{k=0}^{N-1} \left\| \hat{u}(k) - u^i(k) \right\|^2}, \qquad t = 0, 1, \ldots, N . \tag{27}$$

Inequality (27) implies that

$$\lim_{i \to \infty} x^i(t) = \hat{x}(t), \qquad \forall t \in T. \tag{28}$$

Let $\hat{\lambda}(t)$ be the solution of

$$\hat{\lambda}(t) = \frac{\partial H(\hat{x}(t), \ \hat{u}(t), \ \hat{\lambda}(t+1), \ t)}{\partial x} , \qquad \hat{\lambda}(N) = \frac{\partial \theta(\hat{x}(N))}{\partial x} . \tag{29}$$

Then, there is a constant $D > 0$ such that

$$\left\| \hat{\lambda}(t) - \lambda^i(t) \right\| \le D \left| \hat{x}(t) - x^i(t) \right| + D \left\| \hat{u}(t) - u^i(t) \right\| + D \left\| \hat{\lambda}(t+1) - \lambda^i(t+1) \right\| . \tag{30}$$

In the same way as before, we see that

$$\lim_{i \to \infty} \lambda^i(t) = \hat{\lambda}(t), \qquad \forall t \in T. \tag{31}$$

Since the sequence $\{c^i\}$ is bounded and

$$\lim_{i \to \infty} \left\| u^i(t) - u^{i-1}(t) \right\| = 0, \tag{32}$$

from (19) we see that

$$\frac{\partial H(\hat{x}(t), \ \hat{u}(t), \ \hat{\lambda}(t+1), \ t)}{\partial u} (\hat{u}(t) - u) \le 0 , \qquad \forall t \in T, \quad \forall u \in U. \tag{33}$$

Since the function H is a convex function on the convex set U, (33) implies that [6]

$$H(\hat{x}(t), \ \hat{u}(t), \ \hat{\lambda}(t+1), \ t) = \min_{u \in U} H(\hat{x}(t), \ u, \ \hat{\lambda}(t+1), \ t) . \tag{34}$$

Thus, we see that $\hat{u}(t)$ and $\hat{x}(t)$ satisfy the necessary conditions for optimality. $\square$

5. Concluding Remarks

We have presented the algorithm for computing the optimal control of the discrete-time control systems with constraints on the control, but without constraints on the state. It is proved that the values of the cost function continue to decrease at each iteration, and that the converged control satisfies the Pontryagin type necessary conditions for optimality. The algorithm presented here is essentially same as the previously presented algorithm [3] for computing the optimal control of continuous-time control systems. However, the algorithm for the discrete-time systems is more suitable for computation, because the control and the state can be determined sequentially in the discrete-time systems. Consequently there is no problem to determine the state and the control simultaneously as in [3]. Also, since we use the digital computers for computations, the algorithm for the discrete-time systems is much more practical and useful.

The constants c^i should be chosen adaptively depending on the situation of the computation. In general, when the constant c^i is smaller, the obtained variation of the control is larger. Therefore, c^i are desired as small as possible as far as the cost values decrease. According to our computational experience, the following way of choosing the constant c^i is recommended. Choose the initial constant c^1 properly. If $J(\mathbf{u}^i) > J(\mathbf{u}^{i-1})$, then set $c^i := 2c^i$. If $J(\mathbf{u}^i) \le J(\mathbf{u}^{i-1})$, then set $c^{i+1} = \alpha c^i$ where α is a constant such that $0.6 \le \alpha \le 0.9$.

References

[1] E. Polak, *Computational Methods in Optimization, A Unified Approach.* New York : Academic Press,1971.

[2] K. L. Teo, C. J. Goh and K. H. Wong, *A Unified Computational Approach to Optimal Control Problems.* Burnt Mill, Harlow, England: Longman Scientific & Technical, 1991.

[3] Y. Sakawa and Y. Shindo, "On global convergence of an algorithm for optimal control," *IEEE Trans. Automat. Contr.,* vol. AC-25, no. 6, pp. 1149-1153, 1980.

[4] M. D. Canon, C.D. Cullum Jr. and E. Polak, *Theory of Optimal Control and Mathematical Programming.* New York: McGraw-Hill, 1970.

[5] A. D. Ioffe and V. M. Tihomirov, *Theory of Extremal Problems.* Amsterdam: North-Holland, 1979.

[6] M. R. Hestenes, *Optimization Theory: The Finite Dimensional Case,* New York: Wiley, 1975.

WSSIAA 5 (1995) pp. 337–352
© World Scientific Publishing Company

OPTIMAL DECENTRALIZED CONTROL FOR STOCHASTIC DYNAMIC SYTEMS

Sergey V. Savastyuk

and

Dragoslav D. Šiljak

School of Engineering
Santa Clara University
Santa Clara, CA 95053, USA

ABSTRACT

The objective of this paper is to derive sufficient conditions for optimality of decentralized control using the classical method of Lagrange. By formulating the decentralized information structure constraints on the input variables as differential equations we can add the constraints to the equations of motion and minimize the cost functional. To obtain the globally optimal solution we introduce suitable Lagrange–Lyapunov functions of time and state as multipliers and solve the constrained optimization problem. By assuming a Gaussian nature of the state evolution we provide a feedback control structure determined by Riccati-type equations.

1. Introduction

In spite of its relatively long history, the fundamental problem of optimal decentralized control has not been solved satisfactorily. The reason is the difficulty in formulating the decentralized information structure constraints which can be incorporated in a standard optimization framework. Neither Pontryagin's maximum principle [1] nor Bellman's dynamic programming [2] can handle the lack of incomplete state observation, the fact recognized by Fleming [3] far back in the sixties.

The principal idea of this paper is to solve the problem of optimal decentralized control by redefining the decentralized information structure constraints as differential equations, add the constraints to the equations of motion, and solve the problem in the classical framework of Lagrange [4]. A new element in this context is a set of Lagrange multipliers in the form of time-state functions, which we call Lagrange-Lyapunov functions because of their role in the optimization process. This makes it possible to formulate the sufficient conditions of optimality in terms of Hamilton-Jacobi-like equations [3], and prove the conditions using the method of global optimization introduced by Krotov [5].

We consider a linear system composed of interconnected subsystems with local inputs and performance indices, which is described by stochastic linear differential equations of Ito's type [6]. The obtained conditions of optimality provide a feedback structure for decentralized control. When the additional constraint of Gaussian state evolution is added, we obtain a solution to the Linear Quadratic Gaussian (LQG) problem under decentralized constraints. Our solution involves Riccati-type equations in the same manner as in the classical regulator theory of Kalman [7]. The stated theorems are proved by relaying on a number of results established in [8,9]. The results of [10-13] provide a mathematical framework for handling the dynamic constraints in the form of the Fokker–Plank–Kolmogorov equation of motion.

2. Problem formulation

Let us consider a system $\mathcal{S}$ composed of interconnected subsystems, which is described by stochastic differential equations of Ito's type,

$$\mathcal{S} : dx_i = \left[A_{ii}(t)x_i + B_i(t)u_i(t,x) + \sum_{j \neq i}^{N} A_{ij}(t)x_j \right] dt + C_i(t,x)dw_i, \quad i \in \mathbf{N} \quad (2.1)$$

where $x_i \in I\!\!R^{n_i}$ are the local state vectors; $u_i \in \mathbf{U}_i \subset I\!\!R^{m_i}$ are the local control vectors, $\mathbf{U}_i$ are fixed sets; $w_i \in I\!\!R^{\nu_i}$ are independent Wiener processes with zero mean values and incremental covariances $R_i^w dt$; $t \in \mathbf{T} \subset I\!\!R$ is time, $\mathbf{T} = [t_0, t_1]$; $x \in I\!\!R^n$, is the state vector of $\mathcal{S}$, $x = (x_{11}, x_{12}, \ldots, x_{1n_1}, x_{2n_1+1}, x_{2n_1+2}, \ldots, \ldots, x_{Nn})^T$; $u \in I\!\!R^m$, is the control vector of $\mathcal{S}$, $u = (u_1^T, u_2^T, \ldots, u_N^T)^T$; $n = \sum_{i=1}^{N} n_i$, $i \in \mathbf{N} = \{1, 2, \ldots, N\}$; $\mathbf{N}_i = \{k_i, \ldots, k_{i+1}\}$, where $k_i = \sum_{j=1}^{i-1} n_j + 1$, $k_{i+1} = \sum_{j=1}^{i} n_j$; $\mathbf{N}^i = \{1, 2, \ldots, n\} \backslash \mathbf{N}_i$. Matrices A_{ii}, B_i, A_{ij}, C_i, and R_i^w have appropriate dimensions, and $I\!\!R^n$ denotes the n-dimensional Euclidean space.

With the system $\mathcal{S}$ we associate a quadratic cost

$$J = \frac{1}{2}\mathcal{E}\left\{ \sum_{i=1}^{N} \int_{t_0}^{t_1} [x_i^T Q_i(t)x_i + u_i^T R_i(t)u_i]dt \right\}. \quad (2.2)$$

Here $\mathcal{E}\{ \cdot \}$ denotes expectation, Q_i are symmetric nonnegative matrices, and R_i are symmetric positive definite matrices; all matrices being of proper dimensions.

Our central objective is to determine a control $u(t,x)$ which drives system $\mathcal{S}$ optimally with respect to cost J and, at the same time, satisfies the essential information restriction that the control laws u_i at each subsystem $\mathcal{S}_i$ can utilize only the locally available state x_i. More precisely, we state the following:

(2.3) PROBLEM. Determine a control $u(t, x)$ which minimizes the functional J with respect to system $\mathcal{S}$ under the *decentralized information structure constraints*

$$u_i := u_i(t, x_i), \quad i \in \mathbf{N}. \tag{2.4}$$

The control, if one exists, is called the *optimal decentralized control.*

We stated the minimization Problem (2.3) in the terms of stochastic processes, and we want to reformulate it as a problem involving parabolic partial differential equations, so that it can be resolved as a deterministic problem. For these reasons, we need a number of assumptions. Let $\mathbf{L}_2$ be the set of square integrable functions $x \to q(x) : I\!R^n \to I\!R$; let $\mathbf{B}$ be the set of functions $x \to v(x) : I\!R^n \to I\!R^m$ that are Borel measurable; and $\mathbf{M} \subset I\!R$ is a set of measure zero. We introduce the following special notations [3]:

$$(f(y))_y := f_y(y) := \sum_{k=1}^{r} \frac{\partial f_k(y)}{\partial y_k}, \tag{2.5}$$

for vector functions $y \to f(y) : I\!R^r \to I\!R^r$, and

$$(F(y))_{yy} := \sum_{k=1}^{r} \sum_{l=1}^{r} \frac{\partial^2 F_{kl}(y)}{\partial y_k \partial y_l},$$

$$\alpha(y)_{yy} F(y) := \sum_{k=1}^{r} \sum_{l=1}^{r} \frac{\partial^2 \alpha_{kl}(y)}{\partial y_k \partial y_l} F_{kl}(y), \tag{2.6}$$

for matrix functions $y \to F(y) : I\!R^r \to I\!R^{r \times r}$ and a scalar $y \to \alpha(y) : I\!R^r \to I\!R$.

We state the following:

(2.7) ASSUMPTION. The dynamics of the probability density of a diffusion process $x(t)$ satisfies the Fokker–Plank–Kolmogorov equation

$$\frac{\partial p(t, x)}{\partial t} = \sum_{i=1}^{N} \left(-\left[f_i^u(t, x) p(t, x) \right]_{x_i} + \left[F_i(t, x) p(t, x) \right]_{x_i x_i} \right) \tag{2.8}$$

where $F_i := \frac{1}{2}(C_i C_i^T)$, $(t, x) \to C_i : \mathbf{T} \times I\!R^n \to I\!R^{n_i \times \nu_i}$; $f_i^u(t, x) := f_i(t, x, u_i(t, x))$.

(2.9) ASSUMPTION. Function $(t, x) \to p(t, x) : \mathbf{T} \times I\!R^n \to I\!R$ is continuous and is an element of the Sobolev space $\mathbf{W}_2^{1,1}(\mathbf{T} \times I\!R^n)$ [14]; the control law $(t, x) \to u(t, x) : \mathbf{T} \times I\!R^n \to \mathbf{U} \subset I\!R^m$ is Borel measurable; $t \to p_*(t) := p(t, \cdot) : \mathbf{T} \times I\!R^n \to$

340

$\mathbf{L}_2$, $t \to u_*(t) := u(t, \cdot) : \mathbf{T} \times I\!\!R^n \to \mathbf{B}$ are a probability density and a control law at time $t \in \mathbf{T}$.

(2.10) ASSUMPTION. Matrix functions $A_{ii}(t)$, $B_i(t)$, $A_{ij}(t)$, $Q_i(t)$, $R_i(t)$, are Borel measurable, which implies that functions $(t, x, u_i) \to f_i(t, x, u_i) : \mathbf{T} \times I\!\!R^n \times I\!\!R^{m_i} \to I\!\!R^{n_i}$ are Borel measurable, where

$$f_i(t, x, u_i) = A_{ii}(t)x_i + B_i(t)u_i(t, x_i) + \sum_{j \neq i}^{N} A_{ij}(t)x_j. \qquad (2.11)$$

Furthermore, functions $(t, x) \to C_i(t, x) : \mathbf{T} \times I\!\!R^n \to I\!\!R^{n_i \times \nu_i}$ and $(t, x, u_i) \to f_c^i(t, x, u) : \mathbf{T} \times I\!\!R^n \times I\!\!R^{m_i} \to I\!\!R$ are Borel measurable, where $f_c^i(t, x, u) = = \frac{1}{2}(x_i^T Q_i(t)x_i + u_i^T R_i(t)u_i)$.

(2.12) ASSUMPTION. The solution of equation (2.8) is understood in the generalized sense [15], that is,

$$\int_{t_0}^{t_1} \int_{I\!\!R^n} \left[-\eta(t, x)\frac{\partial p(t, x)}{\partial t} + \sum_{i=1}^{N} \left(\eta_{x_i}^T(t, x)f_i^u(t, x) + [\eta_{x_i x_i}(t, x)F_i(t, x)] \right)p(t, x) \right] = 0 \qquad (2.13)$$

for any function $(t, x) \to \eta(t, x) : \mathbf{T} \times I\!\!R^n \to I\!\!R$ from space $\overset{o}{\mathbf{W}}{}_2^{1,2}(t, x)$.

Fundamental to the solution of Problem (2.3) is the set $\mathbf{V}^*$ of control functions satisfying the information structure constraints (2.4). We also introduced the measurability of the admissible control laws on the components of the state vector, which enables us to extend our results to nonlinear systems. Stronger conditions, such as "the function $p(t, x)$ has continuous second derivative and the function $u(t, x)$ is continuous in both arguments" can also be used.

For a given initial state $p_0(x) := p_*(t_0)$, we introduce the set $\mathbf{D}^*$ of control process pairs $z := [u_*(\cdot), p_*(\cdot)]$ satisfying the above assumptions and the following:

(2.14) ASSUMPTION. The sets $\mathbf{Q}$ and $\mathbf{V}^*$ are given, and $p_*(t) \in \mathbf{Q} \subset \mathbf{L}_2$, $u_*(t) \in \mathbf{V}^* \subset \mathbf{B}$ for all $t \in \mathbf{T}$.

(2.15) ASSUMPTION. The functional $z \to J(z) : \mathbf{D}^* \to I\!\!R$ is defined in (2.2), and $\mathbf{D}^* \neq \emptyset$.

Finally, on the basis of the above assumptions, we are in a position to restate Problem (2.3) in technical terms as:

(2.16) PROBLEM. Determine

$$\beta = \inf_{z \in \mathbf{D}^*} J(z), \qquad (2.17)$$

and a minimizing sequence $\{z^\alpha\}$, $z^\alpha := [u_*^\alpha(\cdot), \ p_*^\alpha(\cdot)] \in \mathbf{D}^*$, or a process $\bar{z} := [\bar{u}_*(\cdot), \ \bar{p}_*(\cdot)] \in \mathbf{D}^*$, on which $J(\bar{z}) = \beta$.

(2.18) REMARK. The difference between finding the minimizing sequence and the fixed optimal process is that the former can be considered as obtaining an ϵ-suboptimal solution, while the latter as finding a strict solution; both solutions can be found independently [5].

3. The decentralized information structure constraints

We propose to solve the problem of optimal decentralized control by using the classical Lagrange setting in the spirit of the maximum principle of Pontryagin [1]. A crucial step in our development is the way we introduce the information structure constraints via differential equations, and then use Lagrange multipliers for each constraint participating in the optimization process. Taking into account the nature of these multipliers, we term them *Lagrange–Lyapunov functions*.

(3.1) DEFINITION. The set $\mathbf{V}^*$ of functions $u(t,x)$ is said to obey the decentralized information structure constraints (2.4) if the functions $u_i(t,x)$ do not depend on the components of the vector x^i on $\mathbf{T} \times I\!R^{n^i}$ for all $i \in \mathbf{N}$, and $x^i = (x_1^T, \ldots, x_{i-1}^T, x_{i+1}^T, \ldots, x_N^T)^T \in I\!R^{n^i}$, $n^i = n - n_i$.

The crucial step in solving Problem (2.6) is to substitute the set $\mathbf{V}^*$ for another equivalent set $\mathbf{V}$ specified by the following:

(3.2) DEFINITION. The set $\mathbf{V}$ is comprised of control functions $u(t,x)$ with vector components $u_i(t,x)$ which are continuously differentiable with respect to the components of x^i for all $(t,x^i) \in \mathbf{T} \times I\!R^{n^i}$, and

$$\frac{\partial u_i(t,x)}{\partial x^i} = 0 \ , \qquad i \in \mathbf{N} \ . \tag{3.3}$$

(3.4) LEMMA. The set $\mathbf{V}$ coincides with the set $\mathbf{V}^*$.

Finally, we reformulate Problem (2.16) as follows:

(3.5) PROBLEM. Determine $\beta = \inf_{z \in \mathbf{D}} J(z)$ and a minimizing sequence $\{z^\alpha\}$, $z^\alpha := [u_*^\alpha(\cdot), \ p_*^\alpha(\cdot)] \in \mathbf{D}$, or a process $\bar{z} := [\bar{u}_*(\cdot), \ \bar{p}_*(\cdot)] \in \mathbf{D}$, on which $J(\bar{z}) = \beta$.

(3.6) REMARK. The obvious Lemma (3.4) means that decentralized information structure constraints (2.4) are rewritten effectively as differential equation constraints (3.3), which can now be incorporated into the standard Lagrange optimization framework in a manner suggested by Pontryagin in his maximum principle and by Bellman in his version of the Hamilton-Jacobi equations.

4. Sufficient conditions of optimality

We introduce the class Φ of Lagrange –Lyapunov functions $\varphi = (\varphi^0, \varphi^1, \ldots, \varphi^N)$, where $\varphi^i = (\varphi^{i1}, \varphi^{i2}, \ldots, \varphi^{in})$, $(t, q) \to \varphi^0(t, q) : \mathbf{T} \times \mathbf{L}_2 \to I\!R$, and $(t, x, u_i, q) \to \varphi^i(t, x, u_i, q) : \mathbf{T} \times I\!R^n \times I\!R^{m_i} \times \mathbf{L}_2 \to I\!R^n$, satisfying the following:

(4.1) ASSUMPTION. The function $\varphi^0(t, q)$ is locally Lipschitz on $\mathbf{T} \times \mathbf{L}_2$, differentiable with respect to (t, q) $(\mathbf{T} \backslash \mathbf{M}) \times \mathbf{L}_2$, and has on $\mathbf{T} \times \mathbf{L}_2$ a continuous partial (Frechet) derivative with respect to q everywhere on $\mathbf{T} \times \mathbf{L}_2$.

(4.2) ASSUMPTION. For each element $z := [u_*(\cdot), p_*(\cdot)] \in \mathbf{D}$, we define a function $(t, x) \to \varphi_q^0 := \mathcal{X}(t, x, p_*(t)) : \mathbf{T} \times I\!R^n \to I\!R$, which belongs to $\overset{o}{\mathbf{W}}_2^{1,2}(\mathbf{T} \times I\!R^n)$, where $\overset{*}{\mathbf{W}}_2^{1,2}(\mathbf{T} \times I\!R^n)$ is the subset of functions $\eta(t, x) \in \mathbf{W}_2^{1,2}$ for which constraints (2.4) hold, and $\overset{*}{\mathbf{W}}_2^{1,2} \supset \overset{o}{\mathbf{W}}_2^{1,2}$.

(4.3) ASSUMPTION. Vector functions $(t, x, u_i, q) \to \varphi_{u_i}^{is} : \mathbf{T} \times I\!R^n \times I\!R^{m_i} \times \mathbf{Q} \to I\!R^{m_i}$, $(t, x, u_i, q) \to \varphi_{x^i}^{is} : \mathbf{T} \times I\!R^n \times I\!R^{m_i} \times \mathbf{Q} \to I\!R^{n^i}$, are defined and continuous for $q \in \mathbf{Q}$, $s \in \mathbf{N}_i$.

(4.4) ASSUMPTION. For all $v \in \mathbf{V}$, $q \in \mathbf{Q}$, $t \in \mathbf{T} \backslash \mathbf{M}$, the functions $(t, x_{*k}) \to \gamma^{ik}(t, x_{*k}) : (\mathbf{T} \backslash \mathbf{M}) \times I\!R \to I\!R$, are defined as

$$\gamma^{ik}(t, x_{*k}) := \int_{I\!R^{n-1}} \varphi^{ik}(t, x, v_i(x), q) dx^{*k} , \tag{4.5}$$

where $\lim_{\tau \to \pm\infty} \gamma^{ik}(t, \tau) = 0$, and $x^{*k} = (x_{11}, \ldots, x_{*k-1}, x_{*k+1}, \ldots, x_{Nn}) \in I\!R^{n-1}$, $k \in \mathbf{N}^i$; the simbol $*k$ means that any index can be used at the $*$ place; in addition, we assume $\varphi^{is} \equiv 0$, for all (t, x, u_i, q), $s \in \mathbf{N}_i$.
We will use the following constructs:

$$H_i(t, x, u_i, q) := (\mathcal{X}_{x_i}^T(t, x) f_i(t, x, u_i) + \mathcal{X}_{x_i x_i}(t, x) F_i(t, x) + f_c^i(t, x, u_i)) q(x)$$
$$+ \varphi_x^i(t, x, u_i, q)$$

$$P(t, v, q) := \varphi_t^0(t, q) + \int_{I\!R^n} \sum_{i=1}^N H_i(t, x, u_i(t, x), q) dx$$

$$\bar{P}(t, q) := \varphi_t^0(t, q) + \int_{I\!R_n} \sum_{i=1}^N \min_{u_i \in U_i} H_i(t, x, u_i, q) dx \tag{4.6}$$

$$G(q) := -\varphi^0(t_1, q) .$$

(4.7) THEOREM. Let the following conditions be satisfied:

$$\varphi \in \Phi, \ \mu(\cdot) \in \mathbf{L}_1(\mathbf{T}), \gamma \in I\!R \tag{4.8}$$

$$\bar{P}(t,q) \geq \mu(t), \ for \ all \ (t,q) \in (\mathbf{T}\backslash\mathbf{M}) \times \mathbf{Q} \tag{4.9}$$

$$G(q) \geq \gamma, \ for \ all \ q \in \mathbf{Q} \tag{4.10}$$

Then, the following assertions are true:

(i) $J(z) \geq \lambda(\varphi)$ for all $z \in \mathbf{D}$, where

$$\lambda(\varphi) := \int_{t_0}^{t_1} \mu(t)dt + \gamma + \varphi^0(t_0, p_0); \tag{4.11}$$

(ii) If there exists a sequence $\{z^\alpha\} \subset \mathbf{D}$ such that $lim_{\alpha\to\infty} J(z^\alpha) = \lambda(\varphi)$, then it is minimizing, and $\lambda(\varphi) = \beta$.

The condition of optimality for the process $\bar{z} \in \mathbf{D}$ can be restated in more constructive form as follows:

(4.12) THEOREM. For a pair $\bar{z} \in \mathbf{D}$ to be optimal, it suffices that there exists a function $\varphi \in \Phi$ such that:

$$\bar{P}(t, \bar{p}_*(t)) = \min_{q\in Q} \bar{P}(t,q), t \in (\mathbf{T}\backslash\mathbf{M}) \tag{4.13}$$

$$H_i(t, x, \bar{u}_i(t,x), \bar{p}_*(t)) = \min_{u_i\in\mathbf{U}_i} H_i(t, x, u_i, \bar{p}_*(t)), \quad (t,x) \in (\mathbf{T}\backslash\mathbf{M}) \times \mathbb{R}^n \tag{4.14}$$

$$G(\bar{p}_*(t_1)) = \min_{q\in\mathbf{Q}} G(q) . \tag{4.15}$$

(4.16) REMARK. We call the function φ a Lagrange–Lyapunov function because, first, the function replaces the Lagrange multipliers in their standard role of handling the constraints (2.8) and (3.3) in the optimization process. Second, we exploit the first derivative of φ (for instance, φ^0 with respect to t, and φ^{1i} with respect to x_{*i}), as if it were a Lyapunov function (but not as a vector Lyapunov function in the classical sense, because we have different directions in $\mathbf{T} \times \mathbb{R}^m$). The idea of using the function φ^0 in handling the Fokker–Plank-Kolmogorov equation of motion was considered in the papers [11-13].

So far, we considered the conditions of opimality for a fixed initial condition. Let us now consider another problem of finding a universal function $\varphi \in \Phi$ that solves Problem (3.5) for various initial conditions $(t_0, p_0) \in \mathbf{T} \times \mathbf{Q}$, we consider a family of such problems for all possible $(t_0, p_0) \in \mathbf{T} \times \mathbf{Q}$, replacing at the same time $\mathbf{D}, \beta$, and $J(z)$ by $\mathbf{D}(t_0, p_0)$, $\beta(t_0, p_0)$, and $J(t_0, p_0; z)$. In this way, control becomes a function of the probability density $p_*(t)$ for each $t \in \mathbf{T}$. If we introduce a control strategy $(t, x, q) \to u^+(t, x, q) : \mathbf{T} \times \mathbb{R}^n \times \mathbf{L}_2 \to \mathbf{U} \subset \mathbb{R}^m$, and substitute u^+ into u_* for a given initial condition, we can obtain in principle the trajectory $p_*(\cdot)$ and programmed open–loop control law $\bar{u}_*(t) = u^+(t, \cdot, p_*(t)), \ t \in \mathbf{T}\backslash\mathbf{M}$.

We define the set $\mathbf{Q}_T$ of all possible $(t_0, p_0) \in \mathbf{T} \times \mathbf{Q}$ which satisfy the inclusion $\bar{z} = [\bar{u}_*(\cdot), \bar{p}_*(\cdot)] \in \mathbf{D}(t_0, p_0)$, and prove the following:

(4.17) THEOREM (Hamilton–Jacobi equations). Suppose that the function $\varphi \in \Phi$ and control law u_+ are such that

$$\varphi_t^0(t, q) + \int_{\mathbb{R}^n} \sum_{i=1}^N \min_{u_i \in \mathbf{U}_i} H_i(t, x, u, q) dx = 0 \ , \quad (t, q) \in (\mathbf{T} \backslash \mathbf{M}) \times \mathbf{Q} \qquad (4.18)$$

$$\varphi^0(t_1, q) = 0 \ , \quad q \in \mathbf{Q} \qquad (4.19)$$

$$u_i^+(t, x, q) \in arg \min_{u_i \in \mathbf{U}_i} H_i(t, x, u_i, q) \ , i \in \mathbf{N} \ (t, x, q) \in (\mathbf{T} \backslash \mathbf{M}) \times \mathbb{R}^n \times \mathbf{Q} \ (4.20)$$

Then, the following assertions are true:

(i) For any $(t_0, p_0) \in \mathbf{Q}_T$, the pair $\bar{z} = [\bar{u}_*(\cdot), \bar{p}_*(\cdot)]$ minimizes the functional $J(t_0, p_0; z)$ on $\mathbf{D}(t_0, p_0)$, so that $J(t_0, p_0; \bar{z}) = \beta(t_0, p_0)$;

(ii) Everywhere on the set $\mathbf{Q}_T$ the function $\varphi^0(t, q)$ is equal to the value $\beta(t, q)$ (Bellman's function).

(4.21) REMARK. It is important to underline the "composite nature" of the proposed feedback control: On one hand, the control necessarily depends on the state $x(t)$, while on the other, the system moves to a new state $p_*(t)$, and feedback has to incorporate both arguments, $x(t)$ and $p_*(t)$. The result of [12] was formulated in the case when the control law can depend only on $p_*(t)$, but the conditions of optimality of control $u(t, p_*(t))$ were obtained in the same maner.

(4.22) REMARK. Obviously, the obtained conditions of optimality of the paper can be discussed in the context of control problems with incomplete (current) state information (not necessarily decentralized) in the form $u(t, x^+)$, where x^+ is an observed part of the state vector, for instance, x^+ can obtained as a result of filter. This problem was considered by Katz [10] in the case of total absence of state information, that is, the control function $u(t)$ does not depend on x. Mortensen [12] discussed the case of total absence of state information, but with complete information about the current state distribution, that is, the function $u(t, p_*(t))$ depends on $(t, p_*(t))$. A significant step in this development was taken by Fleming [4], and continued later by Ahmed and Teo [13], who used the control law in the form $u(t, x^+)$. The regular way to handle the constraints in the form of Fokker–Plank–Kolmogorov equation of motion via the method of Lagrange was used in these papers and in many others, but authors did not provide a general scheme for handling these types of information structure constraints. These constraints remained locked inside the formulation of the Maximum Principle, which prevented an algorithmic numerical solution even in the case of LQG. By contrast, we formulated information structure constraints as partial differential equations, added them to the equation of motion, and applied the Lagrange idea to the both sets of the constraints. Finally, we obtained the sufficient conditions for optimality as a set of independent optimization problems at each point (t, x). This crucial feature of our development is of considerable interest in applications.

5. Linear-quadratic-gaussian case

In this section we only assume that probability density function is Gaussian, but we do not assume the control law as a linear function of current state, that is,

$$
\begin{aligned}
p(t, x) &= \Gamma(t)\, exp\{-\frac{1}{2}x^T H(t)x\} \\
u(t, x) &= (u_1^T(t, x), \ldots, u_N^T(t, x))^T
\end{aligned}
\tag{5.1}
$$

and $p(t, x)$ satisfies the Fokker–Plank–Kolmogorov equation (2.8). In (5.1) $H(t)$ is an $n \times n$ matrix, $\Gamma(t)$ is a scalar, and initial condition $p_0(x)$ is defined by the matrix $H(t_0) = H_0$, and the number $\Gamma(t_0) = \Gamma_0$. We also assume that $C(t)_i$ is independent of x, and in Assumption 2.7, $F_i(t) = \frac{1}{2}C_i(t)C_i(t)^T$.

We are going to show that in this consideration the optimal control law can be a linear function, because of the specific way we choose Lagrange–Lyapunov function $\varphi(t, q)$ as

$$
\varphi^0(t, q) = \int_{I\!R^n} \psi^0(t, x)p(t, x)dx \ ,
$$

$$
\varphi^i(t, x, u^i, q) = \psi^i(t, x, u_i) \ ,
$$

$$
\psi^0(t, x) = \frac{1}{2}x^T S(t)x + \nu(t) \ ,
\tag{5.2}
$$

$$
\psi^i(t, x, u_i) = u_i^T W^i(t)\bar{p}(t, x) \ ,
$$

where $n \times n$ matrix $S(t)$, scalar $\nu(t)$, and $m_i \times n^i$ matrices $W^i(t)$ are unknown.

Let M be an $n \times n$ matrix. The notation of the following partitioned matrices is essential throughout this section

$$
M = \begin{bmatrix} M_{11} & M_{12} & M_{13} \\ M_{21} & M_{ii} & M_{23} \\ M_{31} & M_{32} & M_{33} \end{bmatrix} \ , \quad M^i = [\, M_{21} \quad M_{23} \,] \ , \quad M^{ii} = \begin{bmatrix} M_{11} & 0 \\ 0 & M_{33} \end{bmatrix}
\tag{5.3}
$$

where M^i is $n_i \times n^i$ and M^{ii} is $n^i \times n^i$.

Now, we define an optimal trajectory

$$\bar{p}(t,x) = \bar{\Gamma}(t)\,exp\{-\frac{1}{2}x^T\bar{H}(t)x\} \tag{5.4}$$

and omit the derivation of local conditions for optimality for Theorem (4.12), which can be found in [10]. Let us consider the local conditions by obtaining a minimum (4.13) with respect of q, that is,

$$\begin{aligned}
\mathcal{H}_i(t,x,u_i) =&[(A_{ii}x_i + B_iu_i + A^ix^i)^T\psi^0_{x_i}(t,x)\\
&+ \frac{1}{2}(x_i^TQ_ix_i + u_i^TR_iu_i)]\bar{p}(t,x) + \psi^i_{x^i}(t,x,u_i)
\end{aligned} \tag{5.5}$$

and rewrite it in more details as

$$\begin{aligned}
\mathcal{H}_i^*(t,x,u_i) =&(A_{ii}x_i + B_iu_i + A^ix^i)^T(S_{ii}x_i + S^ix^i)\\
&+ \frac{1}{2}(x_i^TQ_ix_i + u_i^TR_iu_i) - u_i^TW^i\bar{H}^{ii}x^i - u_i^TW^i(\bar{H}^i)^Tx_i\;,
\end{aligned} \tag{5.6}$$

where $\mathcal{H}_i(t,x,u_i) = \mathcal{H}_i^*(t,x,u_i)\bar{p}(t,x)$.

Let us group the terms in (5.6) depending on u_i and x^i, and set them equal to zero in order to reach our goal, which is to make u_i independent of x^i. That is,

$$u_i^T(B_i^TS^i - W^i\bar{H}^{ii})x^i = 0, \quad i \in \mathbf{N}\;. \tag{5.7}$$

This produces the unknown matrices W^i as

$$W^i = B_i^TS^i(\bar{H}^{ii})^{-1}\;. \tag{5.8}$$

We can now rewrite (5.6) to get

$$\begin{aligned}
\mathcal{H}_i^*(t,x,u_i) =&x_i^TA_{ii}^TS_{ii}x_i + x_i^TA_{ii}^TS^ix_j + x^{iT}A^{iT}S_{ii}x_i + x^{iT}A^{iT}S^ix^i\\
&+ u_i^TB_i^TS_{ii}x_i + \frac{1}{2}(x_i^TQ_ix_i + u_i^TR_iu_i) - u_i^TW^i\bar{H}^{iT}x_i\;.
\end{aligned} \tag{5.9}$$

Since we can decompose our minimizing operator (4.14) into N parts,

$$\mathcal{H}_i^*(t,x,\bar{u}_i(t,x)) = \min_{u_i\in\mathbf{U}_i}\mathcal{H}_i^*(t,x,u_i) \tag{5.10}$$

we can obtain from (5.9) the conditions

$$R_iu_i + B_i^TS_{ii}x_i - W^i\bar{H}^{iT}x_i = 0\;, \quad i \in \mathbf{N} \tag{5.11}$$

resulting in the optimal decentralized control law

$$u_i(t, x_i) = -R_i^{-1} B_i^T P_i x_i, \quad i \in \mathbf{N} \tag{5.12}$$

which has the familiar LQG form, and

$$P_i = S_{ii} - S^i (\bar{H}^{ii})^{-1} \bar{H}^{iT} . \tag{5.13}$$

The unknowns $S(t)$ and $\nu(t)$ are determined by the following equations:

$$\begin{aligned}
\dot{S}_{ii} + A_{ii}^T S_{ii} + S_{ii} A_{ii} - P_i^T B_i R_i^{-1} B_i^T P_i + Q_i &= 0 \\
\dot{S}_{ij} + A_{ii}^T S_{ij} + S_{ii}^T A_{ij} &= 0 \\
\dot{\nu} + tr(F\bar{H}) &= 0
\end{aligned} \tag{5.14}$$

where the $n_i \times n_j$ matrix S_{ij} corresponds to the interconnection matrix A_{ij}. The closed–loop matrix of the system S has also the familiar form $A_D = A + BK$, with $B = diag\{B_1, B_2, \ldots, B_N\}$, and the gain matrix $K = diag\{K_1, K_2, \ldots, K_N\}$ having the components

$$K_i = -R_i^{-1} B_i^T P_i . \tag{5.15}$$

Associated with equations (5.14) and (2.8) are the two–point boundary conditions

$$\Gamma(t_0) = \Gamma_0, \quad H(t_0) = H_0, \quad S(t_1) = 0, \quad \nu(t_1) = 0 . \tag{5.16}$$

Therefore, we reduced the formulation of the optimal decentralized control law to solution of the two–point boundary value problem involving equations (5.14) with the end–points (5.16).

6. Acknowledgements

The research reported herein was supported by the National Science Foundation under the grant ECS-9114872.

7. References

1. L. S. Pontryagin, V. Boltynskii, R. Gamkrelidze, and E. Mischenko, *The Mathematical Theory of Optimal Processes*, Wiley Interscience, New York, 1961.

2. R. E. Bellman and S. E. Dreyfus, *Applied Dynamic Programming*, Princeton, Princeton University Press, NJ, 1962.

3. W. H. Fleming, Optimal control of partially observable diffusions, *SIAM J. Control*, **6**, (1968), 194–214.

4. G. Leitman, *The Calculus of Variations and Optimal Control*, Plenum, New York, 1981.

5. V. F. Krotov, A Technique of Global Bounds in Optimal Control Theory, *Control and Cybernetics*, **17** (1988), 115-144.

6. D. D. Šiljak, *Decentralized Control of Complex Systems*, Boston, Academic Press, MA, 1991.

7. R. E. Kalman, Contributions to the theory of optimal control, *Bol. Soc. Mat. Mexicana, Second Ser.*, **5** (1960), 102-119.

8. M. M. Khrustalev and S. V. Savastyuk, Optimality conditions of stochastic diffusion systems in problems with constraints on the control-observation process, *Soviet. Math. Dokl.*, **41** (1990), 256-260.

9. S. V. Savastyuk, Method of Lyapunov-Lagrange functions for optimization of dynamic systems with information constraints. *Proceeding of the First World Congress of Nonlinear Analysts,* Paper No WC 757, Tampa, 1992.

10. S. Katz, Best endpoint control of noisy systems, *J.of Electr. and Control,* **12** (1962), 323-343.

11. R. E. Florentin, Optimal control of noisy system, *J. of Control,* 12 (1962), 123-235.

12. R. E. Mortensen, Apriori open loop control of conditions time systems, *Int. J. Control,* **3** (1965), 113-127.

13. N. U. Ahmed nad K. L. Teo, Existence theorem on optimal control of partially observable diffusions, *SIAM J. Control,* **3** (1974), 351-374.

14. S. L. Sobolev, *Some Applications of Functional Analysis in Mathematical Phisics*, Nauka, Moscow, 1988. (in Russian).

15. O. A. Ladyzhinskaya, V. A. Solonnikov, and N. N. Ural'tseva, *Linear and Quasilinear Equations of Parabolic Type*, Nauka, Moscow, 1967. (in Russian).

16. A. N. Kolmogorov and S. F. Fomin, *Elements of Theory of Functions and Functional Analysis*, Nauka, Moscow, 1972. (in Russian).

17. L. A. Lyusternik and V. I. Sobolev, *A Brief Course in Functional Analysis*, Vysshaya Schkola, Moscow, 1982. (in Russian).

Appendix

In order to prove the theorems stated in the paper, we need a number of lemmas, which are based on the results in [8,9].

(A.1) LEMMA. Assume that function $y \to \chi(y) : [a, b] \to \mathbb{R}$ is right-continuous on $[a, b)$. Then, for any $y \in [a, b)$, we have

$$\lim_{h \to 0^+} \int_a^y \frac{\chi(\zeta + h) - \chi(\zeta)}{h} d\zeta = \chi(y) - \chi(a) \qquad (A.2)$$

Proof. From a part of the proof of Theorem 1 in [16, p. 319], we recall that the derivative of the function χ with respect to ζ is limit of the function $\chi^h(\zeta) = \frac{\chi(\zeta+h)-\chi(\zeta)}{h}$ for $h \to 0^+$. Since χ is right–continuous, we have

$$\int_a^y \chi^h(\zeta) d\zeta = \frac{1}{h} \int_a^y \chi(\zeta + h) d\zeta - \frac{1}{h} \int_a^y \chi(\zeta) d\zeta$$
$$= \frac{1}{h} \int_a^{y+h} \chi(\zeta) d\zeta - \frac{1}{h} \int_a^{a+h} \chi(\zeta) d\zeta . \qquad (A.3)$$

Hence, (A.2). Q.E.D.

(A.4) LEMMA. Let $\varphi \in \Phi$, $z \in \mathbf{D}$ hold. Then,

$$\lim_{\substack{\delta \to 0^+ \\ \Delta t \to 0^+}} \int_{\mathbf{T}_\Delta} \frac{\Delta \varphi^0}{\Delta t} dt = \int_{t_0}^{t_1} \left\{ \varphi_t^0 + \sum_{i=1}^N \int_{\mathbb{R}^n} [\chi_{x_i}^T f_i + (\chi_{x_i x_i} F_i)] p dx \right\} dt , \qquad (A.5)$$

where $\Delta \varphi^0 = \varphi^0(t + \Delta t, p_*(t) + \Delta p_*(t)) - \varphi^0(t, p_*(t))$, $\Delta t \in (0, \delta)$, $0 < \delta < (t_1 - t_0)$, $\mathbf{T}_\Delta = [t_0, t_1 - \Delta t]$, $\mathbf{R}_\Delta = \mathbf{T}_\Delta \times \mathbb{R}^n$, $\mathbf{L}_\Delta = \mathbf{L}_2(\mathbf{R}_\Delta)$, $\Delta p_*(t) \in \mathbf{L}_\Delta$.

Proof. Since $\Delta \varphi^0 = \Delta_1 + \Delta_2$, where $\Delta_1 = \varphi^0(t + \Delta t, p_*(t)) - \varphi^0(t, p_*(t))$, $\Delta_2 = \varphi^0(t + \Delta t, p_*(t + \Delta t)) - \varphi^0(t + \Delta t, p_*(t))$, we take the limit of Δ_1 and Δ_2 separately. Assumptions (2.9) and (4.1), along with Theorem 7 of [16, p.87], imply that the function $t \to \Delta_1(t)/\Delta t : \mathbf{T} \to \mathbb{R}$, as a composition of continuous functions $t \to p_*(t) : \mathbf{T} \to \mathbf{Q}$ and $(t, q) \to \varphi^0(t, q) : \mathbf{T} \times \mathbf{L}_2 \to \mathbb{R}$, is continuous for each $\Delta t \in (0, \delta)$. Since $\mathbf{T}_\Delta$ is a compact from $\mathbb{R}$, for each $\Delta t \in (0, \delta)$ the image of $\Delta_1(t)$ is a compact by Theorem 6 of [16, p.95]. Using Lebesque's Theorem 7 of [16, p.285], we obtain

$$\lim_{\substack{\delta \to 0^+ \\ \Delta t \to 0^+}} \int_{\mathbf{T}_\Delta} \frac{\Delta_1(t)}{\Delta t} dt = \int_{t_0}^{t_1} \varphi_t^0 dt \qquad (A.6)$$

Let us now consider $\Delta_2(t)$. By Lagrange's Theorem 2 of [17, p.201] and Assumption (4.2), we obtain

$$\Delta_2(t) = \int_{\mathbb{R}^n} \chi(t + \Delta t, x, p_*(t) + \theta \Delta p_*(t)) \Delta p(t, x) dx \quad \textit{for all} \ \theta \in (0, 1) . \qquad (A.7)$$

350

Then,

$$\int_{\mathbf{T}_\Delta} \frac{\Delta_2(t)}{\Delta t} dt = \int_{\mathbf{T}_\Delta} \int_{I\!R^n} \chi(t,x) \frac{\partial p(t,x)}{\partial t} dx dt + \epsilon(\delta, \Delta t, \Delta p_*(t)) \, , \qquad (A.8)$$

where

$$\epsilon(\delta, \Delta t, \Delta p_*(t)) = \int_{\mathbf{R}_\Delta} \chi(t + \Delta t, x, p_*(t) + \theta \Delta p_*(t)) \frac{\Delta p(t,x)}{\Delta t} dx dt$$
$$- \int_{\mathbf{R}_\Delta} \chi(t,x,p_*(t)) \frac{\partial p(t,x)}{\partial t} dx dt \, . \qquad (A.9)$$

Let us estimate $\epsilon(\delta, \Delta t, \Delta p_*(t))$ for each $\delta > 0$. We have

$$|\epsilon(\delta, \Delta t, \Delta p_*(t))| \le \int_{\mathbf{R}_\Delta} |\chi(t + \Delta t, x, p_*(t) + \theta \Delta p_*(t)) \frac{\Delta p(t,x)}{\Delta t}$$
$$- \chi(t,x,p_*(t)) \frac{\partial p(t,x)}{\partial t}| dx dt$$
$$= \int_{\mathbf{R}_\Delta} |\chi(t + \Delta t, x, p_*(t) + \theta \Delta p_*(t)) \frac{\Delta p(t,x)}{\Delta t}$$
$$- \chi(t + \Delta t, x, p_*(t) + \theta \Delta p_*(t)) \frac{\partial p(t,x)}{\partial t}$$
$$+ \chi(t + \Delta t, x, p_*(t) + \theta \Delta p_*(t)) \frac{\partial p(t,x)}{\partial t}$$
$$- \chi(t,x,p_*(t)) \frac{\partial p(t,x)}{\partial t}| dx dt$$
$$\le \int_{\mathbf{R}_\Delta} |\chi(t + \Delta t, x, p_*(t) + \theta \Delta p_*(t)) \left[\frac{\Delta p(t,x)}{\Delta t} - \frac{\partial p(t,x)}{\partial t} \right]| dx dt$$
$$+ \int_{\mathbf{R}_\Delta} |[\chi(t + \Delta t, x, p_*(t) + \theta \Delta p_*(t)) - \chi(t,x,p_*(t))] \frac{\partial p(t,x)}{\partial t}| dx dt.$$
$$(A.10)$$

Using Theorem 1 of [15, p.104] and Lemma of [14, p.54], we get

$$\lim_{\substack{\delta \to 0^+ \\ \Delta t \to 0^+}} \int_{\mathbf{T}_\Delta} \frac{\Delta \varphi^0}{\Delta t} dt = \int_{t_0}^{t_1} \left[\varphi_t^0 + \int_{I\!R^n} \chi \frac{\partial p(t,x)}{\partial t} dx \right] dt \, . \qquad (A.11)$$

Substituting the corresponding part of identity (2.10) in the last expression, we obtain the lemma. Q.E.D.

(A.12) LEMMA. Let $\varphi \in \Phi$ and $z \in \mathbf{D}$ hold. Then,

$$\int_{I\!R^n} \frac{\partial \varphi^{ik}}{\partial x_{*k}} dx = \int_{I\!R^n} \frac{d\varphi^{ik}}{dx_{*k}} dx = 0 \quad \text{for all} \ t \in \mathbf{T} \backslash \mathbf{M}, \qquad (A.13)$$

for all $k \in \mathbf{N}^i$.

Proof. Because of Assumptions (4.3) and (4.4), we have

$$\int_{I\!\!R^n} \frac{d\varphi^{ik}}{dx_{*k}} dx = \int_{I\!\!R^n} \left[\frac{\partial \varphi^{ik}}{\partial x_{*k}} + \sum_{s=1}^{n} \left(\frac{\partial \varphi^{is}}{\partial u_i} \frac{du_i}{dx_{*k}} \right) \right] dx \quad for\ all\ t \in \mathbf{T} \backslash \mathbf{M}\ . \quad (A.14)$$

Since $u_*(t) \in \mathbf{V}$, we know that $\partial u_i(t,x)/\partial x^i = 0$, and first equality in (A.14) follows. Furthermore, we have

$$\int_{I\!\!R^n} \frac{\partial \varphi^{ik}}{\partial x_{*k}} dx = \lim_{x_{*k} \to +\infty} \int_{I\!\!R^{n-1}} \varphi^{ik} dx^{*k} - \lim_{x_{*k} \to -\infty} \int_{I\!\!R^{n-1}} \varphi^{ik} dx^{*k} \quad for\ all\ t \in \mathbf{T} \backslash \mathbf{M}.$$
$$(A.15)$$

Using Assumption (4.4), we obtain the second equality in (A.14) and, thus, the lemma. Q.E.D.

(A.16) LEMMA. Let $\varphi \in \Phi$, and $z \to \Gamma(z) : \mathbf{D} \to I\!\!R$ be defined as

$$\Gamma(z) = G(p_*(t_1)) + \int_{t_0}^{t_1} P(t, u_*(t), p_*(t))dt + \varphi^0(t_0, p_0)\ . \quad (A.16)$$

Then, $\Gamma(z) = J(z)$ for all $z \in \mathbf{D}$.

Proof. Let us fix an element $z \in \mathbf{D}$. Using Lemmas (A.1), (A.4), and (A.12), we can obtain a new form of functional $\Gamma(z)$, that is,

$$\Gamma(z) = J(z) + \lim_{\substack{\delta \to 0+ \\ \Delta t \to 0+}} \left\{ \int_{\mathbf{T}_\Delta} \frac{\Delta \varphi^0}{\Delta t} dt \right\} + \int_{t_0}^{t_1} \int_{I\!\!R^n} \frac{d\varphi^i}{dx} dx dt - \varphi^0(t_1, p_*(t_1)) + \varphi^0(t_0, p_*(t)).$$
$$(A.17)$$

Then, $\Gamma(z) = J(z)$ for all $z \in \mathbf{D}$.

Proof of Theorem (4.7). Let us estimate functional $\Gamma(z)$ on $\mathbf{D}$ with the help of (4.8)–(4.10) as $\Gamma(z) \geq \lambda(\varphi)$, $z \in \mathbf{D}$, where $\lambda(\varphi)$ is defined by (4.11). Then, by Lemma (A.16), $J(z) \geq \lambda(\varphi)$ holds and $\beta \geq \lambda(\varphi)$. In particular, $J(z^\alpha) \geq \beta \geq \lambda(\varphi)$ for an element $z^\alpha \in \mathbf{D}$ of the sequence $\{z^\alpha\} \subset \mathbf{D}$. Hence, if there exists a sequence $\{z^\alpha\} \subset \mathbf{D}$ such that $lim_{\alpha \to \infty} J(z^\alpha) = \lambda(\varphi)$, then $\lambda(\varphi) = \beta$, and the sequence $\{z^\alpha\} \subset \mathbf{D}$ is minimizing.

Q.E.D.

Proof of Theorem (4.12). Let us introduce a function $\bar{\mu}(\cdot) \in \mathbf{L}_1(\mathbf{T})$ and the constant $\bar{\gamma} \in I\!\!R$ such that

$$\bar{\mu}(t) = \min_{q \in \mathbf{Q}} \bar{P}(t, q), \quad t \in (\mathbf{T} \backslash \mathbf{M}); \quad \bar{\mu} = 0, \quad t \in \mathbf{M} \qquad (A.18)$$

$$\bar{\gamma} = \min_{q \in \mathbf{Q}} \bar{G}(q). \qquad (A.19)$$

352

Obviously, condition (4.18) has the equivalent form

$$
\begin{aligned}
\bar{P}(t, p_*(t)) :&= \varphi^0(t, p_*(t)) + \int_{\mathbb{R}^n} \sum_{i=1}^{N} \min_{u_i \in \mathbf{U}} H_i(t, x, u_i, p_*(t)) dx \\
&= \varphi^0_t(t, \bar{p}_*(t)) + \int_{\mathbb{R}^n} \sum_{i=1}^{N} H_i(t, x, \bar{u}_i(t, x), \bar{p}_*(t)) dx \ .
\end{aligned}
\tag{A.20}
$$

Hence, (A.19), and

$$
\varphi^0_t(t, \bar{p}_*(t)) + \int_{\mathbb{R}^n} \sum_{i=1}^{N} H_i(t, x, \bar{u}_i(t, x), \bar{p}_*(t)) dx = \bar{\mu}(t)
\tag{A.21}
$$

$$
t \in (\mathbf{T} \backslash \mathbf{M}); \quad \bar{\mu} = 0, \quad t \in \mathbf{M} \times \mathbb{R}^n .
$$

Using this fact, we can show that $J(\bar{z}) = \lambda(\varphi)$. Let us consider a pair $\bar{z} \in \mathbf{D}$, and use identity $J(\bar{z}) \equiv \Gamma(\bar{z})$ to get

$$
\Gamma(\bar{z}) = J(\bar{z}) + \int_{t_0}^{t_1} \bar{\mu}(t) dt + \bar{\gamma} + \varphi^0(t_0, p_0) \ .
\tag{A.22}
$$

Since we can set $\mu(t) = \bar{\mu}(t)$ and $\gamma = \bar{\gamma}$, equality $J(\bar{z}) = \lambda(\varphi)$ follows, and the pair $\bar{z} \in \mathbf{D}$ is optimal. $\hfill$ Q.E.D.

Proof of Theorem (4.17). Let us define the function

$$
\Gamma(t_0, p_0; \bar{z}) = G(\bar{p}_*(t)) + \int_{t_0}^{t_1} P(t, \bar{u}_*(t), \bar{p}_*(t)) dt + \varphi^0(t_0, p_0)
\tag{A.23}
$$

for $(t_0, p_0) \in \mathbf{Q}_T$ and the pair $\bar{z} = [\bar{u}_*(\cdot), \bar{p}_*(\cdot)]$, where $\bar{u}_*(t) = u^+(t, \cdot, p_*(t_0, p_0; \bar{z}))$. We recall from Lemma (A.16) that $\Gamma(t_0, p_0; \bar{z}) = J(t_0, p_0; \bar{z})$, and prove the theorem by verifying conditions of Theorem (4.12) for all $(t_0, p_0) \in \mathbf{Q}_T$. Indeed, to establish (4.13)–(4.15), it suffices to require

$$
\begin{aligned}
\bar{P}(t, q) &= 0, \qquad t \in \mathbf{T} \backslash \mathbf{M} \\
G(q) &= 0, \qquad q \in \mathbf{Q} \ ,
\end{aligned}
\tag{A.24}
$$

which is equivalent to (4.18)–(4.20). If for $(t_0, p_0) \in \mathbf{Q}_T$, the pair $\bar{z} = [\bar{u}_*(\cdot), \bar{p}_*(\cdot)]$ with $\bar{u}_*(t) = u^+(t, \cdot, \bar{p}_*(t))$ and $t \in \mathbf{T} \backslash \mathbf{M}$, is such that

$$
\lambda(t_0, p_0; \varphi) = \int_{t_0}^{t_1} \bar{\mu}(t) dt + \bar{\gamma} + \varphi^0(t_0, p_0) \ ,
\tag{A.25}
$$

then we can set $\bar{\mu} \equiv 0$, $\bar{\gamma} \equiv 0$. Hence, $J(t_0, p_0; \bar{z}) = I(t_0, p_0; \bar{z}) = \varphi^0(t_0, p_0) = \lambda(t_0, p_0; \varphi)$, and using Theorem (4.12), we get $J(t_0, p_0; \bar{z}) = \beta(t_0, p_0)$, where $z \in \mathbf{D}(t_0, p_0)$. This implies part (i) of the theorem. We also obtained $\beta(t, q) = \varphi^0(t, q)$ everywhere on $\mathbf{Q}_T$, which is the part (ii). $\hfill$ Q.E.D.

WSSIAA 5 (1995) pp. 353–370

353

PARAMETER ESTIMATION IN DIFFERENTIAL EQUATIONS

K. SCHITTKOWSKI

Mathematisches Institut, Universität Bayreuth

D - 95440 Bayreuth, Germany

Abstract

We describe an approach to estimate parameters in explicit model functions, dynamic systems of equations, Laplace transformations, systems of ordinary differential equations, differential algebraic equations and systems of partial differential equations. Proceeding from given experimental data, i.e. observation times and measurements, the minimum least squares distance of measured data from a fitting criterion is computed, that depends on the solution of the dynamic system. The practical impact of parameter identification is illustrated by a couple of *real life* examples. Also some details about implementation and numerical algorithms used, are presented.

1   Introduction

Parameter estimation plays an important role in many natural science, engineering and other disciplines. The key idea is to estimate unknown parameters in a mathematical model that describes a real life situation, by minimizing the distance of some known experimental data from the theoretical model function values. Thus also model parameters that cannot be measured directly, can be identified by a least squares fit and analysed subsequently in a quantitative way. To sum up, parameter estimation or data fitting, respectively, is extremely important in all practical situations, where a mathematical model and corresponding experimental data are available to describe the behaviour of a dynamic system.

The purpose of the report is to introduce some numerical methods that can be used to compute parameters by a least squares fit in form of a *tool box*. The mathematical model we consider now, has to belong to one of the following categories:

- explicit model functions

- dynamic systems of nonlinear equations

- Laplace transformations of differential equations

- ordinary differential equations with initial values

- algebraic differential equations

- one-dimensional partial differential equations

The general mathematical model to be analysed and the corresponding algorithms could contain additional features with the goal to apply it to a large variety of different applications. Some of them are:

1. More than one fitting criterion is defined, i.e. more than one experimental data set is fitted within a model formulation.

2. The fitting criteria are arbitray functions depending on the parameters to be estimated, the solution of the underlying dynamic system, and the *time* variable.

3. The model may possess arbitrary equality or inequality constraints w.r.t. the parameters to be estimated, and upper and lower bounds for parameters.

4. An additional independent model parameter called *concentration* may be introduced.

5. The model functions can be defined by their Laplace transforms, where the back-transformation is to be done internally.

6. Consistency conditions of initial values for differential algebraic equations are computed internally.

7. In case of partial differential equations, also coupled ordinary differential equations and non-continuous transitions between different areas can be taken into account.

8. Gradients can be evaluated by automatic differentiation, i.e. without additional round-off, truncation or approximation errors, and without compilation of FORTRAN code.

Only for illustration purposes we denote the first independent model variable the *time* variable of the system, the second one the *concentration* variable and the dependend data as *measurement* values of an *experiment*. By this the probably most frequent application is described. On the other hand, these terms may have any other meaning within a model depending on the underlying application problem.

The parameter estimation problem, alternative phrases are data fitting or system identification, is outlined in Section 2. Is is shown, how the dymamic systems have to be adapted to fit into the least squares formulation required for starting an optimization algorithm.

The numerical algorithms are briefly outlined in Section 3. Basically they are FORTRAN codes with the additional option to let model function be evaluated symbolically together with automatic differentiation. An interactive user interface based on a relational database approach, was implemented for organizing data and results.

A couple of case studies are included in Section 4 to motivate the approach and to show the practical impact of the methods investigated. The following application problems are discussed:

- receptor-ligand binding studies
- separation of substances in a distillation column
- multibody system of a truck
- diffusion of a substance through human skin

2 The Parameter Estimation Models

The basic mathematical model is the least squares problem, i.e. the problem of minimization of a sum of squares of nonlinear functions of the following form:

$$p \in \mathbb{R}^n : \quad \begin{aligned} &\min \sum_{i=1}^{l} f_i(p)^2 \\ &g_j(p) = 0 \qquad , j = 1, ..., n_e \\ &g_j(p) \geq 0 \qquad , j = n_e + 1, ..., n_r \\ &p_l \leq p \leq p_u \end{aligned} \qquad (1)$$

Here we assume that the parameter vector p is n-dimensional and that all nonlinear functions are continuously differentiable with respect to p. Upper and lower bounds are treated independently from the remaining constraints.

All least squares parameter estimation algorithms proceed from the above formulation, although in the one or other case different approaches are available to define the objective functions. The assumption, that all problem functions must be smooth, is essential. All efficient numerical algorithm under consideration are based more or less on the Gauss-Newton method, that requires first derivatives.

In the following we restrict all investigations to parameter estimation problems, where one vector valued model function is available, the *fitting criterion function*, with one additional variable called *time*, and optionally with another one called *concentration*. We proceed now from r measurement sets, given in the form

$$(t_i, c_j, y_{ij}^k) , \; i = 1, \ldots, q_1, \; j = 1, \ldots, q_2, \; k = 1, \ldots, r, \qquad (2)$$

where q_1 time values, q_2 concentration values and $q_1 q_2 r$ corresponding measurement values are defined. Together with a vector-valued model function

$$h(p, y, t, c) = (h_1(p, y, t, c), \ldots, h_r(p, y, t, c))^T \qquad (3)$$

that may depend in addition on the solution vector $y(p, t, c)$ of an auxiliary dynamic system, e.g. an ordinary differential equation, we get the above least squares formulation by

$$f_s(p) = w_{ij}^k(h_k(p, y(p, t_i, c_j), t_i, c_j) - y_{ij}^k) \tag{4}$$

where s runs from 1 to $l = q_1 q_2 r$ in any order, and where w_{ij}^k are suitable non-negative weighting factors.

Then the underlying idea is to minimize the distance between the model function at certain *time* and *concentration* points and the corresponding measurement values. This distance is denoted the residual of the problem. In the ideal case the residuals are zero indicating a perfect fit of the model function by the measurements.

By considering now r fitting criteria, the resulting least squares function to be minimized, is

$$\sum_{k=1}^{r} \sum_{i=1}^{q_1} \sum_{j=1}^{q_2} (w_{ij}^k(h_k(p, y(p, y_i, c_j), t_i, c_j) - y_{ij}^k))^2 \tag{5}$$

If there is no dependency of the model function h from any additional dynamic system, we call them explicit model functions.

2.1 Dynamic Systems of Equations

Our special goal is to estimate parameters in systems of nonlinear equations, that depend on the parameters to be estimated, and on the *time* variable. For simplicity we omit the *concentration* variable c, since the variable is treated in acactly the same way as the *time* variable t. In this case, $y(p, t) \in \mathbb{R}^m$ is implicitely defined by the solution of the system

$$y \in \mathbb{R}^m : \quad \begin{aligned} s_1(p, y, t) &= 0 \\ &\cdots \\ s_m(p, y, t) &= 0 \end{aligned} \tag{6}$$

The system functions $s_1, \ldots, s_m$ are assumed to be continuously differentiable with respect to variables p and y. Moreover we require the regularity of the system, i.e. that the system is solvable and that the derivative matrix

$$\left(\frac{\partial s_i(p, y, t)}{\partial y_j} \right)$$

has full rank for all p with $p_l \leq p \leq p_u$ and for all y, for which a solution $y(p, t)$ exists. Consequently the function $y(p, t)$ is differentiable with respect to all p in the feasible domain.

Now let t be fixed and let $y(p, t)$ a solution of the above system. If we denote $S(p, y) := (s_1(p, y, t), \ldots, s_m(p, y, t))^T$ for all p and y, we conclude from the implicit function theorem that the derivative matrix $\nabla y(p, t)$ is the unique solution of the linear system of equations

$$\nabla_p S(p, y(p, t)) + \nabla_y S(p, y(p, t))V = 0 \qquad (7)$$

where V is a $m \times n$-matrix. From this matrix we get easily the derivatives of the fitting function w.r.t. the parameters to be estimated, see Schittkowski [19].

2.2 Laplace Models

In very many practical applications, the model is available in form of a Laplace formulation to simplify the underlying analysis. The numerical algorithms described in this paper, are able to proceed directly from the Laplace transform and to compute its inverse internally by a quadrature formula proposed by Stehfest [24].

The advantage of a Laplace formulation is that the numerical complexity of nonlinear systems can be reduced to a lower level. Linear differential equations, e.g., can be transformed into algebraic equations and partial diffusion problems can be reduced to ordinary differential equations. The simplified systems are often solvable by analytical considerations.

2.3 Systems of Ordinary Differential Equations

Now we assume that the implicit variable y is a solution of the following system of m ordinary differential equations:

$$\dot{y}_1 = F_1(y, p, t, c) \quad , \; y_1(0) = y_1^0(p, c)$$
$$\cdots$$
$$\dot{y}_m = F_m(y, p, t, c) \quad , \; y_m(0) = y_m^0(p, c) \qquad (8)$$

Without loss of generality we assume that, as in many real life situations, the initial time is zero. The initial values of the differential equations $y_1^0(p, c), \ldots, y_m^0(p, c)$ may depend on one or more of the system parameters to be estimated, and on the concentration parameter. In this case, we have to assume in addition that the observation times are strictly increasing.

Again we have to consider the question how to compute the derivative of the solution $y(p, t, c)$ in an efficient way. Basically there are three different approaches:

1. Numerical approximation: The actual parameter vector is perturbed and the derivatives are computed e.g. in form of forward differences.

2. Variational equation: The derivative matrix is the solution of a linear differential equation with initial values, that is solved in parallel, i.e. together with the given equation.

3. <u>Internal differentiation:</u> The underlying numerical integration scheme is differentiated, i.e. an exact derivative of the discrete solution is computed.

Moreover there exists an additional modification known as the *muliple shooting method*, that was developed for the solution of boundary value problems. By defining certain break points, e.g. the *time* values, the system is integrated only within the resulting intervals. By introducing additional optimization variables for the intial values of the piecewise defined differential equation and additional corresponding nonlinear equality constraints, it is guaranteed, that the final trajectory is continuous.

A particular advantage of the shooting method is, that numerical instabilities are prevented, since integration intervals are short. Moreover excellent starting trajectories are derived from the experimental values to be fitted. The disadvantages are, however, that the size of the optimization problems is increased dramatically and that good inital trajectories are provided only if measurements for all individual solution components are available.

2.4   Systems of Algebraic Differential Equations

Parameter estimation problems based on differential algebraic equations, are very similar to those based on ordinary differential equations. The only difference is that we allow additional algebraic equations in the system, i.e. we get additional system variables. Thus the implicit parameter vector $y(p, t, c)$ is the solution of m_o differential and m_a algebraic equations (DAE)

$$
\begin{aligned}
\dot{y}_1 = \quad & F_1(y, p, t, c) \quad &, \; y_1(0) = y_1^0(p, c) \\
& \cdots \\
\dot{y}_{m_o} = \quad & F_{m_o}(y, p, t, c) \quad &, \; y_{m_o}(0) = y_{m_o}^0(p, c) \\
0 = \quad & F_{m_o+1}(y, p, t, c) \quad &, \; y_{m_o+1}(0) = y_{m_o+1}^0(p, c) \\
& \cdots \\
0 = \quad & F_{m_o+m_a}(y, p, t, c) \quad &, \; y_{m_o+m_a}(0) = y_{m_o+m_a}^0(p, c)
\end{aligned}
\tag{9}
$$

This system is called an index 1 problem or an index 1 DAE, if the algebraic equations can be solved w.r.t. the algebraic variables $z := (y_{m_o+1}, \ldots, y_m)^T$ with $m := m_o + m_a$, i.e. if the matrix

$$
\nabla_z F(y, p, t, c)
$$

possesses full rank. In all other cases we get DAE's with a higher index, see e.g. Hairer and Wanner [8] for a suitable definition and more details.

For simplicity we consider now only problems of index 1, since problems with higher index can be transformed to problems of index 1 by successive differentiation of the algebraic equations.

We have to be very careful when defining the inital values of the model, since they must satisfy the consistency equation

$$F_{m_o+j}(y^0(p,c), p, 0, c) = 0$$

for $j = 1, \ldots, m_a$.

Here the intial value $y^0(p, c)$ is a function depending on the parameter vector p to be estimated, and on the concentration variable c.

2.5 One-Dimensional Parabolic Differential Equations

Again we assume without loss of generality, that the initial time is 0. This assumption facilitates the description of the mathematical model and is easily satisfied in practice by a suitable linear transformation of the time variable. To simplify the notation further, we omit the *concentration* variable c.

The model we want to investigate now, is defined by a system of n_p one-dimensional partial differential equations in one or more spatial intervals. These intervals that could describe e.g. certain areas with different diffusion coefficients, are given by the outer boundary values x_L and x_R that define the total integration interval w.r.t. the space variable x, and optionally some additional internal transition points $x_1^a, \ldots, x_{n_a-1}^a$. Thus we get a sequence of $n_a + 1$ boundary and transition points

$$x_0^a := x_L < x_1^a < \ldots < x_{n_a-1}^a < x_{n_a}^a := x_R . \tag{10}$$

For each integration interval, we have to define an arbitrary partial differential equation of the form

$$u_t^i = f^i(x, t, v, u^i, u_x^i, u_{xx}^i, p) , \quad i = 1, \ldots n_a , \tag{11}$$

where $x \in \mathsf{R}$ is the spatial variable with $x_{i-1}^a \leq x \leq x_i^a$ for $i = 1, \ldots, n_a$, $t \in \mathsf{R}$ the time variable with $0 < t \leq T$, $v \in \mathsf{R}^{n_o}$ the solution vector of a coupled system of ordinary differential equations, $u^i \in \mathsf{R}^{n_s}$ the system variable we want to compute, and $p \in \mathsf{R}^n$ the parameter vector to be identified by an outer least squares algorithm.

Any solution of the coupled system depends on the spatial variable x, the time variable t, the parameter vector p, and is therefore written in the form $v(t, p)$ and $u^i(x, t, p)$ for $i = 1, \ldots, n_a$.

For any of both end points x_L and x_R we allow boundary conditions for function or partial derivative values w.r.t. x. In the same way transition conditions between the different areas may be defined. They are allowed at most at transition points and specify function and partial derivative values from one area to the other, see Dobmann and Schittkowski [4] for details.

Since the initial time is assumed to be zero, initial conditions must have the form

$$u^i(x, 0, p) = u_0^i(x, p) , \quad i = 1, \ldots, n_a \tag{12}$$

and are defined for all $x \in \left[x_{i-1}^a, x_i^a \right]$, $i = 1, \ldots, n_a$.

To indicate that the fitting criteria $h_k(p, t)$ depend also on the solution of the differential equation at the boundary and transition points, where k denotes the index of a measurement set, we use the notation

$$h_k(p, y(p, t), t) = \hat{h}_k(p, t, v(t, p), u^{i_k}(x_k, t, p), u_x^{i_k}(x_k, t, p), u_{xx}^{i_k}(x_k, t, p)) \qquad (13)$$

and insert $\hat{h}_k$ instead of h_k into the least squares function. The variable y denotes the implicitely defined solution values at boundary or transition points w.r.t. to an integration interval defined by the index i_k, and corresponding partial derivatives. The spatial variable x_k is the left or right end point, i.e. $x_k \in \{x_{i_k-1}^a, x_{i_k}^a\}$, $k = 1, \ldots, r$, where r denotes the number of measurement sets.

In order to achieve smooth fitting criteria, we assume that all model functions depend continuously differentiable on the parameter vector p. Moreover we assume that the discretized system of differential equations is uniquely solvable for all p with $p_l \leq p \leq p_u$.

3 Numerical Algorithms

The corresponding codes used to get the results of the previous section, realize the algorithms discussed earlier, and allow the numerical identification of parameters in any of the dynamic models under investigation.

First there are four FORTRAN programs called SYSFIT, MODFIT, DAEFIT and PDEFIT, that perform the numerical analysis and that can be used independently of each other. Model functions and data have to be defined by the user in form of two input files, and the codes generate several output files that can be used in many different ways. The usage is documented in Schittkowski [20] (SYSFIT), Schittkowski [21] (MODFIT, DAEFIT), and Dobmann, Schittkowski [4] (PDEFIT).

Nonlinear model functions are either defined in form of FORTRAN code that must be linked to the optimization codes, or are evaluated symbolically. In the latter case, compilation and link of FORTRAN subroutines is not required whenever model functions are specified or altered. A particular advantage of this approach is the automatic differentiation of model functions to avoid numerical truncation errors. The corresponding system is called PCOMP, see Dobmann, Liepelt and Schittkowski [3].

The underlying nonlinear least squares problems are solved by four different algorithms:

DFNLP: By transforming the original problem into a general nonlinear programming problem in a special way, typical features of a Gauss-Newton and quasi-Newton method are retained, see Schittkowski [18]. The resulting optimization problem is solved by a standard sequential quadratic programming code called NLPQL, cf. Schittkowski [17].

DN2GB: The subroutine is a frequently used unconstrained least squares algorithm and was developed by Dennis, Gay and Welsch [2]. The mathematical method is also based on a Gauss-Newton and quasi-Newton approach.

DSLMDF: First successive line searches are performed along the unit vectors by comparing function values only. The one-dimensional minimization is based on a quadratic interpolation. After a search cycle the Gauss-Newton-type method DFNLP is executed with a given number of iterations. If a solution is not obtained with sufficient accuracy, the search cycle is repeated, see Nickel [13] for details.

NELDER: This is a very simple search method developed by Nelder and Mead [12]. Successively edges of a simplex are exchanged by comparing function values only. The edge with highest function value is replaced by another one generated by a move through the center of the actual simplex to the other side.

Note that only DFNLP is able to take linear or nonlinear constraints into account. All algorithms are also capable to solve problems with large residuals. The choice of algorithm NELDER is only useful when the other algorithms fail because of non-differentiable model functions, a very bad starting point or large round-off errors in the function evaluation.

Dynamic systems of nonlinear equations are treated as a general nonlinear programming problems and solved by the FORTRAN code NLPQL, see Schittkowski [17]. Objective function is the sum of squares of the system parameters, and the constraints are identical to the nonlinear system of equations given. Initial values required for the start of an optimization cycle, must be predetermined by the user. They may depend on the parameters of the outer least squares problem. The system of nonlinear equations is solved for each experimental time value. Moreover the gradients of the model function $h(p, z(p,t), t)$ are calculated analytically by the implicit function theorem. In this case a system of linear equations must be solved for each *time* value by numerically stable Householder transformations.

For the solution of ordinary differential equations, it is possible to select among seven different solvers:

DGEAR: Gear's [6] implicit method, cf. Hindmarsh [9].

RK54: Runge-Kutta-Fehlberg method of order 4 to 5, cf. Shampine and Watts [22] or Fehlberg [5].

RK87: Runge-Kutta-Fehlberg method of order 7 to 8, cf. Prince and Dormand [14].

DVERK: Runge-Kutta method of order 5 and 6, cf. Hull, Enright and Jackson [10].

RAD51: Implicit Runge-Kutta method of order 5 for stiff equations, cf. Hairer and Wanner [8]

GRK4T: Runge-Kutta method of order 4, cf. Kaps and Rentrop [11].

IND-DIR: Runge-Kutta-Fehlberg method of order 4 to 5, cf. Shampine and Watts [22] with additional sensitivity analysis implemented by Benecke [1].

The last algorithm IND-DIR is capable to evaluate derivatives of the solution of the ODE internally, i.e. by analytical differentiation of the Runge-Kutta scheme. In all other cases, the gradients are obtained by numerical forward approximations. For solving stiff differential equations, usage of the codes RAD51 or GRK4T is recommended.

Algebraic differential equations are solved by the implicit Runge-Kutta method RAD51, cf. Hairer and Wanner [8]. If consistent initial values cannot be provided by the user, the corresponding nonlinear system of equations is treated as general nonlinear programming problem with equality constraints. A minimum norm solution is computed by the sequential quadratic programming method NLPQL, see Schittkowski [17]. The initial values for the algebraic equations are used as starting values for NLPQL.

Partial differential equations are transformed into a system of ordinary differential equations by discretizing the model functions w.r.t. the spatial variable x. This approach is known as the method of lines, see e.g. Schiesser [16]. We proceed from uniform grid points within each interval and get a discretization of the whole space interval. To approximate the first and second partial derivative of $u(x, t, p)$ w.r.t. the spatial variable at a given point x_j^i, we compute a polynomial interpolation subject to the neighbouring values. Coupled ordinary differential equations are added to the discretized system without any further modification. Sparsity in form a band structure of the Jacobian of the right-hand side, can be taken into account. The discretized system can be solved by any of the ordinary differential equation solvers mentioned above.

4   Application Problems

The practical impact of parameter identification is illustrated by a couple of *real life* examples. In all cases we want to estimate unknown parameters in a mathematical model that describes a dynamic process. The examples represent typical application problems in a somewhat simplified form that arise frequently in biochemistry, chemistry, engineering, and pharmacy.

For the efficient numerical solution of the parameter estimation problems, an *easy-to-use* interactive software system of the author was used. The system is called EASY-FIT and proceeds from an relational database written in MS-Access for storing model information, experimental data and results. Whenever a numerical solution is to be started, suitable input files for the numerical codes MODFIT, DAEFIT, SYSFIT and PDEFIT are generated. Results are stored in the database.

A complete context sensitive help option is included containing additional technical and organisational information e.g. about the input of data and optimization tolerances. All application problems outlined subsequently, have been modelled, analysed and solved by EASY-FIT on a PC with a 90 MHz Pentium processor running under Windows 3.11.

4.1 Receptor-Ligand Binding Studies

The relationship between antibody and antigen or antibody and antigen determinant, i.e. ligand, can be described by a simple mass equilibrium if some assumptions are satisfied, cf. e.g. Rominger and Albert [15]. The simple example to be studied now, consists of one receptor and two ligands, and the mass conservation law yields a system of nonlinear equations

$$
\begin{aligned}
r_1(1 + K_{11}l_1 + K_{12}L_2l_2^{\star}) - R_1 &= 0 \\
l_1(1 + K_{11}r_1) - L_1 &= 0 \\
l_2^{\star}(1 + K_{12}r_1) - 1 &= 0
\end{aligned}
\tag{14}
$$

see Schittkowski [19]. The system parameters are r_1, l_1 and $l_2^{\star}$, and the parameters to be estimated, are K_{11}, K_{12}, R_1 and L_1. L_2 is the independent model or *time* variable to be filled with experimental data. The fitting criterion is $L_1 - l_1$ and we use the starting values $r_1 = R_1$, $l_1 = L_1$ and $l_2^{\star} = 1.0$ for solving the system of nonlinear equations. The last equation was scaled by $l_2^{\star} := l_2/L_2$.

Experimental data and optimal fit can be retrieved from Figure 1. Starting from $K_{11} = 0.05$, $K_{12} = 0.05$, $R_1 = 50$ and $L_1 = 1$, SYSFIT computed a solution with residual 0.00023, where 3.748 function plus 4.068 gradient evaluations are performed. Reason for the relatively large numbers of function and gradient evaluations is the fact that we count them also for individual L_2-values and also for all evaluations needed for solving the nonlinear system. The outer parameter estimation code DN2GB requires 45 iterations to reach the solution.

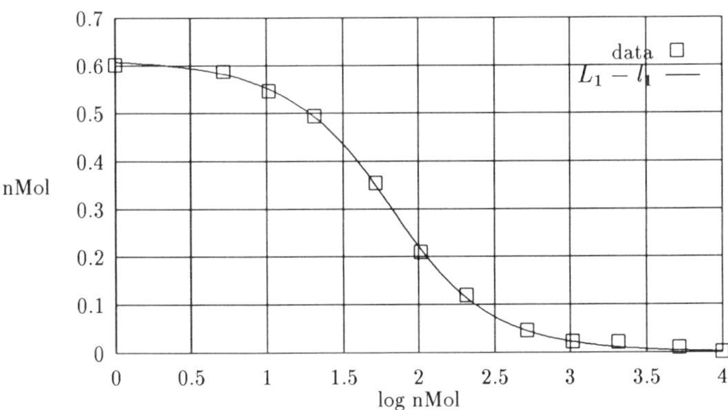

Figure 1: Data and function plot for binding model

364

4.2 Distillation Column

We consider a simple distillation problem for separating two substances, methane and propane. The column consists of $n = 22$ stages with feed at level $n_f = 11$ and condensor at level $n_c = 22$, see Haase [7] for details. The concentration of methane at level i is denoted by $x_i^1(t)$, the concentration of propane by $x_i^2(t)$ for $i = 1, \ldots, n$. Because of the mass balance equation $x_i^1(t) + x_i^2(t) = 1$ for all i, we eliminate $x_i^2(t)$ in our model and define $x_i(t) := x_i^1(t)$ to simplify the notation.

The vapor concentrations of the substance are coupled to the liquid concentrations by the Antoine equation

$$x_i \exp(A_1 - B_1/(C_1 + T_i)) + (1 - x_i) \exp(A_2 - B_2/(C_2 + T_i)) = P \qquad (15)$$

that gives us the temperature T_i for all i and for all time values. Then the vapor concentration is computed from

$$y_i = x_i \exp(A_1 - B_1/(C_1 + T_i))/P \qquad (16)$$

The algebraic equation is solved implicitly by Newton's method whenever a computation of the temperature T_i at level i is requested.

Then we get a system of 22 differential equations in the following form:

$$
\begin{aligned}
\dot{x}_1 &= D(x)(y_2 - x_1) \\
\dot{x}_i &= L_V(x)(x_{i-1} - x_i) + D(x)(y_{i+1} - y_i) \ , \quad i = 2, \ldots, n_f - 1 \\
\dot{x}_{n_f} &= (L_V(x)(x_{n_f-1} - x_{n_f}) + D(x)(y_{n_f+1} - y_{n_f}) + F(x_F - x_{n_f}))/30 \\
\dot{x}_k &= L_A(x)(x_{i-1} - x_i) + D(x)(y_{i+1} - y_i) \ , \quad i = n_f + 1, \ldots, n_c - 1 \\
\dot{x}_{n_c} &= L_A(x)(x_{n_c-1} - x_{n_c}) + D(x)(y_{n_c} - x_{n_c})
\end{aligned}
\qquad (17)
$$

The remaining functions depending on x, are

$$
\begin{aligned}
D(x) &= \frac{3600 Q_{EL}}{x_{n_c} H_1 + (1 - x_{n_c} H_2)} \\
L_V(x) &= R D(x) \\
L_A(x) &= L_V(x) + F
\end{aligned}
$$

where the following constants are used:

$$
\begin{aligned}
Q_{EL} &= 5 & , \quad R &= 0.8 \\
F &= 200 & , \quad H_1 &= 33.071 \\
H_2 &= 42.678 & , \quad x_F &= 0.46 \\
P &= 800 &
\end{aligned}
$$

We want to estimate the Antoine parameters A_k, B_k and C_k for $k = 1, 2$, where we assume that the temperature at feed level 11 is measured. The initial values for the liquid concentrations are all set to one. Measurement data are simulated w.r.t. $A_1 = 18.6$, $B_1 = 3643$, $C_1 = -33.4$, $A_2 = 19.3$, $B_2 = 4117$, $C_2 = -45.7$, and given as follows:

t_k	$T_{n_f}^k$	t_k	$T_{n_f}^k$
0.0	65.863	1.5	72.478
0.1	71.672	2.0	72.479
0.5	72.401	2.5	72.478
1.0	72.476	3.0	72.480

Starting from $A_1 = 20$, $B_1 = 4000$, $C_1 = -30$, $A_2 = 20$, $B_2 = 4000$, $C_2 = -50$, the residual value is reduced by MODFIT from 395 to 0.0022 in 22 iterations.

4.3 Multibody System of a Truck

The general formulation of a mechanical multibody system is given by a differential algebraic equation

$$
\begin{aligned}
\dot{p} &= v \\
M(p, t)\dot{v} &= f(p, v, t) - G(p, t)^T \lambda \\
0 &= g(p, t)
\end{aligned}
\tag{18}
$$

where p are the position coordinates, v the velocities, M the mass matrix, f the applied forces, g the holonomic constraints, λ the generalized constraint forces and G the Jacobi matrix of g w.r.t. p.

We consider now a planar truck model as shown in Figure 2 consisting of 7 bodies, linear and nonlinear suspension elements, and a kinematic joint. The multibody system is described by 22 differential equations and one additional algebraic equation representing the holonomic constraint. Basically the equations of motion are described by

$$
\begin{aligned}
m_1 \ddot{p}_1 &= -F_{10} + F_{13} - m_1 g_e \\
m_2 \ddot{p}_2 &= -F_{20} + F_{23} - m_2 g_e \\
m_3 \ddot{p}_3 &= -F_{13} - F_{23} + F_{35} + F_{34} + F_{43} + F_{53} + F_{37} - m_3 g_e + F_{z_3} \\
l_3 \ddot{p}_4 &= (\alpha_{23} F_{23} - \alpha_{13} F_{13} - \alpha_{37} F_{37} - \alpha_{34} F_{34} - \alpha_{35} F_{35} - \alpha_{43} F_{43} \\
&\quad - \alpha_{53} F_{53}) \cos p_4 + M_{z_4} \\
m_4 \ddot{p}_5 &= -F_{43} - F_{34} - m_4 g_e \\
l_4 \ddot{p}_6 &= (b_{43} F_{43} - b_{34} F_{34}) \cos p_6
\end{aligned}
$$

$$
\begin{aligned}
m_5 \ddot{p}_7 &= -F_{53} - F_{35} + F_{56} - m_5 g_e \\
l_5 \ddot{p}_8 &= \left(c_{53} F_{53} - c_{35} F_{35} - c_{56} F_{56} \right) \cos p_8 \\
m_6 \ddot{p}_9 &= -F_{56} - m_6 g_e \\
m_7 \ddot{p}_{10} &= -F_{37} - m_7 g_e + F_{z_{10}} \\
l_7 \ddot{p}_{11} &= e_{37} F_{37} \cos p_{11} + M_{z_{11}}
\end{aligned}
\tag{19}
$$

see Simeon e.al. [23] for a full description of the mathematical model.

The parameters to be estimated, are the coefficient κ and the parameters d_{13}, d_{23} in the force law of pneumatic springs needed for the computation of F_{13} and F_{23}. Fitting criteria are p_1, p_2, p_3, p_5, p_7, p_9 and p_{10}. Measurement data are simulated subject to $\kappa = 1.4$, $d_{13} = 21593$ and $d_{23} = 38537$ with 14 measurement times between 0 and 3.5. Starting from $\kappa = 1$, $d_{13} = 20000$ and $d_{23} = 30000$, the program DAEFIT approximated an optimal solution in 7 iterations.

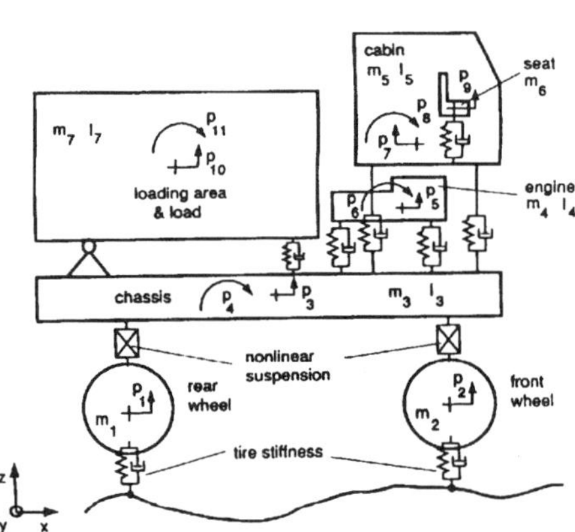

Figure 2: Multibody system of a truck

4.4 Diffusion of a Substance Through Human Skin

By two partial differential equations we describe the diffusion of a substance through human skin taking the Michaelis-Menten effect into account,

$$
\begin{aligned}
u_t^s(x,t) &= D_s\, u_{xx}^s(x,t) - V_{max}\, u^s(x,t)/(K_m + u^s(x,t)) \\
u_t^m(x,t) &= D_m\, u_{xx}^m(x,t) + V_{max}\, u^s(x,t)/(K_m + u^s(x,t))
\end{aligned}
\tag{20}
$$

for $0 \le x \le l$ and $0 \le t \le T$. The corresponding initial conditions are $u^s(x,0) = 0$ and $u^m(x,0) = 0$ for all $x > 0$, and $u^s(0,0) = Y_0 P_s/V_a$, $u^m(0,0) = 0$.

In addition we have coupled ordinary differential equations of the form

$$\dot{v}^s(t) = P_s/V_a F_a D_s \; u_x^s(0,t)$$

$$\dot{v}^m(t) = P_m/V_a F_a D_m \; u_x^m(0,t)$$

$$\dot{w}^s(t) = -P_s/V_a F_a D_s \; u_x^s(l,t)$$

$$\dot{w}^m(t) = -P_m/V_a F_a D_m \; u_x^m(l,t) \qquad (21)$$

for $0 \leq t \leq T$. They describe the behaviour of substance and metabolite at both end points. The initial conditions are $v^s(0) = Y_0$, $v^m(0) = 0$, $w^s(0) = 0$, and $w^m(0) = 0$.

Finally we define four fitting criteria for the concentration of substance and metabolite at both end points, that can be measured by the experiment:

$$h_1(t) = V_a/P_s \; v^s(t)$$

$$h_2(t) = V_a/P_m \; v^m(t)$$

$$h_3(t) = V_a/P_s \; w^s(t)$$

$$h_4(t) = V_a/P_m \; w^m(t)$$

for $0 \leq t \leq T$. Proceeding from data of a real experiment, i.e.

i	t_i	y_i^1	y_i^2	y_i^3	y_i^4
1	1.0E-4	326.60	0.00	0.00	0.00
2	2.0	318.59	0.74	1.73	0.51
3	5.0	308.97	3.15	5.94	2.07
4	7.0	300.14	4.71	8.48	2.88
5	10.0	294.95	6.43	12.89	5.05
6	20.0	268.71	11.14	26.04	11.52
7	30.0	245.76	14.10	38.64	18.19

PDEFIT estimated the optimal parameters

p	inital	optimal
D_s	0.006	0.05582
D_m	0.006	0.000005019
P_s	0.3	0.03448
P_m	0.3	0.8537
V_{max}	1.0	22.13

with a residual 18.005 after 83 iterations. The final contour plots of the concentration distribution are displayed in Figure 3.

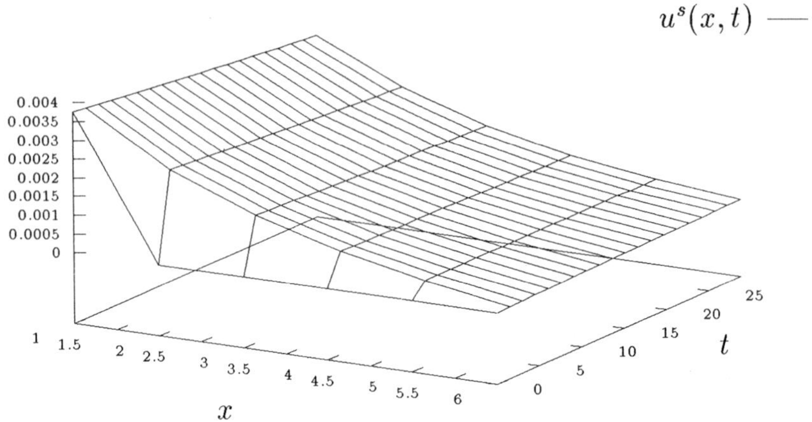

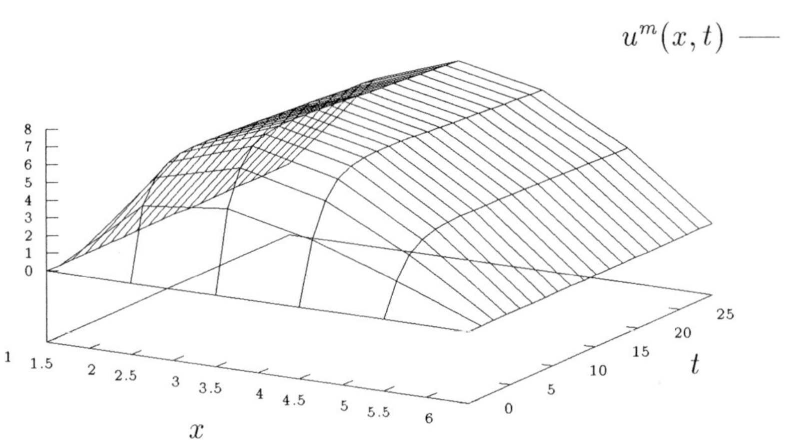

Figure 3: Contour plot for diffusion model

5 Conclusions

It is shown in the paper that a conceptually simple approach can be applied to estimate parameters in dynamic sytems. Basically the solution of the dynamic system is computed by a *black box*. Most important is the question how we can retrieve the gradients of the solution vector w.r.t. the parameters to be estimated. Four different programs are introduced briefly. Some practical application problems show that they can be applied efficiently to solve problems of the size and structure discussed.

References

1. Benecke C. (1993): *Interne numerische Differentiation von gewöhnlichen Differentialgleichungen*, Diplomarbeit, Mathematisches Institut, Universität Bayreuth

2. Dennis J.E.jr., Gay D.M., Welsch R.E. (1981): *Algorithm 573: NL2SOL-An adaptive nonlinear least-squares algorithm*, ACM Transactions on Mathematical Software, Vol. 7, No. 3, 369-383

3. Dobmann M., Liepelt M., Schittkowski K. (1995): *PCOMP: A FORTRAN code for automatic differentiation*, to appear: ACM Transactions on Mathematical Software

4. Dobmann M., Schittkowski K. (1995): *PDEFIT: A FORTRAN code for constrained parameter estimation in partial differential equations*, Report, Mathematisches Institut, Universität Bayreuth

5. Fehlberg E. (1970): *Low-order classical Runge-Kutta formulas with stepsize control*, Computing, Vol. 6, 61-71

6. Gear C.W. (1971): *Numerical Initial Value Problems in Ordinary Differential Equations*, Prentice-Hall

7. Haase G. (1990): *Dynamische Simulation einer Destillationskolonne und Entwurf einer Regelung*, Diplomarbeit, Berufsakademie Mannheim

8. Hairer E., Wanner G. (1991): *Solving Ordinary Differential Equations II. Stiff and Differential-Algebraic Problems*, Springer Series Computational Mathematics, Vol. 14, Springer

9. Hindmarsh A.C. (1974): *GEAR: Ordinary differential equation system solver*, Lawrence Livermore Laboratory, Report UCID-30001, Rev. 3

10. Hull T.E., Enright W.H., Jackson K.R. (1976): *User's guide for DVERK - a subroutine for solving non-stiff ODE's*, TR No. 100, Dept. of Computer Science, University of Toronto, Canada

11. Kaps P., Rentrop P. (1979): *Generalized Runge-Kutta methods of order four with stepsize control for stiff ordinary differential equations*, Numerische Mathematik, Vol. 33, 55-68

12. Nelder J.A., Mead R. (1965): *A simplex method for function minimization*, The Computer Journal, Vol. 7, 308

13. Nickel B. (1995): *Parameterschtzung basierend auf der Levenberg-Marquardt-Methode in Kombination mit direkter Suche*, Diplomarbeit, Mathematisches Institut, Universitt Bayreuth

14. Prince P.J., Dormand J.R. (1981): *High order embedded Runge-Kutta formulae*, Journal on Computational Applied Mathematics, Vol. 7, 67-75

15. Rominger K.L., Albert H.J. (1985): *Radioimmunological determination of Fenoterol. Part I: Theoretical fundamentals*, Arzneimittel-Forschung/Drug Research, Vol.35, No.1a, 415-420

16. Schiesser W.E. (1991): *The Numerical Method of Lines*, Academic Press

17. Schittkowski K. (1985/86): *NLPQL: A FORTRAN subroutine solving constrained nonlinear programming problems*, Annals of Operations Research, Vol. 5, 485–500

18. Schittkowski K. (1988): *Solving nonlinear least squares problems by a general purpose SQP-method*, in: Trends in Mathematical Optimization, K.-H. Hoffmann, J.-B. Hiriart-Urruty, C. Lemarechal, J. Zowe eds., International Series of Numerical Mathematics, Vol. 84, Birkhäuser

19. Schittkowski K. (1994a): *Parameter estimation in systems of nonlinear equations*, Numerische Mathematik, Vol. 68, 129-142

20. Schittkowski K. (1994b): *SYSFIT: A FORTRAN code for estimating parameters in systems of nonlinear equations and coupled mass equilibrium problems*, Report, Mathematisches Institut, Universität Bayreuth

21. Schittkowski K. (1994c): *MODFIT: A FORTRAN code for constrained parameter estimation in differential equations and explicit model functions*, Report, Mathematisches Institut, Universität Bayreuth

22. Shampine L.F., Watts H.A. (1979): *The art of writing a Runge-Kutta code*, Applied Mathematics and Computations, Vol. 5, 93-121

23. Simeon B., Grupp F., Führer C., Rentrop P. (1994): *A nonlinear truck model and its treatment as a multibody system*, Journal of Computational and Applied Mathematics, Vol. 50, 523-532

24. Stehfest H. (1970): *Algorithm 368: Numerical inversion of Laplace transforms*, Communications of the ACM, Vol. 13, 47-49

WSSIAA 5 (1995) pp. 371-391

FOUNDATIONS OF DESIGNING OPTIMAL SYSTEMS AND CONTINGENCY PLANS

Yong Shi
Department of Information Systems and Quantitative Analysis
College of Business Administration
University of Nebraska at Omaha, Omaha, Nebraska 68182, USA

and

Po Lung Yu
School of Business
University of Kansas, Lawrence, Kansas 66045, USA

ABSTRACT

Traditionally, given a system, people try to find an optimal point (solution). This solution only can be used to handle a certain decision situation. In this paper, we introduce the framework of designing optimal systems that can cope with various decision situations under uncertainty to the Optimization Community. We also discuss a number of effective methods for finding the optimal systems as well as their contingency plans. We sincerely invite all interested readers to jointly work on this growing significant research field.

1. Introduction

The problem of optimizing a given system originates in "activity analysis" of linear production systems and in optimal engineering design problems. In the activity analysis, given a production system that transforms a set of resources, *inputs*, into a set of products or services, *outputs* (see Starr [35]), one can use linear programming to identify the best group of activities as an optimal solution for production (see Koopmans [8], Churchman [3]). The traditional optimal engineering design problems are usually formulated by mathematical programming which involves a single objective function and a single availability level of resources (see Wilde [36], Papalambros and Wilde [16]). This approach also finds an optimal point for a given system. Neither the activity analysis nor the engineering design analysis considers two important aspects of formulating and solving optimal system design problems: (i) designing optimal systems and (ii) constructing related contingency plans for each optimal system to cope with various decision situations.

Realizing the importance of designing optimal systems, Zeleny [38, 39] used multiple criteria (MC) de novo programming to design an optimal system. His approach, however, does not explicitly address contingency plans for different decision situations. In a 1990 issue of *Management Science*, Lee, Shi and Yu [12] made the first attempt to consider the above two aspects and proposed a basic procedure to design the optimal systems and construct their corresponding contingency plans by using the MC^2-simplex method derived by Seiford and Yu [17].

According to the basic procedure of [12], with a set of all possible opportunities, one first uses MC^2 linear programming to formulate the design problems. Then one uses the MC^2-simplex method to identify a *set* of *potentially good systems* (PGSs) that contain selected opportunities and can potentially optimize the given design problem under certain ranges of decision parameters, such as the contribution of objectives and resource availability levels. Next, for each PGS, one builds submodels of the design model to construct optimal contingency plans to overcome changes of the decision parameters. The contingency plan for a PGS containing selected opportunities and some slack resources with purchased external resources, if necessary, converts infeasible solutions into feasible ones. Finally, using known techniques of decision making under uncertainty, one selects the optimal linear system(s) from the set of PGSs and their contingency plans as the final decision. Constructing the contingency plans for designing optimal systems *differs* from known postoptimality or sensitivity analysis for linear programming (Gal [5]). The purpose of constructing contingency plans is to ensure the feasibility and optimality over decision situations for the undertaken system design problem; while the sensitivity analysis identifies the optimal solution of a given system as estimates of some system data become available. We proceed this paper as follows.

Section 2 sketches a mathematical model of designing optimal systems and the basic procedure of selecting the optimal system and its contingency plans. Section 3 describes the relationships between the contingency plans, shadow price, and resource price in the designing process. Section 4 explores a method of selecting the optimal systems by using *generalized good systems* (GGSs). These systems contain a large set of selected opportunities and may offer a higher "payoff" than any PGS under certain distributions of decision parameters. Section 5 outlines a number of research topics about the system design problems as well as real-world applications. Section 6 concludes the paper with a few remarks.

2. A Model of Designing Optimal Systems

Given a planning horizon, let $N = \{1, \ldots, n\}$ be n opportunities under consideration. Then, a model of designing optimal linear systems can be formulated by

$$
\begin{aligned}
\text{Max} \quad & \lambda^t Cx \\
\text{s. t.} \quad & Ax \leq D\gamma \\
& x \geq 0,
\end{aligned}
\tag{M1}
$$

where $C \in \mathbf{R}^{q \times n}$, $A \in \mathbf{R}^{m \times n}$, and $D \in \mathbf{R}^{m \times p}$, are matrices of qxn, mxn, and mxp dimensions, respectively; $x \in \mathbf{R}^n$ are decision variables; $\lambda \in \mathbf{R}^q$, $\gamma \in \mathbf{R}^p$ and both (γ, λ) are assumed unknown.

In the above model q situational forces affect the unit contribution, while p situational forces affect resource availability. Both parameters (γ, λ) reflect changes in either internal or external environments of the systems. These environments could be market conditions, economic conditions, social conditions, political conditions, technological conditions, or business philosophy. λ is called the *contribution parameter* and γ the *resource parameter*. Observe that λ and γ may be treated as probability distributions over the situational forces. The interpretation should depend on the individual contexts of linear systems.

If parameters (γ, λ) are known ahead of decision time, we can select the best k opportunities from possible n opportunities as the optimal solution by using linear programming techniques (see e.g., Charnes and Cooper [1], Dantzig [4]). However, if parameters (γ, λ) cannot be known ahead of decision time, the linear programming approach would not be effective, because there are infinitely many possible combinations of (γ, λ). When parameters (γ, λ) are not known before decision time, the change of γ may make the original choice infeasible, while that of λ can render the choice not optimal. Thus, contingency plans need to be constructed to overcome these difficult decision situations.

A *general procedure* of designing the optimal systems is:

(i) to locate a number of good systems associated with feasible (γ, λ) as candidates for the optimal system;
(ii) for each candidate, construct the corresponding optimal contingency plans that offer the optimal contribution under various undesirable (γ, λ) situations;
(iii) use a well-known technique of decision making under uncertainty to select the optimal system(s) from the set of candidates and the corresponding contingency plans as the final decision.

Because of the complex nature of Model (M1), there could be a number of methods to perform the general procedure. When n (the number of possible opportunities), q (number of criteria), and p (the number of resource availability levels) are large, massive computation may be involved in selecting candidates and constructing their contingency plans. Thus, a computer-aided support system is needed to design the optimal systems (Shi, Yu, Zhang and Zhang [32]).

Model (M1) can abstractly be viewed as a "linear program" with parameters (γ, λ). It is well known that if there is an optimal solution, then there is a basic optimal solution that has m basic variables. This suggests the following heuristic assumptions:

Assumption 2. 1. (i) The number of selected opportunities, k, in a good system for Model (M1) should not exceed the number of resources under consideration, m. (ii) The selected k opportunities should be able to "optimize" Model (Ml) under some possible ranges of (γ, λ).

With assumption 2.1, we can use the concept of potential solution of MC^2 linear programming (Yu [37]) to explore a basic procedure of designing the optimal system. Given basic variables $\{x_{j1}, \ldots, x_{jm}\}$ for (M1), we denote the index set of the basic variables by $J = \{j_1, \ldots, j_m\}$. Then , we can write $x(J) = \{x_{j1}, \ldots, x_{jm}\}$. Note that $x(J)$ may contain slack variables. Without confusion, J is called a *basis* for (M1). Let I be an mxm-dimensional identity matrix and 0 be a qxm-dimensional zero matrix. Given a basis J with its basic variables $x(J)$,

374

we define the *associated basis matrix* B_J as the submatrix of $[A, I]$ with column index of J (i.e., column j of $[A, I]$ is in B_J if and only if $j \in J$), and the *associated objective function coefficient* C_J as the submatrix of $[C, 0]$ with column index of J.

Definition 2.1. Given a basis J for Model (M1), define its corresponding
   primal parameter set by $\Gamma_1(J) = \{ \gamma > 0 \mid B_J^{-1} D \gamma \geq 0 \}$; and
   dual parameter set by $\Lambda_1(J) = \{ \lambda > 0 \mid \lambda^t[C_J B_J^{-1} A - C, C_J B_J^{-1}] \geq 0 \}$.

To avoid confusion with parameter sets of the future models, we use the subscript one for both $\Gamma_1(J)$ and $\Lambda_1(J)$ of Model (M1) in definition 2.1.

Statement 2.1. Given a basis J for Model (M1),
   (i) J is called a *primal potential solution* if and only if $\Gamma_1(J) \neq \varnothing$;
   (ii) J is called a *dual potential solution* if and only if $\Lambda_1(J) \neq \varnothing$; and
   (iii) J is called a *potential basis* if and only if $\Gamma_1(J) \times \Lambda_1(J) \neq \varnothing$.
   (iv) The resulting solution $x(J, \gamma) = B_J^{-1} D \gamma \geq 0$ if and only if $\gamma \in \Gamma_1(J)$;
   (v) The objective payoff of $x(J, \gamma)$ is $V(J; \gamma, \lambda) = \lambda^t C_J B_J^{-1} D\gamma$ when (γ, λ) are specified.

Let $\mathcal{J} = \{J_1, \ldots, J_r\}$ be the set of all PGSs derived by using the MC^2-simplex method for Model (M1). Once a PGS J is determined, all other opportunities $j \notin J$ are rejected or not undertaken. The opportunity set of interest reduces to the opportunities in J from N. From statement 2.1, we see that given a PGS J with basic variables $x(J)$, when $\gamma \notin \Gamma(J)$, J is not feasible; and when $\lambda \notin \Lambda(J)$, J is not optimal. Thus, we need to prepare (construct) the corresponding contingency plans for each J to cope with the difficulty (i.e., to make the system feasible and optimal in some sense) by building related submodels of Model (Ml). This leads to research on exploring *various* methods to construct contingency plans for PGSs.

In [12], when some $\gamma \notin \Gamma(J)$, a submodel is built to construct *primal rigid contingency plans* (PRCPs) for a given PGS J. A PRCP for a PGS contains only selected opportunities, slack resources and purchased external resources, if necessary. It converts infeasible solutions to feasible ones and optimizes the submodel related to the PGS under certain ranges of (γ, λ). Note that the PRCP has no *flexibility* to use other possible slack variables (resources) which are not in a PGS.

To construct the PRCPs for a given PGS J, we solve the following submodel:

$$
\begin{aligned}
\text{Max} \quad & \lambda^t C_J x(J) - \alpha^t y \\
\text{s. t.} \quad & B_J x(J) \leq D\gamma + y \qquad\qquad\qquad \text{(M2)}\\
& x(J), y \geq 0,
\end{aligned}
$$

where $\alpha^t = (\alpha_1, \ldots, \alpha_m)$ is the given unit price of purchasing external resources $y = (y_1, \ldots, y_m)^t$, and (γ, λ) are presumed.

The set of all potential bases identified by the MC^2-simplex method for Model (M2) is called the PRCPs for a PGS.

However, using Model (M2) may not be effective to construct all PRCPs for each PGS of $\mathcal{G}$ because there may be r related models which need to be built to complete the task!   In order to facilitate the computation of identifying all PRCPs for each PGS, [12] developed the following *primal augmented model*:

$$\begin{array}{ll} \text{Max} & \lambda^t\,C\,x - \alpha^t\,y \\ \text{s. t.} & A\,x \le D\gamma + y \\ & x,\,y \ge 0. \end{array} \qquad\qquad (M3)$$

Observe that both (M1) and (M2) are submodels of Model (M3) because Model (M1) can be obtained by dropping columns of y from Model (M3) and Model (M2) can be obtained by dropping columns associated with opportunities that are not selected in PGS J from N (i.e., N\J).   To be precise, let us look at the relationship between Model (2) and Model (3).

For a given basis K, without confusion, we decompose its basic variables into $(x(K), y(K))$ with $x(K)$ being selected variables x_j and $y(K)$ being those y_i. Applying the MC^2-simplex method, we have:

Theorem 2.1.   If K is a basis for Model (M2), then K is also a basis for Model (M3).   Conversely, if K is a basis for Model (M3) and contains none of $x(J')$ (the nonbasic variables with respect to a basis J) as basic variables, then K is also a basis for Model (M2).   Furthermore, the basic feasible solution for a given γ, $(x(K, \gamma), y(K, \gamma))$, is identical for the both (M2) and (M3) in either case.

Let the primal parameter sets of K(J) for (M2) and (M3) be $\Gamma_2(K(J))$ and $\Gamma_3(K(J))$, respectively; and the dual parameter sets of K(J) for (M2) and (M3) be $\Lambda_2(K(J), \alpha)$ and $\Lambda_3(K(J), \alpha)$, respectively.   Then, we have:

Corollary 2.1.   (i) $\Gamma_2(K(J)) = \Gamma_3(K(J))$; and (ii) $\Lambda_2(K(J), \alpha) \supseteq \Lambda_3(K(J), \alpha)$, $\alpha \in \mathbf{R}^m$.

Based on the above, all PGSs and their PRCPs can be systematically and effectively located by using the MC^2-simplex method.   This is captured as the following basic procedure [12]:

Procedure   2.1.
Step 1. use the MC^2-simplex method to find a set of PGSs of Model (M1) as candidates for the optimal system.
Step 2. apply theorem 2.1 and corollary 2.1 to solve Model (M3) and identify the PRCPs for each PGS of step 1.
Step 3. use a well-known technique of decision making under uncertainty, such as "maximizing expected payoff", "minimizing the variance of the payoff", "maximin payoff", and "maximizing the probability of achieving a targeted payoff" (Keeney and Raiffa [7], Ziemba and Vickson [40], Yu [37], Shi [18]), to select the "best" system from the set of PGSs and their PRCPs as the final   decision.

The following example, adopted from Shi, Yu and Zhang [30] and Shi [18], is used to illustrate this basic procedure for designing the optimal system:

Example 2.1. Consider a design problem with two criteria and two constraint levels for five possible opportunities:

$$
\text{Max } (\lambda_1, \lambda_2) \begin{pmatrix} 3 & 2 & 1 & 1 & 0 \\ 0 & 1 & 2 & 3 & 3 \end{pmatrix} \begin{pmatrix} x_1 \\ x_2 \\ x_3 \\ x_4 \\ x_5 \end{pmatrix}
$$

$$
\text{s.t.} \begin{pmatrix} 1 & 0 & 2 & 1 & 0 \\ 0 & 1 & 1 & 2 & 2 \\ 1 & 1 & 0 & 0 & 1 \end{pmatrix} \begin{pmatrix} x_1 \\ x_2 \\ x_3 \\ x_4 \\ x_5 \end{pmatrix} \leq \begin{pmatrix} -10 & 20 \\ 40 & 20 \\ 30 & 40 \end{pmatrix} \begin{pmatrix} \gamma_1 \\ \gamma_2 \end{pmatrix}
$$

$$
x_j \geq 0, j = 1, 2, 3, 4, 5.
$$

Let s_i be slack variables for constraints 1, 2, 3 respectively. Then procedure 2.1 is conducted as follows:

Step 1. Using the computer software of the MC^2-simplex method (Chien, Shi and Yu [2]), we find the set of all PGSs, denoted by $\mathcal{J} = \{J_1, J_2, J_3, J_4, J_5\}$ as shown in Table 1. Here, J_1 has (x_1, x_2, s_3) as the basic variables. If J_1 is chosen, then the opportunities $\{1, 2\}$ will be used while s_3 is the slack resource. Assuming $\lambda_1 + \lambda_2 = 1$ and $\gamma_1 + \gamma_2 = 1$, we see that J_1 is optimal whenever $0 < \gamma_1 \leq 2/3$ and $1/5 \leq \lambda_1 \leq 1$. However, if $2/3 < \gamma_1 \leq 1$, J_1 is becomes infeasible and if $0 < \lambda_1 < 1/5$, J_1 is not optimal. Thus, we need to construct the corresponding contingency plans for PGS J_1. The meanings of J_2, J_3, J_4, and J_5 can be similarly explained.

Table 1. Potentially Good Systems

Potentially good systems	Basic variables	$\Gamma_1(J_i)$	$\Lambda_1(J_i)$
J_1	(x_1, x_2, s_3)	$0 < \gamma_1 \leq 2/3$	$1/5 \leq \lambda_1 \leq 1$
J_2	(x_1, x_2, x_4)	$0 < \gamma_1 \leq 2/3$	$1/4 \leq \lambda_1 \leq 1$
J_3	(x_1, x_2, x_5)	$0 < \gamma_1 \leq 2/3$	$1/5 \leq \lambda_1 \leq 1/4$
J_4	(x_1, x_5, s_3)	$0 < \gamma_1 \leq 2/3$	$1/11 \leq \lambda_1 \leq 1/5$
J_5	(x_3, x_5, s_3)	$0 < \gamma_1 \leq 2/3$	$0 \leq \lambda_1 \leq 1/11$

Step 2. In order to effectively construct PRCPs for J_i, $i = 1, 2, 3, 4, 5$, we solve the corresponding augmented problem:

$$\text{Max } (\lambda_1, \lambda_2) \begin{pmatrix} 3 & 2 & 1 & 1 & 0 \\ 0 & 1 & 2 & 3 & 3 \end{pmatrix} \begin{pmatrix} x_1 \\ x_2 \\ x_3 \\ x_4 \\ x_5 \end{pmatrix} - (\alpha_1, \alpha_2, \alpha_3) \begin{pmatrix} y_1 \\ y_2 \\ y_3 \end{pmatrix}$$

$$\text{s.t.} \begin{pmatrix} 1 & 0 & 2 & 1 & 0 \\ 0 & 1 & 1 & 2 & 2 \\ 1 & 1 & 0 & 0 & 1 \end{pmatrix} \begin{pmatrix} x_1 \\ x_2 \\ x_3 \\ x_4 \\ x_5 \end{pmatrix} \leq \begin{pmatrix} -10 & 20 \\ 40 & 20 \\ 30 & 40 \end{pmatrix} \begin{pmatrix} \gamma_1 \\ \gamma_2 \end{pmatrix} + \begin{pmatrix} 1 & 0 & 0 \\ 0 & 1 & 0 \\ 0 & 0 & 1 \end{pmatrix} \begin{pmatrix} y_1 \\ y_2 \\ y_3 \end{pmatrix}$$

$$x_j \geq 0, \ j = 1, 2, 3, 4, 5; \text{ and } y_j \geq 0, \ j = 1, 2, 3.$$

According to theorem 2.1, we first obtain the set of all primal bases, denoted by $\mathcal{K} = \{K_1, \ldots, K_{44}\}$ (see Shi, Yu and Zhang [30] for the details). Then, for each J_i, $i = 1, 2, 3, 4, 5$, we identify the set of all potential bases (PRCPs) for Model (M2) by examining those K of $\mathcal{K}$ such that $x(J) \supseteq x(K)$. Since the parameter set $\Lambda_2(K(J), \alpha)$ is a function of α, for illustration, let $\alpha = (2, 2, 2)^t$ in the process. Table 2 shows the set of all PRCPs for J_1, where basis $K_{40} = J_1$. We see that in terms of the ranges of (γ_1, λ_1), PGS J_1 has four alternative contingency plan sets: $\{K_{40}, K_5, K_{22}\}$, $\{K_7, K_{22}\}$, $\{K_5, K_7, K_{22}\}$, $\{K_{40}, K_7, K_{22}\}$. Similarly, we can find the PRCPs for J_2, J_3, J_4, and J_5.

Table 2. Primal Rigid Contingency Plans for J_1 with $\alpha = (2, 2, 2)^t$

$K_j(J_1)$	$\Gamma_2(K_j(J_1))$	$\Lambda_2(K_j(J_1))$	Payoff $V(K_j(J_1))$
$x(K_{40}) = (x_1, x_2, s_3)$	$0 < \gamma_1 \leq 2/3$	$0 < \lambda_1 \leq 2/3$	$(\lambda_1, \lambda_2) \begin{pmatrix} 50 & 100 \\ 40 & 20 \end{pmatrix} \begin{pmatrix} \gamma_1 \\ \gamma_2 \end{pmatrix}$
$x(K_5) = (x_1, x_2, y_1)$	$0 < \gamma_1 \leq 2/3$	$2/3 \leq \lambda_1 \leq 1$	$(\lambda_1, \lambda_2) \begin{pmatrix} 50 & 100 \\ 40 & 20 \end{pmatrix} \begin{pmatrix} \gamma_1 \\ \gamma_2 \end{pmatrix}$
$x(K_7) = (x_1, x_2, y_3)$	$0 < \gamma_1 \leq 2/3$	$0 < \lambda_1 \leq 1$	$(\lambda_1, \lambda_2) \begin{pmatrix} 50 & 100 \\ 40 & 20 \end{pmatrix} \begin{pmatrix} \gamma_1 \\ \gamma_2 \end{pmatrix}$
$x(K_{22}) = (x_2, y_1, y_3)$	$2/3 < \gamma_1 \leq 1$	$0 < \lambda_1 \leq 1$	$(\lambda_1, \lambda_2) \begin{pmatrix} 50 & 120 \\ 0 & 100 \end{pmatrix} \begin{pmatrix} \gamma_1 \\ \gamma_2 \end{pmatrix}$

Step 3. Suppose we use the "maximizing expected payoff" as the criterion to selecting the final optimal system from $\mathcal{J}$ and the PRCPs. In order to do this, let us assume that γ_1 is independent on λ_1, and (γ_1, λ_1) have the bi-uniform probability distribution $F(\gamma_1, \lambda_1)$. Then, the expected payoff $EV(J_1)$ with respect to the contingency plan set $\{K_{40}, K_5, K_{22}\}$ can be expressed as

378

$$EV(J_1) = \int_{R_1} V(K_{40}(J_1)) \, dF(\gamma_1, \lambda_1) + \int_{R_2} V(K_5(J_1)) \, dF(\gamma_1, \lambda_1)$$

$$+ \int_{R_3} V(K_{22}(J_1)) \, dF(\gamma_1, \lambda_1),$$

where $V(K_{40}(J_1))$, $V(K_5(J_1))$, and $V(K_{22}(J_1))$; and the (γ_1, λ_1) ranges R_j, $j = 1, 2, 3$, are shown in Table 2.

The computing result is $EV(J_1) = 48.45$. If we use other contingency plan sets for J_1, the result is the same. Similarly, we can get $EV(J_2) = 72.21$, $EV(J_3) = 49.83$, $EV(J_4) = 32.88$, and $EV(J_5) = 21.66$. Thus, in terms of the "maximizing expected payoff," J_2 is the "best" (optimal) system we should choose. Note that no matter what decision situation will change in the future, we always has an optimal contingency to deal with. That is, for every (γ_1, λ_1), there is a PRCP of either $\{K_{40}, K_5, K_{22}\}$, $\{K_7, K_{22}\}$, $\{K_5, K_7, K_{22}\}$, or $\{K_{40}, K_7, K_{22}\}$ to optimally undertake the opportunities $\{1, 2, 4\}$.

3. Analysis of External Resource Price

An interesting economic problem in constructing contingency plans for the optimal systems is the analysis of external resource price of Model (M2). Recall that given a basis K for the contingency model (M2) corresponding to PGS J, the dual parameter set $\Lambda_2(K(J), \alpha)$ is a function of α, which is the price of y. Suppose that $y(K, \gamma) > 0$ and that α is very high. Then the decision maker may not purchase y for preparing contingency plans and K may not be a contingency plan for J. Thus, the change of α value can affect the determination of contingency plans.

Let the primal and dual parameter sets of PGS J for (M2) be $\Gamma_2(J)$ and $\Lambda_2(J, \alpha)$, respectively. Then we have the following results (see Shi and Yu [27]):

Theorem 3.1. Given a PGS J for Model (M1), the comparison of Model (M1) to Model (M2) indicates that

(i) $\Gamma_1(J) = \Gamma_2(J)$; and

(ii) if $(\alpha')^t \geq \lambda^t C_J B_J^{-1} \geq 0$ for all $\lambda \in \Lambda_1(J)$, $\alpha' \in \mathbf{R}^m$, then $\Lambda_2(J, \alpha) \supseteq \Lambda_1(J)$.

The condition of $(\alpha')^t \geq \lambda^t C_J B_J^{-1} \geq 0$ has an important economic meaning here. By using the terminology of linear programming [4], we can view $\lambda^t C_J B_J^{-1}$ as the *shadow prices* of PGS J for Model (M1). Then, theorem 3.1 says that if the price of external resource are larger than or equal to the shadow price of PGS J, the dual parameter set of J for Model (M2) contains the dual parameter set of J for Model (M1). Because there exist infinitely many values of α satisfying $(\alpha')^t \geq \lambda^t C_J B_J^{-1} \geq 0$, the theorem also intuitively indicates that if α is large enough, there is no incentive to buy any external resource and/or change the basis for new optimality.

To discuss more details, let Γ be the γ parameter space and Λ be the λ parameter space. Then, the Cartesian space of Γ and Λ is denoted by $\Gamma \times \Lambda$.

Definition 3.1. Given a PGS J of Model (M1),

(i) the space Γ is said to be covered by J if $\Gamma_1(J) \supseteq \Gamma$;

(ii) the space Λ is said to be covered by J if $\Lambda_1(J) \supseteq \Lambda$; and

(iii) the space $\Gamma \times \Lambda$ is said to be covered by J if $\Gamma_1(J) \times \Lambda_1(J) \supseteq \Gamma \times \Lambda$.

From Section 2, we can know that if Γ is covered by PGS J, then J is the contingency plan of (M2) for PGS J because after deleting all $j \notin J$ from N, J covers both Λ and $\Gamma \times \Lambda$ with respect to (M2) (see Shi [18]). Here we are interested in the case when Γ is *not* covered by the contingency plan J selected by Model (M2) for PGS J. What levels of α value can make the contingency plan J cover Λ with respect to Model (M2) so that the purchase of the external resource y is needed only in preparing some PRCPs to cover $(\Gamma \backslash \Gamma_1(J)) \times \Lambda$?

Theorem 3.2. Given a PGS J, suppose that its contingency plan J selected by Model (M2) covers Λ, but not Γ. Then there exists a $\alpha^* \in \mathbf{R}^m$ such that whenever $\alpha \geq \alpha^*$, Λ is covered by the J with respect to (M3). Here, no purchase of y is needed in constructing the contingency plan J for PGS J.

Remark 3.1. For a given PGS J, let $Q_1 = \{\alpha \geq 0 \mid \mathbf{1}_q \alpha^t \geq C_J B_J^{-1}, \alpha \in \mathbf{R}^m, \mathbf{1}_q = (1, \ldots, 1)^t \in \mathbf{R}^m\}$. Then, the "greatest lower bound" of Q_1 is $\alpha^* = (\alpha_1^*, \ldots, \alpha_m^*)^t$, where $\alpha_i^* = \max \{(C_J B_J^{-1})_{ki} \mid 1 \leq k \leq q\}$, $i = 1, \ldots, m$. To verify theorem 3.2, we can choose $\alpha = \alpha^*$.

Let Q be the space of parameter α and $Q_2 = \{\alpha \geq 0 \mid \alpha^t \geq \lambda^t C_J B_J^{-1}, \lambda \in \Lambda_1(J), \alpha \in \mathbf{R}^m\}$. We define $Q_2' = \{\alpha \geq 0 \mid \alpha^t < \lambda^t C_J B_J^{-1}, \lambda \in \Lambda_1(J), \alpha \in \mathbf{R}^m\}$ as the complement set of Q_2. Based on theorem 3.2, we have:

Theorem 3.3. Given a PGS J, the possible cases to purchase the external resource y for preparing contingency plans selected by Model (M2) are:

(i) if $\alpha \in Q_2 \backslash Q_1$ and Γ is covered by J, then y is needed to cover $\Gamma \times (\Lambda \backslash \Lambda_2(J, \alpha))$.

(ii) if $\alpha \in Q_1$ and Γ is *not* covered by J, then y is needed to cover $(\Gamma \backslash \Gamma_2(J)) \times \Lambda$.

(iii) if $\alpha \in Q_2 \backslash Q_1$ and Γ is *not* covered by J, then y is needed to cover $(\Gamma \times \Lambda) \backslash (\Gamma_2(J) \times \Lambda_2(J, \alpha))$.

(iv) if $\alpha \in Q_2'$, then J is not a contingency plan and y is needed to cover $\Gamma \times \Lambda$.

Example 3.1. We use Example 2.1 to illustrate theorem 3.2 and theorem 3.3 (ii) as follows.

Consider the contingency plan J_4 of Model (M2) which does not cover Γ for PGS J_4. From the simplex tableau of basis J_4 (i.e., Table 3), we see that

$$C_{J_4} B_{J_4}^{-1} = \begin{pmatrix} 3 & 0 & 0 \\ 0 & 1.5 & 0 \end{pmatrix}.$$

Table 3. Simplex Tableau of J_4

x_1	x_2	x_3	x_4	x_5	s_1	s_2	s_3	RHS	
1	0	2	1	0	1	0	0	-10	20
0	0.5	-2.5	-2	0	-1	-0.5	1	20	10
0	0.5	0.5	1	1	0	0.5	0	20	10
0	-2	5	2	0	3	0	0	-30	60
0	0.5	-0.5	0	0	0	1.5	0	60	30

Then $\alpha^* = (3, 1.5, 0)^t \in \mathcal{C}_1$. Let $\alpha = \alpha^*$, we have

$$\mathbb{1}_q(\alpha^*)^t - C_{J_4} B_{J_4}^{-1} = \begin{pmatrix} 0 & 1.5 & 0 \\ 3 & 0 & 0 \end{pmatrix}.$$

Thus, $\Lambda_2(J_4, \alpha^*) = \{\lambda > 0 \mid 0 < \lambda_1 \leq 1, \lambda_1 + \lambda_2 = 1\} \supseteq \Lambda$. This means that Λ is covered by the contingency plan J_4 selected by Model (M2) if we choose $\alpha = \alpha^*$. In this case, no purchase of y is needed to cover $\Gamma_2(J_4)X\Lambda$. This illustrates theorem 3.2. However, we still need to purchase y to prepare other PRCPs from Model (M2) to cover $(\Gamma\backslash\Gamma_2(J_4))X\Lambda$. This illustrates theorem 3.3 (ii). Other cases in theorem 3.3 can be similarly demonstrated.

4. Generalized Good Systems and Contingency Plans

Given a system design problem (M1), we can obtain a set of all PGSs, $\mathcal{J} = \{J_1,\ldots, J_r\}$, by using the MC^2-simplex method. Under the restriction of assumption 2.1.(i), the number of selected opportunities, k, in a PGS J can not exceed the number of available resources, m. For some design problems with a large number of possible opportunities, where n farther exceeds k, this assumption may rule many opportunities out of consideration for the optimal system. To *remove* or *relax* the assumption, we can take the *unions* of subsets of $\mathcal{J}$. It is possible to generate some new good systems that can contain more opportunities than any PGS and also have a higher "payoff" than any PGS under certain distribution of (γ, λ). We call such a system the *generalized good system* (GGS), because it is generated by some PGSs.

A natural approach to find all distinct unions, which can be considered as GGSs, of subsets of $\mathcal{J}$ is to check all $2^r - 1 - r$ possible unions of subsets of $\mathcal{J}$. When r is large, this may be a prohibitive task. Also, several different subsets may yield the same union or contain the same variables. To overcome these difficulties, Shi, Yu and Zhang [31] proposed an efficient algorithm of finding all distinct unions from $\mathcal{J}$.

Let Ω be a GGS. Since Ω is generated from some of PGSs, a PGS is a special Ω. We denote the set of all PGSs and GGSs by Ψ. Then Ψ represents all possible candidates for the optimal system. For any given index subset H of opportunity variables and/or slack variables that is not selected as an element of Ψ, we want to know whether there is an element of Ψ, say Ω_0, such that Ω_0 *is preferred to* the H with respect to optimality (see Definition 4.3) for Model (M1). If so, this allows us to ignore the all non-selected subsets of variables from further consideration in designing the optimal system.

Let $N^* = \{1,. . . , n+m\}$ be the index set of all possible opportunity variables x_j and slack variables s_i. Note that $N^* \supseteq N$. Suppose that for given Model (M1), we have obtained Ψ by taking unions of subsets of $\mathcal{J}$. We have:

Definition 4.1. Given a PGS J of Model (M1), we define its *optimal situation set* by $R(J) = \{(\gamma, \lambda) > 0 \mid \gamma \in \Gamma_1(J), \lambda \in \Lambda_1(J)\}$.

Definition 4.2. For any subset H, $H \subseteq N^*$, its *optimal situation set* is defined by $R(H) = \cup \{R(J_k) \mid J_k \subseteq H\}$.

Definition 4.3. For any subsets H_1 and H_2, H_1 and $H_2 \subseteq N^*$, *a optimality preference* with respect to Model (M1) is defined by

(i) $H_1 \succ H_2$ if and only if $R(H_1) \supset R(H_2)$; and

(ii) $H_1 \sim H_2$ if and only if $R(H_1) = R(H_2)$.

By Definition 4.3, the induced preference $\{ \succeq \}$ is defined by: $H_1 \succeq H_2$ if and only if $R(H_1) \supseteq R(H_2)$ for any $H_1, H_2 \subseteq N^*$.

Theorem 4.1. Given a subset H as possible candidate for optimal system, if $H \notin \Psi$, there is a $\Omega_k \in \Psi$ such that $\Omega_k \succeq H$.

Corollary 4.1. For any subset H that contains some opportunity j which is not in any $J_i \in \mathcal{J}$, there is a $\Omega_k \in \Psi$ such that $\Omega_k \succeq H$.

Theorem 4.1 and corollary 4.1 guarantee that Ψ is *the set of all possible candidates* for the optimal system. Therefore, we can concentrate on constructing the corresponding contingency plans for each element of Ψ without paying attention to any non-selected subsets.

As we have discussed how to construct PRCPs for all PGSs in Section 2, we focus on preparing PRCPs for all GGSs as follows.

Since the number of selected opportunities in a GGS Ω may exceed m, the number of resources under consideration, Ω may *not* satisfy assumption 2.1 (i) and (ii). However, by using the MC^2-simplex method, some bases for Model (M1) contained in Ω can optimize (M1) under certain ranges of (γ, λ). To construct PRCPs for a GGS Ω, we are restricted in using the opportunity variables x_j and slack variables s_i involved in Ω and purchase external resources y, if necessary. Given a GGS Ω, let $x(\Omega)$ be the corresponding variables involved in Ω. Then, for a given $x(\Omega)$, we denote A_Ω as the submatrix of $[A, I]$ associated with Ω and C_Ω as the submatrix of $[C, 0]$ associated with Ω.

In constructing the PRCPs for a given GGS Ω, we solve the following submodel:

$$
\begin{aligned}
\text{Max} \quad & \lambda^t C_\Omega x(\Omega) - \alpha^t y \\
\text{s. t.} \quad & A_\Omega x(\Omega) \le D\gamma + y \\
& x(\Omega), y \ge 0,
\end{aligned}
\qquad\qquad \text{(M4)}
$$

where $\alpha^t = (\alpha_1, \ldots, \alpha_m)$ is the given unit price of purchasing external resources $y = (y_1, \ldots, y_m)^t$, and (γ, λ) are presumed.

The set of all potential solutions obtained from Model (M4) is called the PRCPs selected by Model (M4) for a given GGS Ω. Let $x(\Omega')$ be the variables that are not involved in a GGS Ω. Then, we observe that Model (M4) is a submodel of Model (M3). If we drop $x(\Omega')$ from Model (M3), we can identify the PRCPs for each given GGS Ω by applying definition 2.1, statement 2.1, theorem 2.1, and corollary 2.1. We also can develop algorithm similar to procedure 2.1 for designing the optimal system with consideration of GGSs (Shi, Yu and Zhang [30]).

Example 4.1. Based on example 2.1, given the set of all PGSs, $\mathcal{J} = \{J_1, J_2, J_3, J_4, J_5\}$ in Table 1, we generate GGSs by checking 2^5-1-5 (= 26) possible unions of subsets of $\mathcal{J}$. This results in Table 4.

In order to identifying the PRCPs for each Ω_j, $j = 1, 2, 3, 4, 5, 6, 7$, we first assume $\alpha = (2, 2, 2)^t$ and drop $x(\Omega_j')$ from $\mathcal{K} = \{K_1, \ldots, K_{44}\}$. Then, we apply definition 2.1 to find the PRCPs for each Ω_j from the resulting subset of $\mathcal{K}$. For example, let $\Gamma_4(K(\Omega_1))$ be the primal parameter set of Ω_1 and $\Lambda_4(K(\Omega_1), \alpha)$ be the dual parameter set of Ω_1. All of PRCPs for Ω_1 is identified in Table 5. There are four alternative PRCP sets for Ω_1: $\{K_{40}, K_3, K_5, K_{22}, K_{39}\}$, $\{K_{41}, K_3, K_{20}, K_{22}, K_{39}\}$, $\{K_{40}, K_{41}, K_3, K_{20}, K_{22}, K_{39}\}$, and $\{K_{41}, K_3, K_5, K_{20}, K_{22}, K_{39}\}$. The PRCPs for other Ω_j can be similarly found (see Shi [18]).

Now, the set of all possible candidates for the optimal system becomes $\Psi = \{J_1, J_2, J_3, J_4, J_5, \Omega_1, \Omega_2, \Omega_3, \Omega_4, \Omega_5, \Omega_6, \Omega_7\}$. If we also use the "maximizing expected payoff" as the criterion to evaluate $\Omega_1, \Omega_2, \Omega_3, \Omega_4, \Omega_5, \Omega_6$, and Ω_7, then we have $\mathbf{EV}(\Omega_1) = 78.04$, $\mathbf{EV}(\Omega_2) = 55.64$, $\mathbf{EV}(\Omega_3) = 58.21$, $\mathbf{EV}(\Omega_4) = 55.64$, $\mathbf{EV}(\Omega_5) = 58.21$, $\mathbf{EV}(\Omega_6) = 55.64$, and $\mathbf{EV}(\Omega_7) = 35.74$. The optimal system is Ω_1, instead of J_2 in example 2.1. Comparing Ω_1 with J_2, we have the same opportunity variables $\{x_1, x_2, x_4\}$; but, Ω_1 allows us to utilize the slack resource s_3 and have the higher expected payoff than J_2.

Table 4. Generalized Good Systems

Generalized good systems	Basic variables
Ω_1	(x_1, x_2, x_4, s_3)
Ω_2	$(x_1, x_2, x_4, x_5, s_3)$
Ω_3	$(x_1, x_2, x_3, x_4, x_5, s_3)$
Ω_4	(x_1, x_2, x_5, s_3)
Ω_5	$(x_1, x_2, x_3, x_5, s_3)$
Ω_6	(x_1, x_2, x_4, x_5)
Ω_7	(x_1, x_3, x_5, s_3)

Table 5. Primal Rigid Contingency Plans for Ω_1 with $\alpha = (2, 2, 2)^t$

$K_j(\Omega_1)$	$\Gamma_4(K_j(\Omega_1))$	$\Lambda_4(K_j(\Omega_1))$	Payoff $V(K_j(\Omega_1))$
$x(K_{40}) = (x_1, x_2, s_3)$	$0 < \gamma_1 \le 2/3$	$1/7 \le \lambda_1 \le 2/3$	$(\lambda_1, \lambda_2)\begin{pmatrix} 50 & 100 \\ 40 & 20 \end{pmatrix}\begin{pmatrix} \gamma_1 \\ \gamma_2 \end{pmatrix}$
$x(K_{41}) = (x_1, x_2, x_4)$	$0 < \gamma_1 \le 2/3$	$1/7 \le \lambda_1 \le 1$	$(\lambda_1, \lambda_2)\begin{pmatrix} 50 & 100 \\ 40 & 20 \end{pmatrix}\begin{pmatrix} \gamma_1 \\ \gamma_2 \end{pmatrix}$
$x(K_3) = (x_1, x_4, s_3)$	$0 < \gamma_1 \le 1/4$	$0 < \lambda_1 \le 1/7$	$(\lambda_1, \lambda_1)\begin{pmatrix} -70 & 40 \\ 60 & 30 \end{pmatrix}\begin{pmatrix} \gamma_1 \\ \gamma_2 \end{pmatrix}$
$x(K_5) = (x_1, x_2, y_1)$	$0 < \gamma_1 \le 2/3$	$2/3 \le \lambda_1 \le 1$	$(\lambda_1, \lambda_2)\begin{pmatrix} 50 & 120 \\ 40 & 20 \end{pmatrix}\begin{pmatrix} \gamma_1 \\ \gamma_2 \end{pmatrix}$
$x(K_{20}) = (x_2, x_4, s_3)$	$1/4 \le \gamma_1 \le 2/3$	$0 < \lambda_1 \le 1/7$	$(\lambda_1, \lambda_2)\begin{pmatrix} 110 & -20 \\ 30 & 40 \end{pmatrix}\begin{pmatrix} \gamma_1 \\ \gamma_2 \end{pmatrix}$
$x(K_{22}) = (x_2, y_1, y_3)$	$2/3 \le \gamma_1 \le 1$	$3/4 \le \lambda_1 \le 1$	$(\lambda_1, \lambda_2)\begin{pmatrix} 40 & 120 \\ 0 & 100 \end{pmatrix}\begin{pmatrix} \gamma_1 \\ \gamma_2 \end{pmatrix}$
$x(K_{39}) = (x_2, x_4, y_1)$	$2/3 \le \gamma_1 \le 1$	$0 < \lambda_1 \le 3/4$	$(\lambda_1, \lambda_2)\begin{pmatrix} 35 & 130 \\ 15 & 70 \end{pmatrix}\begin{pmatrix} \gamma_1 \\ \gamma_2 \end{pmatrix}$

384

5. Extensions and Applications

This section outlines some extensive studies regarding the model of designing the optimal system (M1) and applications to the real-world problems. The theoretical studies include (i) the construction of primal flexible contingency plans for a candidate (PGS or GGS) for the optimal system, (ii) elimination techniques in solving (M1), and (iii) approaches to dual contingency plans. The applications are optimal selection problems in accounting, production planning, telecommunication, and data file allocation.

5. 1. Primal Flexible Contingency Plans

As described in sections 2 and 4, when we construct a contingency plan for PGS or GGS, three types of variables can be possibly selected. They are opportunity variables x, slack variables s, and external variables y. In selecting a basic basis for PGS or GGS, the ways of utilizing slack variables (unused resources) lead to different approaches to constructing different contingency plans. To contrast with a PRCP, we call the contingency plan that contains not only the variables in a given PGS or GGS, but also other slack variables which are not selected in the PGS or GGS a *primal flexible contingency plan* (PFCP). The PFCP provides flexibility for the decision maker to use all possible slack resources in designing the optimal system.

To construct the PFCPs for a given PGS J, let the corresponding non-basic variables of J be $x(J')$. We decompose $x(J')$ into $x^1(J')$ and $x^2(J')$, where $x^1(J')$ is the set of non-basic variables consisting of those slack variables that are not selected in J, and $x^2(J')$ is the set of non-basic variables consisting of those opportunity variables of N that are not selected in J. Then, we can construct the PFCPs for the J by solving the following submodel:

$$\text{Max} \quad \lambda^t C_J x(J) - \alpha^t y$$
$$\text{s. t.} \quad B_J x(J) + R^1 x^1(J') \leq D\gamma + y \qquad (M5)$$
$$x(J),\ x^1(J'),\ y \geq 0,$$

where R^1 is the submatrix of $[A, I]$ associated with $x^1(J')$, $\alpha^t = (\alpha_1, \ldots, \alpha_m)$ is the given unit price of purchasing external resources $y = (y_1, \ldots, y_m)^t$, and (γ, λ) are presumed.

The set of all potential solutions of (M5) obtained by using the MC^2-simplex method is called the PFCPs for PGS J.

If a GGS Ω is considered, we denote the corresponding variables that are not selected in any bases of Ω by $x(\Omega')$. Similarly to the above, we decompose $x(\Omega')$ into $x^1(\Omega')$ and $x^2(\Omega')$, where $x^1(\Omega')$ is the set of non-selected slack variables and $x^2(\Omega')$ is the set of non-selected opportunity variables with respect to Ω. Then, the PFCPs for the Ω can be constructed by solving the following submodel:

$$\text{Max} \quad \lambda^t C_\Omega x(\Omega) - \alpha^t y$$
$$\text{s. t.} \quad A_\Omega x(\Omega) + A_{\Omega'} x^1(\Omega') \leq D\gamma + y \qquad (M6)$$
$$x(\Omega),\ x^1(\Omega'),\ y \geq 0,$$

where $A_{\Omega'}$ is the submatrix of $[A, I]$ associated with $x^1(\Omega')$, $\alpha^t = (\alpha_1, \ldots, \alpha_m)$ is the given unit price of purchasing external resources $y = (y_1, \ldots, y_m)^t$, and (γ, λ) are presumed.

We call the set of all potential solutions of (M6) obtained by using the MC^2-simplex method the PFCPs for GGS Ω. Both (M5) and (M6) are also the submodels of Model (M3). We can get Model (M5) by dropping the variables $x^2(J')$ from (M3), while have Model (M6) by dropping the $x^2(\Omega')$ from (M3). The details of theoretical discussion of Model (M5) can be found in Shi and Yu [28] and that of Model (M6) is referred to Shi and Zhang [33]. By integrating all of Models (M1)-(M6), we see that (M1)-(M2) and (M4)-(M6) are the different submodels of (M3). This suggests a systematic process that can use Model (M3) to identify PGSs and GGSs and construct the corresponding PRCPs and/or PFCPs. Shi, Yu, Zhang and Zhang [32] explored this process and proposed a computer-aided system to implement it.

5. 2. Elimination Techniques

In designing the optimal system, a key problem is how to efficiently identify the set of all PGSs, $\mathcal{J}$, for Model (M1). Since each PGS of $\mathcal{J}$ must consist of some opportunities from N, we call an opportunity that is not selected in any PGS a *permanently dominated opportunity* (PDO). Note that a permanently dominated opportunity is also not in any GGS, because a GGS is a union of some PGSs. Shi, Yu and Zhang [29] derived a method to eliminate permanently dominated opportunities before $\mathcal{J}$ is identified. This work provides useful means to speed up computation in solving the optimal system problem.

Given a primal potential basis K and a dual potential basis Q for Model (M1), we call (K, Q) a primal-dual pair. Define a half space with respect to Q by $S_j^>(Q) = \{\lambda > 0 \mid \lambda^t [C_Q B_Q^{-1} A_j - C_j] > 0\}$. Then, a sufficient condition for $\lambda^t [C_Q B_Q^{-1} A_j - C_j] > 0$, for all $\lambda \in \Lambda_1(Q)$ is that $\Lambda_1(Q) \subseteq S_j^>(Q)$. To detect whether an opportunity is in a PGS J, we have:

Theorem 5.1. Given a primal-dual pair (K, Q) and a $(\gamma^0, \lambda^0) \in \Gamma_1(K) \times \Lambda_1(Q)$, let $J(\gamma^0, \lambda^0)$ be a nondegenerate PGS. If for given $j \in N$, $\Lambda_1(Q) \subseteq S_j^>(Q)$ and $B_Q^{-1} A_j \geq 0$, then x_j is not a basic variable for $J(\gamma^0, \lambda^0)$. That is, the jth opportunity is not in $J(\gamma^0, \lambda^0)$.

If we used theorem 5.1 and found that a given opportunity $j \in N$ is not in any PGS of $\mathcal{J}$. Then, the j will not be in $\mathcal{J}$ and can be eliminated. This result is given by the following theorem:

Theorem 5.2. Let $\mathcal{D} = \{(K^i, Q^i) \mid i = 1, \ldots, d\}$ be a group of primal-dual pairs such that $\Gamma \times \Lambda \subseteq \cup \{\Gamma_1(K^i) \times \Lambda_1(Q^i) \mid (K^i, Q^i) \in \mathcal{D}\}$. If the jth opportunity is not in any PGS $J^i(\gamma, \lambda)$ with respect to (K^i, Q^i) for any i, then it is not in any PGS $J(\gamma, \lambda)$ of $\mathcal{J}$, and consequently, the jth opportunity corresponding to x_j can be eliminated from further consideration in the process of identifying the entire $\mathcal{J}$.

5. 3. Approaches to Dual Contingency Plans

Both PRCPs and PFCPs for a given PGS or GGS are constructed by adding external resources for the purpose of converting infeasible solutions into feasible and optimal ones. However, for some change of the unit contribution of selected opportunities, these opportunities in a given PGS or GGS may not satisfy the optimality condition under consideration. How to correct these non-optimal situations for designing the optimal system is a question that drives us to explore dual contingency plans.

Shi [19] showed a method of adjusting the unit contribution of selected opportunities in a given PGS to ensure the optimality condition over decision situations. The resulting contingency plans that converts possible non-optimal solutions into optimal ones for the PGS is called *dual rigid contingency plans* (DRCPs), because they only use the selected slack resources, but not all possible slack resources in the process of constructing contingency plans.

We can write the dual model of (M1) as

$$\text{Min } u^t D\gamma$$
$$\text{s. t. } u^t A \geq \lambda^t C \tag{M7}$$
$$u^t \geq 0,$$

where $u^t \in \mathbf{R}^m$ is an unrestricted m-dimensional vector.

Given Model (M1) for the optimal system, we may interpret its economic meaning as to maximize the total contribution of selected opportunities in PGS J by using resources $D\gamma$ "intelligently." The interpretation of Model (M7) may be to minimize the total "implicit value" of the resources $D\gamma$ consumed by producing the opportunities in PGS J [4].

Suppose for some $\lambda \notin \Lambda(J)$, a given PGS J is not optimal. To construct the DRCPs for a given PGS J, we can solve the following submodel:

$$\text{Min } u^t D\gamma - w(J)^t \beta(J)$$
$$\text{s. t. } u^t B_J \geq \lambda^t C + w(J)^t \tag{M8}$$
$$u^t \text{ is unrestricted, and } w(J)^t \geq 0,$$

where $\beta(J) = (\beta_1, \ldots, \beta_m)^t$ is the increments of x(J) and $w(J)^t = (w_1, \ldots, w_m)$ is the unit contribution of $\beta(J)$, and (γ, λ) are presumed.

The set of all potential solutions obtained from Model (M8) is called the set of DRCPs for a PGS J. These DRCPs provide optimality for all possible changes of (γ, λ) when J is considered as a candidate.

A method of constructing dual flexible contingency plans (DFCPs) a PGS J in which all possible slack resources are flexibly adjusted to meet the optimality has been studied by Shi [20]. When the candidate is a GGS, Shi and He [23] provided a procedure for constructing the corresponding dual contingency plans.

5. 4. *Applications*

The flexibility feature of the optimal system models can foster a great potential of applications in many real-world problems. The following application areas are examples:

(i) In business accounting, a transfer pricing problem occurs when a firm processes raw materials into intermediate or final products from one division to another division. This problem involves multiple objective, such as maximizing the overall company's profit, maximizing the market share goal of products in each division, and maximizing the utilized production capacity of the company so that each division manager can avoid any underutilization of normal production capacity. However, existing transfer pricing models formulated by linear programming and goal programming cannot provide a comprehensive scenario of all possible optimal trade-offs between multiple objectives under consideration for a given transfer pricing problem. Using Model (M1), Shi, Kwak and Lee [24] studied these optimal trade-offs.

Similarly, Model (M1) can be used to formulate a capital budget problem that not only incorporates multiple criteria over future periods of investment, but also allows multiple (a group of) decision makers to involve in the decision process (Kwak, Shi, Lee and Lee [9]).

(ii) Production planning is one of the most important functions in the process of production and operations management. In the production planning, managers of a manufacture firm need to make a crucial decision on which specific aggregate levels of production, inventory and work force have to be produced to meet possible demands. Such a decision is traditionally interpreted as to find a best combination of the production, inventory and work force quantities that yields a minimum overall cost.

Production planning has drawn a great deal of attention from both practitioners and academia. Historically, economists initiated research on production planning and product development. Since then, production planning has been extensively studied in the literature. By realizing fluctuations in demand, production and inventory on a seasonal basis, linear programming approach regards production periods that consist of inventory, regular and overtime production rates as "resources," and the corresponding demand periods as "destinations". Then, a transportation method of linear programming is applied to identify an optimal solution that yields the minimum total costs of production and inventory in supplying the product units to the demands. The framework of this model has a great impact on many modeling developments, because it provided a useful channel of not only using mathematical modeling approaches to production planning, but also solving a real aggregate production planning problem by any available commercial linear programming software. A number of authors adopted techniques of goal programming and MC programming to solve aggregate production planning problems with multiple objectives, for instance, treating production, inventory, work force and shortage as individual objectives (For instance, see Singhal and Adlakha [34]).

These current models, however, have explicitly or implicitly assumed that both capacity (supply) and demand levels are fixed. That is, capacity and demand levels in the models are given for a certain situation. It is well known that in reality, capacity and demand fluctuate on both seasonal and situational bases. Using the MC^2 transportation model of Shi [21], which is a special case of Model (M1), Shi and Haase [22] proposed an MC^2 model of aggregate production planning. This production planning model provides manufacture decision makers with a systematic and comprehensive scenario about all possible optimal trade-offs of aggregate production planning depending on the multiple factors.

(iii) Model (M1) can be used to select telecommunication network systems (Lee, Nazem and Shi [10]). Given a set of candidate cities for a telecommunication network, we want to select some subsets of candidate cities as hub city designs for the telecommunication system, such as an information super highway. The selected hub city design (a subset of candidate cities) maximizes the multiple criteria: population, economy, education, health care, and transportation, subject to the location constraints with multiple resource availability levels. The integer format of Model (M1) developed by Shi and Lee [25] can be used to construct optimal hub city designs to reflect policy (decision) makers' goal-seeking and compromise behavior. Similarly, Model (M1) can be applied to the problem of allocating data files over a wide area network (Lee, Shi and Stolen [11]).

6. Conclusions

Designing an optimal system means finding the best subset of opportunities from the set of all possible opportunities for commitment. As decision parameters vary depending on situations, optimal contingency plans must be constructed to overcome difficult decision situations, such as infeasible or non-optimal solutions.

Since both traditional activity analysis of linear production systems and the optimal linear engineering design problems are formulated by linear programming which involves a single objective function and a single resource availability level, they cannot handle the optimal system design problems with multiple objectives and multiple resource availability levels. The model of optimal system design problems proposed by Lee, Shi and Yu [12] is the first model to *explicitly* and *realistically* deal with multiple conflicting criteria and multiple resource availability levels in the real-life design problems. This research project is challenging because the model of the problems involves several decision parameters. It also opens a wide range of research field to the optimization community. In this field, we can build the bridges between the MC^2 framework and various popular techniques to enrich the methodologies of designing the optimal system.

For example, (i) instead of using potential solutions as potential good systems, we may consider the decision makers' goal-seeking and compromise behavior to approach a set of "satisficing solutions" between an upper and a lower aspiration level (Yu [37]). These aspiration levels may be represented by the upper or lower bounds of acceptability for objective payoffs. Being motivated by Zimmermann's [41] work on fuzzy MC linear programming, Liu and Shi [13], Shi and Liu [26], and Liu, Shi and Liu [14] have studied some fuzzy approaches to solving Model (M1).

(ii) Because the models of designing the optimal system involve the parameters γ, λ, α, and β, we may assume that they are random variables with given probability distributions *before* the selection time. Then, we may identify the stochastic solutions for Model (M1) and use them as the potentially good systems (Kall [6]).

(iii) There may be a possibility to solve Model (M1) via neural network techniques (Mangasarian [15]). The resulting solutions, if well-defined, can also serve as the candidates for the optimal system.

We will report any significant results regarding these on-going research problems in the near future.

Acknowledgment

This research has been partially supported by a University Research Fellow award from the University of Nebraska at Omaha for the first author and the National Science Foundation of USA under Grant IST-8418863 for the second author.

References

1. A. Charnes and W. W. Cooper, *Management Models and Industrial Applications of Linear Programming*, Vol. 1 & 2, Wiley, New York, 1961.
2. I. S. Chien, Y. Shi and P. L. Yu, MC2 Program: A Pascal Program run on PC or VAX (revised version), School of Business, University of Kansas, 1989.
3. C. W. Churchman, *The Systems Approach*, Delacorte Press, New York, 1968.
4. G. B. Dantzig, *Linear Programming and Extensions*, Princeton University Press, New Jersey, 1963.
5. T. Gal, *Postoptimal Analysis, Parametric Programming and Related Topics*, McGraw-Hall, New York, 1979.
6. P. Kall, *Stochastic Linear Programming*, Springer-Verlag, Berlin, 1976.
7. R. L. Keeney and H. Raiffa, *Decisions with Multiple Objectives: Preferences and Value Tradeoffs*, Wiley, New York, 1976.
8. T. C. Koopmans, Analysis of production as an efficient combination of activities, in *Activity Analysis of Production and Allocation*, T. C. Koopmans (ed.), Wiley, New York, 1951.
9. W. Kwak, Y. Shi, H. Lee and C. F. Lee, Capital budgeting with multiple criteria and multiple decision makers (Working Paper 94-6), College of Business Administration, University of Nebraska at Omaha, 1994.
10. H. Lee, S. Nazem and Y. Shi, Designing rural area telecommunication networks via hub cities, *Omega: The International Journal of Management Science* **22** (1994), 305-314.
11. H. Lee, Y. Shi and J. D. Stolen, Allocating data files over a wide area network: Goal setting and compromise designs, *Information and Management* **26** (1994), 85-93.
12. Y. R. Lee, Y. Shi and P. L. Yu, Linear optimal designs and optimal contingency plans, *Management Science* **36** (1990), 1106-1119.
13. Y. H. Liu and Y. Shi, A fuzzy programming approach for solving a multiple criteria and multiple constraint level programming problem, *Fuzzy Sets and Systems* **65** (1994), 117-124.

14. Y. Liu, Y. Shi and Y. H. Liu, Duality of fuzzy MC^2 linear programming: A constructive approach, *Journal of Mathematical Analysis and Applications*, forthcoming, 1995.
15. O. L. Mangasarian, Mathematical programming in neural networks, *ORSA Journal on Computing* **5** (1993) 349-360.
16. P. Y. Papalambros and D. J. Wilde, *Principle of Optimal Design*, Cambridge University Press, Massachusetts, 1988.
17. L. Seiford and P. L. Yu, Potential solutions of linear systems: The multicriteria multiple constraint level program, *Journal of Mathematical Analysis and Applications* **69** (1979), 283-303.
18. Y. Shi, Optimal Linear Production Systems: Models, Algorithms, and Computer Support Systems, Ph. D. Dissertation, School of Business, University of Kansas, 1991.
19. Y. Shi, Optimal linear designs and dual contingency plans: A contribution adjustment approach (Working Paper 92-10), College of Business Administration, University of Nebraska at Omaha, 1992.
20. Y. Shi, Constructing flexible dual contingency plans for optimal linear designs with multiple criteria, *Journal of Mathematical Analysis and Applications*, forthcoming, 1995.
21. Y. Shi, A transportation model with multiple criteria and multiple constraint levels, *Mathematical and Computer Modelling*, forthcoming, 1995.
22. Y. Shi and C. Haase, Optimal trade-offs of aggregate production planning with multi-criteria and multi-capacity-demand levels (Working Paper 94-7), College of Business Administration, University of Nebraska at Omaha, 1994.
23. Y. Shi and Z. He, Dual contingency plans in optimal generalized linear designs, *Systems Science* **25** (1994), 1267-1292.
24. Y. Shi, W. Kwak and H. Lee, Optimal trade-offs of multiple factors in international transfer pricing problems (Working Paper 94-3), College of Business Administration, University of Nebraska at Omaha, 1994.
25. Y. Shi and H. Lee, A binary integer linear programming with multicriteria and multiconstraint levels (Working Paper 92-22), College of Business Administration, University of Nebraska at Omaha, 1992.
26. Y. Shi and Y. H. Liu, Fuzzy potential solutions in multi-criteria and multi-constraint level linear programming problems, *Fuzzy Sets and Systems* **60** (1993), 163-179.
27. Y. Shi and P. L. Yu, An introduction to selecting linear optimal systems and their contingency plans, in *Operations Research*, G. Fandel and H. Gehring (eds.), Springer-Verlag, Berlin, 1991.
28. Y. Shi and P. L. Yu, Selecting optimal linear production systems in multiple criteria environments, *Computer and Operations Research* **19** (1992), 585-608.
29. Y. Shi, P. L. Yu, and D. Zhang, Eliminating permanently dominated opportunities in multiple-criteria and multiple-constraint level linear programming, *Journal of Mathematical Analysis and Applications* **183** (1994), 658-705.
30. Y. Shi, P. L. Yu and D. Zhang, Generalized optimal designs and contingency plans in linear systems, *European Journal of Operational Research*, forthcoming, 1995.

31. Y. Shi, P. L. Yu and D. Zhang, Generating new designs using union operations, *Computers and Mathematics with Applications* **27** (1994), 105-117.

32. Y. Shi, P. L. Yu, C. Zhang and D. Zhang, A computer-aided system for linear production designs, *Decision Support Systems* **12** (1994) 127-149.

33. Y. Shi and D. Zhang, Flexible contingency plans in optimal linear designs, *Mathematical and Computer Modelling* **17** (1993), 13-28.

34. K. Singhal and V. Adlakha, Cost and shortage trade-offs in aggregate production planning, *Decision Sciences* **20** (1989), 158-165.

35. M. K. Starr, Evolving concepts in production management, in *Evolving Concepts in Management*, E. Flippo (ed.), Academy of Management Proceedings, 24th Annual Meeting. Reprinted by E. S. Buffa, Readings in Production and Operations. Management, Wiley, New York, 1966, 28-35.

36. D. J. Wilde, *Globally Optimal Design*, Wiley, New York, 1978.

37. P. L. Yu, *Multiple Criteria Decision Making: Concepts, Techniques and Extensions*, Plenum, 1985.

38. M. Zeleny, Optimal system designs with multiple criteria: De novo programming approach, *Engineering Costs and Production Economics* **10** (1986), 89-94.

39. M. Zeleny, Optimizing given systems vs. Designing optimal systems: The de novo programming approach, *International Journal of General Systems* **17** (1990), 295-307.

40. W. T. Ziemba and R. G. Vickson, *Stochastic Optimization Models in Finance*, Academic Press, California, 1975.

41. H. - J. Zimmermann, Fuzzy programming and linear programming with several objective functions, *Fuzzy Sets and Systems* **1**(1978), 45-55.

WSSIAA 5 (1995) pp. 393–408

A SLIDING MODE BASED LEADER-FOLLOWER STRATEGY FOR MULTI-LEVEL HIERARCHIES

MR. MARTIN A. SIKORA
DR. SERGEY V. DRAKUNOV
Department of Electrical Engineering
The Ohio State University
205 Dreese Laboratory, 2015 Neil Avenue
Columbus, OH 43210 USA

DR. JOSE B. CRUZ, JR.
Dean of Engineering
The Ohio State University
142 Hitchcock Hall, 2070 Neil Avenue
Columbus, OH 43210 USA

ABSTRACT

In many hierarchal systems, the lead decision maker does not control all system inputs and thus cannot directly implement a team optimal solution. In this situation the leader must develop incentives to influence the lower level decision makers (followers) into choosing strategies that lead to the leader's desired solution. This paper shows that, for a certain class of systems, affine sliding manifolds may be used as incentives to influence decision makers, level by level, into adopting the team strategy.

I. Introduction

The control of large-scale hierarchal systems has long been of interest both inside and outside of engineering. Within an engineering context these systems occur in many problems, such as shop scheduling in a factory with several departments or controlling electric power generation and transmission among several different utilities in a geographic region. Nonengineering hierarchal problems may also be found in economics (regulatory problems) and management (influencing subordinates).

An important subclass of hierarchal problems are those where the upper hierarchal levels can only influence the lower levels (as opposed to directly controlling them). This leads to the general problem of *incentives*, which has been covered in

394

past literature. This paper will be concerned with the incentive technique known as *leader-follower control* and how such controllers may be formulated using sliding mode techniques.

The specific goal of this paper is to extend the work of Cruz, Drakunov, and Sikora[2] to a multilevel hierarchy, with each hierarchy consisting of a single decision maker. In this formulation there will be $i = 1 \ldots M$ decision makers, with each individual decision maker controlling some corresponding input u_i. The first decision maker is the absolute leader, who will set policy for the group; the Mth decision maker is the absolute follower, who will not be able to influence any in the group. The ith decision maker ($i = 2 \ldots M - 1$) has a dual role, and will act both as the follower for the $(i - 1)^{st}$ decision maker and the leader for the $(i + 1)^{st}$ decision maker.

A. Leader-Follower Control

Leader-Follower control is a variant of the *Stackelberg* strategy. This strategy was proposed by von Stackelberg[8] in 1934 as a way to analyze static economic problems. Chen and Cruz[1] later extended this technique and applied it to dynamic hierarchal problems as well.

The Stackelberg strategy is a biased information strategy. In the strategy, the decision maker with more authority (called the leader) is assumed to know not only his cost functional $J_1(u_1, u_2)$, but also the cost functional $J_2(u_1, u_2)$ of the second decision maker (called the follower). The second decision maker is assumed to know only the cost functional $J_2(u_1, u_2)$. As the leader knows the follower's cost functional he is then able to compute $\gamma : u_1 \mapsto u_2$, the function that defines how the follower will react to any control chosen by the leader. The Stackelberg strategy set of the leader will then be all controls u_1^* which minimize $J_1(u_1, \gamma(u_1))$.

One major weakness of the Stackelberg strategy is that the best performance the leader can achieve is not guaranteed to be the leader's optimal performance. Ho, Luh, and Olsder[7] felt that in following the Stackelberg strategy the leader was actually taking a subservient role. In proposing the leader-follower strategy, they reformulated the problem so that the leader determines an incentive function $g : u_2 \mapsto u_1$ and declares this function openly to the follower. If the follower is rational (meaning that the follower will not choose a suboptimal strategy out of spite) then the follower will restrict his decision space and calculate $u_2^* = arg \min J_2(g(u_2), u_2)$. With this strategy it is the follower which truly takes the subservient position, and with careful choice of incentive it is possible for the leader to obtain his global optimal solution.

Zheng, Basar, and Cruz[11] investigated sufficient conditions which may be used to develop an incentive that will guarantee the leader may obtain his optimal. In the standard incentive design methodology it is assumed that the leader precomputes a desired solution (u_1^*, u_2^*). The desired set of controls for the follower may the be defined as $\Omega_d = \{(u_1, u_2) \in J_2(u_1, u_2) \leq J_2(u_1^*, u_2^*)\}$, which is the set of all control combinations which the follower would prefer to follow instead of (u_1^*, u_2^*). The goal of the leader is to create an incentive that will intersect with Ω_d only at the point (u_1^*, u_2^*).

Three propositions were given concerning incentives[11]. In the first proposition the discontinuous threat strategy was proposed. This incentive strategy takes the form

$$g(u_2) = \begin{cases} u_1^* & \text{if } u_2 = u_2^* \\ \text{any } u_1 \text{ such that } (u_1, u_2) \notin \Omega_d & \text{otherwise} \end{cases}$$

The second proposition demonstrates that for a continuous and strictly convex $J_2(u_1, u_2)$ any desired decision pair (u_1^*, u_2^*) may be induced by a continuously differentiable incentive function. The third proposition adds that if $J_2(u_1, u_2)$ is differentiable and if

$$\nabla_{u_1} J_2(u_1^*, u_2^*) \neq 0$$

then the desired decision pair can be induced by an affine incentive strategy

$$g(u_2) = u_1^* - Q(u_2 - u_2^*)$$

where Q is a linear operator whose adjoint satisfies

$$\nabla_{u_2} J_2(u_1^*, u_2^*) = Q^* \nabla_{u_1} J_2(u_1^*, u_2^*).$$

This affine incentive will tangentially meet the desired decision set for the follower at the point (u_1^*, u_2^*).

Tolwinski[9], however, showed that these affine incentive strategies are not always well-suited for decision making, as the incentive is not robust with respect to nonoptimal behavior by the follower. The threat, when used to punish the follower's nonoptimal behavior, will only further perturb the system away from the desired trajectory. Further, an incentive function based on control values is extremely sensitive to external disturbances or changes in the plant parameters. For this reason, other incentives were examined to add robustness to the leader-follower strategy.

B. *Sliding Mode Control*

Sliding Mode Control has for many years (Flugge-Lotz[6] in the 1950's and Emelyanov[5] in the 1960's) been recognized as one of the key approaches for the

396

systematic design of robust controllers for complex nonlinear dynamic systems operating under uncertainty conditions. The interest in this technique is increasing rapidly. In cases when many classical control methods are not applicable (due to the strong nonlinearity and/or uncertainty of the problem), sliding mode control proved to be an effective tool for rejecting disturbances. An inherent insensitivity of the systems with sliding modes to parameter variations and disturbances eliminates the necessity of exact system modeling (see, for example, DeCarlo[3], Utkin[10] and references therein).

The attractiveness of sliding modes to control designers is explained by many reasons. First of all the design idea is very simple: in order to solve a control problem, such as stabilizing a system or tracking a desired trajector,y one needs only to reformulate the objective as a problem of maintaining a certain relation between the system state variables. Then the control is sought such that the set in the state space where this relation is true forms a sliding manifold, i.e. the integral manifold reachable in finite time. The last is achieved by using discontinuous control action. In real life systems, the element which implements a discontinuous function (for example a relay) switches at high frequency. Its input (the distance to the sliding manifold) may be close the zero while its output (more precisely its average value) takes finite values. Hence the element implements a high (theoretically infinite) gain, which is the conventional tool to suppress the influence of disturbances and uncertainties in the system behavior. Unlike using continuous control, however, the invariance effect can be attained by using finite control actions.

Drazenovic[4] showed that a system in sliding mode will reject certain plant parameter variation and exogenous disturbances. Since then robustness issues have been well studied.

In sliding mode control state space is partitioned into different regions using switching manifolds of the form $s(x) = 0$, where $x \in \Re^n$. Controls are generally discontinuous on the surfaces $s_i(x) = 0$ and take the form

$$u_i = \begin{cases} u_i^+ & \text{if } s_i(x) > 0 \\ u_i^- & \text{if } s_i(x) < 0 \end{cases} \tag{1}$$

for $i = 1 \ldots n$.

A system may be driven to a manifold $s(x) = 0$ if two conditions occur. First, in some region around the switching manifold the derivative of the sliding manifold equation along the system state trajectory must satisfy

$$\dot{s}(x)s(x) < 0, \tag{2}$$

which ensures the sliding surface is locally reachable. The second condition is that

$$\lim_{s(x)\to 0^-} \dot{s} > 0 \quad \text{and} \quad \lim_{s(x)\to 0^+} \dot{s} < 0. \tag{3}$$

This will ensure that the system will reach the sliding surface in finite time, instead of approaching the manifold asymptotically. Eq. (3) is often referred to as the sliding mode existence requirement.

As may be seen from the above conditions, once the system reaches a sliding surface no trajectory will leave the manifold. At this point the system is said to be in the *sliding mode*, and system dynamics may be described according to an equivalent control method developed by Utkin[10]. The sliding mode equivalent control is defined as

$$u_{eq} = - \left[\frac{\partial s}{\partial x}(x) B(t, x) \right]^{-1} f(t, x). \tag{4}$$

Although this continuous equivalent control could be applied directly, the discontinuous control not only guarantees the stability of $s(x) = 0$ but also does not require an exact knowledge of system state equations (hence the invariance to parameter variations). Further, with discontinuous control the system is impervious to disturbances under a sufficient magnitude, as to perturb the system away from the sliding mode a disturbance must be of a large enough magnitude to force the system into a region where Eq. (2) does not hold.

C. *Leader-Follower Control Under Sliding Mode*

As mentioned previously, the classical leader-follower problem leads to non-robust incentive strategies. Further, the computations necessary to develop an incentive are often quite complex. Cruz, Drakunov, and Sikora[2] presented a way of reformulating the leader follower problem that is based on Euler-Lagrange equations. This Euler-Lagrange formulation was used because it leads to a technique that is both more robust and more tractable.

As a review of the process, we consider the system

$$\begin{aligned} \dot{x} &= u, \\ \dot{y} &= v, \end{aligned} \tag{5}$$

where u is the leader's control and v is the follower's control. Further, the leader desires the control and trajectory set (u^*, v^*, x^*, y^*) (found from off-line computation of the leader's cost function) and the leader knows the follower's cost functional

$$J_f(t, x, y, \dot{x}, \dot{y}) = \int_{t_0}^{t_f} g(t, x(t), y(t), \dot{x}(t), \dot{y}(t)) \, dt. \tag{6}$$

398

The variation δJ_f, computed on an extremal, is then given by

$$\delta J_f(t, x, y, \dot{x}, \dot{y}) = \int_{t_0}^{t_f} (H_x(t)\Delta x + H_y(t)\Delta y)\, dt = 0 \qquad (7)$$

where H_x and H_y are the Euler-Lagrange terms

$$H_x(t) = \frac{\partial g}{\partial x}(t, x^*(t), \dot{x}^*(t), y^*(t), \dot{y}^*(t)) - \frac{d}{dt}\left(\frac{\partial g}{\partial \dot{x}}(t, x^*(t), \dot{x}^*(t), y^*(t), \dot{y}^*(t))\right) \quad (8)$$

$$H_y(t) = \frac{\partial g}{\partial y}(t, x^*(t), \dot{x}^*(t), y^*(t), \dot{y}^*(t)) - \frac{d}{dt}\left(\frac{\partial g}{\partial \dot{y}}(t, x^*(t), \dot{x}^*(t), y^*(t), \dot{y}^*(t))\right) \quad (9)$$

and

$$\Delta x = x(t) - x^*(t), \qquad (10)$$
$$\Delta y = y(t) - y^*(t). \qquad (11)$$

Normally, the next step would be to compute the conditions which make $H_x(t) = 0$ and $H_y(t) = 0$; another way to guarantee that $\delta J_f(t, x^*, y^*, \dot{x}^*, \dot{y}^*) = 0$ is for the leader to declare the incentive to be that the system will always be contained on the sliding manifold

$$s(t, x, y) = H_x(t)(x - x^*) + H_y(t)(y - y^*) = 0. \qquad (12)$$

As $x(t_0) = x^*(t_0)$ and $y(t_0) = y^*(t_0)$ the system will be on the manifold from the beginning. One (nonunique) control which will keep the system on the manifold is

$$u = -M(t)sign[H_x(t)s(t, x, y)] \qquad (13)$$

where $M(t) > 0$ is a function of time. To determine the minimum value for $M(t)$ at any time t the control may be substituted into Eq. (2) and any convenient norm may then be applied. In this case if

$$|H_x(t)M(t)| > |\dot{H}_x(t)(x - x^*) - H_x(t)\dot{x}^* + \dot{H}_y(t)(y - y^*) - H_x(t)(v - \dot{y}^*)|$$

then the system will remain on the sliding manifold (and thus the extremal trajectory) regardless of whatever control the follower chooses.

II. Extension to the Multi-hierarchy Case

Let us consider the system of the form

$$\dot{x} = F(t,x) + \sum_{i=1}^{M} B_i(t,x)u_i, \tag{14}$$

where $x \in R^n$ is a state vector and $u_i \in R^{m_i}$, are $M \geq 2$ control vectors associated correspondingly with different levels of the hierarchy. We assume that a decision maker on the level i is the leader for a decision maker on the level $i-1$, who is the follower. On the top of the hierarchy there is a decision maker managing the control u_1 who is an absolute leader.

As in the traditional leader-follower problem statement we assume that there are cost functionals

$$J_i = \int_{t_0}^{t_f} g_i(t, x(t), u_1(t), \ldots, u_M(t))dt, \tag{15}$$

where $i = 1, 2, \ldots, M$ and $g_i(\cdot)$ is convex. The initial and final time instants t_0, t_f as well as the initial and boundary conditions x_0, x_f are supposed to be fixed. The objective of the absolute leader is to minimize J_1. The other decision makers try to make J_i as small as possible given their leader's strategy.

If x^* is unknown, the structure of the decision making policy will be such that the decision maker on the ith level will wait until the incentive of the $(i-1)^{st}$ decision maker has been declared before declaring an incentive for the $(i+1)^{st}$ decision maker (where $i = 2, \ldots, M-1$). Further, it is also assumed the the ith decision maker will not negotiate with lower decision makers to choose a Pareto type solution under the already declared constraints.

We show that under certain conditions the leader can achieve the same optimum as if the follower would cooperate, i.e. obtain the team solution:

$$x^*(\cdot) = arg \min J_1(x(\cdot), u_1(\cdot), \ldots, u_M(\cdot)). \tag{16}$$

Therefore, without loss of generality, we can assume that instead of variational problem with the cost J_1, there is a desired trajectory $x^*(\cdot)$ $(x^*(t_0) = x(t_0))$ which the leader would like to track.

We assume that the functions F, B_i are smooth enough that all partial derivatives used below exist and are continuous. We also shall make the following assumptions on structure:

400

Assumption 1
For every $t \geq t_0$ and $x \in R^n$

$$rank \ B_i(t, x) = m_i, \quad i = 1, \ldots, M \tag{17}$$

Assumption 2
We assume also that for every $t \geq t_0$ and $x \in R^n$

$$range \ B_i \perp range \ B_j. \quad i \neq j \tag{18}$$

Before introducing the main theorem some notations are defined:
If C is a full rank rectangular matrix then let C^+ denote its left inverse (pseudoinverse),

$$C^+ = (C^T C)^{-1} C^T. \tag{19}$$

For every matrix B_i in Eq. (14) there exists a matrix B_z such that

$$B_z(t, x) B_i(t, x) \equiv 0, \tag{20}$$

so

$$B_z(t, x)[\dot{x} - F(t, x)] = 0 \tag{21}$$

is unique. Define $\tilde{g}_i$ as

$$\tilde{g}_i(t, x, \dot{x}) = g_i(t, x, B_1^+(t, x)[\dot{x} - F(t, x)], \ldots, B_M^+(t, x)[\dot{x} - F(t, x)]). \tag{22}$$

and $\bar{g}_M$ as

$$\bar{g}_M(t, x, \dot{x}, p) = \tilde{g}_M(t, x, \dot{x}) + p^T B_z(t, x)[\dot{x} - F(t, x)], \tag{23}$$

where $p \in R^{n - m_1 - m_2 - \cdots - m_M}$ satisfies

$$\dot{p} = -\frac{\partial (B_M^+ F)^T}{\partial x_M}(t, x^*)p - \frac{\partial \tilde{g}_M^T}{\partial x_M}(t, x^*). \tag{24}$$

The cost for decision maker i may be rewritten in the form of

$$J_i = \int_{t_0}^{t_f} \tilde{g}_i(t, x(t), \dot{x}(t)) \, dt. \tag{25}$$

Thus, when Assumptions 1 and 2 are valid then (14) implies (21) and (21) implies (14). The cost function (25) was found from (15) using only the change of variables outlined in (22), so the minimization of (15) is equivalent to the minimization of (25) with respect to x, and thus the original problem (14), (15) is equivalent to the problem (21), (25). The following theorem will detail how the problem (21), (25) may be solved, thus solving the original problem.

Theorem 1 *Under the structural assumptions 1 and 2, the ith decision maker $(i = 1, 2, \ldots, M-2)$ may influence a team solution $\check{x}_i = x^*$ by declaring an incentive of the form*

$$s_i(t, x) = H_{i+1}(t, x^*)^T (x - x^*) \tag{26}$$

where

$$H_{i+1} = \frac{\partial \tilde{g}_{i+1}}{\partial x} - \frac{d}{dt}\left(\frac{\partial \tilde{g}_{i+1}}{\partial \dot{x}}\right) \tag{27}$$

and the $(M-1)^{st}$ decision maker may influence the Mth decision maker by declaring an incentive of the form

$$s_{M-1}(t, x) = H_M(t, x^*, p^*)^T (x - x^*) \tag{28}$$

where

$$H_M = \frac{\partial \bar{g}_M}{\partial x} - \frac{d}{dt}\left(\frac{\partial \bar{g}_M}{\partial \dot{x}}\right). \tag{29}$$

Matrix $B_z(t, x)$ $(\operatorname{rank} B_z(t, x) = n - m$, where $m = \sum_{i=1}^M m_i)$ is such that

$$B_z(t, x) B_i(t, x) \equiv 0, \tag{30}$$

for all $i = 1, \ldots, M$ and the variable $p \in R^{n-m}$ satisfies the equation

$$\dot{p} = -\frac{\partial(B_z^+ F)^T}{\partial x}(t, x^*)p - \frac{\partial \tilde{g}_M^T}{\partial x}(t, x^*). \tag{31}$$

Proof:

Under Assumptions 1 and 2, each control variable u_i may be expressed as a unique function of x and $\dot{x}$ in the form

$$u_i = B_i^+(t, x)[\dot{x} - F(t, x)], \tag{32}$$

Using the M equations developed from Eq. (32) the cost for decision maker i may be rewritten in the form of

$$J_i = \int_{t_0}^{t_f} \tilde{g}_i(t, x(t), \dot{x}(t))\, dt \tag{33}$$

with $\tilde{g}_i$ defined above. There exists a $B_z(t, x)$ such that Eq. (20) holds, so

$$B_z(t, x)[\dot{x} - F(t, x)] = 0. \tag{34}$$

The set of all desired trajectories for the leader, denoted x^*, is assumed to be precomputed. For each decision maker $i = 1, 2, \ldots, M - 2$, the variation of the immediate follower's cost $J_{n+1}(t, x, \dot{x})$ is given by

$$\delta J_{i+1} = \int_{t_0}^{t_f} H_{i+1}(t, x, \dot{x})(x - x^*)\, dt \tag{35}$$

where

$$H_{i+1}(t, x, \dot{x}) = \frac{\partial}{\partial x}\tilde{g}_{i+1}(t, x, \dot{x}) - \frac{d}{dt}\frac{\partial}{\partial \dot{x}}\tilde{g}_{i+1}(t, x, \dot{x}) \tag{36}$$

is the set of unconstrained Euler-Lagrange equations evaluated using the cost of the $(i+1)^{st}$ follower. The theorem may then be proven using recursion.

To influence the second decision maker, the first decision maker must keep $H_2(t, x, \dot{x})$ identically zero along the desired state trajectory x^*. This may be done if the leader declares the incentive manifold

$$s_1(t, x) = H_2(t, x^*, \dot{x}^*)(x - x^*) = 0. \tag{37}$$

Under this manifold, the desired trajectory for both the first and the second decision makers is now x^*.

Now the ith decision maker must influence the $(i+1)^{st}$ decision maker to choose trajectory x^* (for $i = 2 \ldots M - 2$). Declaring

$$s_i(t, x) = H_{i+1}(t, x^*, \dot{x}^*)(x - x^*) = 0 \tag{38}$$

constrains the $(i+1)^{st}$ decision maker to the decision space $s(t, x) = \bigcap_{j=1}^{i} s_j(t, x)$. As the incentive $s_i(t, x)$ is tangent (in functional space) to the convex cost $\tilde{g}_{i+1}(t, x, \dot{x})$ at x^*, minimizing over $s(t, x)$ will lead either to the minimizing solution $x = x^*$ or to some inconsistency. However, $x = x^*$ is a solution for all $s_j(t, x)$, $j = 1, 2, \ldots, i$ and thus is also contained in $s(t, x)$.

Finally, the $(M-1)^{st}$ decision maker must influence the Mth decision maker. The constrained cost for the Mth decision maker is

$$\bar{g}_M(t, x, \dot{x}, p) = \tilde{g}_M(t, x, \dot{x}) - p^T B_z(t, x)[\dot{x} - F(t, x)]. \tag{39}$$

It was proven in Cruz, Drakunov, and Sikora[2] that the incentive

$$s_{M-1}(t, x, p) = H_M(t, x^*, \dot{x}^*, p^*)(x - x^*) = 0 \tag{40}$$

with p^* satisfying

$$\dot{p} = -\frac{\partial (B_z^+ F)^T}{\partial x}(t, x^*)p - \frac{\partial \tilde{g}_M^T}{\partial x}(t, x^*) \tag{41}$$

leads to the minimizing solution $x = x^*$. All decision makers now share as a team solution the leader's desired trajectory. $\qquad\square$

Based on the given information structure, this theorem may be implemented one of two ways. If x^* is unknown to the followers, then the theorem is implemented sequentially. The lead decision maker must first develop a sliding surface of the form Eq. (37) based on the desired trajectory x^* and cost functional of the second follower. Once this is declared the second follower then optimizes under this constraint to determine the desired trajectory is x^*. The second follower then computes an incentive based on x^* and the cost functional of the third follower. The incentive declaration, cost optimization, and incentive computation for the follower cycle is then repeated until decision maker $M - 1$ is reached. The $(M - 1)^{st}$ decision maker must declare an incentive of the form Eq. (39).

If x^* is known to each decision maker, then the incentives may be computed in parallel. The ith decision maker ($i = 1, \ldots, M - 2$) will compute an incentive of the form (38), while the $(M - 1)^{st}$ decision maker will declare an incentive of the form (39).

III.   Example - A Three Level Hierarchy

We examine the system

$$
\begin{aligned}
\dot{x} &= u \\
\dot{y} &= v \\
\dot{z} &= w
\end{aligned}
\tag{42}
$$

where u is the control of the leader, v is the control of the first follower, and w is the control of the second follower. Initial and final conditions are $(x_0, y_0, z_0) = (5, 5, 5)$ and $(x_f, y_f, z_f) = (0, 0, 0)$. The costs for the three decision makers are

$$
\begin{aligned}
J_u &= \frac{1}{2} \int_0^2 [x^2 + 4u^2 + v^2 + w^2] \, dt \\
J_v &= \frac{1}{2} \int_0^2 [y^2 + u^2 + 4v^2 + w^2] \, dt \\
J_w &= \frac{1}{2} \int_0^2 [z^2 + u^2 + v^2 + 4w^2] \, dt.
\end{aligned}
\tag{43}
$$

The leader's desired optimal trajectory set (computed using standard techniques) was found to be

$$
x^*(t) = \frac{5}{e^2 - 1}(e^2 e^{-0.5t} - e^{0.5t})
$$

$$y^*(t) = -\frac{5}{2}t + 5$$
$$z^*(t) = -\frac{5}{2}t + 5. \tag{44}$$

The leader then computes the Euler-Lagrange equations for the first follower's unconstrained cost

$$H^1_x = -\ddot{x}$$
$$H^1_y = y - 4\ddot{y}$$
$$H^1_z = -\ddot{z}$$

and declares the incentive manifold

$$\begin{aligned}
s_1(x,y,z) &= H^{1*}_x(x - x^*) + H^{1*}_y(y - y^*) + H^{1*}_z(z - z^*) \\
&= -\frac{5}{4(e^2 - 1)}(e^2 e^{-0.5t} - e^{0.5t})\left[x - \frac{5}{e^2 - 1}(e^2 e^{-0.5t} - e^{0.5t})\right] \\
&\quad + (-\frac{5}{2}t + 5)\left[y - (\frac{5}{2}t + 5)\right].
\end{aligned} \tag{45}$$

Under this constraint, the leader and first follower now share a team solution. A rational first follower will then attempt to constrain the second follower to obtain the same optimal solution. To begin this, the first follower computes the Euler-Lagrange equations for the second follower's cost

$$H^2_x = -\ddot{x}$$
$$H^2_y = -\ddot{y}$$
$$H^2_z = z - 4\ddot{z}$$

and declares a second incentive manifold

$$\begin{aligned}
s_2(x,y,z) &= H^{2*}_x(x - x^*) + H^{2*}_y(y - y^*) + H^{2*}_z(z - z^*) \\
&= -\frac{5}{4(e^2 - 1)}(e^2 e^{-0.5t} - e^{0.5t})\left[x - \frac{5}{e^2 - 1}(e^2 e^{-0.5t} - e^{0.5t})\right] \\
&\quad + (-\frac{5}{2}t + 5)\left[z - (\frac{5}{2}t + 5)\right].
\end{aligned} \tag{46}$$

The second follower then optimizes only in the region of decision space satisfying $\mathcal{M} = s_1(x,y,z) \cap s_2(x,y,z)$. On this region the second follower's optimal trajectories coincide with those of the leader and of the first follower; the desired solution of the leader has now become the team optimal solution.

Discontinuous controls satisfying Eq. (2) and Eq. (3) were chosen for the leader and first follower. Further, for maximum robustness a discontinuous control was chosen for the second follower, although an open loop control would also work. The controls that keep the system on the desired manifolds were chosen to be

$$u^* = -20\,sign(s_1)$$
$$v^* = -10\,sign(s_2)$$
$$u^* = -5\,sign(z - z^*)$$

where the coefficients were chosen so that the control of the leader dominates the control of the first follower and the control of the first follower dominates the control of the second follower.

Figure 1 demonstrates the effectiveness of the incentives and the choice of control. In the figure it may be seen that the system will track the desired trajectory. Figure 2 shows that real sliding mode controllers are subject to drawbacks; chattering (output noise due to finite switching transition times) was seen to enter the system. This chattering is due to discretizing the system for simulation, and may be minimized by using fast switches and small sampling intervals. Figure 3 shows the robustness of the sliding controls to initial condition variations. Even when perturbed from the initial state the system will rejoin the desired trajectory.

IV.   Summary

This paper provided brief summaries of leader-follower incentives and sliding mode control. A method of developing feedback incentive based on the Euler-Lagrange equations and sliding modes was extended to the case of systems with multiple levels of hierarchies. An example illustrating the design techniques was also given.

One point this paper does not address is the possibility of the followers banding together to develop a Pareto strategy that harms the leader. Current research is being performed to develop conditions under which the leader can directly influence each follower, which will eliminate this problem. Future research will also include extending the technique to systems with multiple decision makes on each level of hierarchy. Further, there will be an examination of the possible nonoptimalities of chattering in the process.

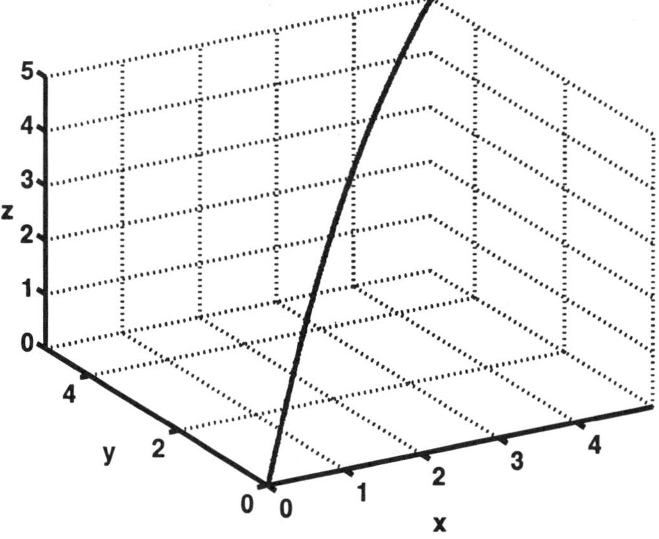

Figure 1: Trajectory of the System under Sliding Incentive Control

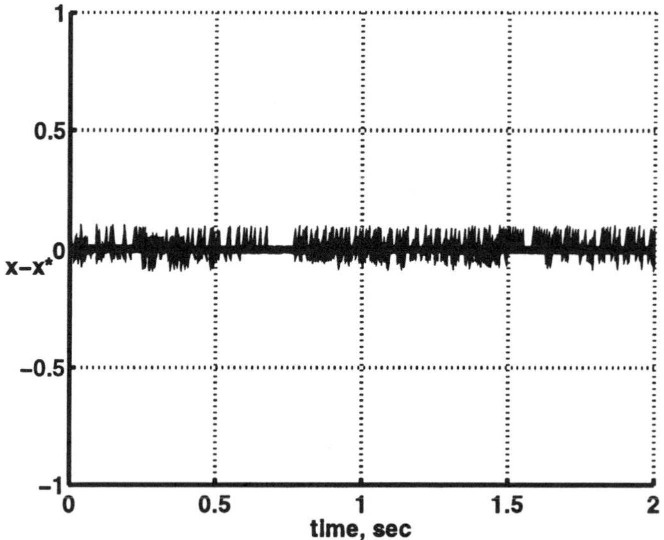

Figure 2: Error in the x Trajectory Due to Chattering

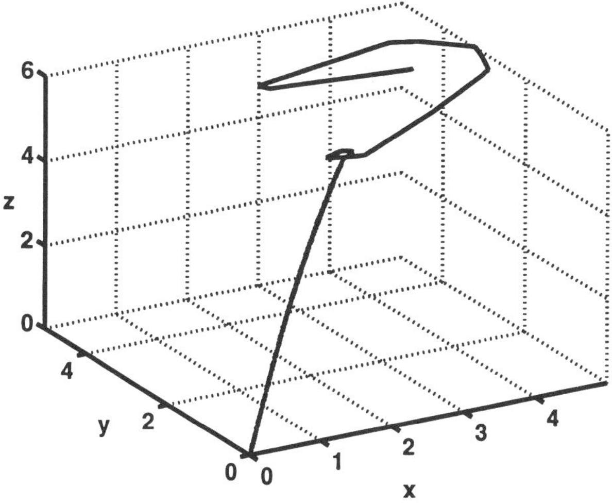

Figure 3: Robustness of System to Initial Condition Disturbances

V.   Acknowledgment

Martin Sikora would like to acknowledge the support of the Engineering Research Center for Net Shaped Manufacturing, located at the Ohio State University, for support in pursuing this research.

VI.   References

[1] C.I. Chen and J.B. Cruz, Jr., Stackelberg Solution for Two-Person Games with Biased Information Patterns, *IEEE Transactions on Automatic Control*, Vol. **17** (1972), pp. 791–797.

[2] J.B. Cruz, Jr., S.V. Drakunov, and M.A. Sikora, Leader-Follower Strategy via a Sliding Mode Approach, *Journal of Optimization Theory and Applications*, to be published in Vol. **88** (1996).

[3] R.A. DeCarlo, S.H. Zak, and G.P. Matthews, Variable Structure Control of Nonlinear Multivariable Systems: A Tutorial, *Proceedings of the IEEE*, Vol. **76** (1988), pp. 212-232.

[4] B. Drazenovic, The Invariance Condition in Variable Structure Systems, *Automatica*, Vol. **5** (1969), pp. 287–295.

[5] S.V. Emelyanov, *Theory of Variable Structure Systems* (in Russian), Nauke, Moscow, Russia, 1970.

[6] I. Flugge-Lotz, *Discontinuous Automatic Control*, Princeton University Press, Princeton, New Jersey, 1953.

[7] Y.-C. Ho, P.B. Luh, and G.J. Olsder, A Control-Theoretic View on Incentives, *Automatica*, Vol. **18** (1982), pp. 167–179.

[8] H. von Stackelberg, *The Theory of Market Economy*, Oxford University Press, Oxford, England, 1952.

[9] B. Tolwinski, Closed-Loop Stackelberg Solution to a Multistage Linear-Quadratic Game, *Journal of Optimization Theory and Applications*, Vol. **34** (1981), pp. 485–499.

[10] V.I. Utkin, *Sliding Modes and Their Application In Variable Structure Systems*, MIR, Moscow, Russia, 1978.

[11] Y.-P. Zheng, T. Basar, and J.B. Cruz, Jr., Stackelberg Strategies and Incentives in Multiperson Deterministic Decision Problems, *IEEE Transactions on Systems, Man and Cybernetics*, Vol. **14** (1984), pp. 10–23.

WSSIAA 5 (1995) pp. 409–421

A General Approach to Deterministic Annealing

Cameron Tovey and Alistair Mees
Center for Applied Dynamics and Optimisation
Department of Mathematics
University of Western Australia
Nedlands 6907

1 Introduction

This paper is about the optimization of difficult discrete problems using continuous methods. Such discrete optimization problems often arise as decision problems in areas such as operational research. Many of these problems, such as the travelling salesman problem, are NP-Hard which means that it is often not practical to find an optimal solution. Instead we seek a feasible approximately optimal solution in a reasonable amount of time.

Often such problems can be formulated as continuous problems with an additional discrete constraint. For example, consider the problem

$$\begin{aligned} \text{Minimize} \quad & 1 - (\sqrt{x} + \sqrt{y} + \sqrt{z}) \\ \text{in} \quad & x, y, z \in \{0, 1\} \\ \text{with} \quad & 0 \leq x + y + z \leq 1. \end{aligned}$$

The variables x, y and z may represent the outcomes of certain decisions which need to be made.

One way of approximately solving such a discrete optimization problem is to create a continuous problem by removing the discrete constraint, perhaps replacing it with a continuous constraint enforcing upper and lower bounds, and to somehow map the resulting continuous solution back into the discrete domain. The problem with this approach is that the continuous problem contains many more states than the discrete problem, and often the solutions of the continuous problem have no discrete counterpart. Some arbitrary way of mapping continuous solutions to discrete solutions must be chosen, and there is no guarantee that a solution found in this manner will be optimal, or even approximately optimal. Often solutions found in this manner are no better that the average objective function value over all the feasible solutions, that is, we expect to do as well by guessing.

For example, in the problem given above we can replace the constraint $x, y, z \in \{0, 1\}$ with the constraint $0 \leq x, y, z \leq 1$. The solution of this new continuous problem is $x = y = z = 1/3$ which is not in the discrete domain. We can map this solution to the discrete domain by rounding off, giving $x = y = z = 0$ which is in domain of the discrete problem but is not optimal, having cost 1. Of the four feasible solutions, three are optimal having cost 0. The average cost of all the feasible solutions is therefore $1/4$ — we expect to do better by guessing than using this method.

The motivation for the method described in this paper comes from the observation that many complex discrete problems exist in nature, which is interesting because the (classical) laws of nature are continuous. This is because any piece of matter in nature is made up of a large number of fluctuating particles, and quantum mechanics theory predicts that the state of each of these particles is discrete. Quantum mechanics also predicts that we can never know exactly what state a particle in the system is in. The so called 'laws of nature' that we observe are actually due to average properties of these fluctuating particles. Physicists study such systems using a technique called statistical mechanics, which is based on the postulate that the valid states of a system of particles are distributed according to the Boltzmann-Gibbs (B-G) distribution. The B-G distribution is a function of the internal energy of the state of a system (so two states with the same internal energy are equally likely) and of temperature. By the state of a system we mean the list of the states of all the particles. It is unnecessary to have any previous knowledge of statistical mechanics to understand this paper.

The results in this paper are based on some properties of systems of particles which can be derived from the B-G distribution. The first of these properties is that as temperature approaches absolute zero the most likely state of a system of particles is the state with the lowest internal energy. The second property is that the state with lowest internal energy is reached more rapidly if the substance annealed, that is, cooled slowly from a high temperature to a low temperature, than if the substance is held at the low temperature originally.

We can impose the B-G distribution on a discrete optimization problem by using the objective function value as internal energy and the points in the domain as states. Since the B-G distribution is parameterized by temperature, an artificial temperature is induced on the problem. We will call such a system an *optimization system*. The calculation of the most likely state at low temperature will then give an approximately optimal solution of the discrete problem. There are two main ways this is done. The first method is to stochastically generate points in the domain that are distributed according to the B-G distribution, and to record the best point found so far. The algorithm that results from doing this as temperature is lowered gradually from a high temperature to a low temperature is the simulated annealing algorithm due to Kirkpatrick et. al. [2]. We do not describe this algorithm in this paper. The second method is derive a new continuous optimization system with the same statistical properties as the discrete system, and to solve this new problem using continuous optimization techniques as temperature is lowered to absolute zero. The solution of the discrete problem can be found from the solution to the continuous problem. This method is called *deterministic annealing* and is the subject of this paper.

The two most important methods of applying deterministic annealing to discrete optimization problems are using the saddle point approximation (Peterson and Anderson [3], Peterson and Soderberg [4]) and using generalized deformable templates (Yuille [13]). Deformable templates are the more efficient of the two methods but can only be applied to problems with a strongly geometry nature. For this reason deformable templates have been used successfully in areas such as computer vision. The saddle point approximation is most easily applied to a general optimization problem. Both these methods require an understanding of statistical mechanics to apply and use efficiently. This is unfortunate because the way in which people working in continuous optimization and operational research approach most problems is quite different to the way in which deterministic annealing is usually applied.

For this reason we have developed a deterministic annealing method which requires little or no knowledge of statistical mechanics to apply (Tovey and Mees [11, 11], Tovey [10]). We describe this method, which we call the *constraint selection method*, in this paper. Although this method is based on the generalized deformable template techniques described by Yuille [13], the method can be applied to non-geometric problems (Tovey & Mees [11]), and in fact the saddle point approximation algorithms and the deformable template algorithms are easily derived using constraint selection without the use of statistical mechanics (Tovey & Mees [12]).

2 The Optimization Systems

In this section we introduce the Boltzmann-Gibbs (B-G) distribution and optimization systems. We start with the optimization problem

$$
\begin{aligned}
\text{Minimize} \quad & f_\beta(\mathbf{x}) \\
\text{in} \quad & \mathbf{x} \in \Xi \subseteq \mathbf{R}^n \\
\text{as} \quad & \beta \to \infty.
\end{aligned}
$$

We require that $f_\infty(\mathbf{x}) \equiv \lim_{\beta \to \infty} f_\beta(\mathbf{x}) \in [l, \infty]$ for some finite number l. We have included parameter β for use as a penalty function multiplier. We think of Ξ as being as discrete set, although the results presented here do not depend on this assumption.

We intend to impose the B-G distribution on the domain and observe expected value of $\mathbf{x}$ as temperature is lowered to absolute zero. It is common to set temperature equal to $1/\beta$ and raise β to infinity instead. For this reason we use the same variable for inverse temperature and as a penalty function multiplier.

The *Boltzmann-Gibbs distribution* is defined by the probability function

$$
P_\beta(\mathbf{x}) = \exp\left[-\beta f_\beta(\mathbf{x})\right] / Z_\beta
$$

where Z_β is some normalization constant that ensures the probabilities 'sum' to one.

Depending on the nature of Ξ 'sum' could involve summation, (Riemann) integration, a combination of summation and integration, or some measure theoretic integration. The first three cases are the important special cases of the latter case, so we will integrate with a general measure to simplify the notation and proofs. We write $\int_\Xi d\mu(\mathbf{x}) P_\beta(\mathbf{x}) = 1$ to mean the integration of the function $P_\beta(\mathbf{x})$ over the set Ξ using measure μ equals 1. From this equation it is clear the Boltzmann distribution depends on the measure chosen. We always formulate our problems so that x_i is either continuous, for which we use Lesbesgue measure λ, or discrete, for which we use counting measure σ.

For readers unfamiliar with measure theory

$$\int_\Xi d\lambda(\mathbf{x}) f(\mathbf{x}) = \int \cdots \int_\Xi f(\mathbf{x}) \, d\mathbf{x}$$

whenever the latter (Riemann) integral is defined, and

$$\int_\Xi d\sigma(\mathbf{x}) f(\mathbf{x}) = \sum_{\mathbf{x} \in \Xi} f(\mathbf{x})$$

whenever the summation is defined.

To illustrate how the B-G distribution can facilitate optimization we have graphed a non-convex function $f(\mathbf{x})$ and the corresponding $P_\beta(\mathbf{x})$ at various values of β in Figure 1. As $\beta \to \infty$ the most likely state is the optimal state. One good question is, "Why use deterministic annealing when we could maximize $P_\beta(\mathbf{x})$ for some large value of β ?" The reason is that $P_\beta(\mathbf{x})$ has exactly the same number of local maxima and as $f(\mathbf{x})$ has local minima and this new problem is therefore just as complex as the old. Further, if $f(\mathbf{x})$ is discrete so is $P_\beta(\mathbf{x})$ so we have just exchanged one difficult discrete problem for another.

However, $\langle \mathbf{x} \rangle_\beta$, the *mean point* or *expected value of* $\mathbf{x}$, defined by

$$\langle \mathbf{x} \rangle_\beta = \int_\Xi d\mu(\mathbf{x}) \, \mathbf{x} P_\beta(\mathbf{x})$$

is a continuous function of β. We will see that $\langle \mathbf{x} \rangle_\beta$ approaches the optimal point of $f_\beta(\mathbf{x})$ as $\beta \to \infty$ (this can be seen from Figure 1).

3 The Statistical Mechanics of Optimization Systems

The statistical mechanics of optimization systems will now be discussed. No previous knowledge of statistical mechanics is assumed. The interested reader should be able to prove the results presented here without much difficulty using the outline proofs presented here.

A central concept in statistical mechanics is the *partition function* which is defined as

$$Z_\beta(\mathbf{h}) = \int_\Xi d\mu(\mathbf{x}) \exp\left[-\beta\left(f_\beta(\mathbf{x}) + \mathbf{h} \cdot \mathbf{x}\right)\right] \tag{1}$$

where μ is the measure we are using for Ξ. We require $0 < Z_\beta(\mathbf{0}) < \infty$. The partition function has a number of important properties. The first of these is that the Boltzmann-Gibbs distribution can be defined as follows

$$P_\beta(\mathbf{x}) \equiv \frac{\exp\left[-\beta f_\beta(\mathbf{x})\right]}{Z_\beta(\mathbf{0})}.$$

Note that $Z_\beta(\mathbf{0})$ is the normalizing factor Z_β discussed in the previous section which ensures that $\int_\Xi d\mu(\mathbf{x}) P_\beta(\mathbf{x}) = 1$.

The second property of the partition function is that the mean value of $\mathbf{x}$ can be calculated from the partition function using

$$\langle \mathbf{x} \rangle_{\beta} = -\frac{\partial}{\partial \mathbf{h}} \log Z_{\beta}(\mathbf{h})|_{\mathbf{h}=\mathbf{0}} /\beta. \tag{2}$$

This is important because it means that systems with the same partition function are statistically equivalent even though they may appear quite different. The following result shows how $\langle \mathbf{x} \rangle_{\beta}$ can be used in optimization.

Theorem 1 *If there is some μ-measurable function g with $|f_{\beta}(\mathbf{x})| < g(\mathbf{x})$ and $f_{\infty}(\mathbf{x}) \equiv \lim_{\beta \to \infty} f_{\beta}(\mathbf{x})$ has a unique global minimum $\mathbf{x}^*$ then $\langle \mathbf{x} \rangle_{\beta} \to \mathbf{x}^*$ as $\beta \to \infty$.*

Outline Proof. Let $\Xi^* = \{\mathbf{x}^*\}$ and let $\Xi^N = \Xi \backslash \{\mathbf{x}^*\}$. First

$$P_{\beta}(\mathbf{x}) = \left(\int_{\Xi} d\mu(\mathbf{u}) \exp\left[-\beta \left(f_{\beta}(\mathbf{u}) - f_{\beta}(\mathbf{x}) \right) \right] \right)^{-1}.$$

Lesbesgue's dominated convergence theorem (see e.g. Dudley [16] p. 101) is required for some of the following results. First

$$\int_{\Xi^N} d\mu(\mathbf{x}) \begin{bmatrix} \mathbf{x} \\ 1 \end{bmatrix} P_{\infty}(\mathbf{x}) = \begin{bmatrix} \mathbf{0} \\ 0 \end{bmatrix}$$

so

$$\int_{\Xi^*} d\mu(\mathbf{x}) P_{\infty}(\mathbf{x}) = 1$$

and then

$$\begin{aligned}
\lim_{\beta \to \infty} \langle \mathbf{x} \rangle_{\beta} &= \mathbf{x}^* \int_{\Xi^*} d\mu(\mathbf{x}) P_{\infty}(\mathbf{x}) + \int_{\Xi^N} d\mu(\mathbf{x}) \mathbf{x} P_{\infty}(\mathbf{x}) \\
&= \mathbf{x}^*.
\end{aligned}$$

$\square$

In general we cannot evaluate the partition function of a discrete optimization system directly without summing over at least as many exp functions as there are points in the domain. In such cases it is more efficient to examine every state. Clearly we must exploit the structure of the domain and the objective function to evaluate the partition function efficiently. One way to achieve this is to derive a new continuous optimization system which is statistically equivalent to the discrete problem.

In order to show how this can be done we need the following results. Suppose we manipulate the partition function (1) of our problem in such a way that it has the form

$$Z_{\beta}(\mathbf{h}) = \int_{\Upsilon} \exp\left[-\beta \left(F_{\beta}(\mathbf{v}) + \mathbf{h} \cdot \mathbf{v} \right) \right] d\nu(\mathbf{v}) \tag{3}$$

where Υ is a connected set and ν is some measure. We see that $Z_{\beta}(\mathbf{h})$ is the partition function of a new optimization system having objective function $F_{\beta}(\mathbf{v})$ and measure $\nu(\mathbf{v})$. Equation (2) shows that

$$\langle \mathbf{x} \rangle_{\beta} = \langle \mathbf{v} \rangle_{\beta} \tag{4}$$

where $\langle \mathbf{v} \rangle_\beta$ is the mean value of $\mathbf{v}$ in the new optimization system. Providing f and F satisfy the conditions of Theorem 1, we see that $\langle \mathbf{x} \rangle_\beta \to \mathbf{x}^*$ as $\beta \to \infty$ and $\langle \mathbf{v} \rangle_\beta \to \mathbf{v}^*$ as $\beta \to \infty$. The function $F_\beta(\mathbf{v})$ is often called an *effective energy function*.

To see why this is useful consider the possibility that f is the objective function of an NP-Hard optimization problem, and F is a continuous function. Then optimization of the continuous function F as $\beta \to \infty$ will also find the optimal solution of the discrete problem.

4 The Constraint Selection Function

In this section we describe a general way of formulating discrete optimization problems so that a suitable effective energy function can be found easily. We do this by enforcing each part of a discrete constraint using a penalty or Lagrange function and using the results of the previous section to decide which constraint is to be enforced.

Penalty and Lagrange functions have had a large degree of success in the area of continuous optimization, but are rarely used in discrete optimization. The reason for this is that penalty functions enforcing discrete constraints are necessarily non-convex, so a global rather than local optimization technique is required. Global optimization of non-convex functions is a difficult task so this situation is to be avoided. To see why penalty functions are necessarily non-convex consider trying to enforce the constraint $x \in \{0, 1\}$ with a penalty function. For a penalty function to enforce this constraint it must have local minima of 0 at both $x = 0$ and $x = 1$ and nowhere else. By the intermediate value this implies the penalty function must have a local maximum somewhere in $(0, 1)$. We refer the interested reader to Tovey and Mees [11] for a more complete discussion of this topic. Lagrange functions suffer from a similar problem.

In contrast, if we want to enforce one constraint *and* another constraint all we need to do is add respective penalty (or Lagrange) functions for each constraint. We would like a way to do this if we want to enforce one constraint *or* another constraint - this would be very useful for discrete optimization. Multiplying respective penalty functions for each constraint will give a penalty function enforcing the correct constraint but it will be non-convex and so of little use. We need to find a way of solving the problem

$$\begin{array}{ll} \text{Minimize} & f(\mathbf{x}) \\ \text{in} & \mathbf{x} \in \mathbf{R}^n \\ \text{with} & \mathbf{x} \in \Xi_1 \quad \text{or} \quad \mathbf{x} \in \Xi_2 \quad \text{or} \ldots \text{or} \quad \mathbf{x} \in \Xi_N. \end{array} \tag{5}$$

where each set Ξ_i is connected. There is no need for f to depend on β here since all our constraints are expressed explicitly not as penalty functions.

We will now introduce the idea of a constraint function. A *constraint function* enforcing the set $X \subset \mathbf{R}^n$ is a function $C_\beta(\mathbf{x})$ for which

$$\begin{array}{ll} \text{Minimize} & f(\mathbf{x}) + \gamma C_\beta(\mathbf{x}) \\ \text{in} & \mathbf{x} \in \mathbf{R}^n \\ \text{as} & \beta \to \infty \end{array} = \begin{array}{ll} \text{Minimize} & f(\mathbf{x}) \\ \text{in} & \mathbf{x} \in X \end{array} \tag{6}$$

for any $\gamma > 0$ and $f(\mathbf{x})$ which is bounded below. A constraint function is a generalization of penalty and barrier functions. The function $\text{Maximize}_\lambda \; \lambda \cdot \mathbf{h}(\mathbf{x})$ where $X = \{\mathbf{x} : \mathbf{h}(\mathbf{x}) = \mathbf{0}\}$ is a constraint function which is independent of β and is the part of a Lagrange function which enforces constraints. Therefore a constraint function also generalizes Lagrange functions.

Suppose we have a constraint function $C_{i,\beta}(\mathbf{x})$ enforcing each constraint $\mathbf{x} \in \Xi_i$ in (5). We re-formulate (5) as

$$
\begin{aligned}
\text{Minimize} \quad & \alpha f(\mathbf{x}) + \sum_{i=1}^{N} b_i C_{i,\beta}(\mathbf{x}) \\
\text{in} \quad & b_i \in \{0,1\}, \mathbf{x} \in V \subset \mathbf{R}^n \\
\text{with} \quad & \sum_{i=1}^{N} b_i = 1 \\
\text{as} \quad & \beta \to \infty
\end{aligned}
\tag{7}
$$

where X is any convex set containing all the Ξ_i and α is a positive scaling constant. For $\sum_{i=1}^{N} b_i C_{i,\beta}(\mathbf{x})$ to be a valid constraint function the minimum value of each $C_{i,\beta}(\mathbf{x})$ must be equal in order that the solution is not biased. If this is not the case it is possible that the solutions of (5) and (7) will not be equal. This is why we insisted that a constraint function must work for all values of f in (6).

Defining $B = \{\mathbf{b} \in \mathbf{R}^N : b_i \in \{0,1\}, \sum_{i=1}^{N} b_i = 1\}$ the partition function of the optimization system corresponding to (7) is

$$
Z_\beta(\mathbf{h_x}, \mathbf{h_b}) = \int_X \sum_{\mathbf{b} \in B} \exp\left[-\beta \left(\alpha f(\mathbf{x}) + \sum_{i=1}^{N} b_i C_{i,\beta}(\mathbf{x}) + \mathbf{h_x} \cdot \mathbf{x} + \mathbf{h_b} \cdot \mathbf{b} \right) \right] d\mathbf{x}.
$$

Equation (4) in the previous section showed the mean point of any problem can be found from the mean point of any other system with the same partition function. We noted that this is particularly useful if the other system is purely continuous.

We require a continuous system with the same partition function as above. We can find such a system by summing $\mathbf{b}$ out of the partition function. We are, therefore, not interested in the average value of $\mathbf{b}$ in the partition function and set $\mathbf{h_b} = \mathbf{0}$ for simplicity. We get

$$
\begin{aligned}
Z_\beta(\mathbf{h_x}) &= \int_X \exp\left[-\beta\left(\alpha f(\mathbf{x}) + \mathbf{h_x} \cdot \mathbf{x}\right)\right] \sum_{\mathbf{b} \in B} \exp\left[-\beta \sum_{i=1}^{N} b_i C_{i,\beta}(\mathbf{x})\right] d\mathbf{x} \\
&= \int_X \exp\left[-\beta\left(\alpha f(\mathbf{x}) + \mathbf{h_x} \cdot \mathbf{x}\right)\right] \sum_{i=1}^{N} \exp\left[-\beta C_{i,\beta}(\mathbf{x})\right] d\mathbf{x} \\
&= \int_X \exp\left[-\beta\left(\alpha f(\mathbf{x}) + \mathbf{h_x} \cdot \mathbf{x} - \frac{1}{\beta} \log \sum_{i=1}^{N} \exp\left[-\beta C_{i,\beta}(\mathbf{x})\right]\right)\right] d\mathbf{x}.
\end{aligned}
$$

This also the partition function of the problem

$$
\begin{aligned}
\text{Minimize} \quad & f(\mathbf{v}) + \gamma \left(-\tfrac{1}{\beta} \log \sum_{i=1}^{N} \exp\left[-\beta C_{i,\beta}(\mathbf{v})\right] \right) \\
\text{in} \quad & \mathbf{v} \in X \subset \mathbf{R}^n \\
\text{as} \quad & \beta \to \infty
\end{aligned}
\tag{8}
$$

416

where γ is a positive scaling constant defined by $\gamma = 1/\alpha$. Equation (4) shows that $\langle \mathbf{x} \rangle_\beta = \langle \mathbf{v} \rangle_\beta$ where $\langle \mathbf{x} \rangle_\beta$ is from the optimization system defined from the discrete problem (7) and $\langle \mathbf{v} \rangle_\beta$ from the optimization system defined by (8). Theorem 1 shows that the two problems have equal solutions, and hence a solution of (8) is a solution of (5) provided the $C_{i,\beta}(\mathbf{v})$ have equal minimum values.

Under this latter condition, the function

$$H_\beta(\mathbf{x}) = -\frac{1}{\beta} \log \sum_{i=1}^{N} \exp\left[-\beta C_{i,\beta}(\mathbf{x})\right] \tag{9}$$

is a constraint function enforcing the constraint $\mathbf{x} \in \Xi_1$ or $\mathbf{x} \in \Xi_2$ or ... or $\mathbf{x} \in \Xi_N$. We call $H_\beta(\mathbf{x})$ the *constraint selection function*. Note that this does not show that $H_\beta(\mathbf{x})$ is any better than a non-convex penalty function enforcing a discrete constraint. We discuss this using an example in the following section.

5 Binary Problems

One common way of formulating discrete optimization problems is using binary variables, that is, variables that take the values $\{0, 1\}$. Such problems have the form

$$\begin{aligned}
\text{Minimize} \quad & f_\beta(\mathbf{x}) \\
\text{in} \quad & x_i \in \{0, 1\} \\
\text{as} \quad & \beta \to \infty.
\end{aligned} \tag{10}$$

A simple example of such a problem was given in the introduction.

Using the penalty term βx_i^2 to enforce $x_i = 0$ and $\beta(x_i - 1)^2$ to enforce $x_i = 1$, a constraint selection function enforcing each $x_i \in \{0, 1\}$ is

$$H_{\text{BIN},\beta} = -\frac{1}{\beta} \sum_{i=1}^{N} \log \left(\exp\left[-\beta^2 x_i^2\right] + \exp\left[-\beta^2 (x_i - 1)^2\right] \right).$$

This constraint selection function was first used to find approximate solutions of discrete multicommodity flow problems (Tovey and Mees [11]).

We note that $H_{\text{BIN},\beta}$ is superior to penalty functions enforcing $x_i \in \{0, 1\}$ because for sufficiently small β (in this case $\beta < \sqrt{2}$) $H_{\text{BIN},\beta}$ is convex. As β is increased the constraint selection function is continuously deformed so that for large values of β it is a non-convex function acting like βx_i^2 near $x_i = 0$ and like $\beta(x_i - 1)^2$ near $x_i = 1$. Similar properties hold for all constraint selection functions provided the $C_{i,\beta}(\mathbf{x})$ are at least weakly convex (Tovey [10]). Figure 2 shows a graph of $H_{\text{BIN},\beta}$.

An important practical point is, therefore, that when solving (8) we start at a small value of β, find the optimal point of the resulting convex function, then track this optimal point by continually minimizing the objective function as β is increased monotonically to infinity. Under this *annealing process* the constraint selection function performs far better than is possible for any penalty function.

The Lagrange terms $\text{Max}_{\lambda_i}\ \alpha\lambda_i x_i$ and $\text{Max}_{\mu_i}\ \alpha\mu_i(x_i-1)/\beta$ can be used instead of penalty functions. Since only one of these constraints is to be enforced we may use $u_i \equiv \lambda_i = \mu_i$. The Lagrange terms are independent of β we may set $\alpha = 1/\beta$. The constraint selection function is then

$$\underset{u_i}{\text{Max}}\ -\frac{1}{\beta}\sum_{i=1}^{N}\log\left(\exp\left[-u_i x_i\right] + \exp\left[-u_i(x_i - 1)\right]\right).$$

Problem (10) can be solved by finding a saddle point of the effective energy function

$$F(\mathbf{u}, \mathbf{x}) = f_\beta(\mathbf{x}) - \frac{1}{\beta}\sum_{i=1}^{N}\log\left(\exp\left[-u_i x_i\right] + \exp\left[-u_i(x_i - 1)\right]\right)$$

by minimizing with respect to $\mathbf{x}$ and maximizing with respect to $\mathbf{u}$. Here constraint selection makes a connection with the saddle point approximation algorithms mentioned in the introduction. Tovey and Mees [12] contains a discussion of this topic. While originally applied to graph matching, saddle point approximation algorithms have been used to solve a variety of problems including the Knapsack problem (Ohlsson et. al. [7]) and time tabling problems (Gislen et. al. [5, 6]).

6   The Elastic Net Algorithm

The elastic net algorithm finds approximate solutions of the travelling salesman problem (TSP). The algorithm was first derived geometrically (Durbin and Willshaw [1]) and has since been derived using a generalized deformable template method (Yuille [13]). The aim of this section is to show the relationship between deformable templates and constraint selection. A deformable template is a shape that is to be mapped to a set of data points.

The TSP is a commonly used example and is one of the most widely accepted benchmark problems for testing new algorithms that optimize (approximately or otherwise) general discrete problems. The TSP can be stated as follows. Suppose there are some cities, which we will represent using the points $z_1, z_2, .., z_N \in \mathbf{R}^2$, and a travelling salesman who must visit each city exactly once, starting from his home city, before returning to his home city after every city has been visited. This is shown in Figure 3. The path the salesman takes on his journey is often called a tour. The problem is to minimize the length of the salesman's journey.

One feature of solutions of the TSP is that paths in a tour never cross (Tovey [9]). The idea of the elastic net is to represent the saleman's tour as an elastic band which is constrained to pass through all the cities. Since the band is elastic it needs to be stretched it in order for it to pass through each city. The elastic band V can be modeled by the points $\mathbf{v}_1, \mathbf{v}_2, .., \mathbf{v}_M \in \mathbf{R}^2$, where $\mathbf{v}_j$ with consecutive indices are connected and the arithmetic of the indices is modulo M so the band points M and 1 are joined. The larger M is, the better the model. It is necessary that $M \geq N$, and desirable that $M \gg N$. The length of the band can be modeled

418

by

$$L_n(V) = \sum_{j=1}^{M} \|\mathbf{v}_j - \mathbf{v}_{j+1}\|^n . \tag{11}$$

The correct length is given when $n = 1$, but in this case $L(V)$ can be difficult to optimize because it is not strictly convex. If $M \gg N$ minimizing $L_2(V)$ gives a reasonable approximation.

The constraint is that each city must be associated with exactly one band point in the final solution. If we use penalty terms of the form $\beta\|\mathbf{z}_i - \mathbf{v}_j\|^2$ to force band point j to pass through city i, then we have a constraint selection problem. That is, for each city i, we must enforce either

$$\mathbf{z}_i = \mathbf{v}_1 \quad \text{or} \quad \mathbf{z}_i = \mathbf{v}_2 \quad \text{or} \ldots \text{or} \quad \mathbf{z}_i = \mathbf{v}_M.$$

The constraint selection function corresponding to city i can be written down from (9) as

$$H_{i,\beta}(V) = -\frac{1}{\beta} \log \left(\sum_{j=1}^{M} \exp \left[-\beta^2 \|\mathbf{z}_i - \mathbf{v}_j\|^2 \right] \right) . \tag{12}$$

The elastic net algorithm can then be stated as

$$\text{Minimize} \ \ L_1(V) + \gamma \sum_{i=1}^{N} H_{i,\beta}(V)$$

as $\beta \nearrow \infty$ from some sufficiently small initial value. An elastic band is shown in Figure 4. We are currently examining a "Lagrangian Elastic Net" which uses Lagrange functions rather than penalty functions to enforce the constraints.

References

[1] R. Durbin, and D. Willshaw (1987) An analogue approach to the travelling salesman problem using an elastic net method. *Nature.* **326**, 689-691.

[2] S. Kirkpatrick, C.D. Gelatt Jr. and M.P. Vecchi (1983) Optimisation by Simulated Annealing. *Science.* **220**, 671-680.

[3] C. Peterson and J.R. Anderson (1987) A mean field learning algorithm for neural networks. *Complex Systems.* **1**, 995-1019.

[4] C. Peterson and B. Soderberg (1989) A new method for mapping optimisation problems onto neural networks. *Int. J. Neural Systems.* **1**, 3-22.

[5] L. Gislen, C. Peterson and B. Soderberg (1989) Teachers and Classes with Neural Networks. Int. J. Neural Systems. **1**, 167-176.

[6] L. Gislen, C. Peterson and B. Soderberg (1992) Complex Scheduling Using Potts Neural Networks. *Neural Computation.* **4**, 805-831.

[7] M.Ohlsson, C. Peterson and B. Soderberg (1993) Neural Networks for Optimization Problems with Inequality Constraints : The Knapsack Problem. *Neural Computation.* **5**, 331-339.

[8] C. Peterson and B. Soderberg (1993) Artificial Neural Networks. In *Modern Heuristic Techniques for Combinatorial Problems.* ed. C. Reeves. Halsted Press, New York, USA.

[9] C. Tovey (1992) Adaptive Cooling Schedules for Deterministic Annealing Algorithms. *Honours Dissertation.* University of Western Australia.

[10] C. Tovey (1995) A New Deterministic Annealing Method with Application to Operational Research and Decision Problems. *Masters Thesis.* The University of Western Australia.

[11] C. Tovey and A. Mees (1994) Approximate Solutions of Multi-Commodity Integer Flow Problems Using Deterministic Annealing. *Journal of Opl. Res. Soc.* Submitted.

[12] C. Tovey and A. Mees (1995) The Connection Between the Saddle Point Approximation and Deformable Templates. In preparation.

[13] A.L. Yuille (1990) Generalized Deformable Models, Statistical Physics, and Matching Problems. *Neural Computation.* **2**, 1-24.

[14] A.L. Yuille and J.J. Kosowsky (1994) Statistical Physics Algorithms That Converge (1994) *Neural Computation.* **6**, 341-356.

[15] D.P. Bertsekas (1982) *Constrained Optimisation and Lagrange Multiplier Methods.* Academic Press, New York, USA.

[16] R.M. Dudley (1989) *Real Analysis and Probability.* Wadsworth & Brooks/Cole, Pacific Grove, California, USA.

[17] M.R Garey and D.S. Johnson (1987) *Computers and Intractability.* W.H. Freeman, San Francisco, USA.

[18] D.J. Amit (1989) *Modeling Brain Function.* Cambridge University Press, Cambridge, England.

[19] G. Parisi (1988) *Statistical Field Theory.* Addison-Wesley, New York, USA.

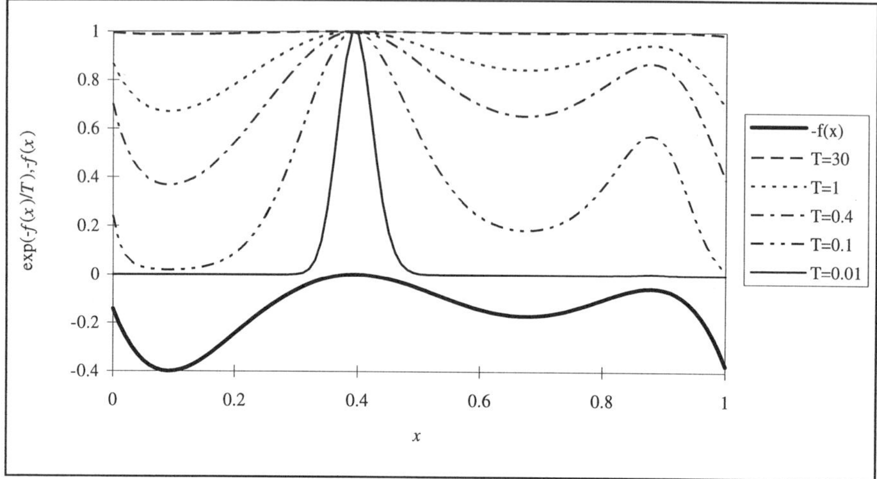

Figure 1 : A graphs of $-f(x)$ and $P(x)=\exp(-f(x)/T)$ for various values of T. Note that $-f(x)$ is shown to make the graph easier to read.

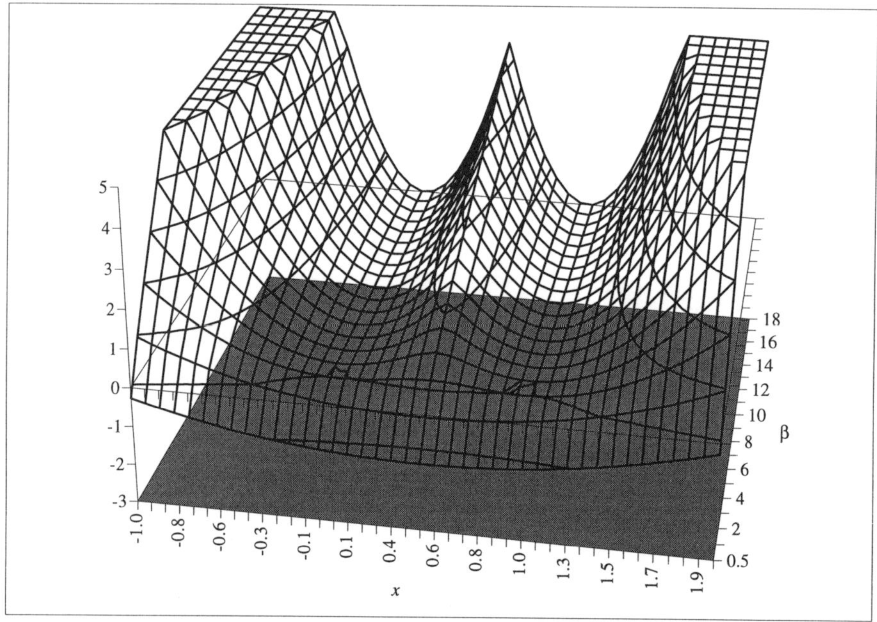

Figure 2 : The binary constraint selection function.

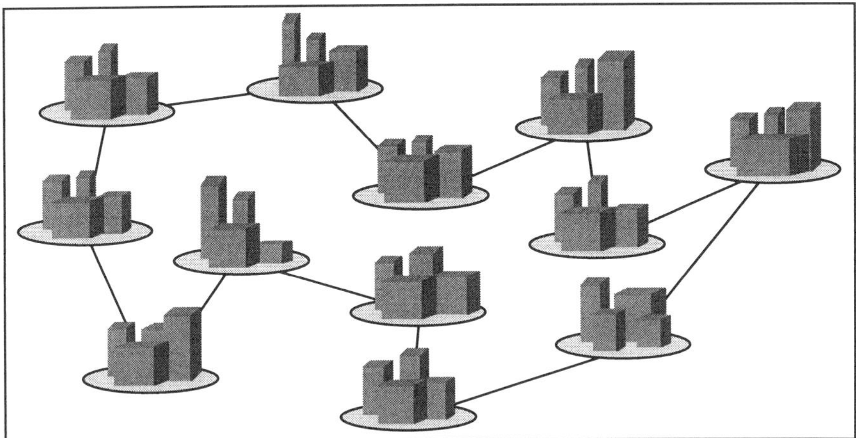

Figure 3 : The travelling salesman problem concerns minimizing the cost of a tour through a number of cities.

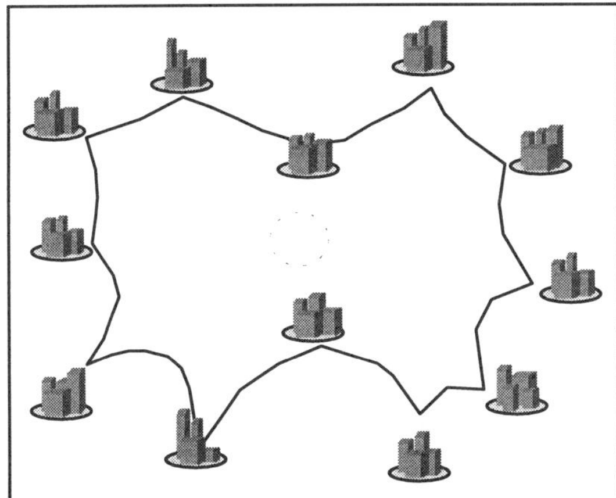

Figure 4 : The elastic net algorithm works by stretching an elastic band around the TSP cities. The dotted line shows the intial configuration of the band. The solid line shows the band at low temperature.

WSSIAA 5 (1995) pp. 423–434
© World Scientific Publishing Company

SIMPLIFIED ANALYSIS OF AN $O(nL)$-ITERATION INFEASIBLE PREDICTOR-CORRECTOR PATH-FOLLOWING METHOD FOR MONOTONE LINEAR COMPLEMENTARITY PROBLEMS

PAUL TSENG

Department of Mathematics, University of Washington
Seattle, Washington 98195, U.S.A.

ABSTRACT

We give a simplified analysis of an infeasible predictor-corrector path-following method for solving monotone linear complementarity problems. This method, like those studied by Mizuno et al. and by Potra and Sheng, (i) requires two factorizations and two backsolves per iteration, (ii) can find a solution in $O(\sqrt{n}L)$ or $O(nL)$ iterations, depending on the quality of the starting point, and (iii) has local quadratic convergence, provided a strictly complementary solution exists. The method decreases the centering parameter and the infeasibility at both predictor step and corrector step, and it is flexible in that either a primal-scaling or dual-scaling or primal-dual scaling can be used for the corrector step without affecting the global and local convergence properties of the method.

1 Introduction

Motivated by their practical performance, there recently have been numerous studies of the global/local convergence rate of infeasible-interior-point methods for linear programming[2,4,5,8–11,13,16,17,20,30] and for solving monotone linear complementarity problem (abbreviated as LCP)[1,12,14,15,18,19,23–26,28,29,31]. The majority of these studies focus on primal-dual path-following methods using predictor-corrector steps, for which they establish global linear convergence and/or, under the assumption that a strictly complementary solution exists, local superlinear/quadratic convergence. However, the analyses tend to be complicated and hence not easily accessible to those outside the research area. (An exception is a method of Mizuno[10] which has a simple global convergence analysis. But this method uses feasible interior-point methods as subroutines and the analysis implicitly assumes an understanding of the latter methods.) In this paper, we present a simplified global and local convergence analysis of an infeasible predictor-corrector path-following method. This method shares the same features as those of Mizuno et al.[12] and of Potra and Sheng[19], namely, it (i) requires two factorizations and two backsolves per iteration, (ii) can start anywhere in the positive orthant and, assuming a solution exists, attain global linear convergence with convergence ratio that, depending on the quality of the starting point, ranges from $1 - O(\sqrt{n})^{-1}$ to $1 - O(n)^{-1}$ (also see related works[2,9,11,18]), (iii) attains local quadratic convergence provided a strictly complementary solution exists. This method differs from existing methods in that the centering parameter and the infeasibility are decreased at both the predictor step and the corrector step. Also, this method is modularized in that either primal-scaling or dual-scaling or primal-dual-scaling can be used for the corrector step without affecting the global/local convergence properties of the method. The predictor step may also be removed without affecting the global convergence property of the method.

We will present our results in the context of the monotone LCP, namely, that of finding an $(x^*, y^*) \in \Re^{2n}$ satisfying

$$x^* \geq 0, \quad y^* \geq 0, \quad Mx^* + q = y^*, \quad (x^*)^T y^* = 0, \tag{1}$$

where $M \in \Re^{n \times n}$ is positive semidefinite and $q \in \Re^n$ are given. However, our results also extend to the horizontal LCP studied by Zhang[29] and to the case where M is a P_*-matrix[6]. We denote by S the set of solutions (possibly empty) of the monotone LCP, i.e.,

$$S = \{ (x^*, y^*) \in \Re^{2n} \mid (x^*, y^*) \text{ satisfies Eq. (1)} \}.$$

In our notation, all vectors are column vectors and superscript T denotes transpose. We denote by $\Re^n$ the n-dimensional real vector space and by $\Re^n_{++}$ the strictly positive orthant in $\Re^n$. We denote by e the vector in $\Re^n$ whose components are all 1. For any $x \in \Re^n$, we denote by x_i the ith component of x, by X the $n \times n$ diagonal matrix whose ith diagonal entry is x_i for all i, and by $\|x\|_1$, $\|x\|$, $\|x\|_\infty$ the 1-norm, the 2-norm and the ∞-norm, respectively, of x. For any $I \subset \{1, ..., n\}$, we denote by x_I the vector with components x_i, $i \in I$.

2 An Infeasible Predictor-Corrector Algorithm

For any $w \in (0, \infty)$, let

$$\phi_w(x, y) \;=\; \|we - Xy\|/w \qquad \forall (x, y) \in \Re^{2n}.$$

The idea of the infeasible predictor-corrector path-following method is to begin each iteration with a $w > 0$ and an $(x, y) > 0$ that is approximately centred (in the sense that $\phi_w(x, y) \leq \beta_0$ for some $\beta_0 \in (0, 1)$), take a damped Newton step (predictor step) for

$$Xy = 0, \qquad Mx + q - y = 0$$

(so to attain fast local convergence) and then followed by a Newton step (corrector step) for

$$Xy = we, \qquad Mx + q - y = -s$$

with s suitably chosen (so to make (x, y) approximately centred again). (See the paper by Mizuno et al.[12] for a survey of the history of this method.) We will focus on the following version of the method which is similar to that studied by Mizuno et al.[12] and Potra and Sheng[19] but differs in that w^t and $Mx^t + q - y^t$ decrease in norm at *both* the predictor step and the corrector step.

Algorithm 2.1 *Choose any* $(\beta_0, \beta_1, \beta_2) \in \Re^3$ *satisfying* $0 < \beta_0 < \beta_1 < \beta_2 < \sqrt{\beta_0} < 1$ *and any* $(x^0, y^0, w^0) \in \Re_{++}^{2n+1}$ *satisfying* $\phi_{w^0}(x^0, y^0) \le \beta_0$. *Let*

$$\eta_1 = (\beta_0 - (\beta_2)^2))/(\beta_0 + \sqrt{n}). \tag{2}$$

For $t = 0, 1, ...,$ *compute* $(x^{t+1}, y^{t+1}, w^{t+1})$ *from* (x^t, y^t, w^t) *according to:*

Predictor Step:

$$\hat{x}^t = x^t + \theta^t \Delta x^t, \qquad \hat{y}^t = y^t + \theta^t \Delta y^t, \qquad \hat{w}^t = (1 - \theta^t)w^t, \tag{3}$$

where $(\Delta x^t, \Delta y^t)$ *is the unique vector in* $\Re^{2n}$ *satisfying*

$$Y^t \Delta x^t + X^t \Delta y^t = -X^t y^t, \qquad M \Delta x^t - \Delta y^t = -(Mx^t + q - y^t), \tag{4}$$

and θ^t *is the largest* $\theta \in (0, 1]$ *satisfying* $(x^t + \theta \Delta x^t, y^t + \theta \Delta x^t) > 0$ *and*

$$\phi_{(1-\theta)w^t}(x^t + \theta \Delta x^t, y^t + \theta \Delta x^t) \le \beta_1. \tag{5}$$

Corrector Step:

$$x^{t+1} = \hat{x}^t + \hat{\theta}^t \Delta \hat{x}^t, \qquad y^{t+1} = \hat{y}^t + \hat{\theta}^t \Delta \hat{y}^t, \qquad w^{t+1} = (1 - \gamma^t \hat{\theta}^t)\hat{w}^t, \tag{6}$$

where γ^t *is the largest* $\gamma \in (0, \eta_1]$ *satisfying*

$$\|\hat{w}^t e - \hat{X}^t \hat{y}^t - \gamma \hat{X}^t (M\hat{x}^t + q - \hat{y}^t)\|/\hat{w}^t \le \beta_2, \tag{7}$$

$(\Delta \hat{x}^t, \Delta \hat{y}^t)$ *is the unique vector in* $\Re^{2n}$ *satisfying*

$$\hat{w}^t (\hat{X}^t)^{-1} \Delta \hat{x}^t + \hat{X}^t \Delta \hat{y}^t = \hat{w}^t e - \hat{X}^t \hat{y}^t, \qquad M \Delta \hat{x}^t - \Delta \hat{y}^t = -\gamma^t (M\hat{x}^t + q - \hat{y}^t), \tag{8}$$

and $\hat{\theta}^t$ *is any* $\theta \in [1, \infty)$ *satisfying*

$$(\hat{x}^t + \theta \Delta \hat{x}^t, \hat{y}^t + \theta \Delta \hat{y}^t) > 0, \qquad \phi_{(1-\gamma^t \theta)\hat{w}^t}(\hat{x}^t + \theta \Delta \hat{x}^t, \hat{y}^t + \theta \Delta \hat{y}^t) \le \beta_0. \tag{9}$$

We use the primal-scaling direction in the corrector step so to simplify the global convergence analysis. This direction may be replaced by either a dual-scaling (corresponding to replacing "$\hat{w}^t(\hat{X}^t)^{-1}$" and the first "$\hat{X}^t$" in Eq. (8) with, respectively, $\hat{Y}^t$ and $\hat{w}^t(\hat{Y}^t)^{-1}$) or a primal-dual-scaling direction (corresponding to replacing "$\hat{w}^t(\hat{X}^t)^{-1}$" in Eq. (8) with $\hat{Y}^t$) without affecting the global and the local convergence properties of the method. In other words, Thms. 3.1 and 4.1 still hold for these modified methods (possibly with different constants). The predictor step, which is used to accelerate local convergence, can be removed (by setting $\theta^t = 0$) without affecting the global convergence property of the method. Note that (θ^t, γ^t), which will be shown to be well defined (see the proof of Thm. 3.1), can be computed very easily. The parameter $\hat{\theta}^t$ is introduced to accelerate the convergence of the method and can always be set to 1 (see the proof of Thm. 3.1).

3 Global Linear Convergence

Our first lemma shows that the direction $(\Delta\hat{x}^t, \Delta\hat{y}^t)$ is a centering direction at $(\hat{x}^t, \hat{y}^t)$. The proof of this lemma is patterned after that of Lem. 1 in an earlier paper[21] for a primal-scaling path-following method (also see Lem. 1 in a related paper[22]).

Lemma 3.1 *Fix any* $(\beta_0, \beta_1, \beta_2) \in \Re^3$ *satisfying* $0 < \beta_0 < \beta_1 < \beta_2 \le \sqrt{\beta_0} < 1$. *Let* η_1 *be given by Eq. (2). For any* $(\hat{x}^t, \hat{y}^t, \hat{w}^t) \in \Re_{++}^{2n+1}$ *satisfying* $\phi_{\hat{w}^t}(\hat{x}^t, \hat{y}^t) \le \beta_1$, *the quantity* $(\Delta\hat{x}^t, \Delta\hat{y}^t)$ *satisfying Eq. (8), with* γ^t *being the largest* $\gamma \in (0, \eta_1]$ *satisfying Eq. (7), also satisfies*

$$(\hat{x}^t + \Delta\hat{x}^t, \hat{y}^t + \Delta\hat{y}^t) > 0, \qquad \phi_{(1-\gamma^t)\hat{w}^t}(\hat{x}^t + \Delta\hat{x}^t, \hat{y}^t + \Delta\hat{y}^t) \le \beta_0.$$

Proof. For simplicity, denote $(x, y, w) = (\hat{x}^t, \hat{y}^t, \hat{w}^t)$, $(\Delta x, \Delta y) = (\Delta\hat{x}^t, \Delta\hat{y}^t)$ and $\gamma = \gamma^t$ respectively. Let $r = we - Xy$, $s = Mx + q - y$ and $d = X^{-1}\Delta x$. Then Eq. (8) may be rewritten as

$$wd + X\Delta y = r, \qquad MXd - \Delta y = -\gamma s.$$

Since $XMX + wI$ is positive definite and hence invertible, we can solve for d to obtain

$$d = (XMX + wI)^{-1}(r - \gamma Xs).$$

This together with the positive semidefinite property of XMX yields

$$w\|d\|^2 \le d^T(XMX + wI)d = d^T(r - \gamma Xs) \le \|d\|\|r - \gamma Xs\|, \tag{10}$$

and it readily follows that

$$\|d\| \le \|r - \gamma Xs\|/w. \tag{11}$$

Let $x' = x + \Delta x$ and $y' = y + \Delta y$. By Eqs. (7) and (11), we have $\|d\| \le \beta_2 < 1$ and so $e + d > 0$. Then $x > 0$ yields $x' = x + Xd = X(e + d) > 0$. Also, $x' = x + Xd$ implies $X' = X + DX$, which together with $y' = y + \Delta y$ yields

$$\begin{aligned}
we - X'y' &= we - (I + D)X(y + \Delta y) \\
&= wd - DX(y + \Delta y) \\
&= wDd,
\end{aligned}$$

where the last two equalities each follows from the first equation in Eq. (8). Thus

$$\begin{aligned}
\|we - X'y'\| &= w\|Dd\| \\
&\le w\|Dd\|_1 \\
&= w\|d\|^2 \\
&\le \|r - \gamma Xs\|^2/w \\
&\le (\beta_2)^2 w,
\end{aligned}$$

where the first inequality follows from properties of the 1-norm and the 2-norm; the second inequality follows from Eqs. (10) and (11); the third inequality follows from Eq. (7). Since $\beta_2 < 1$ and $x' > 0$, the above relation implies that $y' > 0$. Also, using the triangle inequality and the above relation, we obtain

$$
\begin{aligned}
\phi_{(1-\gamma)w}(x', y') &= \|(1-\gamma)we - X'y'\|/((1-\gamma)w) \\
&\leq \|we - X'y'\|/((1-\gamma)w) + \gamma\sqrt{n}/(1-\gamma) \\
&\leq (\beta_2)^2/(1-\gamma) + \gamma\sqrt{n}/(1-\gamma).
\end{aligned}
$$

It follows from $\gamma \leq \eta_1$ that the righthand side of the above relation is below β_0. ∎

The next lemma is needed to bound $\|(x^t, y^t)\|$ from above and certain components of (x^t, y^t) from below. This lemma, due originally to Mizuno[9] (see Lem. 3.3 therein), is widely used in the convergence analysis of infeasible-interior-point methods. For completeness, we include a short proof.

Lemma 3.2 *Assume that $S \neq \emptyset$. For any $\mu \in [0,1]$, any $(x^*, y^*) \in S$ and any $(x, y) \in \Re^{2n}$ satisfying $(x, y) \geq 0$ and $Mx + q - y = \mu(Mx^0 + q - y^0)$, we have*

$$
\mu(x^T y^0 + y^T x^0) \leq x^T y + \mu((x^0)^T y^0 + (x^*)^T y^0 + (x^0)^T y^*), \tag{12}
$$

$$
(1-\mu)(x^T y^* + y^T x^*) \leq x^T y + \mu((x^0)^T y^0 + (x^*)^T y^0 + (x^0)^T y^*). \tag{13}
$$

Proof. We have from $Mx + q - y = \mu(Mx^0 + q - y^0)$ and $Mx^* + q = y^*$ that

$$
M(x - \mu x^0 - (1-\mu)x^*) - (y - \mu y^0 - (1-\mu)y^*) = 0.
$$

Multiplying both sides by $x - \mu x^0 - (1-\mu)x^*$ and using the positive semidefinite property of M yields

$$
(x - \mu x^0 - (1-\mu)x^*)^T(y - \mu y^0 - (1-\mu)y^*) \geq 0,
$$

which can be rearranged to obtain

$$
\begin{aligned}
\mu(x^T y^0 &+ y^T x^0) + (1-\mu)(x^T y^* + y^T x^*) \\
&\leq x^T y + (\mu x^0 + (1-\mu)x^*)^T(\mu y^0 + (1-\mu)y^*) \\
&= x^T y + \mu^2(x^0)^T y^0 + \mu(1-\mu)((x^*)^T y^0 + (x^0)^T y^*),
\end{aligned}
$$

where the equality follows from $(x^*)^T y^* = 0$. Using the nonnegativity of (x, y), (x^0, y^0) and (x^*, y^*) and the fact $0 \leq \mu \leq 1$ completes the proof. ∎

By using Lems. 3.1 and 3.2, we can now establish the global linear convergence of Algorithm 2.1.

Theorem 3.1 *Let β_0, β_1, β_2, η_1, and $\{(x^t, y^t, w^t, \hat{x}^t, \hat{y}^t, \hat{w}^t, \theta^t, \gamma^t, \hat{\theta}^t)\}_{t=0,1,\ldots}$ be generated by Algorithm 2.1. Then,*

$$(x^t, y^t) > 0, \qquad \phi_{w^t}(x^t, y^t) \leq \beta_0, \tag{14}$$
$$w^t = \mu^t w^0, \qquad Mx^t + q - y^t = \mu^t(Mx^0 + q - y^0), \tag{15}$$

for all t, where

$$\mu^t = (1 - \gamma^{t-1}\hat{\theta}^{t-1})(1 - \theta^{t-1}) \cdots (1 - \gamma^0\hat{\theta}^0)(1 - \theta^0). \tag{16}$$

If $\{(\hat{x}^t, \hat{y}^t)\}$ is bounded, then $S \neq \emptyset$. If $S \neq \emptyset$, then $\{(\hat{x}^t, \hat{y}^t)\}$ and $\{(x^t, y^t)\}$ are bounded and, for any $(x^*, y^*) \in S$, we have $\gamma^t \geq \min\{\eta_1, \eta_2\}$ for all t, where

$$\eta_2 = \frac{(\beta_2 - \beta_1)w^0(\min_i y_i^0)/\|Mx^0 + q - y^0\|_\infty}{(1 + \beta_1)nw^0 + (x^0)^T y^0 + (x^*)^T y^0 + (x^0)^T y^*}. \tag{17}$$

Proof. It is easily seen that Eqs. (14)-(15) hold for $t = 0$. Assume Eqs. (14)-(15) hold for some $t \geq 0$. Then, the lefthand side of Eq. (5) is below β_0 when $\theta = 0$, so, by the choice of θ^t and Eq. (3), both θ^t and $(\hat{x}^t, \hat{y}^t, \hat{w}^t)$ are well defined and satisfy $(\hat{x}^t, \hat{y}^t, \hat{w}^t) > 0$ and $\phi_{\hat{w}^t}(\hat{x}^t, \hat{y}^t) \leq \beta_1$. Then, γ^t is well defined (since when $\gamma = 0$, the lefthand side of Eq. (7) is equal to $\phi_{\hat{w}^t}(\hat{x}^t, \hat{y}^t)$ and hence below β_1) and, by Lem. 3.1, Eq. (9) holds for $\theta = 1$. Thus, $\hat{\theta}^t$ is well defined (and in particular can be set to 1) and, since $\theta = \hat{\theta}^t$ satisfies Eq. (9), Eq. (6) implies Eq. (14) holds when t is replaced by $t + 1$. Also, we have from Eqs. (3) and (6) and the second equality in Eqs. (4) and (8) that

$$w^{t+1} = (1 - \gamma^t\hat{\theta}^t)(1 - \theta^t)w^t, \qquad Mx^{t+1} + q - y^{t+1} = (1 - \gamma^t\hat{\theta}^t)(1 - \theta^t)(Mx^t + q - y^t)$$

so Eq. (15) holds when t is replaced by $t + 1$. By induction, Eqs. (14)-(15) hold for all $t \geq 0$.

For any t, we have from $\phi_{\hat{w}^t}(\hat{x}^t, \hat{y}^t) \leq \beta_1$ (shown earlier) that $\hat{X}^t\hat{y}^t \leq (1 + \beta_1)\hat{w}^t e$, so

$$(\hat{x}^t)^T\hat{y}^t \leq (1 + \beta_1)n\hat{w}^t = (1 + \beta_1)n(1 - \theta^t)\mu^t w^0, \tag{18}$$

where the equality follows from Eqs. (3) and (15). Also, we have from the triangle inequality and $\phi_{\hat{w}^t}(\hat{x}^t, \hat{y}^t) \leq \beta_1$ that, for any $\gamma \in [0, \infty)$,

$$\|\hat{w}^t e - \hat{X}^t\hat{y}^t - \gamma\hat{X}^t(M\hat{x}^t + q - \hat{y}^t)\|/\hat{w}^t \leq \phi_{\hat{w}^t}(\hat{x}^t, \hat{y}^t) + \gamma\|\hat{X}^t(M\hat{x}^t + q - \hat{y}^t)\|/\hat{w}^t$$
$$\leq \beta_1 + \gamma\|\hat{X}^t(M\hat{x}^t + q - \hat{y}^t)\|/\hat{w}^t.$$

Let $\bar{\gamma}^t$ be the largest γ for which the lefthand side is below β_2 (so $\gamma^t = \min\{\eta_1, \bar{\gamma}^t\}$). Then $\bar{\gamma}^t$ must exceed the largest γ for which the righthand side is below β_2. This yields

$$\bar{\gamma}^t \geq \frac{(\beta_2 - \beta_1)\hat{w}^t}{\|\hat{X}^t(M\hat{x}^t + q - \hat{y}^t)\|} \geq \frac{(\beta_2 - \beta_1)\hat{w}^t}{\|\hat{x}^t\|_1\|M\hat{x}^t + q - \hat{y}^t\|_\infty} = \frac{(\beta_2 - \beta_1)w^0}{\|\hat{x}^t\|_1\|Mx^0 + q - y^0\|_\infty}, \tag{19}$$

where the equality follows from $\hat{w}^t = (1-\theta^t)\mu^t w^0$ and $M\hat{x}^t + q - \hat{y}^t = (1-\theta^t)\mu^t(Mx^0 + q - y^0)$ (see Eqs. (3), (4) and (15)).

Assume $\{(\hat{x}^t, \hat{y}^t)\}$ is bounded. Then Eq. (19) yields that $\bar{\gamma}^t \geq \eta$ for some constant $\eta > 0$ and so $\gamma^t = \min\{\eta_1, \bar{\gamma}^t\} \geq \min\{\eta_1, \eta\}$ for all t. By Eq. (16) and $\hat{\theta}^t \geq 1$ for all t, $\{\mu^t\} \to 0$. Since $(\hat{x}^t, \hat{y}^t) > 0$, $M\hat{x}^t + q - \hat{y}^t = (1 - \theta^t)\mu^t(Mx^0 + q - y^0)$ and Eq. (18) holds for all t, this implies any cluster point of $\{(\hat{x}^t, \hat{y}^t)\}$ is in S, so $S \neq \emptyset$.

Assume $S \neq \emptyset$ and fix any $(x^*, y^*) \in S$. By Eqs. (14)-(15), we have Eq. (12) holds (with $(x, y) = (\hat{x}^t, \hat{y}^t)$, $\mu = (1 - \theta^t)\mu^t$), so

$$\|\hat{x}^t\|_1(\min_i y_i^0) + \|\hat{y}^t\|_1(\min_i x_i^0) \leq (\hat{x}^t)^T y^0 + (\hat{y}^t)^T x^0$$
$$\leq (\hat{x}^t)^T \hat{y}^t/\mu^t + (x^0)^T y^0 + (x^*)^T y^0 + (x^0)^T y^*$$
$$\leq (1 + \beta_1)nw^0 + (x^0)^T y^0 + (x^*)^T y^0 + (x^0)^T y^*,$$

where the first inequality follows from the nonnegativity of $(\hat{x}^t, \hat{y}^t)$ and (x^0, y^0) and the last inequality follows from Eq. (18). Since $(x^0, y^0) > 0$, this shows $\{(\hat{x}^t, \hat{y}^t)\}$ is bounded. A similar argument shows $\{(x^t, y^t)\}$ is bounded. Combining the above inequality with Eq. (19) yields

$$\bar{\gamma}^t \geq \frac{(\beta_2 - \beta_1)w^0(\min_i y_i^0)/\|Mx^0 + q - y^0\|_\infty}{(1 + \beta_1)nw^0 + (x^0)^T y^0 + (x^*)^T y^0 + (x^0)^T y^*}$$

for every t. The righthand side is exactly η_2, so $\gamma^t = \min\{\eta_1, \bar{\gamma}^t\} \geq \min\{\eta_1, \eta_2\}$. ∎

Thus, if S is nonempty, Algorithm 2.1 has global linear convergence of ratio at most $1 - \min\{\eta_1, \eta_2\}$ and, by using standard techniques, can find an element of S in $O(\min\{\eta_1, \eta_2\}^{-1}L)$ iterations, where L denotes the size of the problem encoding in binary. It is easily seen that $\eta_1^{-1} = O(\sqrt{n})$, so it only remains to estimate η_2^{-1}. In the case where (x^0, y^0, w^0) is chosen so that η_2^{-1} is $O(\sqrt{n})$ (such as when in addition $Mx^0 + q - y^0 = 0$), the iteration count is $O(\sqrt{n}L)$. In the case where (x^0, y^0, w^0) is chosen by the well known choice of (see Cor. 3.6 by Potra and Sheng[19])

$$x^0 = \rho_p e, \quad y^0 = \rho_d e, \quad w^0 = \rho_p \rho_d, \quad \rho_p \geq \frac{\|x^*\|_1}{n}, \quad \rho_d \geq \max\left\{\frac{\|y^*\|_1}{n}, \|\rho_p Me + q\|_\infty\right\},$$

where (x^*, y^*) is any element of S, Eq. (17) yields

$$\eta_2^{-1} \leq 3(4 + \beta_1)n/(\beta_2 - \beta_1) = O(n),$$

so the iteration count is $O(nL)$.

To detect whether S is empty, we can adopt a technique used by Potra and Sheng[19] (see Thm. 3.7 therein): In Algorithm 2.1, for every t let ρ^t be the largest ρ such that

$$(\hat{x}^t)^T y^0 + (\hat{y}^t)^T x^0 > (\hat{x}^t)^T \hat{y}^t/\mu^t + (x^0)^T y^0 + \|(x^0, y^0)\|\rho.$$

Then, by Eq. (12), there is no $(x^*, y^*) \in S$ satisfying $\|(x^*, y^*)\| \leq \rho^t$.

4 Local Quadratic Convergence

The key to local quadratic convergence of $\{(x^t, y^t)\}$ is the following well known result (see Lem. 12 by Mizuno et al.[12], Lem. 4.4 by Potra and Sheng[19], Thm. 4.5 by Wright[25], Thm. 3.6 by Ye and Anstreicher[27]) that shows $\|(\Delta x^t, \Delta y^t)\|$ is in the order of μ^t. However, our proof is simple as it uses only Eqs. (12)-(13) and Hoffman's bound[3] and yields, as a byproduct, a new error bound on the distance $d((x,y), S) := \min_{(x^*, y^*) \in S} \|(x,y) - (x^*, y^*)\|$ (cf. a result of Mangasarian[7]).

Lemma 4.1 *Assume there exists an* $(x^*, y^*) \in S$ *satisfying* $(x_I^*, y_J^*) > 0$ *for some* $I \cup J = \{1, ..., n\}$ *and* $I \cap J = \emptyset$. *Fix any* $\beta_0 \in (0, 1)$, *any* $\bar{\mu} \in (0, 1)$ *and any* $(x^0, y^0, w^0) \in \Re_{++}^{2n+1}$. *There exists a constant* $C_1 > 0$ *(depending on* (x^*, y^*), (x^0, y^0, w^0), β_0, $\bar{\mu}$) *such that, for any* $(x, y) \in \Re_{++}^{2n}$ *and any* $\mu \in (0, \bar{\mu}]$ *satisfying* $\phi_{\mu w^0}(x, y) \le \beta_0$ *and* $Mx + q - y = \mu(Mx^0 + q - y^0)$, *we have* $d((x,y), S) \le C_1 \mu$ *and, moreover, the vector* $(\Delta x, \Delta y)$ *satisfying*

$$Y\Delta x + X\Delta y = -Xy, \qquad M\Delta x - \Delta y = -(Mx + q - y) \tag{20}$$

satisfies $\|(\Delta x, \Delta y)\| \le C_1 \mu$.

Proof. For convenience, we will use "$a = O(b)$" to mean $a \le Cb$ for some constant $C > 0$ depending on (x^*, y^*), (x^0, y^0, w^0), β_0 and $\bar{\mu}$. First, since $\phi_{\mu w^0}(x, y) \le \beta_0$, we have $(1 - \beta_0)\mu w^0 \le x_i y_i \le (1 + \beta_0)\mu w^0$ for all i and so

$$(\min_i x_i y_i)^{-1} = O(\mu^{-1}), \qquad x^T y = O(\mu). \tag{21}$$

Second, by $\mu \le \bar{\mu}$ and the nonnegativity of (x, y) and (x^*, y^*), we have

$$
\begin{aligned}
(1 - \bar{\mu}) \min_{i \in I, j \in J} \{y_j^*, x_i^*\} \|(x_J, y_I)\|_1 &\le (1 - \mu)((x_J)^T y_J^* + (y_I)^T x_I^*) \\
&\le (1 - \mu)(x^T y^* + y^T x^*) \\
&\le x^T y + \mu((x^0)^T y^0 + (x^*)^T y^0 + (x^0)^T y^*),
\end{aligned}
$$

where the last inequality follows from Eq. (13). Since $(x_I^*, y_J^*) > 0$, this together with Eq. (21) shows

$$\|(x_J, y_I)\|_1 = O(x^T y + \mu) = O(\mu). \tag{22}$$

Third, by the nonnegativity of (x, y) and (x^0, y^0) and using Eq. (12), we have

$$\min_i \{y_i^0, x_i^0\}(\|x\|_1 + \|y\|_1) \le x^T y^0 + y^T x^0 \le x^T y / \mu + (x^0)^T y^0 + (x^*)^T y^0 + (x^0)^T y^*$$

so $(x^0, y^0) > 0$ and Eq. (21) imply $\|x\|_1 = O(1)$. Fourth, note the following two linear systems:

$$
\begin{aligned}
M\xi + q - \psi &= \mu(Mx^0 + q - y^0), \quad \xi_J = x_J, \quad \psi_I = y_I, \quad \xi \ge 0, \quad \psi \ge 0, \\
M\xi + q - \psi &= 0, \quad \xi_J = 0, \quad \psi_I = 0, \quad \xi \ge 0, \quad \psi \ge 0, \tag{23}
\end{aligned}
$$

both have solution $((x, y)$ and (x^*, y^*), respectively) so, by a result of Hoffman[3], the system Eq. (23) has a solution $(\bar{x}, \bar{y})$ (which is easily seen to be in S) that satisfies

$$\|(x - \bar{x}, y - \bar{y})\| = O(\|(\mu(Mx^0 + q - y^0), x_J, y_I)\|). \tag{24}$$

Since $(\bar{x}, \bar{y}) \in S$ so $\bar{X}\bar{y} = 0$ and $M\bar{x} + q = \bar{y}$, we can rewrite Eq. (20) as

$$Y(x + \Delta x - \bar{x}) + X(y + \Delta y - \bar{y}) = (X - \bar{X})(y - \bar{y}), \quad M(x + \Delta x - \bar{x}) - (y + \Delta y - \bar{y}) = 0.$$

Substituting the second equation into the first gives

$$(YX + XMX)X^{-1}(x + \Delta x - \bar{x}) = (X - \bar{X})(y - \bar{y}),$$

and multiplying both sides by $X^{-1}(x + \Delta x - \bar{x})$ and using the positive semidefinite property of XMX yields

$$\begin{aligned}
(\min_i x_i y_i)\|X^{-1}(x + \Delta x - \bar{x})\|^2 &\leq (x + \Delta x - \bar{x})^T X^{-1}(YX + XMX)X^{-1}(x + \Delta x - \bar{x}) \\
&= (x + \Delta x - \bar{x})^T X^{-1}(X - \bar{X})(y - \bar{y}) \\
&\leq \|X^{-1}(x + \Delta x - \bar{x})\|\|(X - \bar{X})(y - \bar{y})\|.
\end{aligned}$$

Thus

$$\|X^{-1}(x + \Delta x - \bar{x})\| \leq \|(X - \bar{X})(y - \bar{y})\| / \min_i x_i y_i$$

and since the lefthand side is greater than or equal to $\|x + \Delta x - \bar{x}\|/\|x\|_\infty$, we have

$$\begin{aligned}
\|x + \Delta x - \bar{x}\| &\leq \|x\|_\infty \|(X - \bar{X})(y - \bar{y})\| / \min_i x_i y_i \\
&= O(\|x - \bar{x}\|\|y - \bar{y}\|/\mu) = O(\mu),
\end{aligned}$$

where the first equality follows from Eq. (21) and $\|x\| = O(1)$, and the second equality follows from $\|(x - \bar{x}, y - \bar{y})\| = O(\mu)$ (see Eqs. (22) and (24)). Since $\|x - \bar{x}\| = O(\mu)$, the above relation shows $\|\Delta x\| = O(\mu)$. From the second equality in Eq. (20), we have

$$\|\Delta y\| = \|M\Delta x + Mx + q - y\| = \|M\Delta x + \mu(Mx^0 + q - y^0)\|,$$

and so $\|\Delta y\| = O(\mu)$. Finally, since $(\bar{x}, \bar{y}) \in S$, we have $d((x, y), S) \leq \|(x - \bar{x}, y - \bar{y})\| = O(\mu)$. $\blacksquare$

By using Lem. 4.1, the local quadratic convergence of Algorithm 2.1 readily follows. The proof is fairly standard and is included for completeness.

Theorem 4.1 *Assume there exists an $(x^*, y^*) \in S$ satisfying $x_I^* > 0$, $y_J^* > 0$ for some $I \cup J = \{1, ..., n\}$ and $I \cap J = \emptyset$. Let β_0, β_1, β_2, η_1, and $\{(x^t, y^t, w^t, \hat{x}^t, \hat{y}^t, \hat{w}^t, \theta^t, \gamma^t, \hat{\theta}^t)\}_{t=0,1,...}$ be generated by Algorithm 2.1. Then, there exist constants $C_1 > 0$ and $C_2 > 0$ (depending on (x^*, y^*), (x^0, y^0, w^0), β_0, β_1, η_1, and η_2 given by Eq. (17)) such that*

$$d((x^t, y^t), S) \leq C_1 \mu^t \quad and \quad \mu^{t+1} \leq C_2(\mu^t)^2 \quad \forall t \geq 1.$$

Proof. By Thm. 3.1, we have that the assumption of Lem. 4.1 is satisfied by $\bar{\mu} = 1 - \min\{\eta_1, \eta_2\}$ and $(x, y, \mu) = (x^t, y^t, \mu^t)$ for all $t \geq 1$. Hence, by Lem. 4.1, $d((x^t, y^t), S) \leq C_1\mu^t$ and $\|(\Delta x^t, \Delta y^t)\| \leq C_1\mu^t$ for some constant $C_1 > 0$. The latter implies, for any $\theta \in (0, 1]$,

$$
\begin{aligned}
\phi_{(1-\theta)w^t}(x^t + \theta\Delta x^t, y^t + \theta\Delta y^t) &= \left\| w^t e - X^t y^t - \Delta X^t \Delta y^t \theta^2/(1-\theta) \right\| / w^t \\
&\leq \phi_{w^t}(x^t, y^t) + \|\Delta X^t \Delta y^t\|\theta^2/(1-\theta)w^t \\
&\leq \beta_0 + (C_1)^2\mu^t/(1-\theta)w^0,
\end{aligned}
$$

where the equality uses the first equation in Eq. (4) and the last inequality also uses $\theta^2 \leq 1$ and Eqs. (14)-(15). It is easily checked that the righthand side is below β_1 for all $\theta \in [0, 1 - (C_1)^2\mu^t/((\beta_1 - \beta_0)w^0)]$, so we also have $(x^t + \theta\Delta x^t, y^t + \theta\Delta y^t) > 0$ (since if any component of $(x^t + \theta\Delta x^t, y^t + \theta\Delta y^t)$ reaches 0, the lefthand side would exceed 1). Since θ^t is the largest $\theta \in (0, 1)$ for which the lefthand side is below β_1 and $(x^t + \theta\Delta x^t, y^t + \theta\Delta y^t) > 0$, this yields

$$
\theta^t \geq 1 - (C_1)^2\mu^t/((\beta_1 - \beta_0)w^0)
$$

and it readily follows from Eq. (16) that $\mu^{t+1} \leq (1 - \theta^t)\mu^t \leq C_2(\mu^t)^2$ for some constant $C_2 > 0$. $\blacksquare$

Acknowledgement. This research is supported by National Science Foundation Grant CCR-9311621.

References

1. S. C. Billups and M. C. Ferris, Convergence of infeasible interior-point algorithms from arbitrary starting points, Technical Report, Computer Sciences Department, University of Wisconsin, Madison, October 1993.

2. R. M. Freund, An infeasible-start algorithm for linear programming whose complexity depends on the distance from the starting point to the optimal solution, Working Paper 3559-93-MSA, Sloan School of Management, Massachusetts Institute of Technology, Cambridge, April 1993.

3. A. J. Hoffman, On approximate solutions of systems of linear inequalities, *J. Res. Natl. Bur. Standards*, **49** (1952), 263-265.

4. M. Kojima, Basic lemmas in polynomial-time infeasible-interior-point methods for linear programs, Research Report B-268, Department of Information Sciences, Tokyo Institute of Technology, Tokyo, 1993.

5. M. Kojima, N. Megiddo, and S. Mizuno, A primal-dual exterior point algorithm for linear programming, *Math. Prog.*, **61** (1993), 261-280.

6. M. Kojima, N. Megiddo, T. Noma, and A. Yoshise, *A Unified Approach to Interior Point Algorithms for Linear Complementarity Problems*, Lecture Notes in Computer Science 538, Springer-Verlag, Berlin, 1991.

7. O. L. Mangasarian, Error bounds for nondegenerate monotone linear complementarity problems, *Math. Prog.*, **48** (1990), 437-445.

8. J. Miao, Two infeasible interior-point predictor-corrector algorithms for linear programming, Report RRR 20-93, RUTCOR, Rutgers University, New Brunswick, 1993.

9. S. Mizuno, Polynomiality of infeasible interior point algorithms for linear programming, *Math. Prog.*, **67** (1994), 109-119.

10. S. Mizuno, A simple predictor-corrector infeasible-interior-point algorithm for linear programming, Preprint, The Institute of Statistical Mathematics, Tokyo, October 1993.

11. S. Mizuno and F. Jarre, An infeasible-interior-point algorithm using projections onto a convex set, Preprint 209, Mathematische Institute der Universität Würzburg, Würzburg, August 1993; to appear in *Ann. Oper. Res.*

12. S. Mizuno, F. Jarre, and J. Stoer, A unified approach to infeasible-interior-point algorithms via geometrical linear complementarity problems, Preprint Nr. 213, Mathematische Institute der Universität Würzburg, Würzburg, April 1994.

13. S. Mizuno, M. Kojima, and M. J. Todd, Infeasible-interior-point primal-dual potential-reduction algorithms for linear programming, Technical Report 1023, School of Operations Research and Industrial Engineering, Cornell University, Ithaca, August 1992; to appear in *SIAM J. Optim.*

14. R. D. C. Monteiro and S. Wright, Local convergence of interior-point algorithm for degenerate monotone LCP, Preprint MCS-P357-0493, Mathematics and Computer Science Division, Argonne National Laboratory, Argonne, April 1993.

15. R. D. C. Monteiro and S. Wright, A superlinear infeasible-interior-point affine scaling algorithm for LCP, Preprint MCS-P361-0693, Mathematics and Computer Science Division, Argonne National Laboratory, Argonne, June 1993.

16. F. Potra, A quadratically convergent infeasible interior-point algorithm for linear programming, Technical Report 26, Department of Mathematics, University of Iowa, Iowa City, June 1992.

17. F. Potra, An infeasible interior-point predictor-corrector algorithm for linear programming, Technical Report 28, Department of Mathematics, University of Iowa, Iowa City, July 1992.

18. F. Potra, An $O(nL)$ infeasible-interior-point algorithm for LCP with quadratic convergence, Technical Report 50, Department of Mathematics, University of Iowa, Iowa City, January 1994.

19. F. Potra and R. Sheng, A modified $O(nL)$ infeasible-interior-point algorithm for LCP with quadratic convergence, Technical Report 54, Department of Mathematics, University of Iowa, Iowa City, April 1994.

20. J. Stoer, The complexity of an infeasible-interior-point path-following method for the solution of linear programs, *Optim. Methods Software*, **3** (1994), 1-12.

21. P. Tseng, A simple complexity proof for a polynomial-time linear programming algorithm, *Oper. Res. Letters*, **8** (1989), 155-159.

22. P. Tseng, Complexity analysis of a linear complementarity algorithm based on a Lyapunov function, *Math. Prog.*, **53** (1992), 297-306.

23. S. J. Wright, An infeasible-interior-point algorithm for linear complementarity problems, *Math. Prog.*, **29** (1994), 29-51.

24. S. J. Wright, A path-following infeasible-interior-point algorithm for linear complementarity problems, *Optim. Methods Software*, **2** (1993), 79-106.

25. S. J. Wright, A path-following interior-point algorithm for linear and quadratic problems, Preprint MCS-P401-1293, Mathematics and Computer Science Division, Argonne National Laboratory, Argonne, December 1993.

26. S. J. Wright and Y. Zhang, A superquadratic infeasible-interior-point method for linear complementary problems, Technical Report YZ94-03, Department of Mathematics and Statistics, University of Maryland, Baltimore County; Preprint MCS-P418-0294, Mathematics and Computer Science Division, Argonne National Laboratory, Argonne, February 1994.

27. Y. Ye and K. Anstreicher, On quadratic and $O(\sqrt{n}L)$-iteration convergence of predictor-corrector algorithm for LCP, *Math. Prog.*, **62** (1993), 537-551.

28. D. Zhang and Y. Zhang, A Mehrotra-type predictor-corrector algorithm with polynomiality and Q-subquadratic convergence, Technical Report 93-13, Department of Mathematics and Statistics, University of Maryland, Baltimore County, September 1993.

29. Y. Zhang, On the convergence of a class of infeasible interior-point algorithm for the horizontal linear complementarity problem, *SIAM J. Optim.*, **4** (1994), 208-227.

30. Y. Zhang and D. Zhang, Superlinear convergence of infeasible interior-point methods for linear programming, *Math. Prog.*, **66** (1994), 361-377.

31. Y. Zhang and D. Zhang, On polynomiality of the Mehrotra-type predictor-corrector interior-point algorithms, Technical Report 93-12, Department of Mathematics and Statistics, University of Maryland, Baltimore County, June 1993.

WSSIAA 5 (1995) pp. 435–445

435

SYNTHESIS OF ADAPTIVE CONTROLLERS USING BICRITERIAL OPTIMIZATION AND LYAPUNOV FUNCTIONS

H. UNBEHAUEN

Automatic Control Laboratory, Faculty of Electrical Engineering,
Ruhr University, D-44780 Bochum, Germany

and

N. M. FILATOV

Automatic Control Laboratory, Faculty of Electrical Engineering,
Ruhr University, D-44780 Bochum, Germany

ABSTRACT

Computationally simple multivariable adaptive controllers with improved control quality during the adaptation are synthesised using bicriterial optimization and Lyapunov functions. The suggested design method exploits the dual effect of adaptive control and an uncertainty measure to improve the performance and decrease the adaptation time. Two performance indices are introduced for control optimization which correspond to two goals: to control the plant and to provide an optimal excitation for speeding up the parameter estimation process. Lyapunov functions are used for designing a stable closed-loop control scheme. The interconnections between the suggested approach and linear quadratic control are discussed. A simulated example of a multivariable system demonstrates the potential and superiority of the designed controller over a usually applied controller based on the certainty equivalence assumption.

1. Introduction

It was discovered by Feldbaum[1] that the behaviour of optimal adaptive control is characterized by dual properties: caution and active learning. In other words, it means that the best behaviour of adaptive control must not hastily follow the set point if the plant parameters are not exactly known, (property of cautious control) and, at the same time, the control signal must excite the system to speed up the parameter estimation process and decrease the adaptation time (active learning)[1,2,3]. Difficulties in finding the optimal dual control policy lead to appearance of various suboptimal dual controllers[4,5,6]. Most of them do not find practical application because of complexity and computational difficulties for operation in real time mode or insufficient control quality. Therefore, simple adaptive

436

controllers, which are based on the certainty equivalence (CE) approach, have usually been used up to now. The computational simplicity of the CE adaptive controllers is provided due to the separation of parameter estimation from the control law. The accuracy of estimates is not taken into account in these controllers and the parameter estimations are used as if they are the real values of unknown parameters. Bad control quality and large overshoot are often attributed to this kind of adaptive controllers, especially at the beginning of adaptation, and in cases of inaccurate estimation.

The recently developed bicriterial approach to the synthesis of dual (active adaptive) controllers allows the design of computationally simple control algorithms with improved performance[3,7,8]. This synthesis method is based on a compromised minimization of two performance indices (control losses and uncertainty index) corresponding to the two goals of dual control. However, this approach is developed especially for single-input single-output systems with input-output model representation.

In this paper the bicriterial approach is developed using Lyapunov functions for the design of multivariable adaptive controllers for plants described by linear systems with unknown parameters. The resulting control algorithms are simple in computation and provide improved performance due to the attributed dual properties. In Sec. 2, the general synthesis problem is formulated using bicriterial optimization and Lyapunov functions for linear system models in state space representation. The synthesis procedure of dual controllers is described in Sec. 3. Implementation details and interconnections with the well-known linear quadratic control problem are discussed in Sec. 4. Simulation results of application of the elaborated approach to control a helicopter are used to demonstrate the potential of the method and to compare it with the standard approach based on the CE assumption in Sec. 5. Concluding remarks are given in Sec. 6.

2. Formulation of Synthesis Problem

Consider the following linear dynamic multivariable system in discrete time representation with unknown parameters

$$x(k+1) = A(p)x(k) + B(p)u(k) + \xi(k), \quad k = 0, 1, \ldots, \quad x(0) = x_0, \tag{1}$$

where $x(k) \in \Re^{n_x}$ is the state vector, $u(k) \in \Re^{n_u}$ is the control input vector, $p \in \Re^{n_p}$ is the vector of unknown parameters, $\{\xi(k) \in \Re^{n_\xi}\}$ is the sequence of independent random vectors with zero mean and known covariance matrix Q_ξ, $A(p)$ and $B(p)$ are matrices of corresponding dimensions and are linear functions of the vector of unknown parameters p. The system, according to Eq. (1), is assumed to be controllable and parametrically identifiable and the state vector is exactly observed.

The set of state and control values at time k is denoted as

$$\Im_k = \{x(k), \ldots, x(0), u(k-1), \ldots, u(0)\}, \quad k = 1, \ldots, \quad \Im_0 = \{x(0)\}. \tag{2}$$

Introduce the Lyapunov function of the following form:

$$v_k = x^T(k)P_k x(k), \tag{3}$$

where P_k is a positive definite symmetric matrix. Note, that the first difference of this Lyapunov function is defined by the equation

$$\Delta v_k = x^T(k+1)P_k x(k+1) - x^T(k)P_k x(k). \tag{4}$$

Obviously, the stability of the system will be guaranteed[9] if $v_k > 0$ and $\Delta v_k < 0$ for $x(k) \neq 0$. To derive the control law, which would make the system stable, introduce the following performance index for minimization:

$$J_k = \mathrm{E}\left\{\Delta v_k + u^T(k)Ru(k)\big|\mathfrak{I}_k\right\}, \tag{5}$$

where R is a positive definite symmetric matrix of corresponding dimension and $\mathrm{E}\left\{\cdot\big|\mathfrak{I}_k\right\}$ is the conditional expectation operator when the set $\mathfrak{I}_k$ according to Eq. (2) is available. It is suggested here to choose the matrix P_k of the Lyapunov function to satisfy the following inequality:

$$\mathrm{E}\left\{\Delta v_k + u^T(k)Ru(k)\big|\mathfrak{I}_k,\rho_k\right\} \leq 0, \tag{6}$$

where

$$\rho_k = \left\{A(p) = A(\hat{p}(k)), B(p) = B(\hat{p}(k)), \xi_k = 0\right\} \tag{7}$$

is the CE assumption when all the random values in the system are assumed to be equal to their expectations, and

$$\hat{p}(k) = \mathrm{E}\left\{p\big|\mathfrak{I}_k\right\} \tag{8}$$

is the estimate (conditional expectation) of the unknown parameter vector. A stabilizing feedback of the closed-loop system is guaranteed by inequality (6) after adaptation, because the value of the first difference of the Lyapunov function according to Eq. (4) is negative.

Following the bicriterial approach in dual control[3,7,8], introduce the second performance index

$$J_k^a = -\mathrm{E}\left\{(x(k+1) - \hat{x}(k+1|k))^T W(x(k+1) - \hat{x}(k+1|k))\big|\mathfrak{I}_k\right\}, \tag{9}$$

where

$$\hat{x}(k+1|k) = \mathrm{E}\left\{x(k+1)\big|\mathfrak{I}_k\right\} = A(\hat{p}(k))x(k) + B(\hat{p}(k))u(k), \tag{10}$$

W is a positive semidefinite symmetric matrix. Minimization of Eq. (9) will increase the expectation error of the one step ahead prediction $\hat{x}(k+1|k)$. This error contains the information which is used for parameter estimation (innovational value) and its magnification, in this way, accelerates the parameter estimation process[6]. Minimization of the first performance index according to Eq. (5) results in a control signal which may be named "cautious control" $u_c(k)$ by analogy with the practice adopted in dual control[1-8]. The second performance index according to Eq. (9) is minimized in the compact domain Ω_k and this domain is symmetrically allocated around $u_c(k)$. The value of the optimal excitation will depend on the size of Ω_k. Define the domain Ω_k as follows:

$$\Omega_k = \left\{ u(k): \; u(k) = u_c(k) + u_a(k), \; u_a^T(k) W_a u_a(k) \le f_k(P(k)) \right\}, \tag{11}$$

where $u_a(k)$ is interpreted as excitation signal, W_a is a positive definite matrix of corresponding dimension,

$$P(k) = E\left\{ (p - \hat{p}(k))(p - \hat{p}(k))^T \big| \Im_k \right\} \tag{12}$$

is the covariance matrix of the estimates of unknown parameters and $f_k(\cdot)$ is a positive definite simple scalar function of the covariance matrix. Function f_k, which determines the amplitude of the excitation signal together with the matrix W_a, may be chosen in one of the following forms:

$$f_k(P(k)) = \eta \, \mathrm{tr}\{P(k)\}, \quad \eta \ge 0, \tag{13}$$

$$f_k(P(k)) = \eta (\mathrm{tr}\{P(k)\})^2, \quad \eta \ge 0. \tag{14}$$

In this way, the size of Ω_k and, therefore, the amplitude of the excitation depend on the accuracy of estimation, which is characterized by $P(k)$. The function according to Eq. (14) may be used to provide faster decreasing of the excitation than the previous one described by Eq. (13).

The synthesis problem described above is summarized in the following compromised bicriterial optimization:

$$u(k) = \arg \min_{u(k) \in \Omega_k} J_k^a, \tag{15}$$

$$u_c(k) = \arg \min_{u(k)} J_k, \tag{16}$$

where, according to Eq. (11), Ω_k is an ellipsoid with central point $u_c(k)$. At the same time, matrix P_k in Eq. (4) must be chosen to satisfy inequality (6).

3. Synthesis of Adaptive Controllers

Substitution of Eq. (1) in Eq. (4) leads to the following:

$$\Delta v_k = u^T(k) B^T(p) P_k B(p) u(k) + 2 u^T(k) B^T(p) P_k (A(p) x(k) + \xi(k)) + c_1(k), \tag{17}$$

where $c_1(k)$ does not depend on $u(k)$. Using the last equation, Eq. (5) can be written in the following form:

$$J_k = u^T(k) E\left\{ B^T(p) P_k B(p) \big| \Im_k \right\} u(k) + 2 u^T(k) E\left\{ B^T(p) P_k A(p) \big| \Im_k \right\} x(k)$$

$$+ u^T(k) R u(k) + c_2(k), \tag{18}$$

where $c_2(k)$ does not contain $u(k)$. Taking into account the positive definiteness of R, it is easy to see that the following cautious control minimizes the performance index, Eq. (18)

$$u_c(k) = -\left[\mathrm{E}\{ \boldsymbol{B}^T(\boldsymbol{p})\boldsymbol{P}_k\boldsymbol{B}(\boldsymbol{p})|\mathfrak{I}_k \} + \boldsymbol{R} \right]^{-1} \mathrm{E}\{ \boldsymbol{B}^T(\boldsymbol{p})\boldsymbol{P}_k\boldsymbol{A}(\boldsymbol{p})|\mathfrak{I}_k \}\boldsymbol{x}(k). \tag{19}$$

The second performance index according to Eq. (9) can be represented in the following form after substitution of Eq. (1) and taking the expectation:

$$J_k^a = -\left[\boldsymbol{x}^T(k)\boldsymbol{P}_A(k)\boldsymbol{x}(k) + \boldsymbol{u}^T(k)\boldsymbol{P}_B(k)\boldsymbol{u}(k) + 2\boldsymbol{u}^T(k)\boldsymbol{P}_{BA}(k)\boldsymbol{x}(k) + \mathrm{tr}\{ \boldsymbol{W}\boldsymbol{Q}_\xi \} \right], \tag{20}$$

where

$$\boldsymbol{P}_A(k) = \mathrm{E}\{ \boldsymbol{A}^T(\boldsymbol{p})\boldsymbol{W}\boldsymbol{A}(\boldsymbol{p})|\mathfrak{I}_k \} - \boldsymbol{A}^T(\hat{\boldsymbol{p}}(k))\boldsymbol{W}\boldsymbol{A}(\hat{\boldsymbol{p}}(k)), \tag{21}$$

$$\boldsymbol{P}_B(k) = \mathrm{E}\{ \boldsymbol{B}^T(\boldsymbol{p})\boldsymbol{W}\boldsymbol{B}(\boldsymbol{p})|\mathfrak{I}_k \} - \boldsymbol{B}^T(\hat{\boldsymbol{p}}(k))\boldsymbol{W}\boldsymbol{B}(\hat{\boldsymbol{p}}(k)), \tag{22}$$

$$\boldsymbol{P}_{BA}(k) = \mathrm{E}\{ \boldsymbol{B}^T(\boldsymbol{p})\boldsymbol{W}\boldsymbol{A}(\boldsymbol{p})|\mathfrak{I}_k \} - \boldsymbol{B}^T(\hat{\boldsymbol{p}}(k))\boldsymbol{W}\boldsymbol{A}(\hat{\boldsymbol{p}}(k)). \tag{23}$$

According to Eq. (11), the ellipsoid, which defines the constraint for amplitude of the excitation, has the following form:

$$g_k = (\boldsymbol{u}(k) - \boldsymbol{u}_c(k))^T \boldsymbol{W}_a (\boldsymbol{u}(k) - \boldsymbol{u}_c(k)) - f_k(\boldsymbol{P}(k)). \tag{24}$$

Then, taking into account the convexity of the last constraint and concavity of the performance index according to Eq. (20), the necessary conditions for minimization of the second performance index can be written as

$$\lambda_k \nabla J_k^a = -\nabla g_k, \tag{25}$$

where ∇ is the gradient operator with respect to $\boldsymbol{u}(k)$, $\lambda_k \geq 0$ is a scalar parameter to be defined. Sufficient condition for the minimum is the positive definiteness for the following matrix

$$\frac{\partial^2(\lambda_k J_k^a + g_k)}{\partial \boldsymbol{u}^2(k)} > 0. \tag{26}$$

Substituting Eqs. (20) and (24) in Eq. (25), one has

$$-\lambda_k \boldsymbol{P}_B(k)\boldsymbol{u}(k) + \boldsymbol{W}_a\boldsymbol{u}(k) - \boldsymbol{W}_a\boldsymbol{u}_c(k) - \lambda_k \boldsymbol{P}_{BA}(k)\boldsymbol{x}(k) = \boldsymbol{0}. \tag{27}$$

After resolving the last equation, the sought-for control law can finally be written as

$$\boldsymbol{u}(k) = \left[\boldsymbol{W}_a - \lambda_k \boldsymbol{P}_B(k) \right]^{-1} \left[\boldsymbol{W}_a\boldsymbol{u}_c(k) + \lambda_k \boldsymbol{P}_{BA}(k)\boldsymbol{x}(k) \right] \tag{28}$$

with the conditions

$$(\boldsymbol{u}(k) - \boldsymbol{u}_c(k))^T \boldsymbol{W}_a (\boldsymbol{u}(k) - \boldsymbol{u}_c(k)) - f_k(\boldsymbol{P}(k)) = 0 \tag{29}$$

and

$$\boldsymbol{W}_a - \lambda_k \boldsymbol{P}_B(k) > \boldsymbol{0}. \tag{30}$$

Consider inequality (6) to determine matrix P_k. After using Eqs. (19), (28) and (29) and taking expectation with the CE assumption described by Eq. (7), inequality (6) takes the form

$$u^T(k)B^T(\hat{p}(k))P_kB(\hat{p}(k))u(k)+2u^T(k)B^T(\hat{p}(k))P_kA(\hat{p}(k))x(k)$$

$$+x^T(k)A^T(\hat{p}(k))P_kA(\hat{p}(k))x(k)+u^T(k)Ru(k)-x^T(k)P_kx(k)\leq 0 \qquad (31)$$

or

$$-x^T(k)A^T(\hat{p}(k))P_kB(\hat{p}(k))\left[B^T(\hat{p}(k))P_kB(\hat{p}(k))+R\right]^{-1}B^T(\hat{p}(k))P_kA(\hat{p}(k))x(k)$$

$$+x^T(k)A^T(\hat{p}(k))P_kA(\hat{p}(k))x(k)-x^T(k)P_kx(k)\leq 0. \qquad (32)$$

From the last inequality, it follows that matrix P_k must satisfy the following inequality

$$-A^T(\hat{p}(k))P_kB(\hat{p}(k))\left[B^T(\hat{p}(k))P_kB(\hat{p}(k))+R\right]^{-1}B^T(\hat{p}(k))P_kA(\hat{p}(k))$$

$$+A^T(\hat{p}(k))P_kA(\hat{p}(k))-P_k\leq 0. \qquad (33)$$

Some aspects for implementation of the designed dual controller, according to Eqs. (19), (28) and (29) with the conditions of inequalities (30) and (33) as well as their interconnections with the linear quadratic approach, will be discussed in the next Section.

4. Implementation of the Designed Controller and Interconnections with the Linear Quadratic Control Problem

It is easy to see that inequality (33) will be satisfied if the matrix P_k is defined from the following equation:

$$-A^T(\hat{p}(k))P_kB(\hat{p}(k))\left[B^T(\hat{p}(k))P_kB(\hat{p}(k))+R\right]^{-1}B^T(\hat{p}(k))P_kA(\hat{p}(k))$$

$$+A^T(\hat{p}(k))P_kA(\hat{p}(k))-P_k+V=0, \qquad (34)$$

where V is a positive definite matrix of dimension $n_x\times n_x$. Therefore, Eq. (34) and the additional matrix V can be used to resolve ambiguities resulting from multiple solutions of inequality (33). Moreover, the matrices R and V have close connections with the linear quadratic control problem as shown below.

Consider the system after adaptation when $\hat{p}(k)\equiv p$. In this case, the controller according to Eqs. (19), (28), (29) and (34) takes the following form:

$$u(k)=-\left[B^T(p)P_kB(p)+R\right]^{-1}B^T(p)P_kA(p)x(k), \qquad (35)$$

where $P_k=const$, which, according to Eq. (34), satisfies the discrete (algebraic) matrix Riccati equation

$$-A^T(p)P_kB(p)\left[B^T(p)P_kB(p)+R\right]^{-1}B^T(p)P_kA(p)+A^T(p)P_kA(p)-P_k+V=0. \quad (36)$$

It is easy to see that the controller, according to Eqs. (35) and (36), is the optimal solution of the linear quadratic control problem for the system described by Eq. (1) with the performance index[10]

$$J=\lim_{N\to\infty}\frac{1}{N}\mathrm{E}\left\{\sum_{k=0}^{N}(x^T(k)Vx(k)+u^T(k)Ru(k))\right\}. \quad (37)$$

To calculate the estimate $\hat{p}(k)$ and covariance matrix $P(k)$ during the implementation of the derived dual adaptive controller the extended Kalman filter[11] or other suitable methods of parameter estimation can be used. The controller is implemented by the following sequence of actions:

1. measurement of the state vector $x(k)$
2. calculation of the estimate of the parameter vector $\hat{p}(k)$ and covariance matrix $P(k)$
3. computation of P_k from Eq. (34) and $u_c(k)$ from Eq. (19)
4. finding λ_k and $u(k)$ from Eqs. (28) and (29) using condition (30)
5. if the covariance matrix is small enough, then switching off the adaptation mechanism and using a controller with fixed parameters according to Eqs. (35) and (36) in further steps
6. otherwise, repeating from step 1 for the next time interval $k+1$.

It should be noted that in many cases of time varying parameters in the system described by Eq. (1), the following stochastic model of a Wiener process can be successfully used[7,8]

$$p(k+1)=p(k)+\varepsilon(k), \quad k=1,2,\ldots, \quad p(0)=p_0, \quad (38)$$

where $\varepsilon(k)$ is a discrete-time white noise vector with zero mean and covariance matrix $Q_\varepsilon(k)$. In this case, the extended Kalman filter can also be used for parameter estimation and it is possible to estimate the state vector together with the parameters when measurement noise acts on the system.

5. Simulation Results

The problem of helicopter control[12] is considered for the comparison of the synthesized controller with the standard one based on the CE assumption below. The linearized dynamic equation of the helicopter is

$$\dot{x}(t)=\begin{bmatrix}-0.0366 & 0.0271 & 0.0188 & -0.4555\\ 0.0482 & -1.0100 & 0.0024 & -4.0208\\ 0.1002 & 0.3681+k_1 & -0.7070 & 1.42+k_2\\ 0 & 0 & 1 & 0\end{bmatrix}x(t)+\begin{bmatrix}0.4422 & 0.1761\\ 3.5446+k_3 & -7.5922\\ -5.5200 & 4.4900\\ 0 & 0\end{bmatrix}u(t), \quad (39)$$

442

where the bounds for the uncertain parameters are

$$|k_1| \leq 0.05, \quad |k_2| \leq 0.01 \quad \text{and} \quad |k_3| \leq 0.04. \tag{40}$$

The discrete time model with sampling time 0.1 for the nominal parameters ($k_1 = k_2 = k_3 = 0$) is

$$x(k+1) = \begin{bmatrix} -0.9964 & 0.0026 & -0.0004 & -0.0460 \\ 0.0045 & 0.9037 & -0.0188 & -0.3834 \\ 0.0098 & \underline{0.0339} & 0.9383 & \underline{0.1302} \\ 0.0005 & \underline{0.0017} & 0.0968 & 1.0067 \end{bmatrix} x(k) + \begin{bmatrix} 0.0445 & 0.0167 \\ \underline{0.3407} & -0.7249 \\ -0.5278 & 0.4214 \\ -0.0268 & \underline{0.0215} \end{bmatrix} u(k) + \xi(k)$$

$$= A(p)x(k) + B(p)u(k) + \xi(k), \tag{41}$$

where $\xi(k)$ represents disturbances and unmodelled dynamics, Q_ξ has been taken as $Q_\xi = 0.0001I$, I is a unit matrix. Variation of parameters of the continuous-time model described by Eq. (39) in intervals of uncertainty (40) leads to variation of six parameters of the discrete-time model which generate the following vector of unknown parameters:

$$p = \begin{bmatrix} a_{3,2} & a_{3,4} & a_{4,2} & b_{3,1} & b_{3,2} & b_{4,2} \end{bmatrix}^T, \tag{42}$$

where $a_{i,j}$ and $b_{i,j}$ are the elements of ith row and jth column of matrices $A(p)$ and $B(p)$, respectively. These six parameters are underlined in Eq. (41). Other parameters of the model described by Eq. (41) do not change significantly. The following parameters for simulation have been taken

$$x(0) = \begin{bmatrix} -1 & -1 & -1 & -1 \end{bmatrix}^T, \; P(0) = I, \; p(0) = \begin{bmatrix} 0 & 0 & 0 & 0 & 0 & 0 \end{bmatrix}^T,$$

$$R = 0.5I, \; V = I, \; W = I, \tag{43}$$

where the plant model according to Eq. (41) is taken with changes from the nominal parameter values p as follows:

$$p = \begin{bmatrix} a_{3,2} - 0.004 & a_{3,4} + 0.001 & a_{4,2} - 0.0002 & b_{3,1} - 0.0008 & b_{3,2} + 0.0018 & b_{4,2} + 0.0001 \end{bmatrix}^T. \tag{44}$$

It is assumed here that the state vector is available from exact measurements and the extended Kalman filter is used for estimation of parameter vector p. For the applied dual controller, the function f_k has been chosen in the form according to Eq. (14) with $\eta = 0.0001$. Parameter η can be chosen by simulation to provide a good system behaviour. The designed dual controller is compared with the corresponding controller according to Eq. (35), based on the CE assumption:

$$u(k) = -\left[B^T(\hat{p}(k))P_k B(\hat{p}(k)) + R \right]^{-1} B^T(\hat{p}(k))P_k A(\hat{p}(k))x(k), \tag{45}$$

where P_k is determined from Eq. (34).

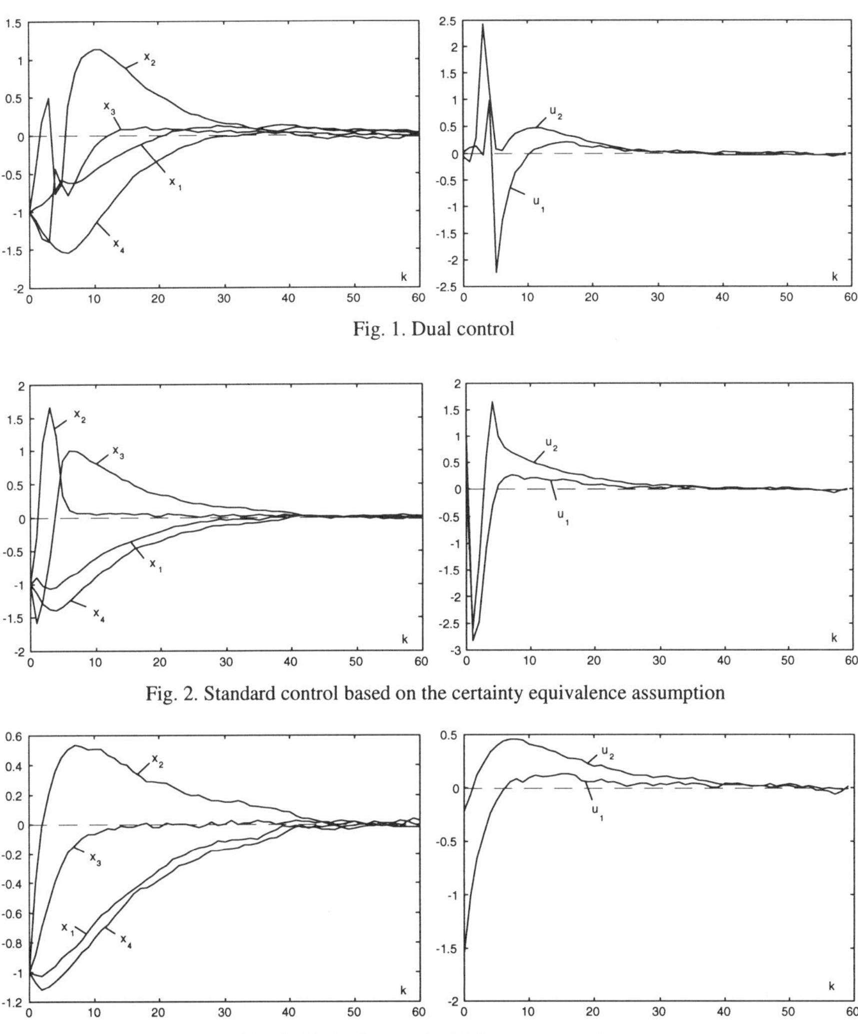

Fig. 1. Dual control

Fig. 2. Standard control based on the certainty equivalence assumption

Fig. 3. Optimal control with known parameters

The simulation results are shown in Fig. 1 (dual control) and Fig. 2 (CE control). In Fig. 3, the simulation results for optimal control with known parameters according to Eqs. (35) and (36) are shown to indicate the best control behaviour, which cannot be achieved in the case of unknown parameters. The following average performance indices have been used for comparison

$$\bar{J}_x = \frac{1}{N_r(N_k+1)} \sum_{i=1}^{N_r} \sum_{k=0}^{N_k+1} x_i^T(k) x_i(k), \quad \bar{J}_u = \frac{1}{N_r N_k} \sum_{i=1}^{N_r} \sum_{k=0}^{N_k} u_i^T(k) u_i(k), \quad (46)$$

$$\bar{J} = \bar{J}_x + 0.5\bar{J}_u. \tag{47}$$

where $N_k = 60$ is the number of sampled date for the simulation and $N_r = 50$ is the number of simulation runs with different parameters of a normally distributed random number generator which was used for simulation. The values of the performance indices according to Eqs. (46) and (47) are indicated in Table 1 and on the diagram in Fig. 4.

Table 1. Average values of performance indices

Performance indices	Dual control	Control based on the CE assumption	Optimal control with known parameters
Total costs $\bar{J}$	0.8601	1.0901	0.5952
State costs $\bar{J}_x$	0.7677	0.8098	0.5387
Control costs $\bar{J}_u$	0.1847	0.5605	0.1129

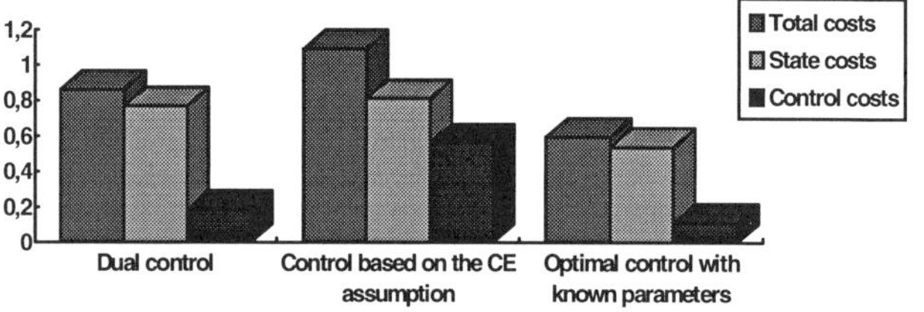

Fig. 4. Diagram of simulation results

The simulation results indicate that the proposed dual controller can improve the control quality and demonstrates superiority over the standard controller based on the CE assumption. Moreover, the synthesized dual control algorithm is simple in computation and has an analytical formulation in contrast to other dual control methods for multivariable systems[4,5,13].

6. Conclusion

In this paper, an adaptive controller, obtained by compromised minimization of two performance indices and using Lyapunov functions, has been developed. The suggested method exploits the dual effect and uses the uncertainty measure to improve the control performance. Lyapunov functions are used to provide stability of the closed-loop control scheme. The interconnections with the well-known linear quadratic controller had been discussed. The potential and superiority of the proposed controller have been demonstrated using a simulated example. Improved control performance and computational simplicity are characteristic for the suggested new controller and, therefore, it can replace usually applied adaptive controllers based on the CE assumption in many practical cases.

7. Acknowledgement

Second author thanks the Alexander von Humboldt Foundation for financial support.

8. References

1. A. A. Feldbaum, *Optimal Control Systems*, Academic Press, New York, 1965.
2. Y. Bar-Shalom, Stochastic dynamic programming: caution and probing, *IEEE Trans. Autom. Control* **26** (1981), 1184-1194.
3. V. P. Zhivoglyadov, G. P. Rao and N. M. Filatov, Application of δ-operator models to active adaptive control of continuous-time plants, *Control - Theory and Advanced Technology* **9** (1993), 127-137.
4. Y. Bar-Shalom and E. Tse, Concepts and methods in stochastic control, in Leondes, C. T. (ed.), *Control and Dynamic Systems*, Academic Press, New York, Volume **12** (1976), 99-172.
5. D. S. Bayard and M. Eslami, Implicit dual control for general stochastic systems, *Opt. Control Appl. Methods* **6** (1985), 265-273.
6. R. Milito, C. S. Padilla and P. Cadorin, An innovation approach to dual control, *IEEE Trans. Autom. Control* **27** (1982), 825-830.
7. N. M. Filatov and H. Unbehauen, Strategies of model reference adaptive control with active learning properties, *Proc. of 2nd IFAC Symposium on Intelligent Components and Instruments in Control Applications*, Budapest, (1994), 989-994.
8. N. M. Filatov and H. Unbehauen, Active adaptation for control systems with implicit reference model, *Proc. of 1st Asian Control Conference*, Tokyo (1994), 287-290.
9. B. C. Kuo, *Digital Control Systems* (2nd Ed.), Saunders College Publishing, Orlando FL(USA), 1992.
10. C. L. Phillips and H. T. Nagle, *Digital Control Analysis and Design*, Prentice Hall, Englewood Cliffs N. Y., 1984.
11. L. Ljung, Asymptotic behaviour of the extended Kalman filter as a parameter estimator for linear systems, *IEEE Trans. Autom. Control* **24** (1979), 36-50.
12. J.-H. Su and I.-K. Fong. Controller robustification via optimization techniques for linear uncertain systems, *Control - Theory and Advanced Technology* **10** (1994), 154-160.
13. K. Birmiwal, A new adaptive LQG control algorithm, *Int. J. Adapt. Control Signal Process.* **8** (1994), 287-295.

WSSIAA 5 (1995) pp. 447–465
© World Scientific Publishing Company

Computational Methods for a Class of Optimization Problems with Functional Inequality Constraints[1]

B. Vo, K. L. Teo and A. Cantoni.[2]

Abstract - Numerical methods for a class of optimization problem with functional inequality constraints are proposed by approximating these with conventional constrained or unconstrained optimization problems. Each of the approximate problems has two interdependent parameters. The sub-optimal solution can be made arbitrarily close to the optimal solution by choosing the parameters of the approximate problem appropriately.

For a narrower class of problems where the cost is quadratic and the functional constraints are convex, a closed form expression for the approximation error can be derived. Globally convergent algorithms are developed for the case where the constraints functions are affine.

1. Introduction

Consider the class of optimization problems, where the continuously differentiable cost function

$$f(x)$$

is to be minimized over a compact subset $\Theta \subset R^n$ subject to the functional inequality constraints

$$g_j(x) \equiv \max \{ \phi_j(x, \omega) : \omega \in \Omega \} \leq 0, j = 1, ..., m.$$

where, $x \in R^n$, Ω is a compact interval in R, $\phi_j : R^n \times \Omega \to R$ is continuously differentiable on R^n for each $\omega \in \Omega$, and Lipschitz continuous on Ω for each $x \in \Theta$. For convenience, this problem will be referred to as problem (P) and its solution is denoted by x^*.

The feasible set $\mathcal{F}$ for problem (P) is defined by

$$\mathcal{F} = \{ x \in \Theta : \phi_j(x, \omega) \leq 0, \quad \forall \omega \in \Omega, j = 1, ..., m \}$$

$$= \{ x \in \Theta : g_j(x) \leq 0, j = 1, ..., m \}$$

Note that $\mathcal{F}$ is closed, since it is an intersection of closed sets, which are the inverse images of the closed semi-infinite interval $(-\infty, 0\,]$ under the continuous maps g_j. Furthermore, let $\mathrm{int}(\mathcal{F})$ be the interior of $\mathcal{F}$ in the sense that,

---

1. This work is supported by Hewlett Packard, the Australian Research Council, and the Cooperative Research Center for Broadband Telecommunications and Networking.

2. Australian Telecommunications Research Institute - Signal Processing Laboratory, Curtin University of Technology, Perth, WA, Australia.

448

$$\mathrm{int}\,(\mathcal{F}) \;=\; \{x \in \mathrm{int}\,(\Theta) : g_j(x) < 0,\, j = 1,\,...,\,m\}\,.$$

It is assumed that the following conditions are satisfied.

(A1) $\mathrm{int}\,(\mathcal{F}) \neq \emptyset$

(A2)   There exists an $\bar{x} \in \mathrm{int}\,(\mathcal{F})$ such that $x_\alpha \equiv \alpha\bar{x} + (1-\alpha)\,x^* \in \mathrm{int}\,(\mathcal{F})$ for all $\alpha \in (0, 1\,]$.

Note that convexity of the interior of the feasible region is sufficient for (A2) to hold.

In Polak and Mayne [1], a reliable algorithm was proposed for solving problem (P) as a non-smooth optimization problem. This was later superseded by a new algorithm in Gonzaga *et al.* [2]. A number of algorithms based on subgradient methods were also developed for this class of problems, see Polak and Wardi [3].

An alternative to these non-smooth methods was proposed in Teo and Goh [4] by introducing a constraint transcription which transforms the functional inequality constraints into conventional equality constraints. This idea is similar to that of Sargent and Sullivan [5]. Unfortunately, the transformed constraints do not satisfy the usual constraint qualifications. Thus convergence is not guaranteed and it has been confirmed experimentally that oscillations do occur in numerical computations.

Jennings and Teo [6] developed a constraint transcription to transform the functional inequality constraints of (P) into non-smooth equality constraints, and a smoothing technique to approximate these non-smooth constraints. Based on this, a sequence of sub-problems, each of which being a standard inequality constrained optimization problem, is constructed. The same constraint transcription and smoothing technique are used to approximate the functional inequalities constraints of (P) in Teo *et al.* [7]. Then an augmented cost function is introduced by incorporating the original cost function and these approximate constraints. This effectively turns problem (P) into a sequence of unconstrained optimization problems.

In this paper, two new generalized methods for constructing sub-problems to approximate (P) are presented. The first of these methods uses conventional inequality constrained optimization problems as sub-problems, while the second uses unconstrained optimization problems. For both methods, results on the convergence of the sub-optimal solutions to a solution of (P) are derived. For the case where the cost is quadratic and the functional constraints are convex, a closed form expression for the approximation error can be found. When the functional inequality constraints are affine, the corresponding augmented costs possess a number of desirable analytical properties which enables globally convergent algorithms to be developed.

The remaining sections of this paper is organized as follows; Section 2 introduce the generalized constraint transcription technique which is used to construct conventional constrained sub-problems to approximate problem (P). In Section 3 conventional unconstrained sub-problems are constructed to approximate problem (P). Section 4 looks at a specific class of problems, those with convex functional constraints and quadratic cost. Algorithms for convex quadratic programming problems with affine functional inequality constraints are proposed in Section 5.

2. Approximations by Conventional Inequality Constrained Problems

This Section introduces a class of conventional inequality constrained optimization problems constructed to approximate problem (P). These approximate problems do not suffer from failure of constraint qualifications as in [4] and can also be chosen to be smooth, thus they can be readily solved by existing methods (for more details see [8] [9] [10]). The major results of this and the following Section can be found in [11].

For each $\upsilon > 0$, let $g_\upsilon : R \to R$ be an integrable function with the following properties:

(P1) $g_\upsilon(x) = 0, \forall x \le -\upsilon$

(P2) $g_\upsilon(x)$ is strictly increasing $\forall x > -\upsilon$

To approximate (P) by the conventional inequality constrained problems, define

$$G_{j,\upsilon}(x) \equiv \int_\Omega g_\upsilon(\phi_j(x,\omega))\, d\omega, \, j = 1, \ldots, m$$

Clearly each of these functions is continuously differentiable in x if g_υ is continuously differentiable.

For each $\upsilon, \tau > 0$ an approximate problem $(P_{\upsilon,\tau})$ is defined as follows:

$$\min\, f(x) \text{ subject to } G_{j,\upsilon}(x) \le \tau, \, j = 1, \ldots, m \tag{1}$$

The smoothness of this problem depend on the choice of g_υ used. Smooth problems are generally easier to handle than their counterpart. The subsequent results developed in this Section allow problem (P) to be solved as smooth constrained optimization sub-problems.

The feasible set $\mathcal{F}_{\upsilon,\tau}$ for problem $(P_{\upsilon,\tau})$ is thus given by

$$\mathcal{F}_{\upsilon,\tau} = \{x \in \Theta : G_{j,\upsilon}(x) \le \tau, \, j = 1, \ldots, m\}$$

The relationship between the approximate feasible regions $\mathcal{F}_{\upsilon,\tau}$, and the actual feasible region $\mathcal{F}$, are presented in the form of the following results.

Lemma 2.1*. Suppose p is an integrable, non-negative and non-decreasing function and q is Lipschitz continuous on a compact interval Ω, i.e.*

$$\left| q(\omega_2) - q(\omega_1) \right| \le \eta \left| \omega_2 - \omega_1 \right|, \, \forall \omega_2, \omega_1 \in \Omega.$$

Then

$$\exists\, an\, \bar{\omega} \in \Omega \ni q(\bar{\omega}) > 0 \quad \Rightarrow \forall \xi > 0, \int_\Omega p(q(\omega))\, d\omega \ge p(-\xi) \min\left(|\Omega|, \frac{\xi}{\eta}\right)$$

Proof*. Supposing that there exists an $\bar{\omega} \in \Omega$ such that $q(\bar{\omega}) > 0$, then, for any $\xi > 0$, let $I \subset \Omega$ be the largest interval containing $\bar{\omega}$ such that $q(\omega) \ge -\xi$, for all $\omega \in I$. First, let's establish that $|I| \ge \min(|\Omega|, \xi/\eta)$ (see Figure 1 for illustration).

450

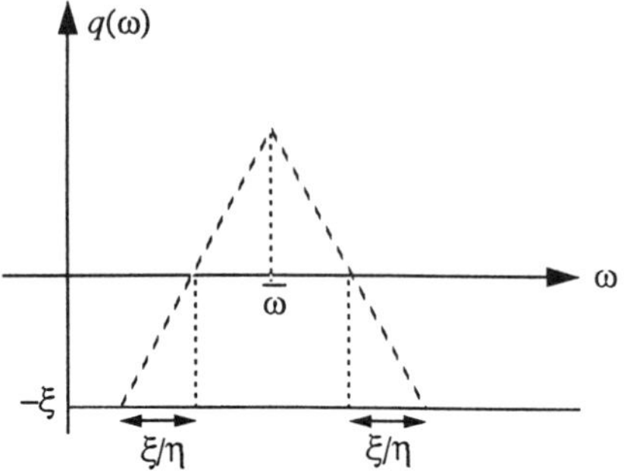

Figure 1

Assume that there exists an $\omega_0 \in I$ such that $q(\omega_0) = -\xi$ (otherwise, from the definition of I and the continuity of q, $I = \Omega$). The Lipschitz condition gives

$$\left| q(\omega_0) - q(\overline{\omega}) \right| = \eta \left| (\omega_0 - \overline{\omega}) \right| \le \eta |I| \tag{2}$$

Since $q(\overline{\omega}) > 0$, it follows from (2) that $-\xi > q(\omega_0) - q(\overline{\omega}) \ge -\eta |I|$. Consequently, $\xi/\eta < |I| \le |\Omega|$. Now, since p is non negative,

$$\int_\Omega p(q(\omega)) \, d\omega \ge \int_I p(q(\omega)) \, d\omega \ge \min \{ p(q(\omega)) : \omega \in I \} \, |I|$$

Furthermore, since p is non-decreasing and $q(\omega) \ge -\xi$, for all $\omega \in I$, it follows that

$$\min \{ p(q(\omega)) : \omega \in I \} \ge p(-\xi) .$$

Thus,

$$\int_\Omega p(q(\omega)) \, d\omega \ge p(-\xi) \min(|\Omega|, \xi/\eta) . \quad \square$$

Theorem 2.1. $0 < \tau < \tau(\upsilon) \equiv \sup \{ g_\upsilon(-\xi) \min(|\Omega|, \xi/\eta) : \xi > 0 \} \Rightarrow \mathcal{F}_{\upsilon, \tau} \subset \mathcal{F}$, where η is such that $\eta |\omega_2 - \omega_1| \ge \left| \phi_j(x, \omega_2) - \phi_j(x, \omega_1) \right|$, $\forall \omega_2, \omega_1 \in \Omega, x \in \Theta, j \in \{1, ..., m\}$.

Proof. Since g_υ is integrable non negative and non-decreasing, and for each $x \in R^n$, $j \in \{1, ..., m\}$, $\phi_j(x, \omega)$ is Lipschitz continuous on Ω, applying Lemma 2.1 leads to

if $\exists$ a $j \in \{1, ..., m\}$ and an $\overline{\omega} \in \Omega$ such that $\phi_j(x, \overline{\omega}) > 0$ then

$$G_{j, \upsilon}(x) \equiv \int_\Omega g_\upsilon(\phi_j(x, \omega)) \, d\omega \ge \sup \left\{ g_\upsilon(-\xi) \min\left(|\Omega|, \frac{\xi}{\eta} \right) : \xi > 0 \right\} \equiv \tau(\upsilon)$$

This is equivalent to $G_{j, \upsilon}(x) \le \tau < \tau(\upsilon) \Rightarrow \phi_j(x, \omega) \le 0, \forall \omega \in \Omega, j = 1, ..., m$. Thus, from the definitions of $\mathcal{F}_{\upsilon, \tau}$ and $\mathcal{F}$, it follows that $x \in \mathcal{F}_{\upsilon, \tau} \Rightarrow x \in \mathcal{F}. \quad \square$

The above result suggests that if τ is sufficiently small, then the solution to the conventional constrained problem $(P_{\upsilon,\tau})$ is a feasible point for problem (P).

The parameter τ can be chosen by calculating $\tau(\upsilon)$ directly or by finding a positive lower bound for it. Figure 2 graphically depicts $g_\upsilon(-\xi)$ and $\min(|\Omega|, \xi/\eta)$ as functions of ξ.

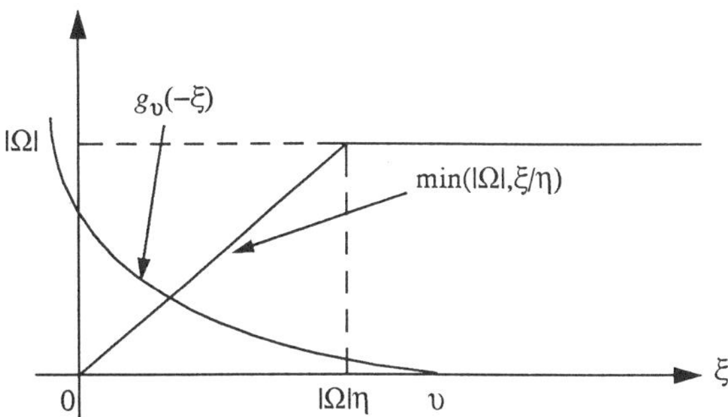

Figure 2 : $g_\upsilon(-\xi)$, $\min(|\Omega|, \xi/\eta)$

Note that Figure 2 assumes υ is chosen such that $|\Omega|\eta < \upsilon$. Thus, it is intuitive that for any $\rho \in (0, 1)$, $g_\upsilon(-\rho|\Omega|\eta)\,\rho|\Omega|$ is a positive lower bound for $\tau(\upsilon)$. In general, the parameter τ can be estimated according to the following result.

Corollary 2.1. *Let* $\bar{\upsilon} = \min(|\Omega|\eta, \upsilon)$ *where* η *is given in Theorem 2.1. Then for any* $\rho \in (0, 1)$

$$\tau = g_\upsilon(-\rho\bar{\upsilon})\,\rho\bar{\upsilon}/\eta \Rightarrow \mathcal{F}_{\upsilon,\tau} \subset \mathcal{F}$$

Proof. If $|\Omega|\eta \geq \upsilon$, then

$$g_\upsilon(-\xi)\min(|\Omega|, \xi/\eta) = \begin{cases} g_\upsilon(-\xi)\,\xi/\eta, & \xi \in (0, \upsilon) \\ 0, & \xi \in [\upsilon, \infty) \end{cases},$$

hence

$$\sup\{g_\upsilon(-\xi)\min(|\Omega|, \xi/\eta) : \xi > 0\} > g_\upsilon(-\rho\upsilon)\,\rho\upsilon/\eta.$$

Otherwise, if $|\Omega|\eta < \upsilon$, then

$$g_\upsilon(-\xi)\min(|\Omega|, \xi/\eta) = \begin{cases} g_\upsilon(-\xi)\,\xi/\eta, & \xi \in (0, |\Omega|\eta) \\ g_\upsilon(-\xi)\,|\Omega|, & \xi \in [|\Omega|\eta, \upsilon) \\ 0, & \xi \in [\upsilon, \infty) \end{cases},$$

hence

$$\sup \{ g_\upsilon (-\xi) \min (|\Omega|, \xi/\eta) : \xi > 0 \} > g_\upsilon (-\rho|\Omega|\eta) \rho|\Omega| .$$

Thus the conclusion follows from Theorem 2.1. ◻

For the case where $|\Omega|\eta \geq \upsilon$, a sensible value for ρ is approximately 0.5. On the other hand, when $|\Omega|\eta < \upsilon$, it is advisable to choose ρ close to unity.

The relationship between the solutions of $(P_{\upsilon, \tau})$ and (P), are illustrated by the following results.

Lemma 2.2. *For any* $\upsilon > 0$*, let* $\mathcal{F}_\upsilon = \{ x \in \Theta : G_{j, \upsilon} (x) = 0, j = 1, ..., m \}$ *. Then*

(i) $\mathcal{F}_\upsilon \subset \mathcal{F}_{\upsilon, \tau}$

(ii) $0 < \mu < \upsilon \Rightarrow \mathcal{F}_\upsilon \subset \mathcal{F}_\mu \subset \mathcal{F}$ *and*

(iii) for any $\alpha \in (0, 1]$ $\exists$ $\delta > 0, x_\alpha \equiv \alpha \bar{x} + (1 - \alpha) x^* \in \mathcal{F}_\delta .$

Proof.

(i) $\mathcal{F}_\upsilon \subset \mathcal{F}_{\upsilon, \tau}$ follows straightforward from the definition of $\mathcal{F}_{\upsilon, \tau} .$

(ii) To prove parts (ii) and (iii) of the lemma, consider an alternative expression for $\mathcal{F}_\upsilon$, given as follows:

$$\mathcal{F}_\upsilon = \{ x \in \Theta : \phi_j (x, \omega) \leq -\upsilon, \quad \forall \omega \in \Omega, j = 1, ..., m \}$$

Let $0 < \mu < \upsilon$, and suppose that $x \in \mathcal{F}_\upsilon$. Then $\phi_j (x, \omega) \leq -\upsilon, \forall \omega \in \Omega, j = 1, ..., m$ and hence

$$\phi_j (x, \omega) \leq -\upsilon < -\mu < 0, \forall \omega \in \Omega, j = 1, ..., m .$$

Consequently, $x \in \mathcal{F}_\mu$. Moreover, for any $x \in \mathcal{F}_\mu$, it also follows that $x \in \mathcal{F}$. Thus $\mathcal{F}_\upsilon \subset \mathcal{F}_\mu \subset \mathcal{F}$.

(iii) By (A2), for each $\alpha \in (0, 1]$, $x_\alpha \equiv \alpha \bar{x} + (1 - \alpha) x^* \in \text{int} (\mathcal{F})$, i.e.

$$\phi_j (x_\alpha, \omega) < 0, \forall \omega \in \Omega, j = 1, ..., m .$$

Hence there exists a $\delta > 0$ such that $\phi_j (x_\alpha, \omega) \leq -\delta, \forall \omega \in \Omega, j = 1, ..., m$ and thus $x_\alpha \in \mathcal{F}_\delta .$ ◻

Theorem 2.2. *Let* $x^*_{\upsilon, \tau}$ *denote a solution to* $(P_{\upsilon, \tau})$*, where* $\tau < \tau (\upsilon)$*. Then,* $\lim_{\upsilon \to 0} f(x^*_{\upsilon, \tau}) = f(x^*)$*, and any accumulation points of* $\{ x^*_{\upsilon (n), \tau} \}_{n = 1}^\infty$ *is a solution of* (P)*, where* $\{ \upsilon (n) \}_{n = 1}^\infty$ *is any positive sequence which converges to zero.*

Proof. From (A2) and the continuity of f, it follows that

$$\forall \varepsilon > 0, \exists \; \theta(\varepsilon) \in (0, 1] \; : 0 < \beta < \theta(\varepsilon) \Rightarrow 0 \leq f(x_\beta) - f(x^*) \leq \varepsilon \qquad (3)$$

For a fixed value of β, there exists, from Lemma 2.2 (iii), a $\delta > 0$ such that $x_\beta \in \mathcal{F}_\delta$. It follows from Lemma 2.2 (ii) and (i) that $0 < \upsilon \leq \delta \Rightarrow x_\beta \in \mathcal{F}_\upsilon \subset \mathcal{F}_{\upsilon,\tau}$. Moreover, by Theorem 2.1, $\mathcal{F}_{\upsilon,\tau} \subset \mathcal{F}$. Thus $f(x^*) \leq f(x^*_{\upsilon,\tau}) \leq f(x_\beta)$. Combining this with (3), yields

$$\forall \varepsilon > 0, \exists \; \delta(\varepsilon) > 0 \; : 0 < \upsilon \leq \delta(\varepsilon) \Rightarrow 0 \leq f(x^*_{\upsilon,\tau}) - f(x^*) \leq \varepsilon.$$

That is, $\lim_{\upsilon \to 0} f(x^*_{\upsilon,\tau}) = f(x^*)$.

Note that the sequence $\{x^*_{\upsilon(n),\tau}\}_{n=1}^{\infty} \in \Theta$. Since Θ is compact, the sequence has an accumulation point $\hat{x} \in \Theta$, and a subsequence, again denoted by $\{x^*_{\upsilon(n),\tau}\}_{n=1}^{\infty}$, such that

$$\lim_{n \to \infty} \left\| x^*_{\upsilon(n),\tau} - \hat{x} \right\| = 0,$$

where the norm is the Euclidean norm for finite dimensional vectors. Thus by continuity of f,

$$\lim_{n \to \infty} f(x^*_{\upsilon(n),\tau}) = f(\hat{x}).$$

Combining this with the first part of the theorem gives $f(\hat{x}) = f(x^*)$. From Theorem 2.1, $x^*_{\upsilon(n),\tau} \in \mathcal{F}_{\upsilon(n),\tau} \subset \mathcal{F}$, and since $\mathcal{F}$ is closed, $\hat{x} \in \mathcal{F}$. Hence, $\hat{x}$ is a solution to (P). $\square$

The above result shows that the solutions to problem $(P_{\upsilon,\tau})$ approach the solution of (P) as the parameter υ is made sufficiently small (and τ is chosen appropriately). Note that the smooth sub-problems introduced in [6] can be considered as special cases of the larger classes of smooth sub-problems presented in this Section.

3. Approximations by Unconstrained Problems

In this Section, a penalty function associated with the constraints of problem (P) is added to the original cost to form an augmented cost. The resulting class of conventional unconstrained problems can then be used to approximate (P). The advantage of this method is that with existing software, unconstrained problems are much easier to solve than constrained problems.

Problem (P) can be approximated by a sequence of unconstrained optimization problem $(P_{\upsilon,\gamma})$ defined for each $\upsilon, \gamma > 0$ as follows:

$$\min \; f_{\upsilon,\gamma}(x) \qquad (4)$$

where $f_{\upsilon,\gamma}(x) \equiv f(x) + \gamma \sum_{j=1}^{m} G_{j,\upsilon}(x)$.

The relations between $(P_{\upsilon,\gamma})$ and (P) are given in the following theorems.

Theorem 3.1. *Let $\tilde{x} \in \Theta$ be such that $f(\tilde{x}) \leq f(x)$, $\forall x \in \Theta$, let υ be such that $\mathcal{F}_\upsilon$ is non-empty and let $x_\upsilon \in \mathcal{F}_\upsilon$. If $x^*_{\upsilon,\gamma}$ is a solution to $(P_{\upsilon,\gamma})$, where $\gamma > \gamma(\upsilon) \equiv (f(x_\upsilon) - f(\tilde{x})) / \tau(\upsilon)$, then $x^*_{\upsilon,\gamma} \in \mathcal{F}$.*

454

Proof. Since Θ is compact and f is continuous, $\tilde{x}$ exists. On the other hand, from the definition of $\mathcal{F}_\upsilon$, it follows that $\forall x \in \mathcal{F}_\upsilon, G_{j,\upsilon}(x) = 0$, subsequently

$$f_{\upsilon,\gamma}(x^*_{\upsilon,\gamma}) \leq f_{\upsilon,\gamma}(x_\upsilon) = f(x_\upsilon) .$$

Hence, using the definition of $f_{\upsilon,\gamma}(x)$ and rearranging yields

$$\gamma \sum_{j=1}^m G_{j,\upsilon}(x^*_{\upsilon,\gamma}) \leq f(x_\upsilon) - f(x^*_{\upsilon,\gamma}) \leq f(x_\upsilon) - f(\tilde{x})$$

$$\Rightarrow \sum_{j=1}^m G_{j,\upsilon}(x^*_{\upsilon,\gamma}) \leq \frac{f(x_\upsilon) - f(\tilde{x})}{\gamma} < \frac{f(x_\upsilon) - f(\tilde{x})}{\gamma(\upsilon)} = \tau(\upsilon) .$$

Let $\tau = (f(x_\upsilon) - f(\tilde{x}))/\gamma$. Then it follows that $x^*_{\upsilon,\gamma} \in \mathcal{F}_{\upsilon,\tau}$, moreover $\tau < \tau(\upsilon)$, but from Theorem 2.2, $\mathcal{F}_{\upsilon,\tau} \subset \mathcal{F}$. Thus, $x^*_{\upsilon,\gamma} \in \mathcal{F}. \square$

The above result asserts that for a sufficiently large penalty parameter γ, the solution $x^*_{\upsilon,\gamma}$ of the unconstrained problem $(P_{\upsilon,\gamma})$ is a feasible point for problem (P). Note that the parameter υ must be sufficiently small so that $\mathcal{F}_\upsilon$ is non-empty.

Corollary 3.1. *Let* $\upsilon = \min(|\Omega|\eta, \upsilon)$ *where* η *is given in Theorem 2.1. Then for any* $\rho \in (0, 1)$

$$\gamma = \frac{(f(x_\upsilon) - f(\tilde{x}))\eta}{g_\upsilon(-\rho\upsilon)\rho\upsilon} \Rightarrow x^*_{\upsilon,\gamma} \in \mathcal{F}$$

The proof follows from Corollary 2.1 and Theorem 3.1.

Theorem 3.2. *Let* $x^*_{\upsilon,\gamma}$ *denote a solution to* $(P_{\upsilon,\gamma})$, *where* $\gamma > \gamma(\upsilon)$. *Then,* $\lim_{\upsilon \to 0} f(x^*_{\upsilon,\gamma}) = f(x^*)$, *and any accumulation point of* $\{x^*_{\upsilon(n),\gamma}\}_{n=1}^\infty$ *is a solution of* (P), *where* $\{\upsilon(n)\}_{n=1}^\infty$ *is any positive sequence which converges to zero.*

Proof. From (A2) and continuity of f, we have (3), which is restated here,

$$\forall \varepsilon > 0, \exists\ \theta(\varepsilon) \in (0, 1] : 0 < \beta < \theta \Rightarrow 0 \leq f(x_\beta) - f(x^*) \leq \varepsilon$$

For a fixed β, from Lemma 2.2 (ii) and (iii), there exists a $\delta > 0$ such that $0 < \upsilon \leq \delta \Rightarrow x_\beta \in \mathcal{F}_\upsilon$. Note from the definition of $\mathcal{F}_\upsilon$ that $\forall x \in \mathcal{F}_\upsilon, f_{\upsilon,\gamma}(x) = f(x)$. Hence,

$$f_{\upsilon,\gamma}(x^*_{\upsilon,\gamma}) \leq f_{\upsilon,\gamma}(x_\beta) = f(x_\beta) .$$

Since $G_{j,\upsilon}(x^*_{\upsilon,\gamma}) \geq 0$, the penalty term is non-negative. Thus $f(x^*_{\upsilon,\gamma}) \leq f(x_\beta)$, and hence the first part of the theorem follows, i.e.

$$\forall \varepsilon > 0, \exists\ \delta(\varepsilon) > 0 : 0 < \upsilon \leq \delta \Rightarrow 0 \leq f(x^*_{\upsilon,\gamma}) - f(x^*) \leq \varepsilon .$$

From Theorem 3.1, $\{x^{*}_{\upsilon(n),\gamma}\}^{\infty}_{n=1} \subset \mathcal{F} \subset \Theta$. Consequently, the conclusion to the second part follows in the same line of reasoning as the second part of Theorem 2.1. $\Box$

The parameter υ is called the accuracy parameter. Collectively, υ and γ are referred to as approximation parameters. Note that as υ tends to zero, the parameter γ tends to infinity.

Since the cost function is smooth, the augmented cost function is smooth if g_{υ} is also smooth. Smooth unconstrained problem can be readily solved by existing gradient based algorithms. An example of a smooth g_{υ} is given below

$$g_{\upsilon}(x) = \begin{cases} 0, & x \le -\upsilon \\ (x+\upsilon)^2/4\upsilon, & -\upsilon \le x \le \upsilon \\ x, & x \le \upsilon \end{cases} \tag{5}$$

This particular function generates the methods introduced in [6] and [7]. It is smooth, but not twice continuously differentiable. Thus the augmented cost may not have a well defined Hessian. Consequently, optimization methods based on Newtonian directions cannot be applied thereby not making use of advantages offered by this method. An example of a twice continuously differentiable g_{υ} is given as follows

$$g_{\upsilon}(x) = \begin{cases} 0, & x \le -\upsilon \\ -\dfrac{\upsilon}{\pi}\cos\left(\dfrac{\pi x}{2\upsilon}\right) + \dfrac{(x+\upsilon)}{2}, & -\upsilon \le x \le 0 \\ \dfrac{\pi x^2}{8\upsilon} + \dfrac{x}{2} + \upsilon\left(\dfrac{1}{2} - \dfrac{1}{\pi}\right), & x \ge 0 \end{cases} \tag{6}$$

For problems where the original cost is twice continuously differentiable, a continuously differentiable g_{υ} yields a twice continuously differentiable augmented cost.

4. Convex Functional Inequality Constrained Quadratic Programming

The materials presented in the current and following Sections are straightforward generalizations of the work found in [12]. The main focus of this Section is on problems for which the cost is a quadratic function, while the functional constraints are convex functions of x. That is, problem (P) now takes the form

$$\min \ x^{T}Lx \ \text{subject to} \ \phi_j(x, \omega) \le 0, \ \forall \omega \in \Omega, j = 1, ..., m$$

where $L \in R^{n \times n}$ is positive definite, and for each $\omega \in \Omega$, $j \in \{1, ..., m\}$,

$$\phi_j(\alpha u + (1-\alpha)v, \omega) \le \alpha\phi_j(u, \omega) + (1-\alpha)\phi_j(v, \omega), \ \forall u, v \in R^n \ \text{for any} \ \alpha \in [0, 1].$$

Note that $x^{T}Lx$ defines a norm on R^n, for convenience, this norm is denoted by $\|x\|^2_L$. The augmented cost for this case becomes

$$f_{\upsilon,\gamma}(x) = \|x\|^2_L + \gamma\sum_{j=1}^{m}\int_{\Omega}g_{\upsilon}(\phi_j(x, \omega))\,d\omega$$

Lemma 4.1. *The feasible set $\mathcal{F}$ of problem (P) is convex.*

The proof is quite straightforward but is provided here for completeness. Let $u, v \in \mathcal{F}$. Then or each $j \in \{1, ..., m\}$, $\omega \in \Omega$, $\phi_j(u, \omega) \leq 0$ and $\phi_j(v, \omega) \leq 0$. Now for any $\alpha \in [0, 1]$, $\phi_j(\alpha u + (1 - \alpha) v, \omega) \leq \alpha \phi_j(u, \omega) + (1 - \alpha) \phi_j(v, \omega)$. Thus, $\phi_j(\alpha u + (1 - \alpha) v, \omega) \leq 0$ and hence $\alpha u + (1 - \alpha) v \in \mathcal{F}$. Therefore, $\mathcal{F}$ is convex. $\Box$

Convexity of the cost and the feasible region implies that problem (P) has a unique solution. This uniqueness implies that not only $\lim_{\upsilon \to 0} f(x^*_{\upsilon, \gamma}) = f(x^*)$ but $\lim_{\upsilon \to 0} x^*_{\upsilon, \gamma} = x^*$ as in the following result.

Theorem 4.1. *Let $x^*_{\upsilon, \gamma}$ denote the solution to problem $(P_{\upsilon, \gamma})$. If for each υ such that $\mathcal{F}_{\upsilon} \neq \emptyset$, $\gamma > \|x_{\upsilon}\|^2_L / \tau(\upsilon)$ for some $x_{\upsilon} \in \mathcal{F}_{\upsilon}$, then, $\lim_{\upsilon \to 0} x^*_{\upsilon, \gamma} = x^*$.*

Proof. Suppose that $\lim_{\upsilon \to 0} x^*_{\upsilon, \gamma} \neq x^*$, i.e.

$$\exists \varepsilon > 0 \ni \forall \delta > 0, \quad \exists \upsilon \in (0, \delta) \ni \|x^*_{\upsilon, \gamma} - x^*\| \geq \varepsilon.$$

Then a real positive sequence $\{\upsilon(n)\}_{n=1}^{\infty}$ which converges to zero can be constructed, by letting

$$\upsilon(n) \in \{\upsilon : 0 < \upsilon < \frac{1}{n}, \|x^*_{\upsilon, \gamma} - x^*\| \geq \varepsilon\} .$$

Clearly, the sequence $\{x^*_{\upsilon(n), \gamma}\}_{n=1}^{\infty} \subset \Theta$. Since Θ is compact, the sequence has an accumulation point y. Theorem 3.2 asserts that any accumulation points of the sequence $\{x^*_{\upsilon(n), \gamma}\}_{n=1}^{\infty}$ is a solution to (P). Hence, y is a solution to (P), but $y \neq x^*$, because x^* is not an accumulation point of $\{x^*_{\upsilon(n), \gamma}\}_{n=1}^{\infty}$ (since $\|x^*_{\upsilon(n), \gamma} - x^*\| \geq \varepsilon$ for all $n \in \{1, 2, ...\}$). This contradicts the uniqueness of the solution to (P). $\Box$

Theorem 4.2. *If g_{υ} is continuous and convex, then the augmented cost function $f_{\upsilon, \gamma}$ is also continuous and strictly convex.*

Proof. Continuity is trivial from the definition of $f_{\upsilon, \gamma}$. To show convexity, for each $\omega \in \Omega$, $j \in \{1, ..., m\}$ the convexity of $\phi_j(x, \omega)$ and the non-decreasing property of g_{υ} implies that for any $u, v \in R^n$ and $\alpha \in [0, 1]$

$$g_{\upsilon}(\phi_j(\alpha u + (1 - \alpha) v, \omega)) \leq g_{\upsilon}(\alpha \phi_j(u, \omega) + (1 - \alpha) \phi_j(v, \omega)) ,$$

Since g_{υ} is also convex, it follows that

$$g_{\upsilon}(\phi_j(\alpha u + (1 - \alpha) v, \omega)) \leq \alpha g_{\upsilon}(\phi_j(u, \omega)) + (1 - \alpha) g_{\upsilon}(\phi_j(v, \omega)) .$$

Hence for any $u, v \in R^n$ and $\alpha \in [0, 1]$

$$\int_{\Omega} g_{\upsilon}(\phi_j(\alpha u + (1 - \alpha) v, \omega)) \, d\omega \leq \int_{\Omega} \alpha g_{\upsilon}(\phi_j(u, \omega)) + (1 - \alpha) g_{\upsilon}(\phi_j(v, \omega)) \, d\omega$$

$$= \alpha \int_{\Omega} g_{\upsilon}(\phi_j(u, \omega)) \, d\omega + (1 - \alpha) \int_{\Omega} g_{\upsilon}(\phi_j(v, \omega)) \, d\omega$$

That is each $G_{j,\,\upsilon}(x)$ is a convex function of x and since $\|x\|_L^2$ is strictly convex, it follows that $f_{\upsilon,\,\gamma}(x)$ is also strictly convex. $\square$

The augmented cost has unique minimum and thus unconstrained techniques can be applied directly.

By exploiting the convexity of inequality constraints and the quadratic cost function, it can be shown that, for any given error bound ε, the accuracy parameter υ can be chosen (without using any information on the solutions $x_{\upsilon,\,\gamma}^*$ and x^*) so that the approximation error $\|x_{\upsilon,\,\gamma}^*\|_L^2 - \|x^*\|_L^2$, is less than the error bound ε. This is presented in the following theorem.

Theorem 4.3. *Let $x_{\upsilon,\,\gamma}^*$ denote the solution to problem $(P_{\upsilon,\,\gamma})$ and let $\delta(\bar{x}) \equiv \min_j \ \min_{\omega \in \Omega} \{-\phi_j(\bar{x},\,\omega)/2\|\bar{x}\|_L^2\}$ for any $\bar{x} \in \mathrm{int}(\mathcal{F})$. If $\upsilon > 0$ is such that $\mathcal{F}_\upsilon$ is non-empty, and $\gamma > \|x_\upsilon\|_L^2/\tau(\upsilon)$ for some $x_\upsilon \in \mathcal{F}_\upsilon$, then,*

$$0 < \upsilon \le \delta(\bar{x}) \min\{\varepsilon,\, 2\|\bar{x}\|_L^2\} \ \Rightarrow 0 \le \|x_{\upsilon,\,\gamma}^*\|_L^2 - \|x^*\|_L^2 \le \varepsilon.$$

Proof. Let

$$\beta = \min\{\varepsilon,\, 2\|\bar{x}\|_L^2\}/2\|\bar{x}\|_L^2.$$

Then,

$$0 < \upsilon \le \delta(\bar{x}) \min\{\varepsilon,\, 2\|\bar{x}\|_L^2\} \ \Leftrightarrow 0 < \upsilon \le -\beta\phi_j(\bar{x},\,\omega),\ \forall \omega \in \Omega, j \in \{1,\,...,\,m\}\ ,$$

$$\Leftrightarrow \beta\phi_j(\bar{x},\,\omega) \le -\upsilon < 0,\ \forall \omega \in \Omega, j \in \{1,\,...,\,m\} \tag{7}$$

Note that each $\phi_j(x,\,\omega)$ is a convex function of x, i.e.

$$\phi_j(x_\beta,\,\omega) = \phi_j(\beta\bar{x} + (1-\beta)x^*,\,\omega) \le \beta\phi_j(\bar{x},\,\omega) + (1-\beta)\phi_j(x^*,\,\omega)\ . \tag{8}$$

Since $\phi_j(x^*,\,\omega) \le 0,\ \forall \omega \in \Omega, j \in \{1,\,...,\,m\}$, and $0 < \beta \le 1$, $(1-\beta)\phi_j(x^*,\,\omega) \le 0$, i.e.

$$\beta\phi_j(\bar{x},\,\omega) + (1-\beta)\phi_j(x^*,\,\omega) \le \beta\phi_j(\bar{x},\,\omega)\ .$$

Hence it is clear from (8) that $\phi_j(x_\beta,\,\omega) \le \beta\phi_j(\bar{x},\,\omega),\ \forall \omega \in \Omega, j \in \{1,\,...,\,m\}$. Moreover, using (6) yields

$$\phi_j(x_\beta,\,\omega) \le \beta\phi_j(\bar{x},\,\omega) \le -\upsilon < 0,\ \forall \omega \in \Omega, j \in \{1,\,...,\,m\}$$

$$\Rightarrow G_{j,\,\upsilon}(x_\beta) = \int_\Omega g_\upsilon(\phi(x_\beta,\,t))\,dt = 0$$

$$\Rightarrow \|x_{\upsilon,\,\gamma}^*\|_L^2 \le f_{\upsilon,\,\gamma}(x_{\upsilon,\,\gamma}^*) \le f_{\upsilon,\,\gamma}(x_\beta) = \|x_\beta\|_L^2$$

Also, as $\gamma > \|x_\upsilon\|_L^2/\tau(\upsilon)$, by Theorem 3.3, $x_{\upsilon,\,\gamma}^* \in \mathcal{U}$, and hence,

458

$$0 \leq \left\| x^*_{\upsilon, \gamma} \right\|^2_L - \left\| x^* \right\|^2_L \leq \left\| x_\beta \right\|^2_L - \left\| x^* \right\|^2_L = \left(\left\| x_\beta \right\|_L - \left\| x^* \right\|_L \right) \left(\left\| x_\beta \right\|_L + \left\| x^* \right\|_L \right)$$

$$= \left(\left\| (\beta \bar{x} + (1 - \beta) \, x^*) \right\|_L - \left\| x^* \right\|_L \right) \left(\left\| (\beta \bar{x} + (1 - \beta) \, x^*) \right\|_L + \left\| x^* \right\|_L \right)$$

$$\leq \beta \left(\left\| \bar{x} \right\|_L - \left\| x^* \right\|_L \right) \left(\beta \left\| \bar{x} \right\|_L + (2 - \beta) \left\| x^* \right\|_L \right)$$

$$= \beta (2 - \beta) \left(\left\| \bar{x} \right\|_L - \left\| x^* \right\|_L \right) \left(\frac{\beta}{(2 - \beta)} \left\| \bar{x} \right\|_L + \left\| x^* \right\|_L \right)$$

$$< \beta (2 - \beta) \left(\left\| \bar{x} \right\|_L - \left\| x^* \right\|_L \right) \left(\left\| \bar{x} \right\|_L + \left\| x^* \right\|_L \right)$$

$$< 2\beta \left\| \bar{x} \right\|^2_L = \min \{ \varepsilon, 2 \left\| \bar{x} \right\|^2_L \} \leq \varepsilon. \quad \Box$$

When an interior point is known, given any tolerance on the cost, the accuracy parameter can be easily selected according to Theorem 4.3 so that $\left\| x^*_{\upsilon, \gamma} \right\|^2_L - \left\| x^* \right\|^2_L$ is less than the tolerance. This is very useful in implementing interior point algorithms for problem (P).

5. Affine Functional Inequality Constrained Quadratic Programming

Throughout this Section, ordering relations applies component-wise for vectors unless otherwise stated. Matrix norm is taken to be the induced Euclidean norm for vectors, i.e. for an $n \times n$ matrix A, its norm is defined as

$$\| A \| = \max_{x \neq 0} \frac{\| A x \|}{\| x \|}.$$

With this in mind, an affine functional inequality constrained quadratic programming problem can be regarded as a problem of the form

$$\min \ x^T L x \ \text{subject to} \ \underset{\sim}{\phi} (x, \omega) \equiv Y (\omega) \, x - z (\omega) \leq 0, \ \forall \omega \in \Omega$$

where $L \in R^{n \times n}$ is a positive definite matrix,

$$\underset{\sim}{\phi} (x, \omega) = \begin{bmatrix} \phi_1 (x, \omega) \\ \vdots \\ \phi_m (x, \omega) \end{bmatrix} \in R^m, \ Y (\omega) = \begin{bmatrix} y_1 (\omega)^T \\ \vdots \\ y_m (\omega)^T \end{bmatrix} \in R^{m \times n} \ \text{and} \ z (\omega) = \begin{bmatrix} z_1 (\omega) \\ \vdots \\ z_m (\omega) \end{bmatrix} \in R^m$$

It is obvious that $\underset{\sim}{\phi} (x, \omega)$ is an affine function of x, hence the name affine functional inequality constraints. This class of problems arises in signal processing applications such as [12]. In fact, the study of quadratic programming involving affine functional constraints is motivated by this application.

The following lemma summarises some useful analytical properties of the augmented cost function, see Appendix for proofs.

Lemma 5.1.

    (i)   *If g_υ is once continuously differentiable, then $f_{\upsilon, \gamma}$ is once continuously differentiable and its gradient is given by*

$$\nabla f_{\upsilon,\gamma}(x) = 2Lx + \gamma \sum_{j=1}^{m} \int_{\Omega} g'_{\upsilon}(\phi_j(x,\omega)) y_j(\omega)\, d\omega.$$

(ii)   *If g_{υ} is twice continuously differentiable, then $f_{\upsilon,\gamma}$ is twice continuously differentiable and its Hessian is given by*

$$\nabla^2 f_{\upsilon,\gamma}(x) = 2L + \gamma \sum_{j=1}^{m} \int_{\Omega} g''_{\upsilon}(\phi_j(x,\omega)) y_j(\omega) y(\omega)_j^{T}\, d\omega.$$

(iii)   *If g_{υ} has bounded second derivative i.e. $\left| g''_{\upsilon}(x) \right| \le h(\upsilon), \forall x \in R$ then $f_{\upsilon,\gamma}$ also has bounded Hessian, i.e.*

$$2\|L\| \le \left\| \nabla^2 f_{\upsilon,\gamma}(x) \right\| \le 2\|L\| + \gamma h(\upsilon) \sum_{j=1}^{m} \int_{\Omega} \left\| y_j(\omega) \right\|^2 d\omega.$$

(iv)   *If g_{υ} has Lipschitz continuous second derivative, i.e. there exists some $L(\upsilon)$, such that $\left| g''_{\upsilon}(x) - g''_{\upsilon}(y) \right| \le L(\upsilon)|x - y|, \forall x, y \in R$, then*

$$\left\| \nabla^2 f_{\upsilon,\gamma}(u) - \nabla^2 f_{\upsilon,\gamma}(v) \right\| \le \gamma L(\upsilon) \|u - v\| \sum_{j=1}^{m} \int_{\Omega} \left\| y_j(\omega) \right\|^3 d\omega.$$

This implies that $\nabla^2 f_{\upsilon,\gamma}(x)$ is also Lipschitz continuous.

By choosing g_{υ} appropriately, globally-convergent algorithms for minimizing the augmented cost can be developed.

5.1 Iterative Algorithms

To obtain sub-optimal solutions, the following iterative process is used

$$x_{k+1} = x_k + \mu_k d_k$$

where x_k is kth iterate, t_k is the step size, and d_k satisfying $\nabla f_{\upsilon,\gamma}(x_k)^T d_k < 0$ is called the descent direction.

Assuming that g_{υ} is at least once continuously differentiable, the simplest of the descent directions is the steepest descent (SD) direction, $d_k = -\nabla f_{\upsilon,\gamma}(x_k)$. At present the method of steepest descent is one of the best known minimization methods. The popularity of the method has been favoured by the comparatively simple implementation and applicability to a wide range of functions.

Theorem 5.1. *Suppose that g_{υ} is twice continuously differentiable for each $\upsilon > 0$ with $\left| g''_{\upsilon}(x) \right| \le h(\upsilon), \forall x \in R$. Then, starting from any initial point x_0, the sequence $\{x_k\}_{k=0}^{\infty}$ generated by the SD algorithm with constant step size t converges to the solution $x^*_{\upsilon,\gamma}$ of problem $(P_{\upsilon,\gamma})$ if*

$$0 < t < \frac{2}{2\|L\| + \gamma h(\upsilon) \sum\limits_{j=1}^{m} \int_{\Omega} \|y_j(\omega)\|^2 d\omega} \,.$$

Proof. Using Lemma 5.1 (iii), the bounds on the Hessian of the augmented cost function is obtained. Then the conclusion follows readily from Theorem 1.3 of [13].

The SD algorithm is attractive to a large extent due to its computational simplicity, however the method has several serious practical drawbacks. Most importantly, convergence is usually slow [13], [14].

Another popular method is the Newton-Ralphson (NR) algorithm, which uses second derivative information to calculate the descent direction,

$$d_k = -\left(\nabla^2 f_{\upsilon,\gamma}(x_k)\right)^{-1} \nabla f_{\upsilon,\gamma}(x_k) \,.$$

Here, it is assumed that g_υ is chosen so that the augmented cost functional is at least twice continuously differentiable, and hence its Hessian is positive definite (twice continuously differentiable convex functions have positive definite Hessian). For algorithms using constant step size, a larger step size than the SD algorithm is allowed as illustrated in the following theorem.

Theorem 5.2. *Suppose that g_υ is twice continuously differentiable for each $\upsilon > 0$ with $\left|g_\upsilon''(x)\right| \leq h(\upsilon)$, $\forall x \in R$. Then, starting from any initial point x_0, the sequence $\{x_k\}_{k=0}^{\infty}$ generated by the NR algorithm with constant step size t converges to the solution $x_{\upsilon,\gamma}^*$ of problem $(P_{\upsilon,\gamma})$ if*

$$0 < t < \frac{4}{2\|L\| + \gamma h(\upsilon) \sum\limits_{j=1}^{m} \int_{\Omega} \|y_j(\omega)\|^2 d\omega} \,.$$

Proof. Again, using Lemma 5.1 (iv), the bounds on Hessian of the augmented cost function is obtained. Then, the conclusion follows readily from Theorem 1.6 of [13].

Considerably faster convergence can be achieved with the NR method by performing some line search at each iteration to determine the step size. It is imperative to point out that an iteration in this context refers to a search direction evaluation. For algorithms with constant step size, each search direction evaluation is followed by a single filter update, while algorithms with line search requires at least one update to determine the step size before the next search direction is calculated. A popular form of line search involve the Goldstein condition [10], [13], [14],

$$f_{\upsilon,\gamma}(x_{k+1}) - f_{\upsilon,\gamma}(x_k) < \alpha t_k \nabla f_{\upsilon,\gamma}(x_k)^T d_k, \alpha \in (0, 0.5) \,.$$

and the Wolfe-Powell condition [10], [14]

$$\nabla f_{\upsilon,\gamma}(x_k + t_k d_k)^T d_k \geq \beta \nabla f_{\upsilon,\gamma}(x_k)^T d_k, \beta \in (\alpha, 1)$$

This type of line search can be implemented as follows

```
line_search

    t_upp = INF; t_low = 0; t = 1; ADMISSIBLE = FALSE;

    while ADMISSIBLE == FALSE,
```

$$\text{if } f_{\upsilon,\gamma}(x_k + t d_k) - f_{\upsilon,\gamma}(x_k) < \alpha t \nabla f_{\upsilon,\gamma}(x_k)^T d_k, \alpha \in (0, 0.5),$$

$$\text{if } \nabla f_{\upsilon,\gamma}(x_k + t d_k)^T d_k \geq \beta \nabla f_{\upsilon,\gamma}(x_k)^T d_k, \beta \in (\alpha, 1),$$

```
                ADMISSIBLE = TRUE;
            else
                t_low = t

                if t_upp = INF,

                    t = Increase(t);
                else
                    t = Refine(t, t_low, t_upp);
                end
            end
        else
            t_upp = t;

            if t_low == 0,

                t = Decrease(t);
            else
                t = Refine(t, t_low, t_upp);
            end
        end
    end
```

There are four possible scenarios in this line search. If $t = 1$ satisfies both the Goldstein and Wolfe-Powell conditions, then $x_{k+1} = x_k + d_k$, and no further line search calculations are performed.

If $t = 1$ is not admissible because it fails the Goldstein condition, then t will be decreased. This is most often and effectively done by setting the next step size to max (t_m, ct), where c is typically 0.1 and t_m is the one-dimensional minimizer of the one-dimensional quadratic approximation $q(t)$ to $f_{\upsilon,\gamma}(x_k + t d_k)$ that interpolates $f_{\upsilon,\gamma}(x_k)$, $\nabla f_{\upsilon,\gamma}(x_k)^T d_k$ and $f_{\upsilon,\gamma}(x_k + t d_k)$, i.e.

$$t_m = \frac{-t^2 \nabla f_{\upsilon,\gamma}(u_k)^T d_k}{2\left(f_{\upsilon,\gamma}(u_k + t d_k) - f_{\upsilon,\gamma}(u_k) - t \nabla f_{\upsilon,\gamma}(u_k)^T d_k \right)}$$

This Decrease(t) routine can be repeated if the new step size continues to fail the Goldstein condition.

Alternatively if the step size t satisfies the Goldstein condition but not the Wolfe-Powell condition, it will be increased. Generally a simple routine such as `Increase` $(t) = t/\rho$, where $\rho \in (0, 1)$ is used although more sophisticated strategies are possible.

After one or more repetitions of either the `Increase(t)` or `Decrease(t)`, either an admissible step size is found or the last two values of t which have been tried t_{low} and t_{upp} bracket an acceptable step size. That is, t_{low} satisfies Goldstein condition but not the Wolfe-Powell, while, t_{upp} fails the Goldstein condition. In this case an admissible step size must be in the open interval (t_{low}, t_{upp}) and is selected by the refine phase. A simple routine such as `Refine` $(t, t_{low}, t_{upp}) = (t_{low} + t_{upp})/2$ is used here, although there are others.

If g_υ is continuously differentiable then $f_{\upsilon, \gamma}$ is continuously differentiable and bounded below, hence there are $0 < t_1 < t_2$ such that for any $t_k \in [t_1, t_2]$, $x_{k+1} = x_k + t_k d_k$ satisfies both the Goldstein and the Wolfe-Powell condition, see Dennis and Schnabel [15].

Theorem 5.3. *Suppose that g_υ is twice continuously differentiable for each $\upsilon > 0$ with $\left| g_\upsilon''(x) \right| \leq h(\upsilon)$. If $\left| g_\upsilon''(x) - g_\upsilon''(y) \right| \leq L(\upsilon) |x - y|$, $\forall x, y \in R$, i.e. $g_\upsilon''(x)$ is Lipschitz continuous, then, starting from any initial point x_0 the sequence $\{x_k\}_{k=0}^{\infty}$, generated by the NR algorithm with step size t_k chosen by the above line search procedure, converges to the solution $u_{\upsilon, \gamma}^*$ of problem $(P_{\upsilon, \gamma})$ at a quadratic rate.*

Proof. From Lemma 5.1 (iii), (iv), the Hessian of the augmented cost function is bounded and satisfies the Lipschitz condition. Thus, applying Theorem 2.2. of [13] completes the theorem.

6. Conclusions

In this paper, numerical methods for solving a class of functional inequality constrained problems has been proposed. The technique involves approximating these problems by conventional constrained or unconstrained problems. In either case, it has been established that appropriate choice of approximation parameters yields a feasible solution. Moreover, both the conventional constrained and unconstrained approximations to this class of functional inequality constrained problems converge to the optimal solution in the limit.

A closed form expression for the approximation error of the unconstrained approximation has been derived when the functional constraints are convex and the cost function is quadratic. For the same cost function with affine constraints, a number of globally convergent algorithms have been proposed and it has been shown that one of these algorithms has a quadratic rate of convergence.

7. References

[1] E. Polak and D. Q. Mayne, "An algorithm for optimization problems with functional inequality constraints", *IEEE Trans. Aut. Control*, **AC-21**, 184-193, 1976.

[2] G. Gonzaga, E. Polak and R. Trahan, "An improved algorithm for optimization problems with functional inequality constraints", *IEEE Trans. Aut. Control*, **AC-25**, 49-54, 1980.

[3]   E. Polak and Y. Wardi, "Nondifferential optimization algorithm for designing control systems having singular value inequalities", *Automatica*, **18**(3), 267-283, 1982.

[4]   K. L. Teo and C. J. Goh, "A simple computational procedure for optimization problems with functional inequality constraints", *IEEE Trans. Aut. Control*, **AD-32**, 940-941, 1987.

[5]   R. W. H. Sargent and G. R. Sullivan, "The development of an efficient optimal control package", IN Proc. *8th Conf. Optimiz. Techniques*, Wurzberg, Germany, Springer, Berlin, 1977.

[6]   L. S. Jennings and K. L. Teo, "A computational algorithm for functional inequality constrained optimization problems", *Automatica*, **26**, 371-375, 1990.

[7]   K. L. Teo, V. Rehbock and L. S. Jennings, "A new computational algorithm for functional inequality constrained optimization problems", *Automatica*, **29**, 789-792, 1993.

[8]   E. Polak, *Computational methods in optimization*, Academic Press, New York, 1971.

[9]   Y. G. Evtushento, *Numerical optimization techniques*, Optimization Software, New York, 1985.

[10]   R. Fletcher, *Practical methods of optimizations*, 2nd edition, John Wiley & Sons, 1987.

[11]   B. Vo, W. X. Zheng, A. Cantoni, K. L. Teo, "Approximation of functional inequality constrained problem", *Technical Report*, TM002, ATRI, Curtin University of Technology, Western Australia (1994).

[12]   B. Vo, A. Cantoni, K. L. Teo, "Envelope constrained filter with linear interpolation", *Technical Report*, TM006, ATRI, Curtin University of Technology, Western Australia (1994).

[13]   B. N. Pshenichni and Y. M. Danilin, *Numerical methods in Extremal Problems*, Mir Publishers, Moscow, 1978.

[14]   J. E. Dennis Jr. and R. B. Schnabel, "A view of unconstrained optimization", *Handbooks in operations research and management science*, G. L. Nemhauser *et al.*, Eds., Elsevier Science Publishers B. V. (North-Holland) **Vol. 1**, pp. 1-72, 1989.

[15]   J. E. Dennis Jr. and R. B. Schnabel, *Numerical methods for unconstrained optimization and nonlinear equations*, Prentice-Hall, Inc., Englewood Cliffs, New Jersey, 1983.

Appendix :   Proof of Lemma 5.1.

(i) Since $\nabla \phi_j(x, \omega) = y_j(\omega)$, hence using the chain rule and Leibnitz's differentiation rule yields the desired result.

(ii) Using the product rule for differentiation on (i) and substituting for $\nabla^2 \phi_j(x, \omega) = 0$ and gives the desired expression.

(iii) Using (ii) and the triangle inequality yields

$$\left\| \nabla^2 f_{\upsilon, \gamma}(x) \right\| \le 2\|L\| + \gamma \sum_{j=1}^{m} \int_{\Omega} \left| g''_{\upsilon}(\phi_j(x, \omega)) \right| \left\| y_j(\omega) y_j(\omega)^T \right\| d\omega$$

$$\le 2\|L\| + \gamma h(\upsilon) \sum_{j=1}^{m} \int_{\Omega} \left\| y_j(\omega) y_j(\omega)^T \right\| d\omega$$

Now consider the norm of a matrix of the form $a_j a_j^T$

$$\left\| a_j a_j^T \right\| = \max \frac{\left(x^T a_j a_j^T a_j a_j^T x \right)^{1/2}}{\|x\|} = \max \frac{\|a_j\| \left\| a_j^T x \right\|}{\|x\|}$$

by the Cauchy-Schwartz inequality

$$\max \frac{\|a_j\| \left\| a_j^T x \right\|}{\|x\|} = \|a_j\|^2$$

Hence

$$\left\| \nabla^2 f_{\upsilon, \gamma}(x) \right\| \le 2\|L\| + \gamma h(\upsilon) \sum_{j=1}^{m} \int_{\Omega} \left\| y_j(\omega) \right\|^2 d\omega$$

To show that $\left\| \nabla^2 f_{\upsilon, \gamma}(x) \right\| \ge 2\|L\|$, observe that the matrix norm induced by the Euclidean vector norm is in fact the spectral norm

$$\|A\| = \sqrt{\lambda_{max}\left(A^T A \right)}$$

since

$$\max \frac{\|Ax\|}{\|x\|} = \max \sqrt{\frac{x^T A^T A x}{x^T x}} = \sqrt{\lambda_{max}\left(A^T A \right)} .$$

Let

$$\gamma \sum_{j=1}^{m} \int_{\Omega} g''_{\upsilon}(\phi_j(x, \omega)) y_j(\omega) y_j(\omega)^T d\omega = M ,$$

then it is clear from the positive definiteness of L and M that

$$\left\| \nabla^2 f_{\upsilon,\gamma}(\boldsymbol{u}) \right\| = \left\| 2L + M \right\| \geq 2 \|L\|$$

Thus, the conclusion follows.

(iv) Again using (ii)

$$\nabla^2 f_{\upsilon,\gamma}(\boldsymbol{u}) - \nabla^2 f_{\upsilon,\gamma}(\boldsymbol{v}) = \gamma \sum_{j=1}^{m} \int_{\Omega} g''_{\upsilon}(\phi_j(\boldsymbol{u}, \omega)) \, y_j(\omega) \, y_j(\omega)^T d\omega -$$

$$\gamma \sum_{j=1}^{m} \int_{\Omega} g''_{\upsilon}(\phi_j(\boldsymbol{v}, \omega)) \, y_j(\omega) \, y_j(\omega)^T d\omega$$

Hence, the triangle inequality and the Lipschitz condition $\left| g''_{\upsilon}(x) - g''_{\upsilon}(y) \right| \leq L(\upsilon) \left| x - y \right|$ gives

$$\left\| \nabla^2 f_{\upsilon,\gamma}(\boldsymbol{u}) - \nabla^2 f_{\upsilon,\gamma}(\boldsymbol{v}) \right\| \leq \gamma L(\upsilon) \sum_{j=1}^{m} \int_{\Omega} \left| \phi_j(\boldsymbol{u}, \omega) - \phi_j(\boldsymbol{v}, \omega) \right| \left\| y_j(\omega) \right\|^2 d\omega$$

Since $\left| \phi_j(\boldsymbol{u}, \omega) - \phi_j(\boldsymbol{v}, \omega) \right| = \left| y_j(\omega)^T (\boldsymbol{u} - \boldsymbol{v}) \right|$

$$\left\| \nabla^2 f_{\upsilon,\gamma}(\boldsymbol{u}) - \nabla^2 f_{\upsilon,\gamma}(\boldsymbol{v}) \right\| \leq \gamma L(\upsilon) \sum_{j=1}^{m} \int_{\Omega} \left| y_j(\omega)^T (\boldsymbol{u} - \boldsymbol{v}) \right| \left\| y_j(\omega) \right\|^2 d\omega$$

$$\leq \gamma L(\upsilon) \left\| \boldsymbol{u} - \boldsymbol{v} \right\| \sum_{j=1}^{m} \int_{\Omega} \left\| y_j(\omega) \right\|^3 d\omega.$$

WSSIAA 5 (1995) pp. 467–482

467

MULTIOBJECTIVE CONTROL SYSTEM DESIGN — A MIXED OPTIMIZATION APPROACH

J F WHIDBORNE

Department of Mechanical Engineering, King's College London, Strand
London WC2R 2LS, United Kingdom

I POSTLETHWAITE and D-W GU

Department of Engineering, University of Leicester
Leicester LE1 7RH, United Kingdom

ABSTRACT

Control system design problems are generally multiobjective, in that they require several, generally conflicting, requirements to be simultaneously met. One procedure for multiobjective computer-aided control system design is the method of inequalities (MOI). Here, the MOI is combined with an H^∞-optimization loop-shaping design procedure in a mixed-optimization approach which designs for both robustness and explicit closed-loop performance. This mixed-optimization approach has been incorporated into a MATLAB-based interactive multiobjective computer-aided control system design environment.

1 Introduction

The majority of engineering design problems are multiobjective, in that there are several conflicting design aims which need to be simultaneously achieved. If these design aims are expressed quantitatively as a set of n design objective functions $\phi_i(p) : i = 1 \ldots n$, where p denotes the design parameters chosen by the designer, the design problem could be formulated as a multiobjective optimization problem:

$$\min_{p \in \mathcal{P}} \left\{ \phi_i(p), \text{ for } i = 1 \ldots n \right\}. \tag{1}$$

where $\mathcal{P}$ denotes the set of possible design parameters p. In most cases, the objective functions are in conflict, so the reduction of one objective function leads to the increase in another. Subsequently, the result of the multiobjective optimization is known as a Pareto-optimal solution[1]. A Pareto-optimal solution has the property that it is not possible to reduce any of the objective functions without increasing at least one of the other objective functions.

A point $p^* \in \mathcal{P}$ is defined as being Pareto-optimal if and only if there exists no other point $p \in \mathcal{P}$ such that

$$
\begin{aligned}
&\text{a)} && \phi_i(p) \leq \phi_i(p^*) \text{ for all } i = 1, \ldots, n \quad \text{ and} \\
&\text{b)} && \phi_j(p) < \phi_j(p^*) \text{ for at least one } j.
\end{aligned}
\tag{2}
$$

The problem with multiobjective optimization is that there is generally a very large set of Pareto-optimal solutions. Subsequently there is a difficulty in representing the set of Pareto-optimal solutions and in choosing the solution which is the best design.

Many multiobjective approaches have been suggested to overcome this difficulty[2]. One particularly successful approach is the method of inequalities[3] (MOI), which is a general-purpose, multiobjective design procedure for interactive computer-aided design where the design problem is expressed as a set of algebraic inequalities which need to be simultaneously satisfied for a successful design. The problem is to find a $p \in \mathcal{P}$ such that

$$\phi_i(p) \leq \varepsilon_i \quad \text{for} \quad i = 1 \ldots n, \tag{3}$$

where ε_i are real numbers, called the design goals, and are the largest tolerable values of ϕ_i. Numerical search algorithms are used to find a solution to Eq. 3. The MOI can be combined with analytical optimization techniques, such as McFarlane and Glover's loop-shaping design procedure[4, 5] (LSDP), to provide a mixed-optimization approach which can design for both robustness and explicit closed-loop performance.

The mixed-optimization approach has been implemented in a MATLAB toolbox called MODCONS[6] to create an interactive multiobjective computer-aided control system design (CACSD) environment for designing robust control systems. The toolbox contains several different numerical search routines which can be used for solving Eq. 3. It contains automatic function generators, which allow the user to specify the design inequalities in a user-friendly manner, the specifications are then automatically generated into MATLAB code for use with the numerical search procedures. It contains an interactive graphical interface for the solution of the MOI, which aids the designer in making informed decisions during the design process. It contains a number of additional utility routines to calculate traditional control problem objective indices such as rise-time, settling time, phase-margin and bandwidth.

In the next section, the MOI is introduced, and in Section 3, some numerical search techniques for solving Eq. 3 are summarized. In Section 4, the use of the MOI for multiobjective control system design using the mixed-optimization approach is described. In Section 5, the MODCONS toolbox is summarized.

2 The Method of Inequalities

Performance specifications for control and other engineering systems are frequently given in terms of algebraic or functional inequalities, rather than in the minimization of some objective function. For example, a control system may be required to have a rise-time of less than 1 second, a settling time of less than 5 seconds and an overshoot of less than 10%. In such cases, it is obviously more logical and convenient if the design problem is expressed explicitly in terms of such inequalities.

The method of inequalities[3] (MOI) is a computer-aided multiobjective design approach, where desired performance is represented by such a set of algebraic inequalities, and where the aim of the design is to simultaneously satisfy these inequalities. The design problem is expressed as

$$\phi_i(p) \leq \varepsilon_i \quad \text{for} \quad i = 1 \ldots n, \tag{4}$$

where ε_i are real numbers, $p \in \mathcal{P}$ is a real vector $(p_1, p_2, \ldots, p_q)$ chosen from a given set $\mathcal{P}$ and ϕ_i are real functions of p. The functions ϕ_i are the *objective functions*, the components of p represent the *design parameters* and ε_i are the *design goals* which are chosen by the designer and represent the largest tolerable values of ϕ_i. The aim is the satisfaction of the set of inequalities in order that an acceptable design p is reached.

Each inequality $\phi_i(p) \leq \varepsilon_i$ of the set of inequalities in Eq. 4 defines a set $\mathcal{S}_i$ of points in the q-dimensional space $\mathbb{R}^q$ and the co-ordinates of this space are $p_1, p_2, \ldots, p_q$, so

$$\mathcal{S}_i = \{p : \phi_i(p) \leq \varepsilon_i\}. \tag{5}$$

The boundary of this set is defined by $\phi_i(p) = \varepsilon_i$. A point $p \in \mathbb{R}^q$ is a solution to the set of inequalities in Eq. 4 if and only if it lies inside every set $\mathcal{S}_i$, $i = 1, 2, \ldots, n$ and hence inside the set $\mathcal{S}$ which denotes the intersection of all the sets $\mathcal{S}_i$,

$$\mathcal{S} = \bigcap_{i=1}^{n} \mathcal{S}_i. \tag{6}$$

$\mathcal{S}$ is called the *admissible set* and any point p in $\mathcal{S}$ is called an *admissible point* denoted p_s.

The objective is thus to find a point p such that $p \in \mathcal{S}$. Such a point satisfies the set of inequalities in Eq. 4 and is said to be a solution. In general, a point p_s is not unique unless the subset $\mathcal{S}$ is a point in the space $\mathbb{R}^q$. In some cases, there is no solution to the problem, i.e. $\mathcal{S}$ is an empty set. It is then necessary to relax the boundaries of some of the inequalities, i.e. increase some of the numbers ε_i, until an admissible point p_s exists.

For control system design, the functions $\phi_i(p)$ may be functionals of the system step response, for example the rise-time, overshoot or the integral absolute error, or functionals of the frequency response, such as the bandwidth. They can also represent measures of the system stability, such as the maximum real part of the closed-loop poles. Additional inequalities which arise from the physical constraints of the system can also be included, to restrict for example, the maximum control signal. In practice, the constraints on the design parameters p which define the set $\mathcal{P}$ are also included in the inequality set, e.g. to constrain the possible values of some of the design parameters, or to limit the search to stable controllers only.

3 Numerical Search Algorithms

The actual solution to the set of inequalities in Eq. 4 may be obtained by means of numerical search algorithms. Generally, the design process is interactive, with the computer providing information to the designer about conflicting design requirements, and the designer adjusting the inequalities to explore the various possible solutions to the problem. The progress of the search algorithm should be monitored, and, if a solution is not found, the designer may either change the starting point, relax some of the design goals ε or change the design configuration. Alternatively, if a solution is found easily, to improve the quality of the design, the design goals could be tightened or additional design objectives could be included in Eq. 4. The design process is thus a two way process, with the MOI providing information to the designer about conflicting design requirements, and the designer making decisions about the 'trade-offs' between design requirements based on this information as well as on the designer's knowledge, experience and intuition about the particular problem. The designer can be supported in this role by various graphical displays[2] which provide information about the progress of the search algorithm and about the conflicting design requirements.

The original algorithm for the MOI, proposed by Zakian and Al-Naib[3], is known as the moving boundaries process (MBP). This algorithm uses Rosenbrock's hill-climbing[7] to perform a local search to try and improve on at least one of the unsatisfied performance indices. This algorithm is simple, robust and effective and has worked well over the years, however, it does rely on a great deal of user-interaction to provide assistance for when local minima are found. The success of the algorithm is very dependent on being provided with a good starting point. This does have the advantage of forcing the user to analyze carefully the problem before the design is completed, and hence guarding against 'unreasonable' solutions.

Ng[2] has proposed another algorithm, which is also based on a hill-climbing method, namely Nelder and Mead's modified simplex method[8]. It provides a dynamic minimax formulation which makes all indices with unsatisfied bounds equally active at the start of each iteration of the Nelder Mead algorithm, so that at the kth iteration, *one step* of the following minimax problem is solved:

$$\min_{p} \psi(p) \tag{7}$$

where

$$\psi(p) = \max_{i} \left\{ \bar{\phi}_i = \frac{\phi_i(p) - \phi_i^g}{\phi_i^b - \phi_i^g}, \ i = 1, 2, \ldots, n \right\} \tag{8}$$

where

$$\phi_i^g = \begin{cases} \varepsilon_i & \text{if } \phi_i(p^k) > \varepsilon_i \\ \phi_i(p^k) - \delta & \text{if } \phi_i(p^k) \leq \varepsilon_i \end{cases}, \tag{9}$$

$$\phi_i^b = \begin{cases} \phi_i(p^k) & \text{if } \phi_i(p^k) > \varepsilon_i \\ \varepsilon_i & \text{if } \phi_i(p^k) \leq \varepsilon_i \end{cases}, \tag{10}$$

and δ is set to a small positive number. This algorithm also works well, but is also very dependent on being provided with a good starting point. Here, we call it the Nelder Mead Dynamic Minimax (NMDM) method.

Another algorithm is the *multiobjective genetic algorithm* (MOGA), developed by Fonseca and Fleming [9, 10, 11]. The design philosophy is slightly different from the other two approaches, in that a set of simultaneous solutions is sought, and the designer then selects the best solution from the set. The idea behind the MOGA is to develop a population of Pareto-optimal or near Pareto-optimal solutions. To restrict the size of the near Pareto-optimal set and to give a more practical setting to the MOGA, Fonseca and Fleming have formulated the problem in a similar way to the MOI. This formulation maintains the genuine multiobjective nature of the problem. The aim is to find a set of solutions which are non-dominated and which satisfy a set of inequalities. An individual j with a set of objective functions $\phi^j = (\phi_1^j, \ldots, \phi_n^j)$ is said to be *non-dominated* if for a population of N

individuals, there are no other individuals $k = 1, \ldots, N, k \neq j$ such that

$$
\begin{array}{lll}
\text{a)} & \phi_i^k \leq \phi_i^j \text{ for all } i = 1, \ldots, n & \text{and} \\
\text{b)} & \phi_i^k < \phi_i^j \text{ for at least one } i.
\end{array}
\tag{11}
$$

The MOGA is set into a multiobjective context by means of the fitness function. The individuals are ranked on the basis of the number of other individuals they are dominated by for the unsatisfied inequalities. Each individual is then assigned a fitness according to their rank. The MOGA is to be included in the next version of the Genetic Algorithms Toolbox[12].

There are a number of other search algorithms for the MOI. An algorithm based on simulated annealing has been developed by the authors[13]. The goal attainment method[14] can also be used for solving inequalities, this algorithm has been included in the Optimization Toolbox[15]. Other algorithms have been developed by Polak, Mayne and co-workers[16] for the solution of functional inequalities.

4 The MOI for Multiobjective Control System Design

The MOI has been used successfully for a great many control system design problems. These include several different classes of control system/control system design such as mixed-optimization[17, 18, 19], critical control systems[20] and eigen-structure assignment[21]; as well as for more straight-forward design using fixed structure controllers with performance based on the step response[3] or the frequency response[22].

In most of the previous applications of the MOI, the design parameter p has parameterised a controller with a particular structure. For example, $p = (p_1, p_2)$ could parameterise a PI controller $p_1 + p_2/s$. This has meant that the designer has had to choose the structure of the control scheme and the order of the controllers. The performance is generally expressed in terms of functionals on the closed-loop system step response or frequency response, such as rise time, settling time, overshoot, undershoot, band-width and phase margin. This approach is of particular value when there are constraints on the structure of the controller.

4.1 Mixed-Optimization

Traditionally, control system designers use design methods based on either analytical optimization or parameter optimization. Both methods have advantages and disadvantages; briefly, analytical optimization techniques (e.g. $\mathcal{H}_\infty$, LQG) generally (i) have non-explicit closed-loop performance, (ii) are single-objective, (iii) are robustly stable, (iv) provide high-order controllers, (v) are not very flexible, (vi) provide a global optimum, (vii) can deal with relatively large MIMO problems; whereas parameter optimization based methods (e.g. MOI) generally (i) have explicit closed-loop performance, (ii) are multiobjective, (iii) are not implicitly robustly stable, (iv) provide simple controllers, (v) are flexible, (vi) are often non-convex resulting in local minima, (vii) can deal with small problems only, (viii) may have difficulty stabilizing the system. A combination of analytical optimization and parameter search methods may be able to overcome some of the limitations of using just one approach.

The MOI can be combined with analytical optimization techniques by using the parameters of the weighting functions generally required by such techniques as the design parameters. Thus, for a nominal plant $G(s)$ augmented by a set of n_w weighting functions

$W(s) = (W_1(s), W_2(s) \ldots W_{n_w})$, a controller $K_{\min}(s, G, W)$, which is optimal in some sense, can generally be synthesized, and a set of closed-loop performance functions ϕ of the optimal control system can be calculated. If the weighting functions are simply parameterized by the design vector p, the problem can be formulated as for Eq. 4, and the MOI used to design the weights for the analytical optimization problem. The designer thus chooses the optimization technique and the structure of the weighting functions $W(s, p)$. He/she defines suitable performance functions (e.g. rise time, settling time, bandwidth, system norms etc.) along with the design goals ε_i, and the MOI can be used to search for suitable values of the weighting function parameters p such that Eq. 4 is satisfied. Details of this procedure applied to McFarlane and Glover's LSDP are given in the next section. Details of the mixed-optimization approach applied to other other analytical optimization methods may be found in Whidborne *et al* [18].

4.2   The Loop-Shaping Design Procedure with the MOI

McFarlane and Glover's loop-shaping design procedure[4, 5] is a procedure for the design of robust MIMO systems based on shaping the open-loop transfer function with weighting functions and on H_∞-optimization of a normalized coprime factorization description of the weighted nominal plant. Certain aspects of the LSDP make it very suitable for combining it with the method of inequalities to design directly for both closed-loop performance and stability robustness.

A plant model $G = \tilde{M}^{-1}\tilde{N}$, is a normalized left coprime factorization (NLCF) of G if $\tilde{M}, \tilde{N} \in RH_\infty$; there exists $V, U \in RH_\infty$ such that $\tilde{M}V + \tilde{N}U = I$; and $\tilde{M}\tilde{M}^* + \tilde{N}\tilde{N}^* = I$ where for a real rational function of s, X^* denotes $X^T(-s)$.

Using the notation

$$G(s) = D + C(sI - A)^{-1}B \stackrel{s}{=} \left[\begin{array}{c|c} A & B \\ \hline C & D \end{array} \right],$$

(12)

then[4]

$$\begin{bmatrix} \tilde{N} & \tilde{M} \end{bmatrix} \stackrel{s}{=} \left[\begin{array}{c|cc} A + HC & B + HD & H \\ \hline R^{-1/2}C & R^{-1/2}D & R^{-1/2} \end{array} \right],$$

(13)

is a normalized coprime factorization of G where $H = -(BD^T + ZC^T)R^{-1}$, $R = I + DD^T$, and the matrix $Z \geq 0$ is the unique stabilizing solution to the algebraic Riccati equation (ARE)

$$(A - BS^{-1}D^TC)Z + Z(A - BS^{-1}D^TC)^T - ZC^TR^{-1}CZ + BS^{-1}B^T = 0,$$

(14)

where $S = I + D^TD$.

A perturbed model G_P is defined as

$$G_P = (\tilde{M} + \Delta_M)^{-1}(\tilde{N} + \Delta_N)$$

(15)

where $\Delta_M, \Delta_N \in RH_\infty$. To maximize the class of perturbed models defined by Eq. 15 such that the configuration of Fig. 1 is stable, we need to find the controller K which stabilizes the nominal closed-loop system and which minimizes γ where

$$\gamma = \left\| \begin{bmatrix} K \\ I \end{bmatrix} (I - GK)^{-1}\tilde{M}^{-1} \right\|_\infty.$$

(16)

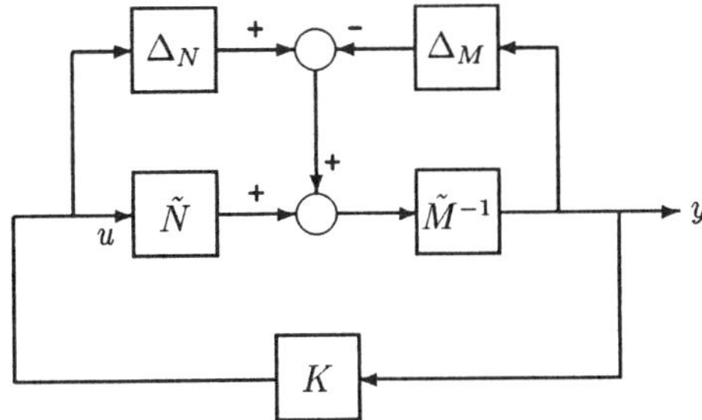

Figure 1: Robust stabilization with respect to coprime factor uncertainty

This is the problem of robust stabilization of normalized coprime factor plant descriptions[23]. From the small gain theorem, the closed-loop system will remain stable if

$$\| [\Delta_N \quad \Delta_M] \|_\infty < \gamma^{-1}. \tag{17}$$

The minimum value of γ for all stabilizing controllers K is

$$\gamma_0 = \inf_{K \text{ stabilizing}} \left\| \begin{bmatrix} K \\ I \end{bmatrix} (I - GK)^{-1} \tilde{M}^{-1} \right\|_\infty \tag{18}$$

and is given by

$$\gamma_0 = \left(1 - \left\| [\tilde{N} \quad \tilde{M}] \right\|_H^2 \right)^{-1/2}, \tag{19}$$

where $\| \cdot \|_H$ denotes the Hankel norm. Now

$$\left\| [\tilde{N} \quad \tilde{M}] \right\|_H^2 = \lambda_{\max}\left(ZX(I + ZX)^{-1} \right), \tag{20}$$

where $\lambda_{\max}(\cdot)$ represents the maximum eigenvalue, and $X \geq 0$ is the unique stabilizing solution of the ARE

$$(A - BS^{-1}D'C)'X + X(A - BS^{-1}D'C) - XBS^{-1}B'X + C'R^{-1}C = 0. \tag{21}$$

Hence from Eq. 19, it can be shown that

$$\gamma_0 = (1 + \lambda_{\max}(ZX))^{1/2}. \tag{22}$$

All controllers[23] optimizing γ are given by $K = UV^{-1}$, where U and V are stable and are right coprime factorizations of K, and where

$$\left\| \begin{bmatrix} -\tilde{N}^* \\ \tilde{M}^* \end{bmatrix} + \begin{bmatrix} U \\ V \end{bmatrix} \right\|_\infty = \left\| [\tilde{N} \quad \tilde{M}] \right\|_H. \tag{23}$$

This is a Hankel approximation problem and can be solved using an algorithm developed by Glover[24].

A controller which achieves a $\gamma > \gamma_0$ is[4]

$$K \overset{s}{=} \left[\begin{array}{c|c} A + BF + \gamma^2 (Q^T)^{-1} Z C^T (C + DF) & \gamma^2 (Q^T)^{-1} Z C^T \\ \hline B^T X & -D^T \end{array} \right], \qquad (24)$$

where $F = -S^{-1}(D^T C + B^T X)$, and $Q = (1 - \gamma^2)I + XZ$. However, if $\gamma = \gamma_0$, then $Q = XZ - \lambda_{\max}(XZ)I$ which is singular, and thus Eq. 24 cannot be implemented. This problem can be resolved using the descriptor system[25, 26], whereby

$$\hat{E}\dot{x} = \hat{A}x + \hat{B}u \qquad (25)$$
$$y = \hat{C}x + \hat{D}u, \qquad (26)$$

which can be written as the system $\hat{G}(s)$

$$\hat{G}(s) = \hat{D} + \hat{C}(s\hat{E} - \hat{A})^{-1}\hat{B} \overset{s}{=} \left[\begin{array}{c|c} -\hat{E}s + \hat{A} & \hat{B} \\ \hline \hat{C} & \hat{D} \end{array} \right] \qquad (27)$$

The descriptor system allows $\hat{E}$ to be singular, and can be converted to the usual state space system[25, 26] via the singular value decomposition of $\hat{E}$. Thus, from Eq. 24, a controller which achieves a $\gamma \geq \gamma_0$ is given in the descriptor form by

$$K \overset{s}{=} \left[\begin{array}{c|c} -Q^T s + Q^T(A + BF) + \gamma^2 Z C^T (C + DF) & \gamma^2 Z C^T \\ \hline B^T X & -D^T \end{array} \right]. \qquad (28)$$

In practice, to design control systems using normalized coprime factorizations, the plant needs to be weighted to meet closed-loop performance requirements. A design procedure has been developed[4, 5], known as the loop-shaping design procedure, to choose the weights by studying the open-loop singular values of the plant, and augmenting the plant with weights so that the weighted plant has an open-loop shape which will give good closed-loop performance.

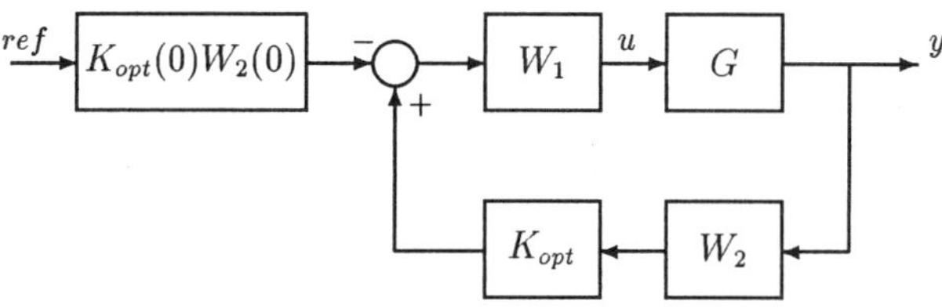

Figure 2: Controller configuration for LSDP

The nominal plant G is augmented with pre- and post-compensators W_1 and W_2 respectively, so that the augmented plant G_s is $G_s = W_2 G W_1$. Using the procedure outlined earlier, an optimum feedback controller K_{opt} is synthesized which robustly stabilizes the NLCF of G_s given by $(\tilde{N}_s, \tilde{M}_s)$ where $G_s = \tilde{M}_s^{-1} \tilde{N}_s$. The final feedback controller K is then constructed by simply combining K_{opt} with the weights to give

$$K = W_1 K_{opt} W_2. \qquad (29)$$

Note that[4],

$$\gamma_0 = \inf_{K \ stabilizing} \left\| \begin{bmatrix} W_1^{-1} K \\ W_2 \end{bmatrix} (I - GK)^{-1} [W_2^{-1} \quad GW_1] \right\|_\infty . \tag{30}$$

Essentially, with the LSDP, the weights W_1 and W_2 are the design parameters which are chosen both to give the augmented plant a 'good' open-loop shape and to ensure that γ_0 is not too large. γ_0 is a design indicator of the success of the loop-shaping as well as a measure of the robustness of the stability property.

The mixed-optimization problem can thus be expressed as:

Problem: *For the system of Fig. 2, find (W_1, W_2) such that*

$$\gamma_0(W_1, W_2) \leq \epsilon_\gamma, \tag{31}$$

and

$$\phi_i(W_1, W_2) \leq \varepsilon_i \quad for \quad i = 1 \ldots n, \tag{32}$$

where

$$\gamma_0(W_1, W_2) = \inf_{K \ stabilizing} \left\| \begin{bmatrix} W_1^{-1} K \\ W_2 \end{bmatrix} (I - GK)^{-1} [W_2^{-1} \quad G] \right\|_\infty , \tag{33}$$

and $\phi_i(W_1, W_2)$ are functionals of the closed-loop system, ϵ_γ, ε_i are real numbers representing desired bounds on γ_0 and ϕ_i respectively, (W_1, W_2) is a pair of fixed order weighting functions with real parameters $p = (p_1, p_2, \ldots, p_q)$. □ □

A design procedure to solve the above problem is:

i) Define the plant G, and define the objective functions ϕ_i.

ii) Define the values of ε_γ and ε_i.

iii) Define the form and order of the weighting functions W_1 and W_2. Bounds should be placed on the values of p_i to ensure that W_1 and W_2 are stable and minimum phase to prevent undesirable pole/zero cancellations. The order of the weighting functions, and hence the value of q, should initially be small.

iv) Define initial values of p_i based on the open-loop frequency response of the plant.

v) Implement a search algorithm in conjunction with Eq. 24 and Eq. 33 to find a W which satisfies the inequalities in Eq. 31 and Eq. 32, i.e. locate an admissible point. If a solution is found, the design is satisfactory. If no solution is found, change the initial values of p by returning to step (iv); change the structure of the controller by returning to step (iii), or relax one or more of the design goals ε_γ and ε_i by returning to step (ii).

vi) With satisfactory weighting functions W_1 and W_2, a satisfactory controller is obtained from Eq. 29.

5 MODCONS Implementation

A number of MATLAB routines which implement the mixed-optimization approach have been collected together into the MODCONS Toolbox[6]. There are three main components at the core of the toolbox, they are; (i) the search algorithm, (ii) the user-interface and (iii) the objective functions calculator. The relationship between these three components and the user is shown in Fig. 3. The search algorithms comprise the core of the design package. These are generic to most design situations, not just for CACSD, and are the main design machine within the process. The user-interface is a very important component of the process, and this provides the interface between the search algorithms and the designer/user, and enables the designer to make intelligent decisions about the design process. The final component is the objective functions calculator. This is specific to CACSD, in fact, some functions are specific to the particular class of control system (fixed controller, mixed-optimization etc.).

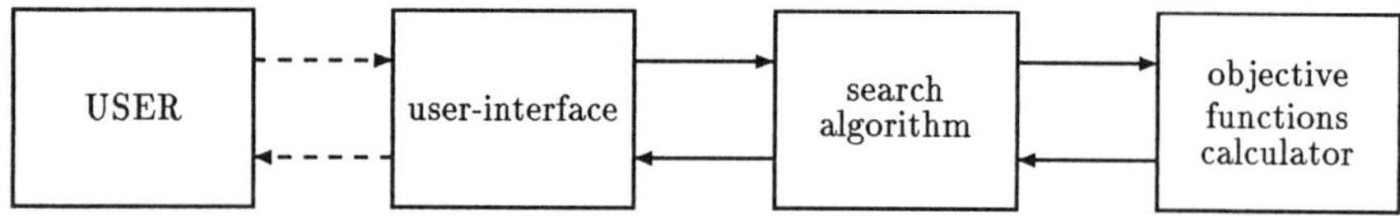

Figure 3: Main component relationship

MODCONS also contains several periphery categories of functions. These consist of the code generators, which generate application-specific user-interfaces and objective functions M-files based upon specifications from the user. There are also some routines to assist in report writing.

5.1 The Search Algorithms

The search algorithms implemented so far are the three algorithms described in Section 3, namely, the moving boundaries process, the Nelder Mead dynamic minimax method and the multiobjective genetic algorithm.

The search algorithm M-files can be called from the command line or from within a user-interface routine. The search algorithms require the definition of an M-file which is the objective functions calculator, and which returns the function set $\phi(p)$. The name of the function M-file must be provided as an argument to the search algorithms M-file, along with initial values of the design parameter p_0, and the design goals ε. Upper and lower bounds on the values of the design parameters may also be provided. The search algorithms terminate when an admissible point is found, when a local minima is reached and the search becomes 'trapped', or when the maximum number of iterations is reached. The algorithms may provide the user with information on the progress of the algorithm, both in text and graphical form.

The MOGA implementation uses the GA Toolbox[12] for the genetic operations, the MOGA will in fact be included in the next version of the GA Toolbox. A simulated annealing algorithm[13] will be included at a later date. The goal attainment method[14] can also be used for solving inequalities, this algorithm can be found in the Optimization Toolbox[15].

5.2 User-Interface

The user-interface provides the mechanism for the designer decision making described in Section 3. More specifically, it is the interface between the search algorithms and the user. The user needs to make decisions on the initial design parameters, parameter bounds and design goals required by the search algorithm. The search algorithm, in turn, must provide the user with information regarding the progress of the search in order that the user can make these decisions. The user can also provide the search algorithms with specific information required for the mechanics of the algorithm, such as the maximum number of iterations, the initial step length, local minima termination criteria and so on.

The user has the option of two user-interfaces, either the MATLAB command line, or a graphical user-interface (GUI) provided by MODCONS. The GUI provides sliders, graphics, menu choice of algorithm etc. which aid the user in decision making. An example of a GUI is shown in Fig. 4.

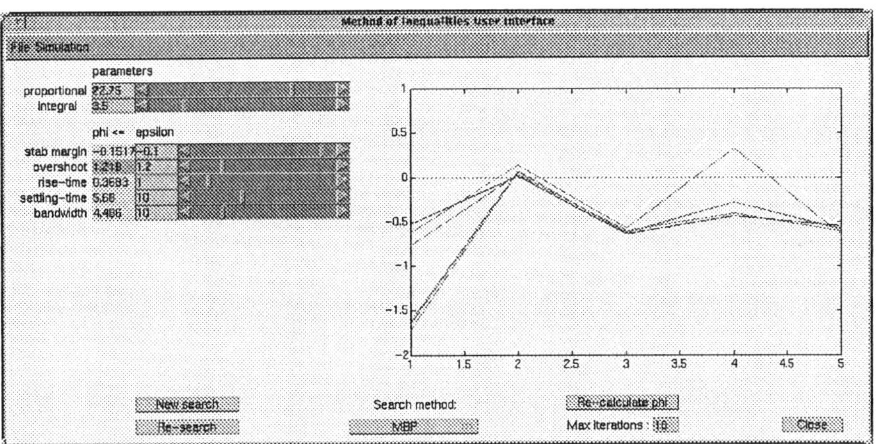

Figure 4: Graphical user-interface example

5.3 Objective Functions Calculator

The objective functions calculator M-file is called by the search algorithm with the current design parameters and returns the objective function vector $\phi(p)$. For control system design, within the objective functions calculator, the following steps are generally taken; (i) synthesize or construct the controller from the design parameters, (ii) compute the closed-loop transfer functions, (iii) compute relevant time and frequency responses, (iv) compute the performance and robustness functionals from the responses. The objective functions calculator M-file can be written by the user, or it can be generated automatically from within MODCONS (see Section 5.5). A number of function M-files are provided by MOD-CONS to calculate commonly-used functionals from step and frequency responses, such as the rise-time, settling time, bandwidth etc. Additional information on the progress of the search may also be directed to the user-interface from the objective functions calculator.

5.4 Example — Mixed-Optimization Implementation

Fig. 5 shows a more detailed relationship between the components for a mixed-optimization design using the LSDP. The user interacts with the user-interface, which calls and monitors the progress of the search algorithms. At each iteration, i, the progress of the objective functions is displayed graphically by the user-interface. The objective functions calculator implements the steps outlined in Section 5.3. Additional information may be output to the user-interface.

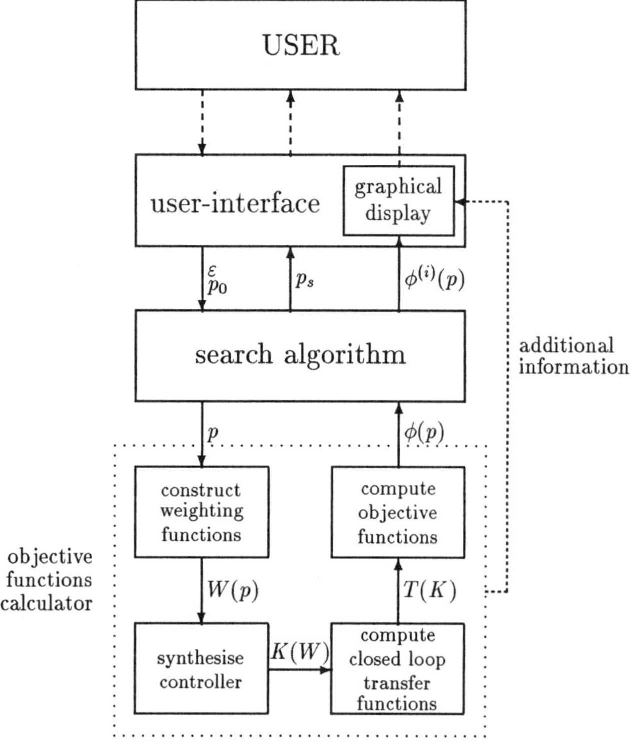

Figure 5: Mixed-optimization implementation

5.5 Code Generators for Objective Function M-Files

Incorporated into MODCONS are a number of automatic code generators for application-specific objective function M-files. For example, for mixed-optimization using the LSDP, a MODCONS routine will bring up the window shown in Fig. 6. Clicking on the "G" box will bring up the window shown in Fig. 7, where the nominal plant can be defined as (A, B, C, D) state-space matrices. The objective function set can be defined from the option

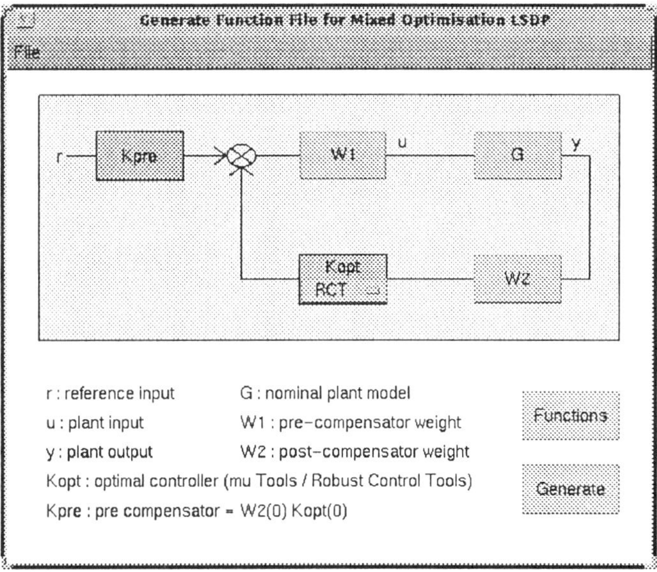

Figure 6: Window for generating mixed-optimisation objective function M-files

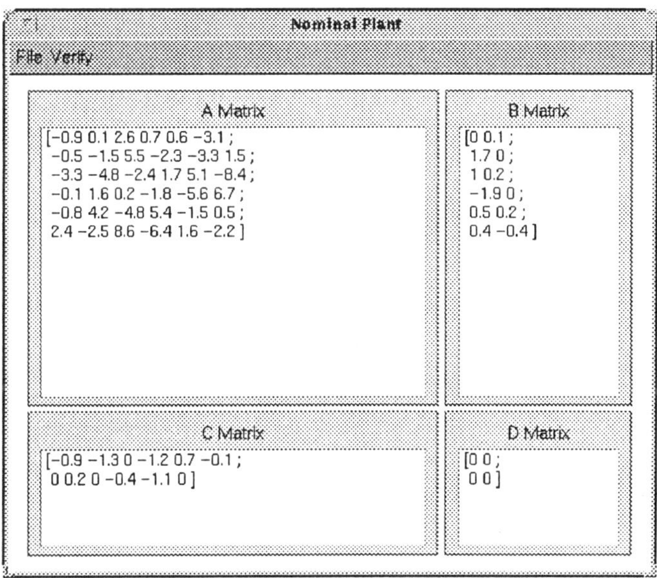

Figure 7: Window for defining nominal plant

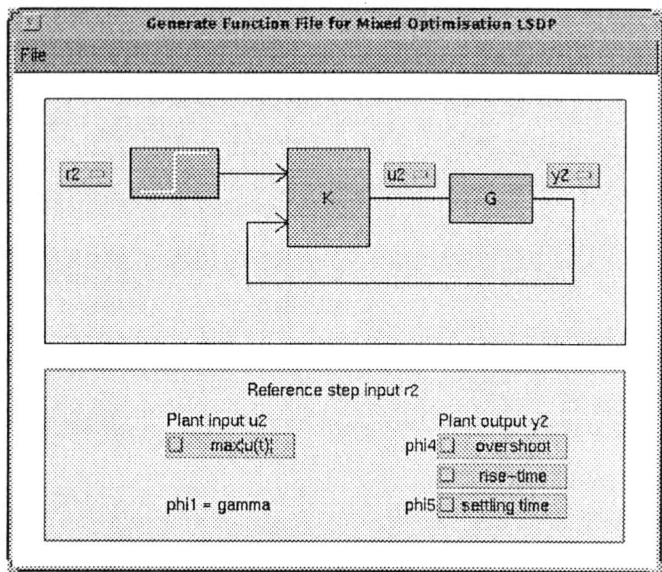

Figure 8: Window for defining objective function set

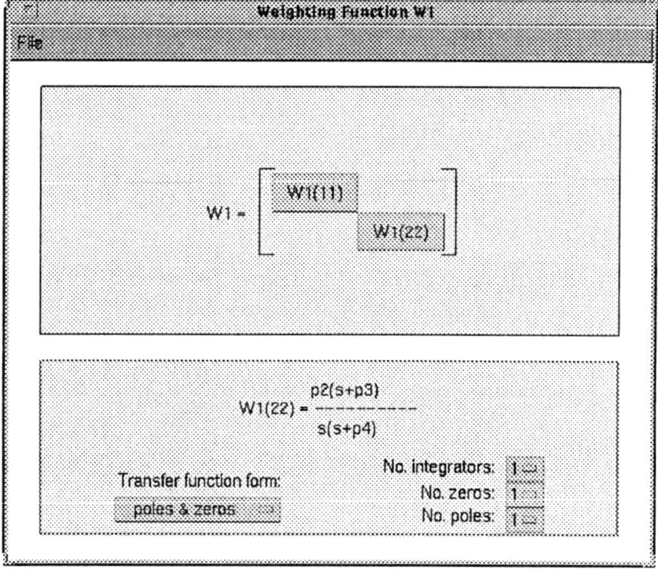

Figure 9: Window for defining structure of weighting function W_1

"functions", which will bring up the window shown in Fig. 8. Clicking on the "W1" box will bring up the window in shown in Fig. 9, where the structure of the pre-plant weighting function W_1 can be defined. The structure of W_2 defined similarly with the "W2" box. The option "Generate" will generate the function M-file according to the specifications given. Similar code generators are being written for other mixed-optimization approaches.

6 Conclusions

The mixed-optimization approach provides a powerful design tool for robust control systems, particularly when stringent time-domain performance specifications need to be met. The interactive nature of the approach aids the designer in making the necessary trade-offs between conflicting design requirements. The approach is well-suited for implementation in MATLAB to provide an interactive CACSD environment. The facilities provided by toolboxes such as MODCONS increases the availability of such advanced design techniques to the engineering community.

7 Acknowledgements

The authors would like to acknowledge N. P. Foster and N. K. Rutland for their suggestions, and D. J. Walker for the code to implement the LSDP with the Robust Control Toolbox.

8 References

1. V. Pareto. *Manuale di Economia Politica*. Societa Editrice Libraria, Milan, Italy, 1906.

2. W.Y. Ng. *Interactive Multi-Objective Programming as a Framework for Computer-Aided Control System Design*, volume 132 of *Lect. Notes Control & Inf. Sci.* Springer-Verlag, Berlin, 1989.

3. V. Zakian and U. Al-Naib. Design of dynamical and control systems by the method of inequalities. *Proc. IEE*, 120(11):1421–1427, 1973.

4. D.C. McFarlane and K. Glover. *Robust Controller Design Using Normalized Coprime Factor Plant Descriptions*, volume 138 of *Lect. Notes Control & Inf. Sci.* Springer-Verlag, Berlin, 1990.

5. D.C. McFarlane and K. Glover. A loop shaping design procedure using H_∞ synthesis. *IEEE Trans. Autom. Control*, AC-37(6):759–769, 1992.

6. J.F. Whidborne, D.-W. Gu, and I. Postlethwaite. MODCONS – a MATLAB toolbox for multi-objective control system design. Technical Report 94-26, Leicester University Engineering Dept., Leicester, U.K., 1994.

7. H.H. Rosenbrock. An automatic method for finding the greatest or least value of a function. *Comp. J.*, 3:175–184, 1960.

8. J.A. Nelder and R. Mead. A simplex method for function minimization. *Comp. J.*, 7(4):308–313, 1965.

9. C.M. Fonseca and P.J. Fleming. Genetic algorithms for multiobjective optimization: formulation, discussion and generalization. In *Genetic Algorithms: Proceeding of the Fifth International Conference*, San Mateo, CA, 1993.

10. C.M. Fonseca and P.J. Fleming. Multiobjective genetic algorithms. In *IEE Colloquium on Genetic Algorithms for Control Systems Engineering*, number 1993/130, pages 6/1–6/5, London, England, 1993.

11. C.M. Fonseca and P.J. Fleming. Multiobjective optimal controller design with genetic algorithms. In *Proc. Control 94*, pages 745–749, Coventry, England, 1994.

12. A. Chipperfield, P.J. Fleming, H. Pohlheim, and C.M. Fonseca. *Genetic Algorithm Toolbox: User's Guide*. Dept. Automatic Control and Systems Engineering, University of Sheffield, U.K., 1994.

13. J.F. Whidborne, D.-W. Gu, and I. Postlethwaite. Simulated annealing for multi-objective control system design. Technical Report 96- , Leicester University Engineering Dept., Leicester, U.K., 1996. In preparation.

14. F.W. Gembicki and Y.Y. Haimes. Approach to performance and sensitivity multiobjective optimization: the goal attainment method. *IEEE Trans. Autom. Control*, AC-20(8):821–830, 1975.

15. A. Grace. *Optimization Toolbox: User's Guide*. The MathWorks, Inc., MA, USA., 1990.

16. R.G. Becker, A.J. Heunis, and D.Q. Mayne. Computer-aided design of control systems via optimization. *IEE Proc.-D*, 126(6):573–578, 1979.

17. I. Postlethwaite, J.F. Whidborne, G. Murad, and D.-W. Gu. Robust control of the benchmark problem using H_∞ methods and numerical optimization techniques. *Automatica*, 30(4):615–619, 1994.

18. J.F. Whidborne, G. Murad, D.-W. Gu, and I. Postlethwaite. Robust control of an unknown plant – the IFAC 93 benchmark. *Int. J. Control*, 1995. To appear.

19. J.F. Whidborne, I. Postlethwaite, and D.-W. Gu. Robust controller design using H_∞ loop-shaping and the method of inequalities. *IEEE Trans. on Contr. Syst. Technology*, 29(4):455–461, 1994.

20. J.F. Whidborne and G.P. Liu. *Critical Control Systems: Theory, Design and Applications*. Research Studies Press, Taunton, U.K., 1993.

21. R.J. Patton, G.P. Liu, and J. Chen. Multiobjective controller design using eigenstructure assignment and the method of inequalities. *J. Guidance*, 17(4):862–864, 1994.

22. K.E. Bollinger, P.A. Cook, and P.S. Sandhu. Synchronous generator controller synthesis using moving boundary search techniques with frequency domain techniques. *IEEE Trans. Power Apparatus & Systems*, PAS-98(5):1497–1501, 1979.

23. K. Glover and D.C. McFarlane. Robust stabilization of normalized coprime factor plant descriptions with H_∞-bounded uncertainty. *IEEE Trans. Autom. Control*, AC-34(8):821–830, 1989.

24. K. Glover. All optimal Hankel-norm approximations of linear multivariable systems and their L^∞-error bounds. *Int. J. Control*, 39(6):1115–1193, 1984.

25. R.Y. Chiang and M.G. Safonov. *Robust Control Toolbox: User's Guide*. The MathWorks, Inc., MA, USA., 1992.

26. M.G. Safonov, R.Y. Chiang, and D.J.N. Limebeer. Hankel model reduction without balancing – a descriptor approach. In *Proc. 26th IEEE Conf. Decision Contr.*, pages 112–117, Los Angeles, CA., 1987.